CHEMISTRY RESEARCH AND APPLICATIONS

BENZIMIDAZOLE

PREPARATION AND APPLICATIONS

Chemistry Research and Applications

Additional books and e-books in this series can be found on Nova's website under the Series tab.

CHEMISTRY RESEARCH AND APPLICATIONS

BENZIMIDAZOLE

PREPARATION AND APPLICATIONS

AKSEL A. VESTERGAARD
EDITOR

nova
science publishers
New York

Library of Congress Cataloging-in-Publication Data

ISBN: 978-1-53616-986-7

Published by Nova Science Publishers, Inc. † New York

Contents

PREFACE

Benzimidazole: Preparation and Applications first defines the structure, physical and chemical properties of the benzimidazole compound class. Further conventional and alternative synthesis methods and important reactions of these compound derivatives are illustrated with examples.

The authors provide an overview of benzimidazole-based synthetic medicine and their significant applications in treating various ailments. The chemical reactions of benzimidazole with other heterocyclic compounds/chemical reagents produce a lot of new substituted molecules, which have potential pharmaceutical applications.

A literature review is presented focusing on the synthesis of chiral compounds and their subsequent applications as therapeutic compounds in order to better evaluate the results so far.

Additionally, the authors discuss how nenzimidazole derivatives play a vital role in organic light emitting diodes as emissive materials, host materials, electron transport materials, hole blocking materials, etc., and have pharmaceutical applications such as antimicrobial, antihypertensive, anti-HIV and anticancer.

Following this, this collection examines classes of potential chemosensors possessing benzimidazole moieties which are capable of both visual and optical estimation of target analytes over a broad concentration range and without interference of contemporary analytes.

Important methodologies used for the synthesis of variedly substituted benzimidazoles are presented along with their multi-target therapeutic uses.

The latest research on anti-tumor and antimicrobial benzimidazole compounds, as well as structure-activity correlations, drug design, clinical and preclinical studies is also presented.

Benzimidazole-based polymers with various architecture and copolymers are discussed with a variation of properties such as mechanical, thermal stability, thermo-oxidative stability and enhanced performance.

The penultimate chapter covers a critical analysis of the recent developments in benzimidazole framework research and the future scope of benzimidazole framework in ligand designing.

In closing, the synthesis and biological properties of coordination compounds containing benzimidazole derivatives are presented and discussed.

Chapter 1 - In this study, firstly the structure, physical and chemical properties of benzimidazole compound class are defined. Further, conventional and alternative synthesis methods and important reactions of these compound derivatives illustrated by examples. Also benzimidazole ring have a lot of different biological activities. Drugs such as albendazole and mebendazole, which have benzimidazole structure, are also included in the drug guidelines by the *World Health Organization*, reveal the importance of this system. In the last part, the studies that express the pharmacological properties of important derivative compounds of this class of compounds are evaluated. In addition, new drug studies belonging to this class of compounds written.

Chapter 2 - This book chapter gives an overview of Benzimidazole based synthetic medicine and their significant application in treating various ailments. The chemical reactions of Benzimidazole with other heterocyclic compounds/chemical reagents produce a lot of new substituted molecules, which have plentiful biological actions and potential pharmaceutical applications. Benzimidazole is considered a major class of heterocyclics and worldwide researchers have put great efforts in studying this nucleus in order to design and synthesize various new derived/hybrid molecules and identified their significant role in many classes of drugs such as antibacterial, antifungal, anthelmintic, antitumor/antiproliferative/anticancer, anti-inflammatory, antiviral including anti-HIV, anticonvulsant, antihistaminic, antidepressant and antioxidant, and recently some researchers have got highly promising results of Benzimidazole derivatives in antiasthamatic and antidiabetic activities also. Benzimidazole has an inspiring molecular geometric structure that offers a suitable skeleton to build newer chemical entities which has encouraged medicinal chemists to design and synthesize novel and effective molecules as potential therapeutic agents. However, there is an urgent need to study the mechanism of action at molecular level of several pharmacological actions resulting from Benzimidazole scaffold through modern methods to furnish safer and effective new molecules for the treatment of various diseases.

Chapter 3 - The chemistry of chiral benzimidazoles is a relatively young branch of stereochemistry which has been required to be more developed over the past decades because of the superior therapeutic properties of the pure isomers as compared to racemic mixtures. Recent studies refer to methods of chiral functionalization of benzimidazoles as well as to the medicinal activities of chiral benzimidazoles. In this chapter benzimidazole syntheses with central, axial and planar chirality, with the specific reaction conditions were presented briefly. A wide range of therapeutic activities for benzimidazole chiral compounds is known, after almost 60 years of research in medicinal chemistry, such as:

anticancer, antimicrobial, antiviral, antithrombotic, antihypertensive, anti-inflammatory, analgesic, hypoglycemic, antimalarial, FRX and CRF1 antagonists and renin inhibitors. In this chapter the authors propose a presentation of the synthetic routes of chiral benzimidazoles and their applications in the field of medicinal chemistry. The literature review presented in this chapter, with the synthesis of chiral compounds and the subsequent application of these as therapeutic compounds, will open new horizons in medicinal chemistry, in order to better evaluate the results so far.

Chapter 4 - Benzimidazole derivatives (BI) play a vital-role in organic light emitting diodes (OLEDs) as emissive materials, host materials, electron transport materials, hole blocking materials, etc., and have pharmaceutical applications such as antimicrobials, antihypertensive, anti-HIVs, anticancer, etc. The non-doped blue emitters with BI derivatives, N,N-diphenyl-4′-(1-phenyl-1H-benzo[d]imidazol-2-yl)-[1,1′-biphenyl]-4-amine (TPABBI) / N,N-diphenyl-4′-(1-phenyl-1H-benzo[d] imidazol-2-yl)-[1,1′-biphenyl]-3-amine (*m*-TPABBBI) [1] show low efficiency roll-off with high luminance, current efficiency (η_c) -3.56/5.1 cdA^{-1}; power efficiency (η_c)- 3.11/5.77 lmW^{-1}; external quantum efficiency (η_{ex}) -4.23/5.66%. The BI used as green emitters shows η_c -7.8 cd/A [2] and 3,7-bis(1-(biphenyl-4-yl)-benzimidazol-2-yl)-10-methylphenothiazine (PhBBMP) exhibits η_c -4.5 cd/A, η_{ex}-0.9% and luminance (L)-12000 cd/m^2 with CIE (0.20, 0.55) [3]. The Pt(II) complex based 2-(2’-pyridyl)benzimidazolyl functionalized molecular stars afford red emission [3b]. The white-emitting device with BI derivative show L-260500 cdm^{-2} and η_p-12.8 lmW^{-1} [4]. 2, 2’, 2’’-(1,3,5-benzinetriyl)-tris(1-phenyl-1- H-benzimidazole) (TPBi) used as hole blocking layer and electron transport layer in blue OLEDs with CIE (0.15, 0.09) and high η_{ex} - 4.3% due to effective carrier recombination [5, 6]. 2,2’-bis(-phenyl-1H-benzo[d]imidazol-2-yl)biphenyl (BImBP) used as host material exhibit η_c- 57.2 cdA^{-1} with low efficiency roll-off, η_p -50.4 lmW^{-1} and η_{ex} -25.7% with CIE (0.15, 0.38) and BImBP basedPHOLED show high η_c-53 cd/A [7]. Among the green host materials, 9-phenyl-3-(4-(1-phenyl-1H-benzo[d]imidazol-2-yl)phenyl)-9H-carbazole (PBIC) and 9-(naphthyl-2-yl)-3-(4-(1-phenyl-1H-benzo[d]imidazol-2-yl)phenyl)-9H-carbazole (NBIC), PBIC host exhibit higher efficiency with η_c and η_p enhanced by 1.6 and 2.0 times compared with CBP host device [8]. Among the BI based hosts, *o*-mCPBI, *m*-mCPBI and *p*-mCPBI, *o*-mCPBI PHOLEDs exhibit, η_{ex} of 20% [9]. The blue PHOLEDs with BI based Ir(III) emitter demonstrated longer lifetime than 300 h and show η_{ex} of 20.1% [10]. Among the two bipolar molecules, tris(3′-(1-phenyl-1*H*-benzimidazol-2-yl)biphenyl-4-yl)amine(TBBI) and tris(2-methyl-3′-(1-phenyl-1*H*-benzimidazol-2-yl)biphenyl-4-yl)amine (Me-TBBI), Me-TBBI device exhibit excellent device efficiency (L-16400 cd m^{-2}, η_c- 27.4 cdA^{-1}, η_p- 4.5 lmW^{-1}) than TBBI [11]. The two novel and stable red luminescent radicals namely, bis[4-(1H-benzimidazolyl)-2,6-dichlorophenyl](2,4,6-trichlorophenyl) methyl radical (TTM-2Bi) and tris[4-(1H-benzimidazolyl)-2,6-dichlorophenyl]methyl radical (TTM-3Bi) reported, TTM-2Bi exhibits better performance [12]. Jayabharathi et al., reported synthesis, characterization and electroluminescent studies of BI derivatives along with their Ir(III) complexes [13]. Herein at first, the BI based emitters with multi-functional abilities have been reviewed comprehensively. The molecular design- strategy and synthetic routes for BI materials and BI based host materials for RGB PHOLEDs are discussed.

Chapter 5 - Considering the significance of natural receptors in biology, chemistry, medicine and environmental studies, numerous synthetic receptors like cryptands, spherands, crown ethers, calixarenes, thiacalixarenes, porphyrins, cyclodextrins and benzo-fused heterocyclic ring systems etc. have been reported in the literature. Using design strategies similar to nature, synthetic assemblies have been developed to encapsulate a variety of guest molecules, stabilize reactive intermediates and even facilitate chemical reactions. These synthetic receptors play an imperative role in ion recognition. In a related advanced *supramolecular concept*, recognition sites can be coupled to particular groups that are capable of "signaling" the ion coordination process. In this case, the binding process is transduced into a signaling event. These receptor molecules exhibit discriminatory response to specific ions or neutral species to be used as chemosensors. This information is used to analyze the stimulant, both qualitatively and quantitatively. This concept has long been an interest of study in the research field because through the integration of different functionalities into the molecular structures, these systems would respond differently upon the introduction of external stimuli (or target analyte) and organize into interesting supramolecular structures. Such distinctive ingenuity endows these systems with versatile potential to be used in the field of analyte detection. Another chief aspect for molecular recognition is the signaling mechanism involved in the detection process. Chemical sensors utilize specific transduction techniques to yield analyte information. The most extensively used techniques employed in chemical sensors are optical absorption, luminescence and redox potential etc.

In the present chapter, class of potential chemosensors possessing benzimidazole moieties have been discussed which are capable of both visual and optical estimation of target analyte over a broad concentration range and without interference of contemporary analytes.

Chapter 6 - Benzimidazole has evolved as an important heterocyclic system for the development of diversely active molecules over several decades of research worldwide. Starting from its exploration for developing physiologically active molecules (such as vitamin B12 derivative) to single-target drugs (such as omeprazole, astemizole, enviradene, telmisartan and others), it has emerged as a nucleus of choice for the development of multifunctional drugs for treatment and/or management of multifactorial diseases such as cancer, rheumatoid arthritis, multiple sclerosis, etc. Various reviews explaining the therapeutic journey of benzimidazole derivatives as single-target drugs have appeared in literature during last few years. However, despite the exploration of benzimidazole scaffold for development of multifunctional molecules during last decade by different research groups, no review on this multifaceted ability of benzimidazole has appeared so far. Also, numerous specialized synthetic methods to introduce different chemical descriptors at various positions of benzimidazole are suggested in literature that assist a medicinal chemist to synthesize complex multi-target drugs. Herein, the authors have searched, reviewed, analyzed and compiled various reports on the development of benzimidazole-based multifunctional molecules to provide medicinal chemists a comprehensive information on the subject till date. The present chapter is focused on important methodologies used for the synthesis of variedly substituted benzimidazoles along with their multi-target therapeutic uses.

Chapter 7 - Benzimidazole moiety is one of the prominent heterocycle in medicinal chemistry, with outstanding results in the last decades of research. The pharmaceutical market comprises a series of benzimidazole - based medicines such as bendamustine, selumetinib - antitumorals; pantoprazole, lansoprazole - proton pump inhybitors; bezitramide - analgesic; mebendazole, albendazole - antihelmintics; ridinazole - antibacterial; enviradine - antiviral and candesartan as antihypertensive.

Cancer continues to be one of the worst diseases of the century in which we live, for which research is most advanced for the discovery of new structures with anti-tumor action. Also, antimicrobial resistance is one of the major problems faced by today's medicine that requires the synthesis of new drugs with a wider range of biological activities with a known mechanism of action and less harmful side effects.

This chapter reviews the latest research on anti-tumor and antimicrobial benzimidazole compounds, as well as structure-activity correlations, drug design, clinical and preclinical studies. The review also covers synthetic strategies for the obtaining of the target structures.

Literature shows that there is a need for modification of existing benzimidazole agents that will bring about a big change for improvement of activity or change in activity, due to the problems like resistance, toxicity, or poor bioavailability.

Chapter 8 - High-temperature polymer electrolyte membrane fuel cell has been demonstrated as a sustainable green technology for power generation. In fact, some of the problems associated with other membranes, such as CO tolerance, improved kinetic and enhanced water management etc., have been solved by above type of membrane. Polybenzimidazole is considered one of the best membranes, which is extensively studied due to excellent properties most suitable for high temperature applications.

In this chapter benzimidazole based polymers with various architecture and copolymers have been discussed with variation of properties, such as mechanical properties, thermal stability, thermo-oxidative stability and enhanced performance. Most of the studies are based on phosphoric acid dopant. Few variations have also been incorporated. This is followed by discussion on blends and composites, which are expected to augment the properties required for fuel cell performance at high temperature. Nanocomposites are also incorporated in the chapter. Various nanocarbons, nanometal salts etc. have been discussed in the section. The design of nanoporous membranes have been discussed as they provide higher space for acid doping to enhance the performance.

Few special cases have been discussed to bring out the efforts in a particular direction. Finally, various characterization techniques have been discussed in brief.

Chapter 9 - Benzimidazole framework has got considerable attention due to their importance in biology and ligand designing. The substitution at one or both “N” with functional groups such as –SR, $–PPh_2$, –SeR, –TeR, –OR, $–NR_2$ etc., results in the formation of multi-dentate ligands suitable for catalysis and transition metal complexation. Several benzimidazole based ligand systems have been designed by using this strategy and their transition metal complexes (including those of group VIII) have been investigated. The single-crystal structures of the metal complexes have many interesting features due to a variety of secondary interactions resulted from ligand itself. Diverse applications of benzimidazole based ligands and their metal complexes have been reported in the literature. The present chapter covers a critical analysis of the recent developments in this area and future scope of benzimidazole framework in ligand designing.

Chapter 10 - Benzimidazole derivatives are known as heterocyclic compounds involved in a several biological processes and have been proved to possess biological activity and pharmacological properties. In this chapter, their coordination behavior towards transition metal ions such as copper(II), zinc(II), cobalt (II), cadmium(II) and cobalt (II) are investigated. The structures of the resulted compounds have been elucidated using crystallographic studies.

The synthesis and the biological properties of the coordination compounds containing benzimidazole derivatives are also presented and discussed.

In: Benzimidazole
Editor: Aksel A. Vestergaard
ISBN: 978-1-53616-986-7

Chapter 1

OVERVIEW OF BENZIMIDAZOLE COMPOUND CLASS

Bahittin Kahveci[1], PhD and İnci Selin Doğan[2,*], PhD
[1]Department of Nutrition and Dietetics, Faculty of Health Sciences, Karadeniz Technical University, Trabzon, Turkey
[2]Department of Pharmaceutical Chemistry, Faculty of Pharmacy, Karadeniz Technical University, Trabzon, Turkey

ABSTRACT

In this study, firstly the structure, physical and chemical properties of benzimidazole compound class are defined. Further, conventional and alternative synthesis methods and important reactions of these compound derivatives illustrated by examples. Also benzimidazole ring have a lot of different biological activities. Drugs such as albendazole and mebendazole, which have benzimidazole structure, are also included in the drug guidelines by the *World Health Organization*, reveal the importance of this system. In the last part, the studies that express the pharmacological properties of important derivative compounds of this class of compounds are evaluated. In addition, new drug studies belonging to this class of compounds written.

Keywords: Benzimidazole, synthesis, conventional method, MW irradiation, biological activity

* Corresponding Author's Email: selinci@gmail.com, isdogan@ktu.edu.tr.

INTRODUCTION

Today, major parts of the drugs in the treatment of disease are obtained by synthesis. The fact that they are more economical than the ones obtained naturally has led to the rapid development of synthetic drugs. In this context heterocyclic chemistry have an important place. Synthetic heterocyclic compounds have a wide range of uses [1].

Benzimidazoles are an important member of the family of heterocyclic compounds in terms of drug properties.

Structure

Benzimidazole ring is a bicyclic heteroaromatic structure consisting of fusion of benzene and imidazole rings (Figure 1).

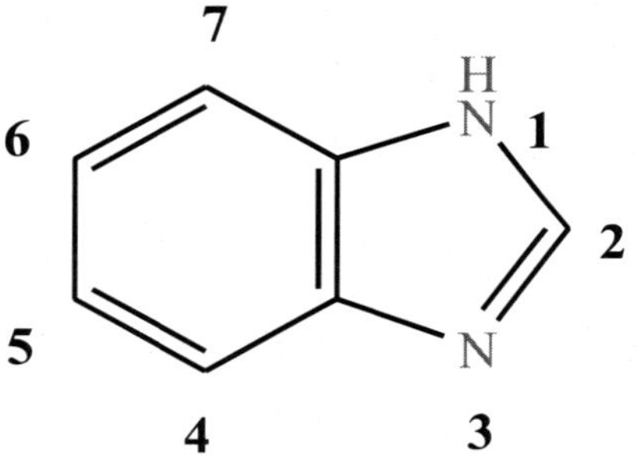

Figure 1. Benzimidazole structure.

The benzimidazole ring has two different nitrogen atoms. One of these nitrogen atoms carries a hydrogen atom and is defined as "imino nitrogen" (figure 1, number one nitrogen atom). The other nitrogen which does not carry hydrogen and has a tertiary structure is defined as "tertiary nitrogen" (figure 1, at the number three nitrogen atom).

Physical Properties

Benzimidazoles are generally crystalline, have low solubility in water and high solubility in alcohol. They dissolve in polar solvents, are not well soluble in apolar solvents and are easily associated with an acidic proton (number one nitrogen atom with hydrogen). They also give displacement reaction due to the presence of acidic hydrogens in imino nitrogen, which results in a reduction in melting and boiling points.

As a result, number one *N*-substituted benzimidazoles cannot associate. Benzimidazoles resemble imidazoles in many respects. Both structures show acidic and basic character. The N-H group in the benzimidazole is acidic. Most benzimidazoles are soluble in aqueous base and acid solutions. The acidic properties of benzimidazoles allow it to form salts with metals. The most attract attention of the metal salts of benzimidazoles is the formation of water-soluble silver salts.

Tautomeric Character

For substituted benzimidazoles that do not have a plane of symmetry, it is possible to write two isomeric structures in which the imino hydrogen is at different positions. For example, since 4-methylbenzimidazole is the tautomer of 7-methylbenzimidazole, this compound is named 4(7) methylbenzimidazole (Figure 2).

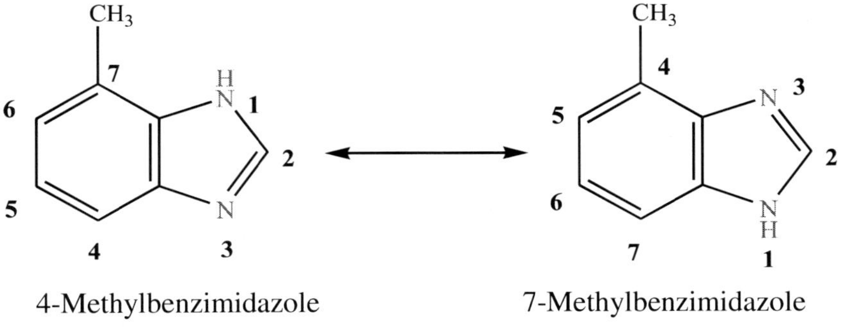

Figure 2. Structure of 4(7)methylbenzimidazoles.

The tautomeric forms of the benzimidazoles containing the symmetry plane are identical. For example, 2-methyl, 5,6-dimethyl and 4,7-dimethyl benzimidazoles have the plane of symmetry (Figure 3) [2].

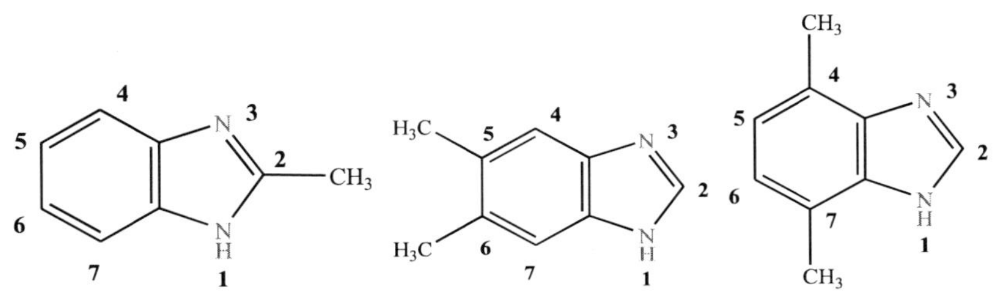

Figure 3. Structures of 2-methyl, 5,6-dimethyl and 4,7-dimethyl benzimidazoles.

Chemical Properties

The most obvious chemical property of benzimidazoles is their chemical stability. They are resistant to treatment even with acids, bases and oxidizing agents. When the benzimidazole is reacted with potassium permanganate, the benzene ring is oxidized to decompose to form 1*H*-imidazole-4,5-dicarboxylic acid (Scheme 1).

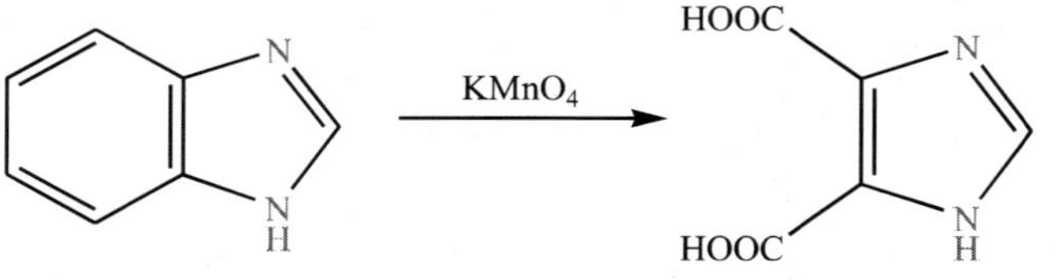

Scheme 1. Oxidation reaction with $KMnO_4$.

Substitution Reaction of Imino Hydrogen Atom of the Benzimidazoles

Because of the acidity of the imino hydrogen atom in the benzimidazole derivative compounds, it is possible to replace this imino hydrogen with bases such as Na_2CO_3, K_2CO_3, NaH, NaOH and NaOEt. The 1-substituted benzimidazoles obtained as a result of this reaction often have a lower melting point than the non-substituted benzimidazoles. The chemical reaction is given below (Scheme 2) [3].

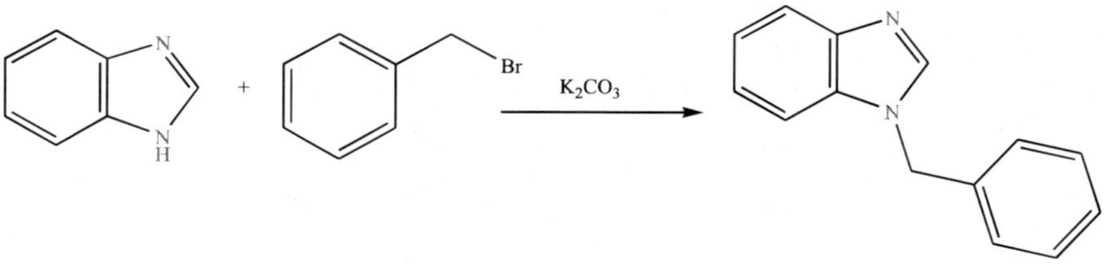

Scheme 2. Synthesis of 1-benzyl-1H-benzimidazole.

Synthesis of Benzimidazole via Conventional Methods

The first synthesis of benzimidazole was gained by the reduction of 2-nitro-4-methylacetanilide on 1872 (Scheme 3) [4].

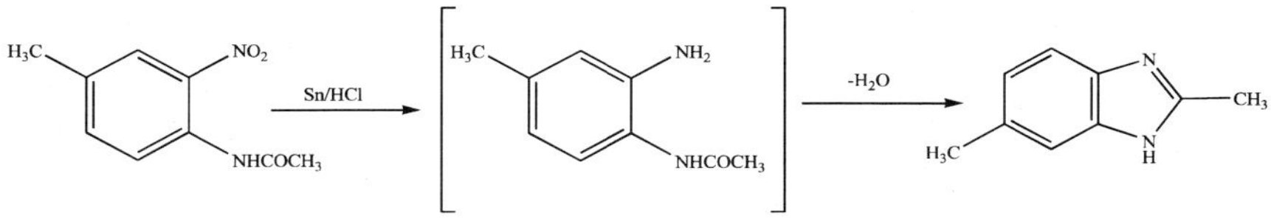

Scheme 3. Synthesis of benzimidazole derivative.

One of the commonly used methods of synthesis of benzimidazole compounds is the reaction of *ortho*-phenylene diamines with carboxylic acid or derivatives. These reactions were carried out under conventional heating conditions. One example is given below (Scheme 4) [5].

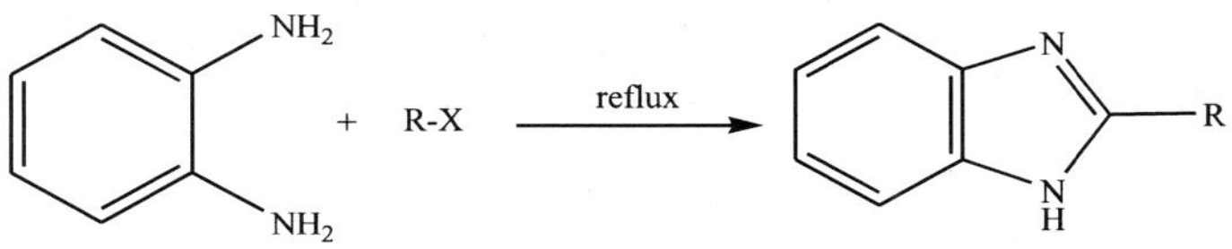

Scheme 4. Synthesis of benzimidazole from *o*-phenylenediamine.

Reduction of *ortho*-dinitro benzene derivatives with Tin(II)chloride in the presence of hydrochloric acid and acetic acid without isolation leads to the formation of the benzimidazole structure (Scheme 5) [6].

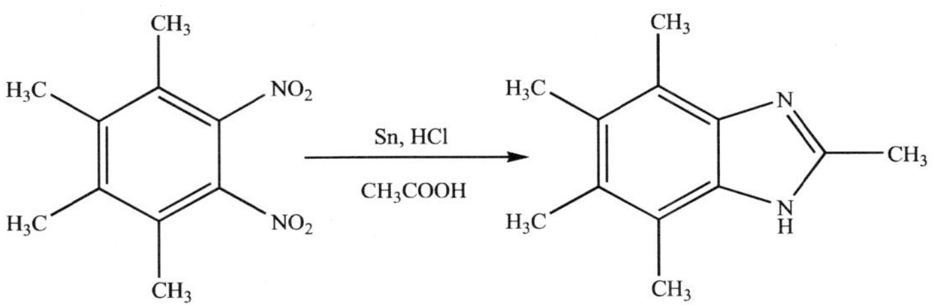

Scheme 5. Synthesis of benzimidazole from *o*-dinitrobenzene.

Traditionally, the most used method of benzimidazole synthesis is the reaction of *ortho*-phenylenediamine with dilute hydrochloric acid (usually 4N), carboxylic acid or anhydride. This method is known as "*Phillips' benzimidazole synthesis*" (Scheme 6) [7].

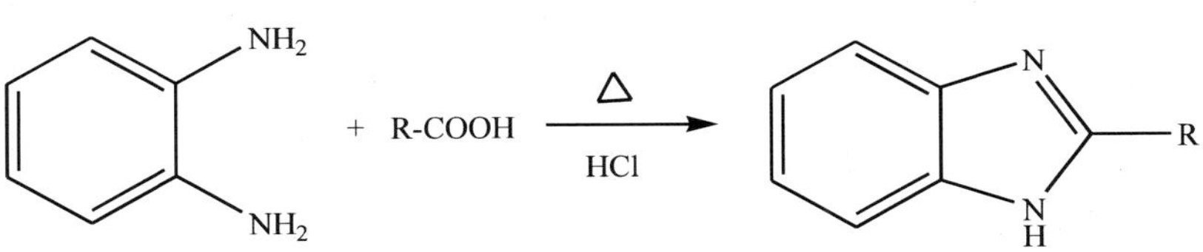

Scheme 6. Synthesis of benzimidazole via *Phillips' method.*

By reaction of 3,4-diaminobenzophenone with cyanogen bromide, benzimidazole compounds were synthesized (Scheme 7) [8].

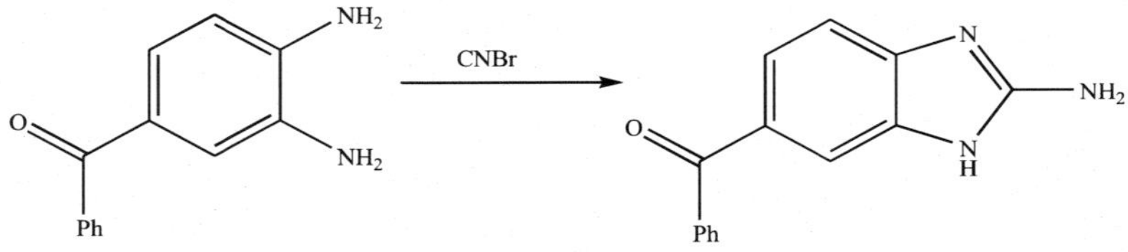

Scheme 7. Synthesis of benzimidazole from 3,4-diaminobenzophenone.

Synthesis of benzimidazoles from aldehydes was carried out by conventional heating using water as an environmentally friendly solvent in a study (Scheme 8) [9]. This reaction was also carried out in the presence of catalyst such as $SiO_2/ZnCl_2$ in another study [10].

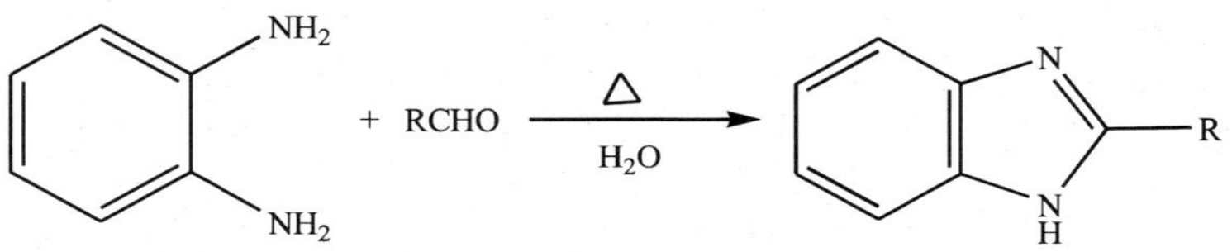

Scheme 8. Synthesis of benzimidazole from aldehyde.

Within the traditional method, in 2010, various benzimidazole derivatives were synthesized using p-TsOH catalyst under different conditions (Scheme 9) [11, 12].

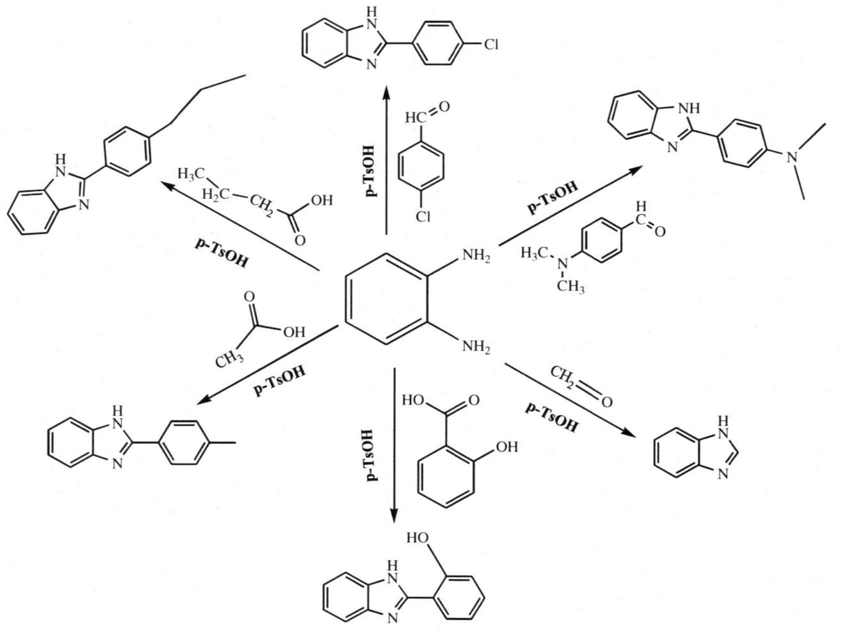

Scheme 9. Synthesis of benzimidazole by p-TsOH catalyst.

In another conventional study, benzimidazole-derived compounds were readily obtained by using environmentally friendly water as a solvent and boric acid as a catalyst (Scheme 10) [13].

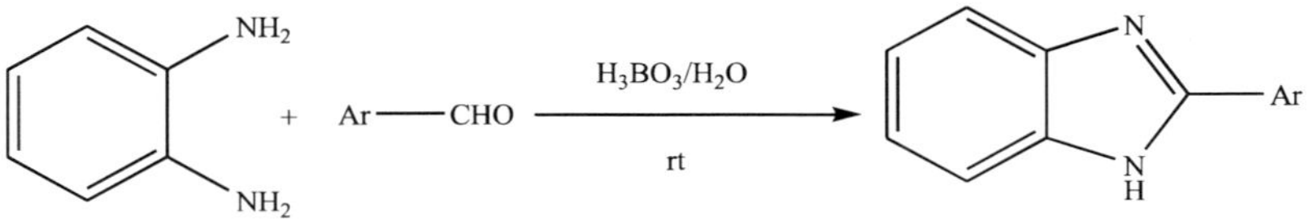

Scheme 10. Synthesis of benzimidazole using boric acid as catalyst, H_2O as a solvent.

Synthesis of Benzimidazole via Microwave Irradiation Technics

Reactions with conventional heating methods require a very long time, and reactive or product degradation due to local heating causes a significant decrease in reaction efficiency. For these reasons, the need for alternative heating sources is increasing.

Recently, microwave (MW) heating systems have been also developed in parallel with the development of technology. In the early days, domestic type microwave heating systems were gradually used in organic synthesis. However, there is no sustainability in using for this purpose due to the lack of safety, temperature and pressure controls.

In 1986, Gedye et al. conducted the first study using microwave irradiation for the purpose of organic synthesis [14]. Following this initial reaction, microwave-assisted organic synthesis became widespread in parallel with the development of microwave heating systems, resulting in a new field of study.

Studies have shown that the most effective method of alternative energy sources developed for carrying out chemical reactions is the use of microwave energy. Compared with conventional heating, microwave heating has been shown to significantly reduce reaction time, reduce the occurrence of undesirable side reactions, and increase product yield and purity.

Because of all these, this technology has been used remarkably in recent years, especially in pharmaceutical chemistry, polymer chemistry, material science, multi-step multicomponent synthesis, nanotechnology and biochemical processes.

Thus, microwave irradiation or heating has become a widely used technique in the world of organic synthesis. It is now recognized by scientists that in principle, any chemical reaction with heat can also occur under microwave conditions. As a result, the use of microwave method has become widespread in pharmaceutical chemistry and drug design applications. Parallel synthesis by microwave method resulted in a rapid increase in the number of new compounds. In addition, it contributed to the discovery of new reaction methods [15].

In order to accelerate organic reactions, the first study of using microwave heating method reported by Richard Gedye et al. in 1986. In this study, the reaction performed in 1 hour with 90% efficiency with conventional heating and it was performed with microwave heating in 10 minutes with 99% yield (Scheme 11) [14].

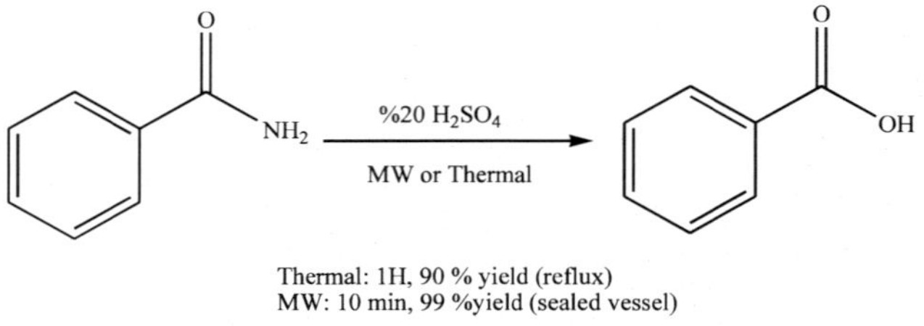

Scheme 11. Synthesis of benzimidazole via MW irradiation.

In the solvent-free conditions, environmentally friendly microwave-assisted synthesis studies were carried out in the 1990s. An interesting example of this is 4,5-diphenyl-4-imidazolin-2-one was synthesized in a short time such as 4 minutes by heating benzoin and urea in a domestic type microwave oven (Scheme 12) [16].

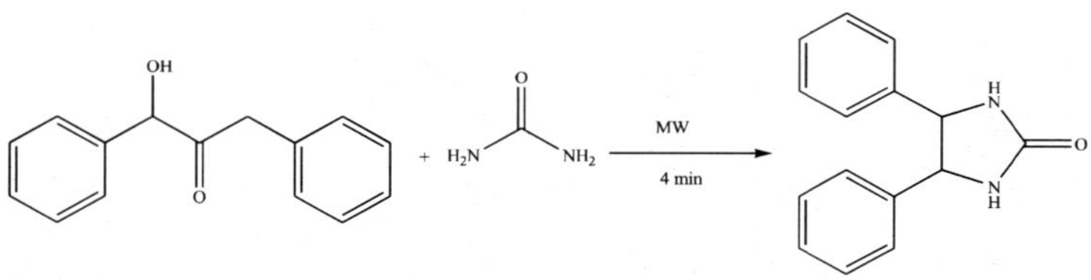

Scheme 12. Synthesis of imidazolinone derivative compounds via MW technique.

An exemplary study of synthesis of benzimidazole by microwaveheating is recorded in the literature as reactions of aldehydes with *o*-phenylenediamines (Scheme 13) [17].

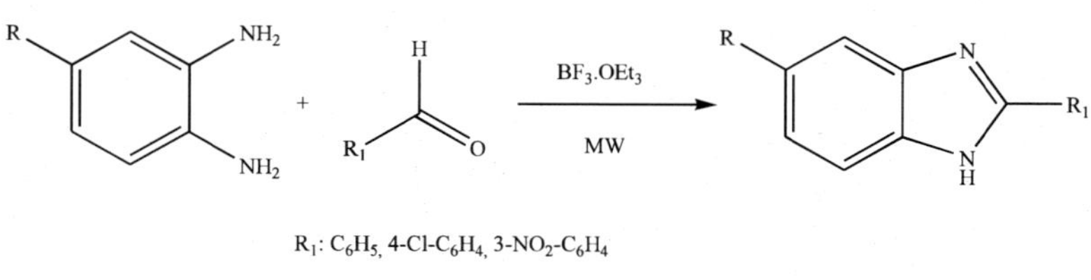

Scheme 13. Synthesis of benzimidazole using *o*-phenylenediamine via microwave irradiation.

A study on the synthesis of bisbenzimidazole derivatives using water as an environmentally friendly solvent was also carried out by microwave heating (Scheme 14) [18].

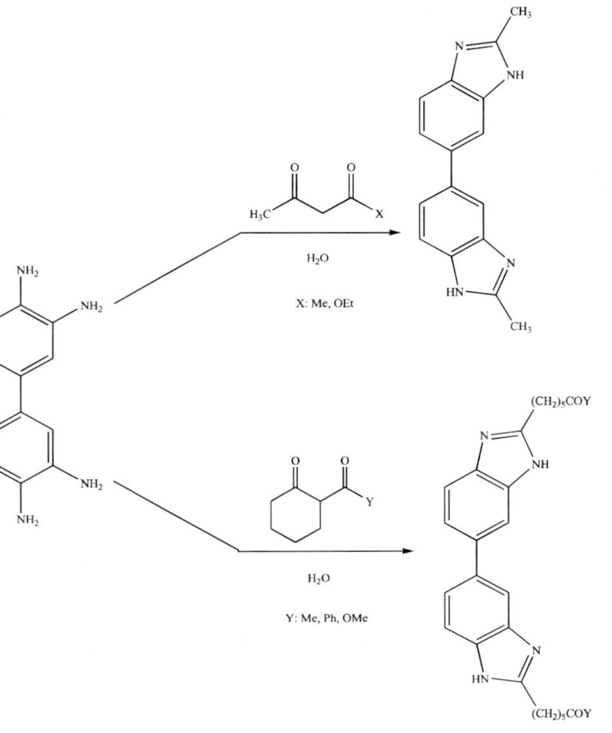

Scheme 14. Synthesis of bisbenzimidazole derivatives in water medium.

It has recently been reported in studies on the synthesis ofbenzimidazole derivatives from *o*-phenylendiamine and *iminoester hydrochlorides* using by microwave method in both solvent and non-solvent medium in a short time and in high yields (Scheme 15) [19].

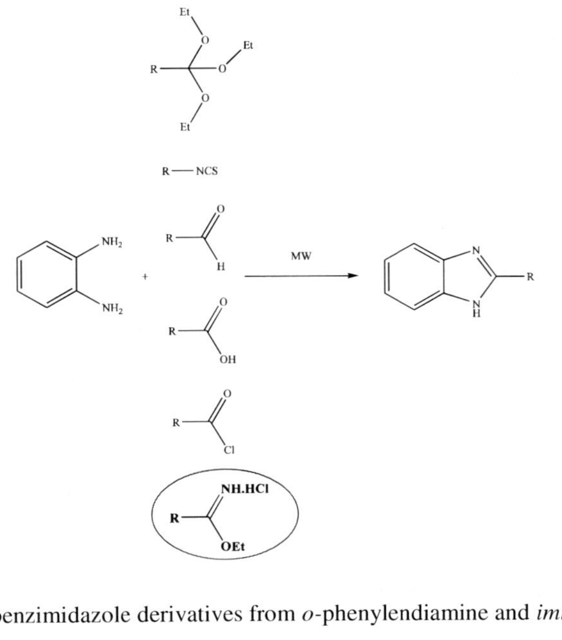

Scheme 15. Synthesis of benzimidazole derivatives from *o*-phenylendiamine and *iminoester hydrochlorides*.

Via microwaveirradiation, one pot synthesis of 2-ethylamino benzimidazole, benzoxazole and benzothiazole derivatives has been reported in literature (Scheme 16) [20].

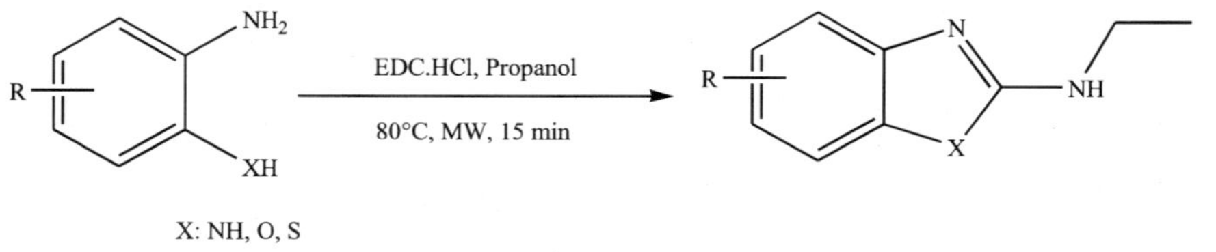

Scheme 16. Synthesis of 2-ethylamino benzimidazole.

In another recent study, the results of the synthesis of 2,5-disubstituted benzimidazoles by microwave heating using various catalysts are given in the literature (Scheme 17) [21].

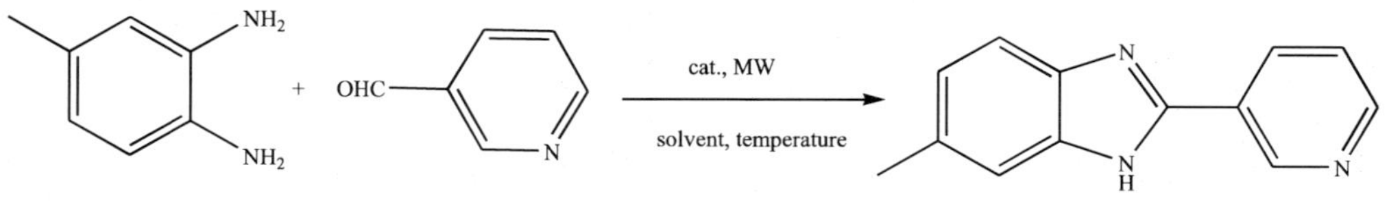

Scheme 17. Synthesis of 2,5-disubstituted benzimidazoles.

Biological Activity of Benzimidazole Derivative Compounds

Goodman et al., have studied deficiency of biotin and purine bases like adenine and guanine on animals [22]. They explained that biotin and benzimidazole have similar structure and the production of biotin deficiency was attempted in several microorganisms by addition of benzimidazole. They recognize that benzimidazole and purine have close structural similarity. And going on the study, Wooley et al., synthesized several benzimidazole compounds, which shown antifungal and antibacterial activity (yeast and bacteria growth-inhibiting action). This growth-inhibiting action was completely negated by aminopurines. The structural similarity of purines to benzimidazole is already known. They found that only adenine and guanine (more active) were effective in reversing the action of benzimidazole [23].

Drugs such as albendazole and mebendazole, which have benzimidazole structure, are also included in the drug guidelines by the *World Health Organization* (WHO); reveal the importance of this system [24-26]. The compounds having benzimidazole structure are classified under several ATC groups by WHO [27]. The benzimidazole can be substituted on different positions; 1st, 2nd and 5th position also together or alone [28]. The first group is antihelmintic benzimidazole compounds which having a small substituent on 2nd position, and sometimes on 5th position. The second is proton pump inhibitor benzimidazole compounds which have bulky 2-substituents. The third is H_1 antihistaminics with bulky 1- and 2-substituents [29]. The other one is antiviral acting benzimidazole compounds with having substituents at various positions [30].

And there were a lot of study about the activities of the compounds with benzimidazole moiety [31] such as antiinflamatory [32], analgesic [33], antifungal, antibacterial [34], antiviral [35], anthelmintic, antiproliferative [36, 37], antihypertensive [38], H_3 antagonistic [39], anticancer [40], antiulcer [41] and antiinfective [42] activities.

In addition, some of the examples of the drugs in clinical practice that have benzimidazole structure were showing on Table 1 [43].

In nature, vitamin B_{12} (N-riosyldimethylbenzimidazole) has a benzimidazole moiety and this part is a ligand for cobalt (Scheme 18) [44].

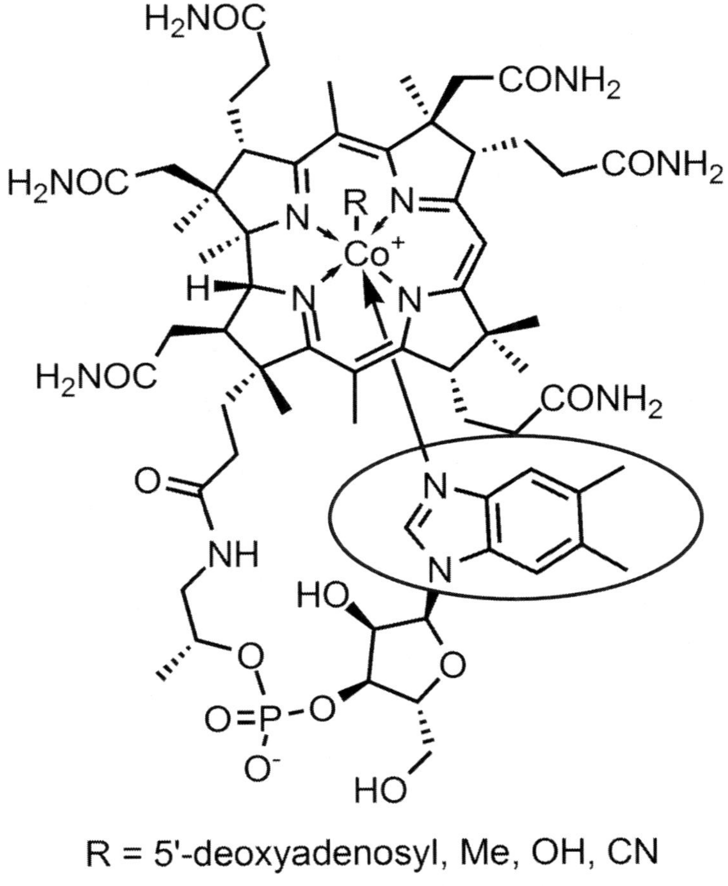

Scheme 18. Vitamin B_{12} molecule which have benzimidazole moiety.

Table 1. Some drugs carrying benzimidazole ring used in treatment [43]

Structure	Name	Pharmacological Activity
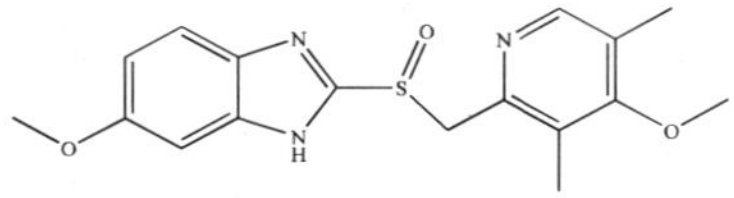	Omeprazole	Proton pump inhibitor [28]
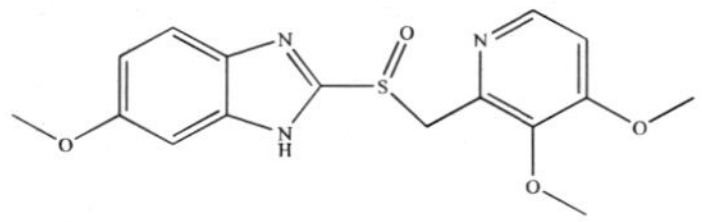	Pantoprazole	Proton pump inhibitor [28]
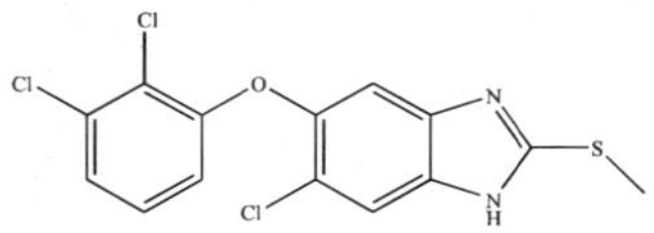	Ticlabendazole	Antihelmintic [36]
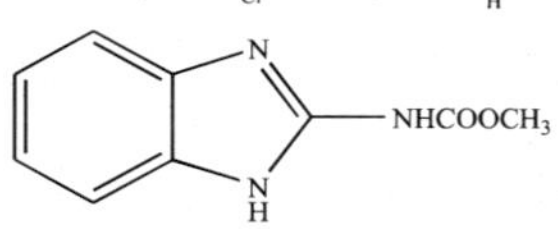	Carbendazime	Fungicide [34]
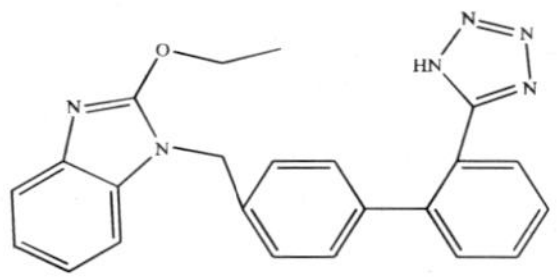	Candesartan	Antihypertensive [38]
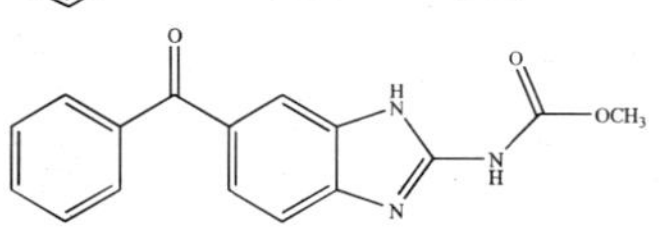	Mebendazole	Worm infestation [42]
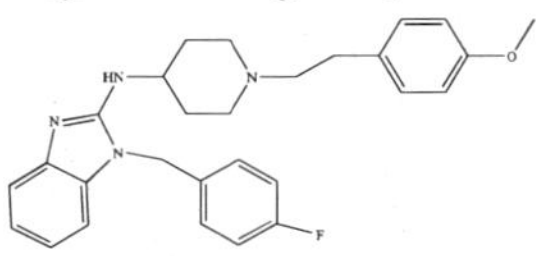	Astemizole	Antihistamine [39]
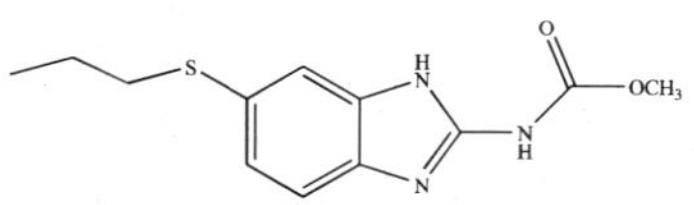	Albendazole	Antihelmintic [37]
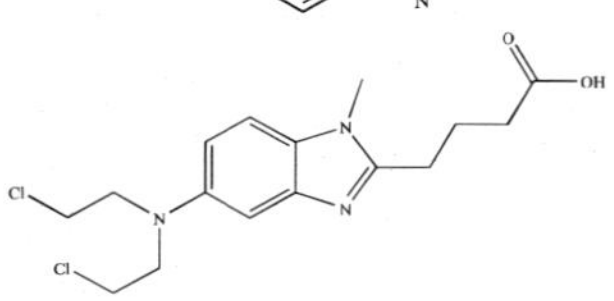	Bendamustine	Anticancer [40]
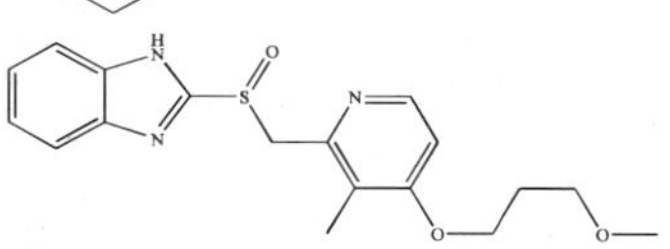	Rabeprazole	Antiulcer [41]

Mariappan et al. [45], synthesized (1*H*-benzimidazol-2-yl-methyl)phenyl-amine derivative compounds and screened for analgesic and antiinflammatory activities. They found that almost all the compounds showed very potent analgesic activity against to standard pentazocine. And most of the tested compounds showed significant antiinflammatory activity against to standard diclofenac sodium. And the result showed that all the compounds have significant effect at 100 mg/kg p.o. (Figure 4).

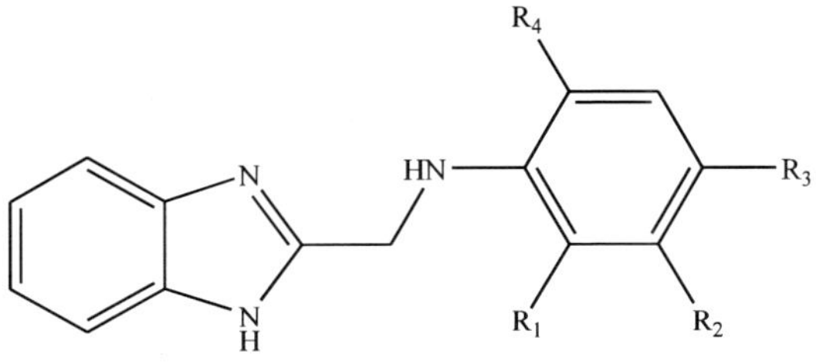

Figure 4. An example for benzimidazole derivatives having antiinflammatory and analgesic activity.

Sontakke et al. [46], synthesized 2-aryl substituted benzimidazole by using silica supported periodic acid (H_5IO_6-SiO_2) as a catalyst and the compounds evaluated for anticancer activity against two cell lines MCF7 (human breast adenocarcinoma) and HL60 (human promyelocytic leukemia) by MTT colorimetric assay using cisplatin as a standard anticancer drug (Figure 5). The results showed that newly synthesized 5-substituted derivatives exhibited excellent activity in micromolar range.

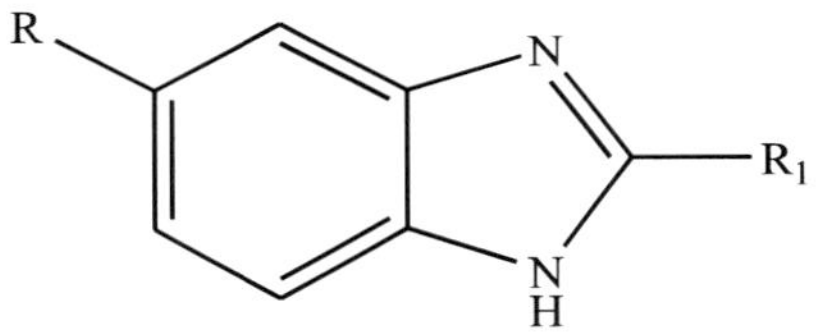

5 R = H
6 R = COOH
7 R = COPh

Figure 5. Some examples of benzimidazole derivatives having anticancer activity.

Nannapaneni et al. [47], synthesized 2-substituted benzimidazole compounds and screened for acute and chronic antianxiety activity in wistar rats by using an elevated plus maze model (Figure 6). Results shows that the compounds [4-(1*H*-benzimidazol-2-yl)phenol], [2-(4-methylphenyl)-1*H*-benzimidazole], [4-(6-nitro-1*H*-benzimidazol-2-yl)phenol], [2-(4-methylphenyl)-6-nitro-1*H*-benzimidazole], and [4-(6-methyl-1*H*-benzimidazol-2-yl)phenol] showed potent activity when compared to the standard drug diazepam. The compound [4-(6-methyl-1*H*-benzimidazol-2-yl)phenol] exhibited the highest antianxiety activity when compared to the other prepared benzimidazole compounds.

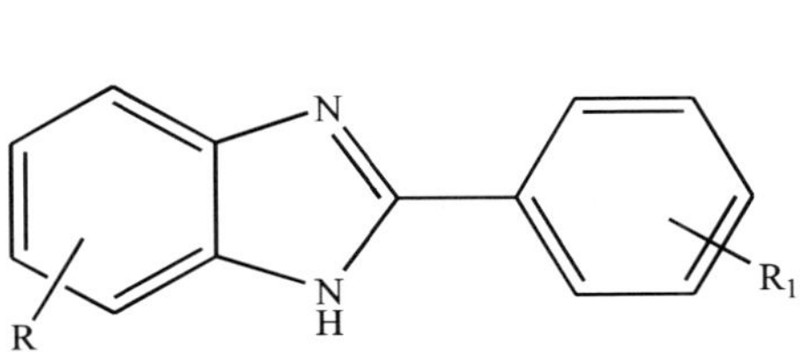

Figure 6. Some examples of benzimidazole derivatives having antianxiety activity.

Shaharyar et al. [48], synthesized 2-[2-(phenoxymethyl)-1*H*-benzimidazol-1-yl]acetohydrazide derivative compounds and evaluated their anticonvulsant activity by maximal electroshock seizure test, pentylenetetrazole-induced seizure test. The compound 7g and j has shown potent anticonvulsant activity and lesser neurotoxicity as compared with the standard drugs because NO_2 and F have more electron withdrawing capacity than the other substituent. The protective index of these two compounds was found to be better than standard drugs. Thus the study revealed that these compounds have potential anticonvulsant activity (Figure 7).

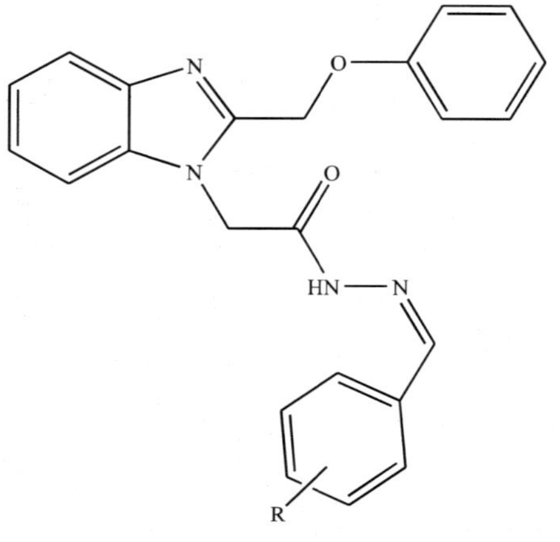

Figure 7. Some examples of benzimidazole derivatives having anticonvulsant activity.

Demirel et al. [49], synthesized 2-(2-phenyl)-1*H*-benzo[d]imidazol-1-yl)-*N*'-(arylmethylene) acetohydrazide derivatives and measured their epidermal growth factor receptor (EGFR) kinase inhibitor activity by comparing them with erlotinib (Figure 8). At 1 μM concentration, erlotinib inhibited EGFR kinase activity by nearly 90%. The results indicated that almost all of the compounds' kinase inhibitor activities were somewhat limited. Compounds that have 3-nitrophenyl (**4**) and 4-benzyloxyphenyl (**7**) substituents as aryl groups were found to have slight inhibitory potencies in the kinase assay at 13.71% and 12.42%, respectively.

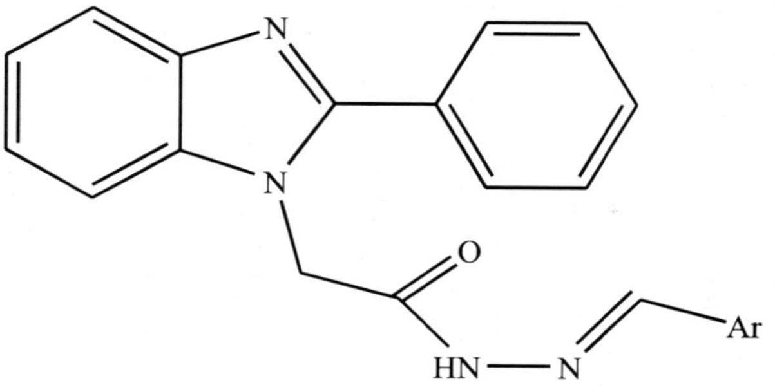

Figure 8. An examples for benzimidazole derivative having EGFR kinase inhibitor activity.

Pérez-Villanueva et al. [50], try to evaluate the antiprotozoal activity of the proton pomp inhibitor drugs omeprazole, lansoprazole, rabeprazole, pantoprazole (table 1) which have benzimidazole nucleus against *Trichomonas vaginalis, Giardia intestinalis and Entamoeba histolytica*. Rabeprazole and pantoprazole were the most active compounds, having IC50 values in the nanomolar range, which were even better than metronidazole, the drug of choice for these parasites.

Anand and Wakode [51] synthesized a series of 2-substituted benzimidazole derivatives and the compounds were evaluated for antimicrobial and antioxidant activity (Figure 9). Cefodoxamine, ascorbic acid and fluconazole were used as reference for antibacterial, antioxidant and antifungal activity, respectively. The results showed that for *in vitro* antibacterial activity against gram positive strain of bacteria (*Staphylococcus aureus*) and gram negative strain of bacteria (*Escherichia coli),* the benzimidazole compounds substituted with (3,4,5-trimethoxyphenyl)-, (2-chloro-4-nitrophenyl)-, (2,4,5-trifluorophenyl)-, (4-chloro-3,5-dinitrophenyl)- are comparatively more active than all other compounds. And the antioxidant activity results which was measured by DPPH and ABTS assay method, the compounds with (2-iodophenyl)- and (2,4,5-trifluorophenyl)-substituents showed best scavenging activity at concentration of 40, 60, 80 and 100 μg/ml. It is found that the results of *in vitro* antifungal activity compounds substituted with (2-chloro-4-nitrophenyl)-, (2-iodophenyl)-, (4-ethylphenyl)-, (2,4,5-trifluorophenyl)- were comparatively more active than all other compounds against *Candida albicans.*

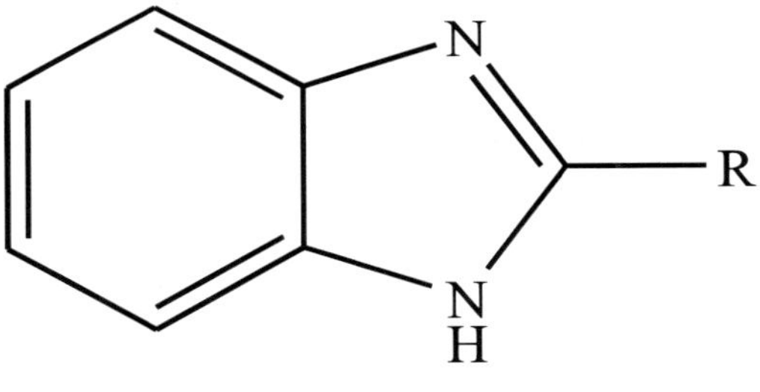

Figure 9. An example for benzimidazole derivative having antimicrobial and antioxidant activity.

Barot et al. [52], synthesized a series of 1,3,4-thiadiazole **(14),** 1,2,4-triazole-5-thione **(16)** and 1,3-thiazolan-4-one **(15)** derivatives of benzimidazole and evaluated for antibacterial and antifungal activities (Figure 10). Some of the synthesized compounds showed good antibacterial and antifungal activities with 2.0 and 2.5 μg/mL MIC, respectively. And according to activity results, it is concluded that as the number of carbon atom increases in side chain at 2-position of benzimidazole, it causes an increase in the activity against *S.aureus, Bacillus subtilis and C. albicans* which supports the chemotherapeutic activity.

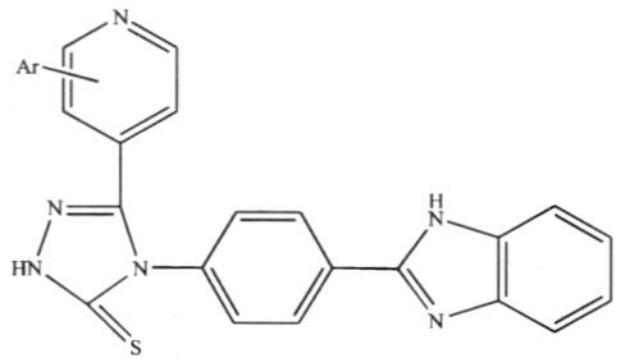

16

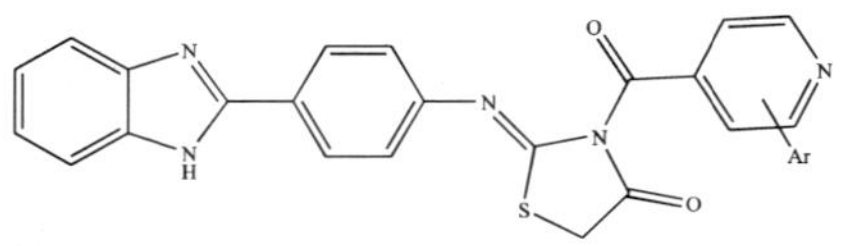

15

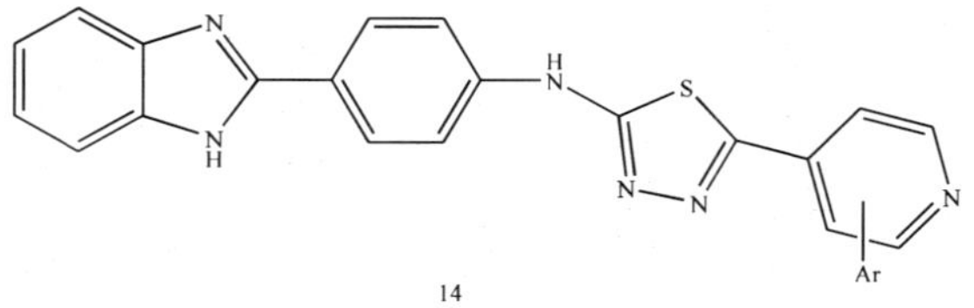

14

Ar: Substituted hydrazide derivative

Figure 10. Some examples of benzimidazole derivatives having antibacterial and antifungal activity.

El-masry et al. [53], synthesized a serial 1-substitute-2-methyl-benzimidazole hydrazide/triazole/oxadiazole/pyrazole derivative compound and evaluated their antimicrobial activities by disk diffusion method using Gentamycine and Ampicilline as a reference drug (Figure 11). The prepared compounds were tested against one strain of Gram (+) bacteria (*Bacillus cereus*), Gram (-) bacteria (*E. coli*), Yeast (*Saccharomyces cerevisae*) and Fungi (*Aspergillus niger*). It is clear that compound 2 was found to be highly active against *B. cereus* but, slightly active against *E.coli*. Compounds 4 and 9a were found to be moderately active against *B. cereus* while compounds 3a and 10c were found to be slightly active against *B. cereus*. But all the tested compounds were found to be inactive against *S. cerevisae* and *A. niger*. Also, compounds 1, 5b, 8, 10a and 11a were found to be inactive against all microorganisms used.

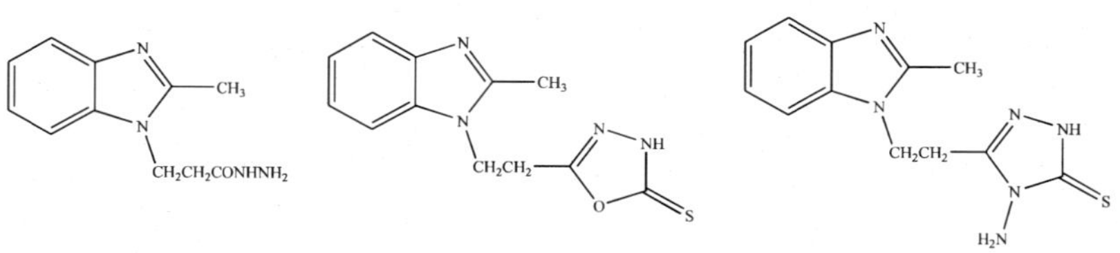

Figure 11. Some examples of benzimidazole derivatives having antibacterial and antifungal activity

Ajani et al. [54], synthesized a serial 2-substituted and 1,2-disubstituted benzimidazole derivatives and screened them for their antibacterial activity (Figure 12). The activity results showed that 2-benzyl-1-(phenylsulfonyl)-1*H*-benzimidazole as the most active derivative with lowest MIC value of 15.63 μg/mL. When they evaluated structure activity relationship, the highest potency was observed on 2-(1*H*-benzimizadol-2-yl)aniline and 2-benzyl-1-(phenylsulfonyl)-1*H*-benzimidazole compounds and they saw that substituent on 1-position and 2-positions of the benzimidazole nucleus had significant effect on the antibacterial activity, and pharmacophoric core is 2-phenylbenzimidazole.

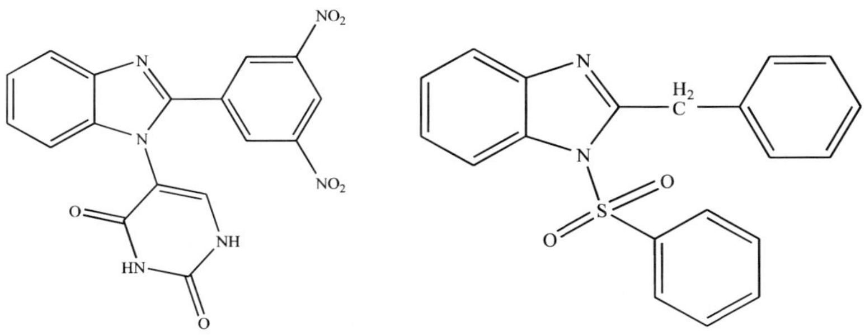

MIC: 15.63 μg/mL
MBC: 31.25 μg/mL

MIC: 15.63 μg/mL
MBC: 31.25 μg/mL

Figure 12. Some examples of benzimidazole derivatives having antibacterial activity

Padalkar et al. [55], synthesized a series of 2-(1*H*-benzimidazol-2-yl)-5-(diethylamino)phenol, 2-(1,3-benzoxazol-2-yl)-5-(diethylamino)phenol, 2-(1,3-benzothiazol-2-yl)-5-(diethylamino)phenol and evaluated for in vitro antibacterial activities against *E. coli* and *S. aureus* strains and in vitro antifungal activity against *C. albicans* and *A. niger* strains by dilution method (Figure 13). When structure activity relationship (SAR) evaluated, new compounds contain benzimidazoles exhibited a broad spectrum of antibacterial profile against tested organism as compared to 2-substituted benzoxazole and benzothiazole. The electron donating and electron withdrawing groups do not affect the growth inhibitory activity against tested bacterial strains. And when observed antifungal activity SAR, new compounds contain benzothiazole exhibited a broad. Electron donating and electron withdrawing groups on benzimidazole, benzoxazole and benzothiazole do not affect the growth inhibitory activity against tested fungal strains. They have noted benzimidazole derivatives give better antibacterial potential.

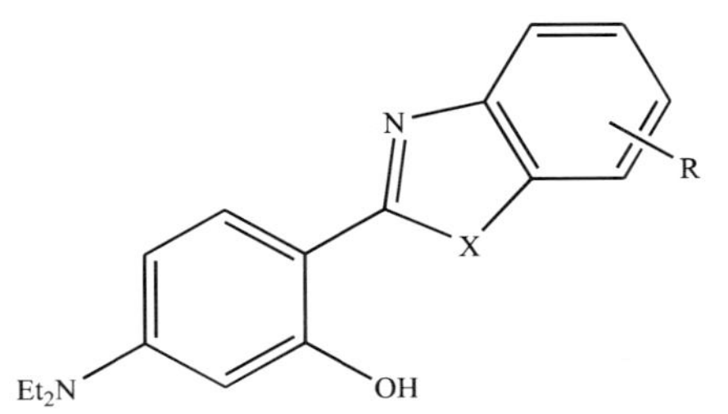

X: NH, S, O R: H, NH2

Figure 13. Some examples of benzimidazole derivatives having antibacterial activity

Kahveci et al. [19], synthesized a serial benzimidazole compound and evaluate their antimicrobial activity (Figure 14). Compounds **2** (5(6)-bromo-2-(4-bromobenzyl)-1*H*-benzimidazol), it's ethyl acetate derivative Compound **3**, hydrazide derivative compound **4** and the hydrazone derivative compounds **5c, 5e** and **5h** showed antimicrobial activity against some tested organisms (*E. coli, Y. pseudotuberculosis, P. aeruginosa, S. aureus, E.*

faecalis, B. cereus, M. smegmatis, C. albicans, S. cerevisiae) and reference drugs are ampicilline, streptomycine, fluconazole.

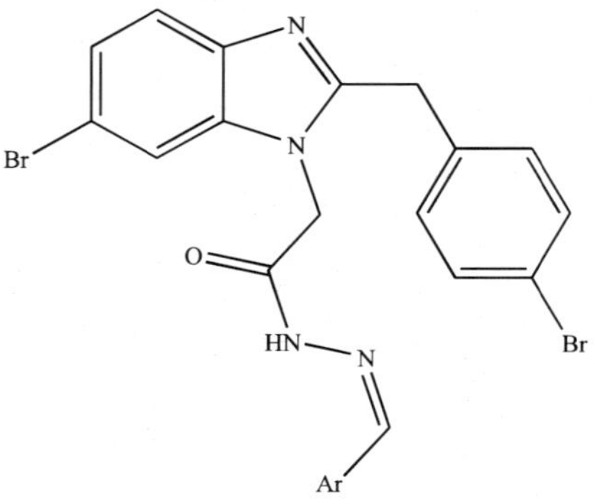

Figure 14. An example for benzimidazole derivative having antimicrobial activity

Doğan et al. [56], synthesized some novel 2-substitutebenzyl-(4)7-phenyl-1*H*-benzo[d]imidazoles **(2a-k)** in mild conditions as potent anti-tyrosinase and antioxidant agents (Figure 15). All of the synthesized compounds displayed certain inhibitory effects on tyrosinase, with IC_{50} values ranging from 37.86 ± 0.24 to 75.81 ± 2.49 μM. Among the compounds, **2j** exhibited similar tyrosinase inhibitory effect (IC_{50} = 37.86 ± 0.24 μM) to the positive control, kojic acid (IC_{50} = 21.93 ± 0.11 μM). Kinetic studies revealed it to act as non-competitive tyrosinase inhibitor with a K_i value of 50.2 μM. The antioxidant activities of the compounds were investigated by using *in vitro* antioxidant assays, including 2,2-diphenyl-1-picrylhydrazyl (DPPH) and ferric reducing antioxidant power (FRAP). All of these results indicated that the compounds might have potential application as tyrosinase inhibitors.

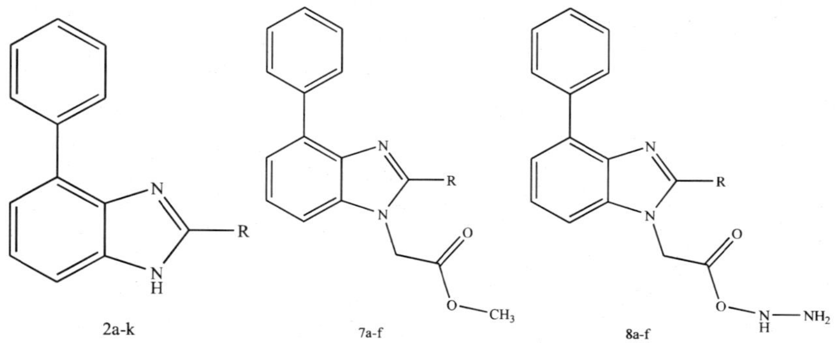

Figure 15. Some examples of benzimidazole derivatives having anti-tyrosinase and antioxidant activity.

And also Gümüş et al. [57], synthesized the derivative of these benzimidazole compounds, 2-[2-(substituted-benzyl)-4(7)-phenyl-1*H*-benzimidazol-1-yl]acetohydrazides **(8a-f)** and methyl 2-[2-(substituted-benzyl)-4(7)-phenyl-1*H*-

benzimidazol-1-yl]acetates **(7a-f)** and evaluate their acetylcholinesterase, butyrylcholinesterase and tyrosinase inhibition, and antioxidant activity (Figure 15). Biological activity results revealed that compounds 8f and 7c showed the highest tyrosinase inhibition with 52.46 ± 2.67% and 52.32 ± 0.70%, respectively. Compound 8f had the highest radical scavenging activities with 24.83 ± 1.28% among the compounds whilst compound 8c showed the lowest activities with 9.62 ± 0.85%. Compound 8f gave the highest PRAP and FRAP absorbance values with 0.180 ± 0.013 and 0.358 ± 0.009 at 100 μM, respectively.

RESULT

Benzimidazoles have an important role in drug discovery and these derivatives represent an excellent scaffold for the development of novel drugs, polymers and ligands. For this reason, a good knowledge about the structure and properties of the benzimidazole compound class makes a significant contribution to the implementation of new and important studies. In this regard, this section will provide convenience to those who work in this field.

REFERENCES

[1] Davies, David T. 1992. *Aromatic Heterocylic Chemistry*. Oxford University Press. ISBN-10: 9780198556602.

[2] Hoffman, K. 1953. *Imidazole and its derivatives*. New York: Interscience Publishers, p 247-261.

[3] Iemurar, H., Hori, M., Othaka, H. (1989). Syntheses of the metabolites of 1-(2-ethoxyethyl-2-(hexahydro-4-methyl-1*H*-1,4-diazepin-1-yl)-1*H*-benzimidazole difumarate and related compounds. *Chem. Pharm. Bul.,* 37: 962-966.

[4] Wright, J. B. (1951). The chemistry of benzimidazoles. *Chemical Reviews,* 48: 397-541.

[5] Harazi, A., Zantour, H. (2002). Ammonium sulphate-magnesium selective reduction of *N*-2-nitrophenylimidates: Synthesis of 2-substitutes benzimidazoles. *Synthetic Communications*, 32(3): 387-392.

[6] Kopenska, K., Najda, A., Zebrowski, J. Chomiez, L., Pielarcyzk, J., Myjak, P., Bretner, M. (2004). Synthesis and activity of 1*H*-benzimidazole and 1*H*-benzotriazole derivatives as inhibitors *Acanthamoeba Castellanii. Bioorg. & Med. Chem.*, 12: 2617-2624.

[7] Phillips, M. A. (1928). The formation of 2-substituted benzimidazoles. *J. Chem. Soc.,* 2392-2399.

[8] Ohemeng, K. A., Roth, B. (1991). Receptor-based design of novel dihydrofolate reductase inhibitors: Benzimidazole and indole derivatives. *J. Med. Chem.*, 34: 1383-1394.

[9] Lin, S., Yang, L. (2005). Simple and efficient procedure for the synthesis of benzimidazoles using air as the oxidant. *Tetrahedron Lett.*, 46: 4315-4319.

[10] Heravi, M. M., Sadjadi, S., Oskooie, H. A. (2008). Heteropolyacids as heterogeneous and recyclable catalysts for the synthesis of benzimidazoles. *Catalysis Comm.,* 9:504-507.

[11] Dawood, Kamal M., Abdel-Wahab, Bakr F. (2010). Synthetic routes to benzimidazole-based fused polyheterocycles. *Arkivoc,* (i):33-389.

[12] Kadhim, A. J. and Kazim, A. C. (2018). Synthesis and characterization of benzimidazole by using o-phenylenediamine with different aldehydes and carboxylic acids in the presence of ρ-TSOH as a catalyst. *Orient. J. Chem.,* 34(4): 2131-2136.

[13] Heravi, M. R. P. and Ashori, M. (2013). Boric acid catalyzed convenient synthesis of benzimidazoles in aqueous media. *J. of Chem.*, Article ID 496413, 5 pages. http://dx.doi.org/10.1155/2013/496413.

[14] Gedye, R., Smith, F., Westaway, K., Ali, H., Baldisera, L., Laberge, L., Rousell, J. (1986). The use of microwave ovens for rapid organic synthesis. *Tetrahedron Lett.,* 27: 279-282.

[15] Kappe, C. O., Stadler, A., Dallinger, D., Mannhold, R., Kubinyi, H., Folkers, G. (2013). *Microwaves in Organic and Medicinal Chemistry*. Wiley-VCH. ISBN: 9783527647842.

[16] Nuchter, M., Müller, U., Ondruschka, B., Tied, A., Lautenschlager, W. (2003). Microwave-asisted chemical reactions. *Chem. Engin. & Technol.,* 26(12): 1207-1216.

[17] Navarrete-Vazquez, G., Moreno-Diaz, H., Estrado-Soto, S., Torres piedra, M. (2007). Microwave assisted one pot synthesis of 2-(substituted phenyl)-1*H*-benzimidazole derivatives. *Synth. Comm.,* 37: 2815-2825.

[18] Wang, Z. X., Qin, H. L. (2005). Reaction of 1,3-dicarbonyl compounds with o-phenylendiamine or 3,3'diaminobenzidine in water or under solvent-free conditions via microwave irradiation. *J. of Hetero. Chem.*, 42:1001-1005.

[19] Kahveci B., Mentese, E. (2017). Microwave-assisted synthesis of benzimidazoles and their derivatives from 1994 to 2016-A review. *Curr. Microwave Chem.*, 4(1): 73-101.

[20] Rapolu, T., Pavan Kumar, K. V. P., Babu, K. R., Kumar Dende S., Nimmareddy, R. R., Reddy, L. K. (2019). Microwave assisted one pot synthesis of 2-ethylamino benzimidazole, benzoxazole and benzothiazole derivatives. *Synth. Comm.*, 49(10): 1308–1315.

[21] Shi, Y., Jiang, K., Zheng, R., Fu, J., Yan, L., Gu, Q., Zhang, Y., Lin, F. (2019). Design, microwave-assisted synthesis and in vitro antibacterial and antifungal activity of 2,5-disubstituted benzimidazole. *Chem. Biodiversity, 16*: e1800510.

[22] Goodman, L., Gilman, A., Hart, N. (1943). Preliminary investigations on the pharmacology of benzimidazole. *Federation Proc.*, 2:8O.

[23] Woolley, D. W., White, A. G. C. (1943). Production of thiamine deficiency disease by the feeding of a pyridine analogue of thiamine. *J. Biol. Chem.*, 149:285-289.

[24] Asteinza, J., Camacho-Carranza, R., Reyes-Reyes, R. E., Dorado Gonzales, V., Espinosa-Aguirre, J. J. (2000). Induction of cytochrome P450 enzymes by albendazole treatment in the art. *Environ. Toxicol. Chem.*, 9:31-37.

[25] Baliharova, V., Skalova, L., Maas, R. F. M., De Vrieze, G., Bul, S., Fink-Gremmels, J. (2003). The effects of benzimidazole anthelmintics on P450 1A in rat hepatocytes and Hepg2 cells. *Res. in Veterinary Sci.,* 75:61-69.

[26] Jiang, S., Meadows, J., Anderson, S. A., Mukkada, A. (2002). Antileismanial activity of the antiulser agent omeprazole. *Antimicrob. Agents*, *Chemother*, 46: 2569-2574.

[27] WHO, 2001. *ATC Classification and DDD Assignment*, fourth edition. Oslo, WHO Collaborating Centre for Drug Statistics Methodology.

[28] Velik, J., Baliharova, V., Fink-Gremmels, J., Bull, S., Lamka, J., Skalova, L. (2004). Benzimidazole drugs and modulation of biotransformation enzymes. *Res. in Veterinary Sci.,* 76: 95-108.

[29] Nicolas, J. M., Whomsley, R., Collart, P., Roba, J. (1999). In vitro inhibition of human liver drug metabolising enzymes by second generation antihistamines. *Chemico-Bio. Inter.,* 123:63–79.

[30] Tebbe, M. J., Jensen, C. B., Spitzer, W. A., Franklin, R. B., George, M. Ch., Phillips, D. L. (1999). The effects of antirhino- and enteroviral vinylacetylene benzimidazoles on cytochrome P450 function and hepatic porphyrin levels in mice. *Antiviral Res.,* 42: 25–33.

[31] Akhtar, W., Khan, M. F., Verma, G., Shaquiquzzaman, M., Rizvi, M. A., Mehdi, S. H., Akhter, M., Alam, M. M. (2017). Therapeutic evolution of benzimidazole derivatives in the last quinquennial period. *Eur. J. of Med. Chem.,* 126:705-753.

[32] El-Feky, S. A. H., El-Samii, Z. K. A., Osman, N. A., Lashine, J., Kamel, M. A., Thabet, H. K. (2015). Synthesis, molecular docking and anti-inflammatory screening of novel quinoline incorporated pyrazole derivatives using the Pfitzinger reaction II. *Bioorg. Chem.*, 58: 104-116.

[33] Mohammed, A. F., Abdel-Moty, S. G., Hussein, M. A., Abdel-Alim, A. A. (2013). Design, synthesis and molecular docking of some new 1,2,4-triazolobenzimidazol-3-yl acetohydrazide derivatives with anti-inflammatory-analgesic activities. *Arch. Pharm. Res.*, 36: 1465-1479.

[34] Kankate, R. S., Gide, P. S., Belsare, D. P. (2015). Design, synthesis and antifungal evaluation of novel benzimidazole tertiary amine type of fluconazole analogues. *Arab. J. Chem.*

[35] http://dx.doi.org/10.1016/j.arabjc.2015.02.002.

[36] Evans, C. W., Atkins, C., Pathak, A., Gilbert, B. E., Noah, J. W. (2015). Benzimidazole analogs inhibit respiratory syncytial virus G protein function. *Antivir. Res.* 121:31-38.

[37] Mavrova, A. T., Wesselinova, D., Vassilev, N., Tsenov, J. A. (2013). Design, synthesis and antiproliferative properties of some new 5-substituted- 2-iminobenzimidazole derivatives. *Eur. J. Med. Chem.*, 63: 696-701.

[38] Hranjec, M., Pavlovic, G., Zamola, G. K. (2012). Synthesis, crystal structure determination and antiproliferative activity of novel 2-amino-4-aryl-4,10-dihydro[1,3,5]triazino[1,2-a]benzimidazoles. *J. Mol. Struct.,* 1007: 242-251.

[39] Sharma, M. C., Sharma, S., Sahu, N. K., Kohli, D. V. (2013). 3D-QSAR kNN-MFA studies on 6-substituted benzimidazoles derivatives as nonpeptide angiotensin II receptor antagonists: a rational approach to anti-hypertensive agents. *J. Saudi. Chem. Soc.,* 17: 167-176.

[40] Rizzati, V., Briand, O., Guillou, H., Gamet-Payrastre, L. (2016). Effects of pesticide mixtures in human and animal models: an update of the recent literature. *Chem. Biol. Interact.,* 254:231-246.

[41] Hubera, S., Antonia, F., Schickanederb, C., Schickanederc, H., Bernhardta, G., Buschauer, A. (2015). Stabilities of neutral and basic esters of bendamustine inplasmacompared to the parent compound: kinetic investigations by HPLC. *J. Pharm. Biomed. Anal.,* 104: 137-143.

[42] Chunduri, R. H. B., Dannana, G. S. (2016). Development and validation of a high throughput UPLC-MS/MS method for simultaneous quantification of esomeprazole, rabeprazole and levosulpiride in human plasma. *J. Pharm. Anal.*, 6, 190-198.

[43] Zhang, Z., Ojo, K. K., Johnson, S. M., Larson, E. T., Geiger, P. H. J. A., Gonzalez, A. C., White, A. C., Parsons, M., Merritt, E. A., Maly, D. J., Verlinde, C. L. M. J., Voorhis, W. C. V., Fan, E. (2012). Benzoylbenzimidazole based selective inhibitors targeting cryptosporidium parvum and toxoplasma gondii calcium-dependent protein kinase-1. *Bioorg. Med. Chem. Lett.* 22, 5264-5267.

[44] Digwal, C. S., Yadav, U., Sakla, A. P., Ramya, P. V. S., Aaghaz, S., Kamal, A. (2016). $VOSO_4$ catalyzed highly efficient synthesis of benzimidazoles, benzothiazoles, and quinoxalines. *Tetrahedron. Lett.*, 57:4012-4016.

[45] O'Neil, M. J., Smith, M., Heckelman. (2001). *The Merck Index*, 13th ed. Merck & Co Inc., P-1785, 10074.

[46] Mariappan, G., Hazarika, R., Alam, F., Karki, R., Patangia, U., Nath, S. (2015). Synthesis and biological evaluation of 2-substituted benzimidazole derivatives. *Arabian J. of Chem.,* 8:715–719.

[47] Sontakke, V. A., Ghosh, S., Lawande, P. P., Chopade, B. A., Shinde, V. S. (2013). A simple, efficient synthesis of 2-aryl benzimidazoles using silica supported periodic acid catalyst and evaluation of anticancer activity. *ISRN Org. Chem.,* Article ID 453682, 7 pages.

[48] Nannapaneni, D. T., Gupta Atyam, V. S. S. S., Reddy, M. I., Sarva Raidu, Ch. (2010). Synthesis, characterization, and biological evaluation of benzimidazole derivatives as potential anxiolytics. *J. Young Pharm.* 2(3): *273-279.*

[49] Shaharyar, M., Mazumder, A., Salahuddin, Garg, R., Pandey, R. D. (2016). Synthesis, characterization and pharmacological screening of novel benzimidazole derivatives. *Arabian J. of Chem.,* 9(1): S342-S347.

[50] Demirel, S., Ayhan Kilcigil, G., Kara, Z., Güven, B., Onay Beşikci, A. (2017). Synthesis and pharmacologic evaluation of some benzimidazole acetohydrazide derivatives as EGFR inhibitors**.** *Turk J. Pharm. Sci.*, 14(3):285-289.

[51] Pérez-Villa nueva, J., Romo-Mancillas, A., Hernández-Campos, A., Yépez-Mulia, L., Hernández-Luis, F., Castillo, R. (2011). Antiprotozoal activity of proton-pump inhibitors. *Bioorg. Med. Chem. Lett.,* 21:7351-7354.

[52] Anand, K., Wakode, S. (2018). Synthesis, characterization and biological evaluation of benzimidazole derivatives. *Int. J. Pharm. Sci. Res*, 9(2):617-624.

[53] Barot, K. P., Manna, K. S., Ghate, M. D. (2017). Design, synthesis and antimicrobial activities of some novel 1,3,4-thiadiazole, 1,2,4-triazole-5-thione and 1,3-thiazolan-4-one derivatives of benzimidazole. *J. of Saudi Chem. Society,* 21, S35–S43.

[54] El-masry, A. H., Fahmy, H. H., Abdelwahed, S. H. A. (2000). Synthesis and antimicrobial activity of some new benzimidazole derivatives. *Molecules*, *5*:1429–1438.

[55] Ajani, O. O., Tolu-Bolaji, O. O., Olorunshola, S. J., Zhao, Y., Aderohunmu, D. V. (2017). Structure-based design of functionalized 2-substituted and 1,2-disubstituted benzimidazole derivatives and their in vitro antibacterial efficacy. *J. of Advanced Res.,* 8: 703–712.

[56] Padalkar, V. S., Borse, B. N., Gupta, V. D., Phatangare, K. R., Patil, V. S., Umape, P. G., Sekar, N. (2016). Synthesis and antimicrobial activity of novel 2-substituted benzimidazole, benzoxazole and benzothiazole derivatives. *Arabian J. of Chem.* 9: S1125–S1130.

[57] Doğan, İ. S., Ozel, A., Birinci, Z., Barut, B., Sellitepe, H. E., Kahveci, B. (2016). Synthesis of some novel 2-substitutebenzyl-(4)7-phenyl-1*H*-benzo[d]imidazoles in mild conditions as potent anti-tyrosinase and antioxidant agents. *Arch. Pharm. Chem. Life Sci.,* 349(11): 881–888.

[58] Gümüş, M. K., Doğan, İ. S., Barut, B., Ozel, A., Kahveci, B. (2017). Synthesis, Acetylcholinesterase, Butyrylcholinesterase and Tyrosinase Inhibition, and Antioxidant Studies of 2-[2-(substituted-benzyl)-4(7)-phenyl-1*H*-benzimidazol-1-

yl]acetohydrazides and methyl 2-[2-(substituted-benzyl)-4(7)-phenyl-1*H*-benzimidazol-1-yl]acetates. *Bulgarian Chem. Comm.*, I: 107-114.

Biographical Sketch

İnci Selin DOĞAN

Affiliation: Assis. Prof.

Education: Pharmacy, Pharmaceutical Chemistry

Business Address: Department of Pharmaceutical Chemistry, Faculty of Pharmacy, Karadeniz Technical University, Trabzon, Turkey

Research and Professional Experience: Pharmaceutical Chemistry, Medicinal Chemistry, Pharmacology, Pharmacy

Professional Appointments: -Head of the Department of Pharmacy Technology
-Head of the Pharmaceutical Chemistry Department

Publications from the Last 3 Years:

1. İnci Selin Doğan*, Hasan Erdinç Sellitepe, Nuran Kayıkçı, Rengin Reis, Hande Sipahi, Nurettin Yaylı. Synthesis and anticancer (MCF-7, PC-3) activities of new 2-hydroxy-2,2-bis(4-substitutedphenyl)-N'-[(1E)-(3/4-substitutedphenyl) methylene]-acetohydrazides. *Organic Communications*, 11:3; 142–148, 2018. http://doi.org/10.25135/acg.oc.47.18.06.107.
2. Hasan Erdinç Sellitepe, İnci Selin Doğan*, Gamze Eroğlu, Burak Barut, Arzu Özel. Synthesis and Investigation of Antioxidant, Metal Chelating and Enzyme Inhibition Activities of Some 4-Aryl-1,4-Dihydropyridine Derivatives., Marmara Pharmaceutical *Journal (Journal of Research in Pharmacy-J. Res Pharm.)* 2019; 23(4): 608-616. https://doi.org/10.12991/jrp.2019.168.
3. Suat Sari, Didem Kart, Suna Atasayar, İnci Selin Doğan, Zeynep Özdemir, İrem Bozbey, Melis Gencel, Şebnem Eşsiz Gökhan, Jóhannes Reynisson, Arzu Karakurt, Selma Saraç, Sevim Dalkara. Antifungal screening and in silico mechanistic studies of an in-house azole library. *Chemical Biology & Drug Design*, 2019. doi: 10.1111/cbdd.13587.
4. Doğan, İ. S.*, Ozel, A., Birinci, Z., Barut, B., Sellitepe, H. E., Kahveci, B. Synthesis of some novel 2-substitutebenzyl-(4)7-phenyl-1H-benzo[d]imidazoles

in mild conditions as potent anti-tyrosinase and antioxidant agents, *Archiv der Pharmazie-Chemistry in Life Sciences 2016*, 349(11), 881–888.

5. Doğan, İ. S., Saraç, S., Sarı, S., Kart, D, Gokhan, Ş. E., Vural, İ., Dalkara,S. New azole derivatives showing antimicrobial effects and their mechanism of antifungal activity by molecular modeling studies, *European Journal of Medicinal Chemistry* 130 (2017) 124-138. http://dx.doi.org/10.1016/j.ejmech.2017.02.035.
6. Gümüş, M. K., Doğan, İ. S.*, Barut, B., Ozel, A., Kahveci, B. Synthesis, Acetylcholinesterase, Butyrylcholinesterase and Tyrosinase Inhibition, and Antioxidant Studies of 2-[2-(substituted-benzyl)-4(7)-phenyl-1H-benzimidazol-1-yl]acetohydrazides and methyl 2-[2-(substituted-benzyl)-4(7)-phenyl-1H-benzimidazol-1-yl]acetates. *Bulgarian Chemical Communications*, 2017, 49(I), 107-114.
7. Fandaklı, S., Doğan, İ. S., Sellitepe, H. E., Yaşar, A., Yaylı,N. Synthesis, theoretical calculation and α-glucosidase inhibition of new chalcone oximes. *Organic Communications*, 2018, 11:1, 23-34.
8. İnci Selin Doğan*, Zeynep Özdemir, Suat Sari, İrem Bozbey, Arzu Karakurt, Selma Saraç, Synthesis, anticonvulsant activity, and molecular modeling studies of novel 1-phenyl/1-(4-chlorophenyl)-2-(1H-triazol-1-yl)ethanole ester derivatives, *Medicinal Chemistry Research,* 27:2171–2186, 2018, doi: 10.1007/s00044-018-2225-6.

Bahittin Kahveci

Affiliation: Prof.

Education: Chemistry, Organic Chemistry

Business Address: Department of Nutrition and Dietetics, Faculty of Health Sciences, Karadeniz Technical University, Trabzon, Turkey

Research and Professional Experience: Chemistry, Organic Chemistry, Medicinal Chemistry, Nutrition and Dietetics

Professional Appointments: Head of the Department of Nutrition and Dietetics

Publications from the Last 3 Years:

1. Menteşe E., Yilmaz F., Emirik M., Ülker S., Kahveci B., "Synthesis, molecular docking and biological evaluation of some benzimidazole derivatives as potent pancreatic lipase inhibitors," *Bioorganic Chemistry*, vol. 76, pp. 478-486, 2018.

2. Kantar G. K., Menteşe E., Beriş F. Ş., Şaşmaz S., Kahveci B., "Synthesis And Antimicrobial Activity Of Some New Triazole Bridged Benzimidazole Substituted Phthalonitrile And Phthalocyanines," *Revue Roumaine De Chimie*, vol. 63, pp. 59-65, 2018.
3. Menteşe E., Kahveci B., Mentese M., "Green protocol: catalyst-free synthesis of 2-substituted benzimidazoles under ultrasound irradiation," *Journal of Chemical Research*, pp. 329-331, 2018.
4. Mentese E., Bektaş H., Sokmen B. B., Emirik M., Çakır D., Kahveci B., "Synthesis and molecular docking study of some 5,6-dichloro-2- cyclopropyl-1H-benzimidazole derivatives bearing triazole, oxadiazole, and imine functionalities as potent inhibitors of urease," *Bıoorganıc & Medıcınal Chemıstry Letters*, vol. 27, pp. 3014-3018, 2017.
5. Doğan İ. S., Özel A., Birinci Z., Bulut B., Sellitepe H. E., Kahveci B., "Synthesis of Some Novel 2-Substitutedbenzyl-(4)7-phenyl-1H-benzo[d]imidazoles in Mild Conditions as Potent Anti-Tyrosinase and Antioxidant Agents," *Archıv Der Pharmazıe*, vol. 349, pp. 1-8, 2016.
6. Özil M., Yılmaz Etlik S., Ülker S., Kahveci B., "A Simple and efficient of novel inhibitors of alpha-glucosidase based on benzimidazole skeleton and molecular docking studies," *Bıoorganıc Chemıstry*, vol. 68, pp. 226-235, 2016.
7. Kahveci B., Menteşe E., Özil M., Yılmaz F., Serdar M., "A Convenient Synthesis of Bisbenzimidazoles from Iminoester Hydrochlorides Via Microwave Technique," *Journal Of Heterocyclıc Chemıstry*, vol. 53, pp. 975-980, 2016.

In: Benzimidazole
Editor: Aksel A. Vestergaard

ISBN: 978-1-53616-986-7

Chapter 2

CHEMISTRY AND MEDICINAL IMPORTANCE OF BENZIMIDAZOLE

Mohammad Rashid[1,*] and Asif Husain[2]
[1]Department of Medicinal Chemistry and Pharmacognosy,
College of Pharmacy and Dentistry, Buraydah Colleges,
Buraydah, Al-Qassim, Kingdom of Saudi Arabia
[2]Department of Pharmaceutical Chemistry, School of Pharmaceutical Education and Research, Jamia Hamdard, New Delhi, India

ABSTRACT

This book chapter gives an overview of Benzimidazole based synthetic medicine and their significant application in treating various ailments. The chemical reactions of Benzimidazole with other heterocyclic compounds/chemical reagents produce a lot of new substituted molecules, which have plentiful biological actions and potential pharmaceutical applications. Benzimidazole is considered a major class of heterocyclics and worldwide researchers have put great efforts in studying this nucleus in order to design and synthesize various new derived/hybrid molecules and identified their significant role in many classes of drugs such as antibacterial, antifungal, anthelmintic, antitumor/antiproliferative/ anticancer, anti-inflammatory, antiviral including anti-HIV, anticonvulsant, antihistaminic, antidepressant and antioxidant, and recently some researchers have got highly promising results of Benzimidazole derivatives in antiasthamatic and antidiabetic activities also. Benzimidazole has an inspiring molecular geometric structure that offers a suitable skeleton to build newer chemical entities which has encouraged medicinal chemists to design and synthesize novel and effective molecules as potential therapeutic agents. However, there is an urgent need to study the mechanism of action at molecular level of several pharmacological actions resulting from Benzimidazole scaffold through modern methods to furnish safer and effective new molecules for the treatment of various diseases.

* Corresponding Author's Email: rashidpharm2008@gmail.com; Rashid.Khalid@bpc.edu.sa.

Keywords: Benzimidazole, imidazole, heterocyclic compounds, chemical properties, synthetic reactions and pharmacological application

1. INTRODUCTION

The synthesis of exiting drug derivatives has been an important part and is aimed at modifying the action of drugs, particularly to reduce the side effects and to potentiate the drug action. Today more than 60% of drugs used in practice are synthetic derivatives and day by day the scope of synthetic medicinal chemistry is broadening.

Heterocyclic compounds are those cyclic compounds whose ring contains, besides carbon, one or more atom of other elements. The non-carbon atoms in such rings are referred to as hetero atoms. The most common hetero atoms are nitrogen, sulphur and oxygen. The heterocyclic ring is comprises of very core of the active moiety or the pharmacophore [1]. Azoles are a class of five membered ring, nitrogen heterocyclic ring compounds containing at least one other heteroatom (or non-carbon atom) of nitrogen, sulfur or oxygen and are considered to be derived from pyrrole, furan and thiophene by replacement of methane groups (—CH=) by pyridine type nitrogen (—N=) atoms from the different positions [1-2]. If azole contained: Oxygen is referred as oxazole, Sulphur is referred as thiazole, and Nitrogen is referred as pyrazole, imidazoles or imidazolines. This chapter covers various aspects of benzimidazole including chemistry, synthetic reaction, pharmacological prospects and its role as a multitargeting agent in multifactorial diseases.

Different classes of heterocyclic and fused heterocyclic compounds have been identified through molecular biology, empirical screening and rational drug development in search of potent and effective new molecules [3-4]. Especially nitrogen containing heterocyclic systems like azoles nucleus playing a vital role in the discovering of novel drugs with potential activity. In terms of searching it could be considered that the benzimidazole heterocyclics are of great importance in their biological as well as synthetic approach of medicinal chemistry. Scientists from around the world have reported remarkable anticancer/antitumor/ antiproliferative [5-9], anti-inflammatory [7-10], antiviral including anti-HIV [11-12], antibacterial [13-14], antifungal [15], antioxidant [16-17] and cysticidal activities [18-19] of different benzimidazole derivatives. Benzimidazoles are a class of heterocyclic compounds that have attracted significant interest in medicinal chemistry owing to their wide range of pharmacological application.

2. BENZIMIDAZOLE CHEMISTRY

Benzimidazole, a heterocyclic aromatic organic compound, is bicyclic in nature which consists fusion of benzene to the 4, 5-positions of imidazole ring [20-22]. Numbering of imidazole and benzimidazole nucleus is shown in Figure 1.

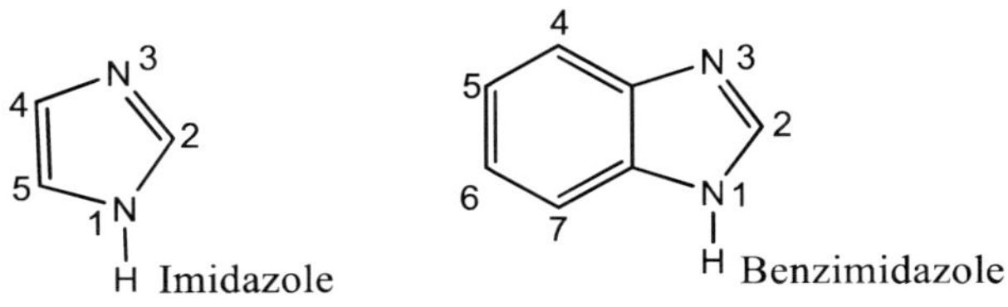

Figure 1. Numbering of imidazole and benzimidazole nucleus.

Benzene ring present in 1H-benzimidazole functions as an electron attracting group on imidazole and decreases imidazole electronic density which is of important role in chemical interaction and having varied biological activities and still of great scientific interest now a days [23] represented in Figure 2.

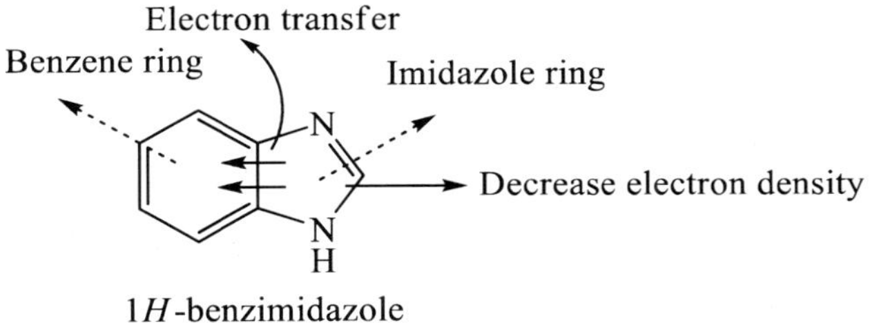

Figure 2. Electronic distribution of benzimidazole nucleus.

Benzimidazoles which contain a hydrogen atom attached to nitrogen in the 1-position readily tautomerize. Because of this tautomerism in benzimidazole certain derivatives which appear at first to be isomers are in reality tautomer, although two nonequivalent structures can be written; only one compound is known [24]. Both structures (1&2) represent the same compound as shown in Figure 3.

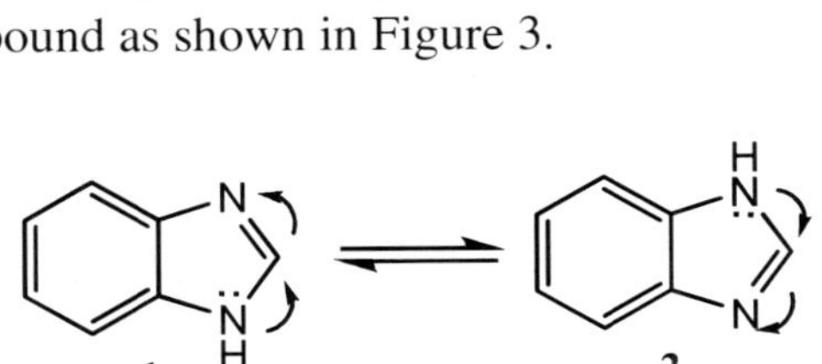

1 **2**

Figure 3. Tautomerism form of benzimidazole nucleus.

The overall reactivity of imidazole and benzimidazole is referred from sets of resonance structure in which the dipolar contributors have finite importance. These predict electrophilic attack in imidazole at N-3 or any ring carbon atom, nucleophilic attack at C-2 or C-1 and also the amphoteric nature of the molecule. In benzimidazole the nucleophilic attack is predicted at C-2. The reactivity of benzimidazole ion at the C-2 position with nucleophiles is enhanced as compared with the neutral molecule [25].

General methods for the synthesis of benzimidazole ring systems involve cyclization reactions. For which 1, 2-disubstituted benzenes of the appropriate type are the most common starting materials. This synthesis can usually be adapted to incorporate additional substituents in the benzenoid ring system, if required, for this reason electrophilic substitution reactions have not been widely used, although it is possible to carry out nitration, halogenation and similar reactions. Benzimidazole is attacked preferentially at the 5-position, imidazole at the 3 and 5-position and benzotriazole at the 4-position. The position of benzimidazole is susceptible to nucleophilic attack. Chloride is displaced from the 2-position of benzimidazole and N-alkyl benzimidazole by alkoxides, amines, thiols and other nucliophiles their activity towards nucleophiles is lower than corresponding benzoxazole and benzothiazole. 1-Substituted benzimidazoles can be directly aminated at the 2-position by heating with sodamide [26].

3. Benzimidazole Physical Properties

Benzimidazole with the imide nitrogen (hydrogen in the 1-position) is usually more soluble in polar solvents and less soluble in organic solvents. With the introduction of other nonpolar substituents in various positions of the benzimidazole ring, the solubility in nonpolar solvents is increased; for example, 2-methylbenzimidazole is easily soluble in ether. Benzimidazole is also distillable and distills unchanged above 300°C.

Benzimidazole is weakly basic, being somewhat less basic than the imidazole. Accordingly, they are in general soluble in dilute acids. Benzimidazole is also sufficiently acidic to be generally soluble in aqueous alkali and form N-metallic compounds. The acidic properties of the benzimidazole, like those of the imidazole [27], seem to be due to stabilization of the ion by resonance. The dipole moment of benzimidazole has been determined, the values that have been obtained being 3.93 D (in dioxane) and 4.08 D [28]. The infrared spectra of "pseudo bases" have been studied and indicate the presence of an amide linkage in these substances [29].

4. Benzimidazole Chemical Properties

The benzimidazole ring possesses a high degree of stability. Benzimidazole, for example, is not affected by concentrated sulfuric acid when heated under pressure to 270°C, or by vigorous treatment with hot hydrochloric acid or with alkalis. Oxidation cleaves the benzene ring of benzimidazole only under vigorous conditions [30]. The benzimidazole ring is also quite resistant to reduction however tetra hydro and hexahydrobenzimidazoles in which the benzene ring is reduced may be prepared by catalytic reduction under certain conditions [31].

Benzimidazole contains a phenyl ring fused to an imidazole ring. Historically the first benzimidazole was prepared in 1872 by Hoebrecker who obtained 2, 5 or 2, 6-dimethylbenzimidazole by the reduction of 2-nitro-4-methylacetanilide. Several years later Ladenburg brained the same compound by refluxing 3, 4-diaminotoluene with acetic acid. The benzimidazoles are known also as benziminazoles or benzoglyoxalines. Thus, benzimidazole according to this nomenclature would be called methenyl-o-phenylenediamine and 2-methylbenzimidazo. Also, they have been named as derivatives of the grouping composing the imidazole portion of the ring [32]. Thus, for example, benzimidazole has also been called 2 (*3H*)-benzimidazolone (3) and 2(*3H*)-benzimidazolethione (4) are also known as o-phenylene urea and o-phenylenethiourea respectively shown in Figure 4.

Figure 4. 2(3H)-benzimidazolone (3) and 2(3H)-benzimidazolethione (4).

Benzimidazoles which contain a hydrogen atom attached to nitrogen in the 1- position readily tautomerize (5 and 6) as shown in Figure 5. This tautomerism is analogous to that found in the imidazoles and amidines. In fact, the benzimidazoles may be considered as cyclic analogs of the amidines. Because of this tautomerism in benzimidazoles certain derivatives which appear at first to be isomers are in reality tautomers although two non-equivalent structures can be written only one compound is known. This may be illustrated with 5 or 6-methylbenzimidazole [33].

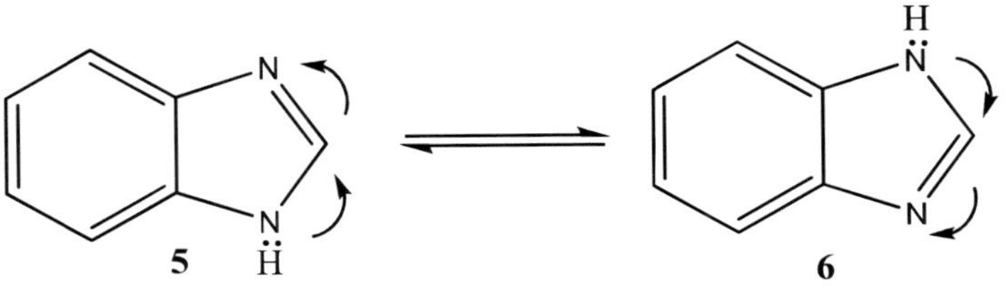

Figure 5. Benzimidazoles tautomerism.

Thus, 5-methylbenzimidazole (7) is a tautomer of 6-methylbenzimidazole (8) and both structures (7 & 8) represent the same compound as shown in Figure 6. When the group attached to the nitrogen in the l-position is larger than hydrogen such tautomerism is not indicated and isomeric forms exist. Thus 1, 5-dimethyl benzimidazole and 1, 6-dimethylbenzimidazole are separate and distinct compounds.

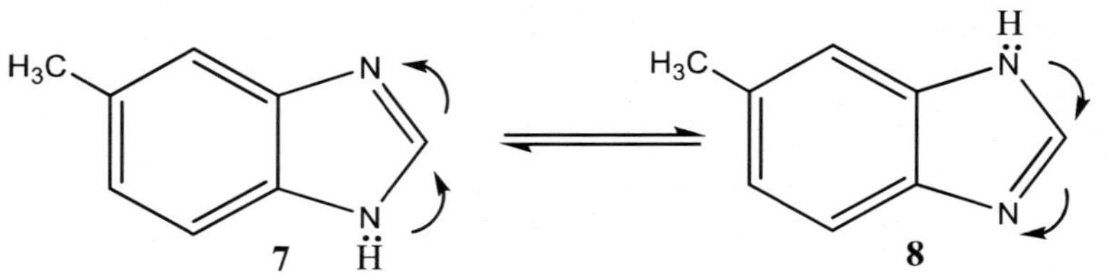

Figure 6. 5-Methylbenzimidazole (7) a tautomer of 6-methylbenzimidazole (8).

5. Benzimidazole Nucleus Value

5.1. Substitution at the N-1 Position

Ionic liquids have been reported to accelerate slow N-benzylation reactions of benzimidazole (8) utilizing dibenzyl carbonate (9) as an alkylating reagent [34]. An additional rate enhancement was observed when microwave irradiation applied in this reaction to afford the N-benzyl benzimidazole (10) (Scheme 1).

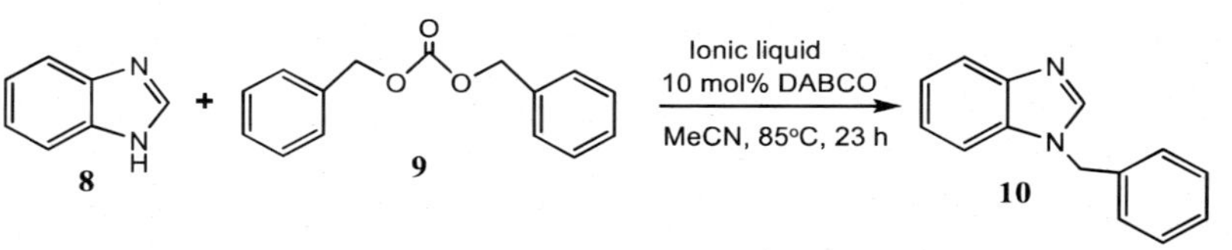

Scheme 1. Benzimidazole substitution reaction at the N-1 Position.

Thus several derivatives of N-substituted benzimidazole (11) and (12) (Scheme 2) have been prepared as antihepatitis B virus activity and several N-alkyl benzimidazole derivatives (13) behave as selective androgen receptor antagonists [35-36].

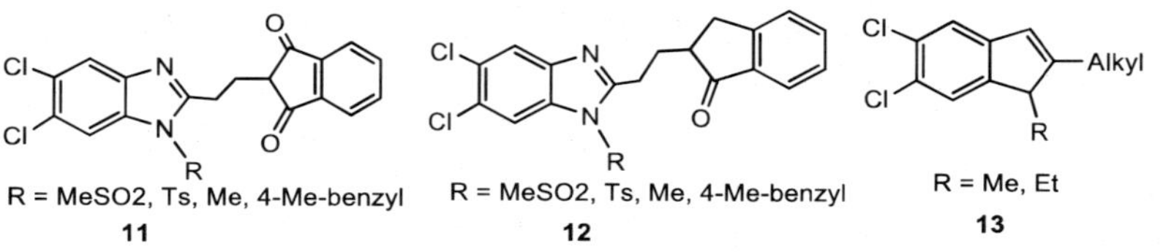

Scheme 2. N-Substituted benzimidazole derivatives.

5.2. Direct Coupling at the C-2 Position

Functionalization of C–H bonds of heterocycles to C-acylation (16) is an important synthetic reaction that is used to build important bioactive structures (Scheme 3). Recently palladium and copper mediated C-2-arylations of benzimidazole (14) with aryl iodides (15) under ligand less and base free conditions have been described [37].

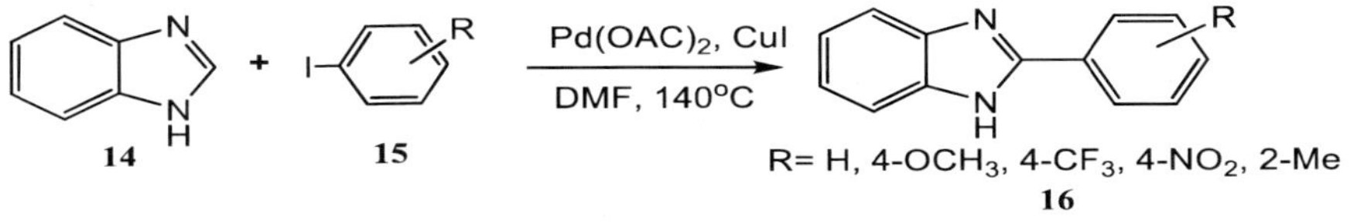

Scheme 3. Benzimidazole direct coupling reaction at the C-2 position.

5.3. Doebner-Von Miller reaction at the C-7 position

Imidazoquinolines (18) have been synthesized in good yield by coupling benzimidazole (17) with dimethyl trans-2-ketoglutaconate under Doebner-Von Miller reaction conditions in dichloromethane. The compound (18) was demethylated using hydro bromic acid in glacial acetic acid, reesterifed with methanolic hydrochloric acid and oxidized to afford benzimidazole-quinone (19) [38] (Scheme 4).

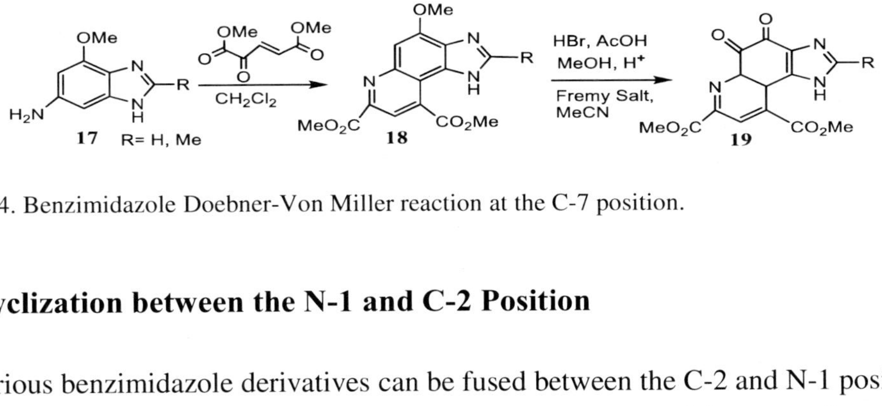

Scheme 4. Benzimidazole Doebner-Von Miller reaction at the C-7 position.

5.4. Cyclization between the N-1 and C-2 Position

Various benzimidazole derivatives can be fused between the C-2 and N-1 positions in order to build novel heterocyclic ring systems. For example, polyheterocyclic structures such as benzimidazoquinazolines made up of two fused heterocyclic rings often possess potent biological activity, like antiproliferative and DNA-intercalated activity [39], antifertility activity [40], anticonvulsant activity and myorelaxant activity [41]. These benzimidazoquinazoline compounds (22) have been obtained by the condensation of 2-cyanobenzothiazoles orbenzoxazoles (20) with 2-(2-aminophenyl) benzimidazole (21) under microwave conditions in the presence of graphite as a catalyst [42] (Scheme 5).

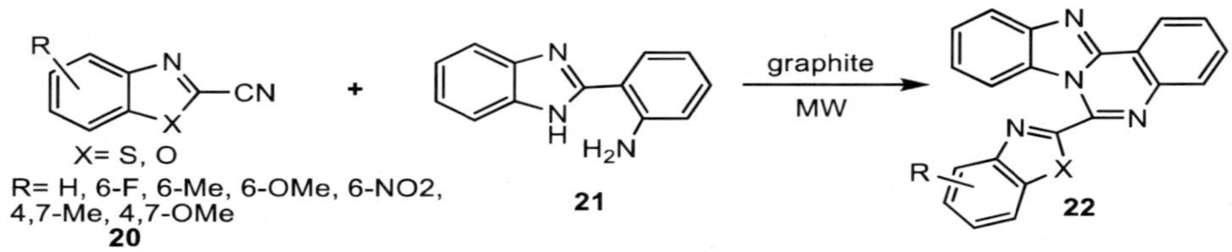

Scheme 5. Benzimidazole cyclization reaction between the N-1 and C-2 position.

5.5. Benzimidazole Derived Metal Complexes

A series of nickel (II) complexes (23) and (24) ligated by 2-(2-benzimidazole)-pyridine derivatives containing N and O donor atoms and nickel dichloride hexahydrate have been prepared [43] (Scheme 6). Interestingly, benzimidazole derived copper (II) and nickel (II) complexes have revealed antibacterial, antifungal and DNA intercalated activities [44], whereas lanthanide (III) complexes exhibited seed germination inhibition activity [45].

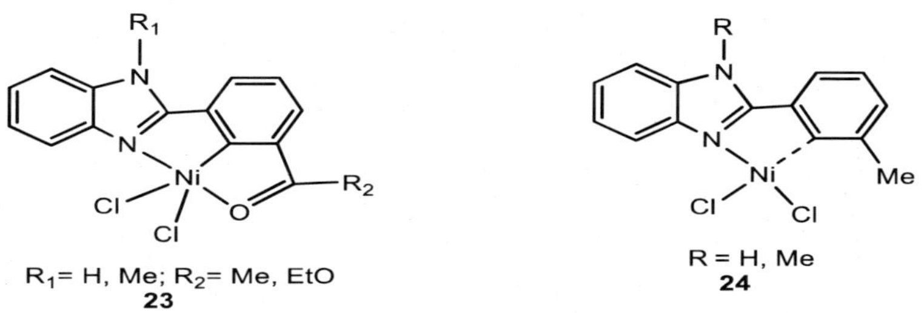

Scheme 6. Benzimidazole derived metal complexes.

6. Benzimidazole Ring Reactions

6.1. 1 and 3-Nitrogens Concerning Reaction

Benzimidazoles react with acids to form salts freely, so that made monohydric chloride, monopicrate, mononitrate and monoacetate etc. Benzimidazole also forms a salt with 2-nitro-l, 3-indanedione and copper azide [46]. The reaction of Benzimidazole ring occurs by one of following methods.

6.1.1. Alkylation Reaction

Benzimidazoles (25), upon alkylation with alkyl halides (26) yield 1-alkylbenzimidazoles (27) and under more vigorous conditions to form 1, 3-dialkylbenzirnidazolium halides [47] (Scheme 7).

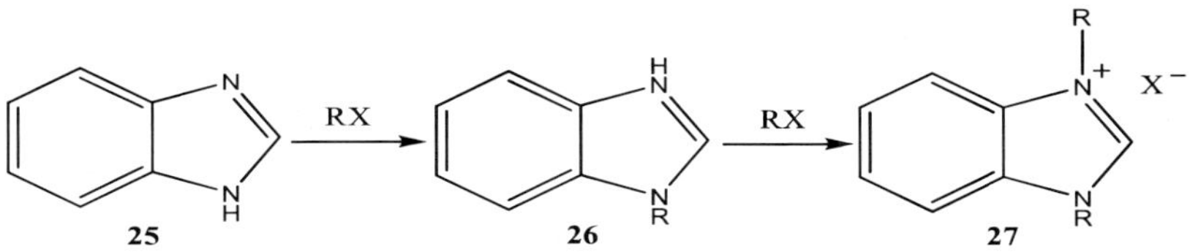

Scheme 7. Benzimidazole alkylation reaction with alkyl halides.

6.1.2. Acylation Reaction

N-Acylbenzimidazole (29) may be prepared by the action of acid chlorides or anhydrides on Benzimidazoles (28). The reaction is usually carried out in the absence of water. In the presence of water and especially in alkaline solution cleavage the imidazole ring. 1-Benzoyl benzimidazole and l-acetylbenzimidazole prepared by the action of benzoyl chloride and acetyl chloride, respectively on benzimidazole (Scheme 8). L-Benzoyl and l-acetyl derivatives have been prepared by the action of these acid chlorides on the silver salts of Benzimidazole [48-49].

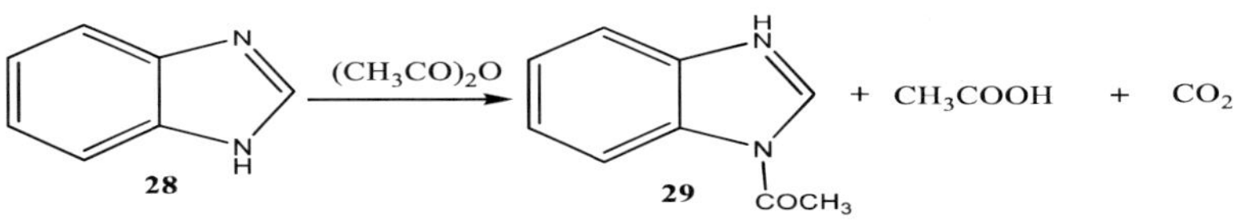

Scheme 8. Benzimidazole acylation reaction with acid chlorides or anhydrides.

6.2. Reducing Reaction

Until very recently it was thought that the benzimidazole ring was stable to reduction. Treatment of benzimidazole, 2-methylbenzimidazole, 2, 5 or 2, 6-dimethylbenzimidazole with red phosphorus and hydro iodic acid even to 300°C gives no reduced product. Red phosphorus and hydro iodic acid have been used in fact to cleave aryl amino (31) groups from Benzimidazole (30) without affecting the benzimidazole ring in the process (Scheme 9). Sodium in refluxing absolute alcohol also causes no reduction of the benzimidazole ring [50-51]. Catalytic reduction of benzimidazole even under high pressure with nickel as the catalyst is reported to give negative results [52-53].

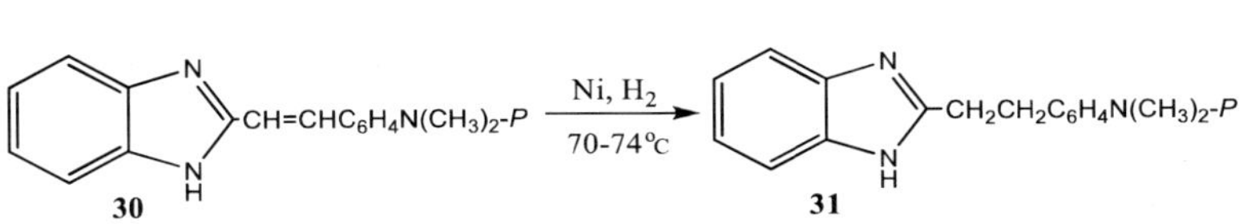

Scheme 9. Benzimidazole reducing reaction with red phosphorus.

Hartmann and Panizzonhowever first showed that certain benzimidazoles could be reduced to 4, 5, 6, 7-tetrahydrobenzimidazoles (32) in good yield with platinum (Adam's catalyst) in acetic acid solution at atmospheric pressure (Scheme 10). The reduction is successful only with those benzimidazoles which contain a substituent in the 2-position. Thus 2-methyl, 2-ethyl and dimethylbenzimidazoles give the corresponding tetra hydro benzimidazoles and 2-phenyl benzimidazole yields 2-cyclohexyl tetrahydrobenzimidazole. Benzimidazole and those benzimidazoles containing a substituent in the l-position (but not in the 2- position) could not be reduced with platinum according to this method [54].

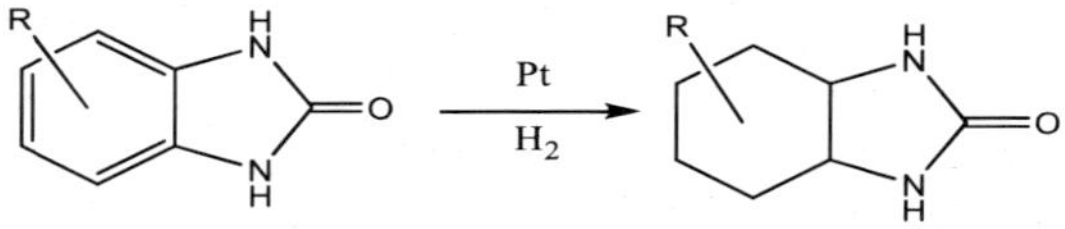

R = - (CH_2)nCOOH (where n = 0, 3,4). **32**

Scheme 10. Benzimidazole reducing reaction with Adam's catalyst.

7. Imidazolering Reaction

The imidazole ring of Benzimidazole may be readily cleaved by one of several methods:

7.1. Aroyl-Halide Reaction

Benzimidazoles when treated with benzoyl chloride yield first N-benzoylbenzimidazoles. In the presence of additional acid chloride and water further reaction may occur and N, N'-dibenzoyl-*o*-phenylenediaminea (33) are obtained [55] (Scheme 11).

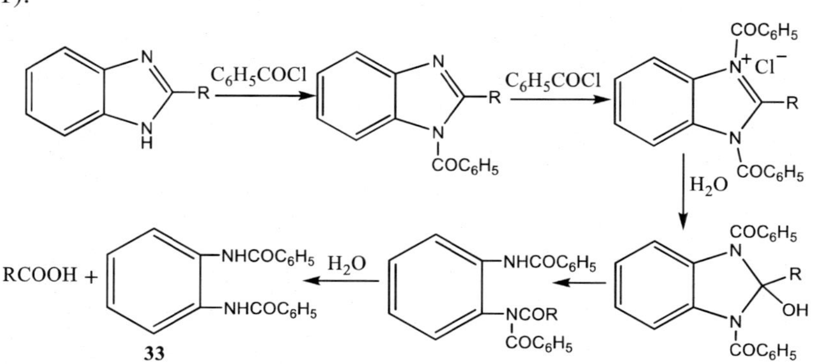

33

Scheme 11. Benzimidazole reducing reaction with benzoyl chloride.

7.2. Halogenation Reaction

When 2,5 or 2,6-dimethylbenzimidazolein an aqueous acid solution **is** treated with a saturated solution of bleaching powder at 35°C to form l-chloro-2,5 or 2,6-dimethylbenzimidazole (34) is obtained [56]. The N-chloro compound loses chlorine quite readily even at relatively low temperatures. When heated under reflux in benzene solution a rearrangement of the chlorine atom to the benzene ring takes place (35). The N-chloro derivative (36) of this compound may then be prepared by treatment again with bleaching powder (Scheme 12).

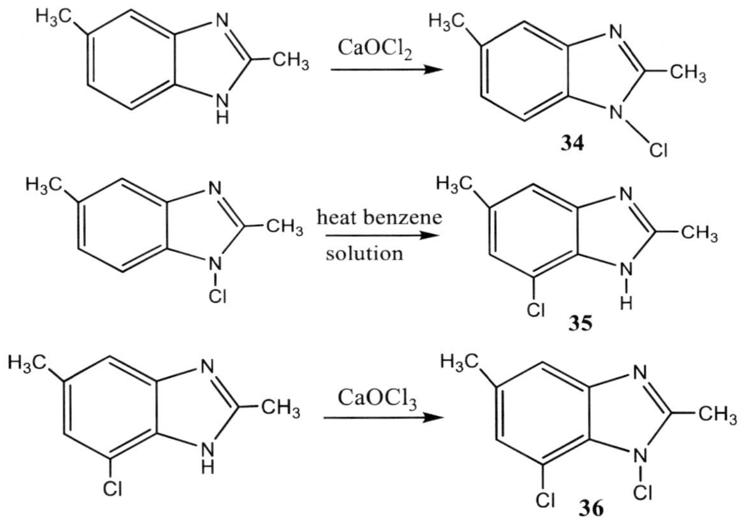

Scheme 12. Benzimidazole chlorination reaction.

The bromination of 2-methylbenzimidazole has been studied by Baczynski and Von Niementowski reaction. The addition of two moles of bromine with the cold acetic acid solution in 2-methylbenzimidazole gives a tetrabromo derivative (37) (Scheme 13). 2-Methyl-6-bromo benzimidazole and 2-methyl-4-bromobenzimidazole under the same conditions give 2-methyl-2, 3, 4, 6-tetrabromobenzimidazolinhey. 2-Methyl-3, 4, 6-tribromo benzimidazole at steam bath temperature adds bromine across the 2, 3- double bond [57].

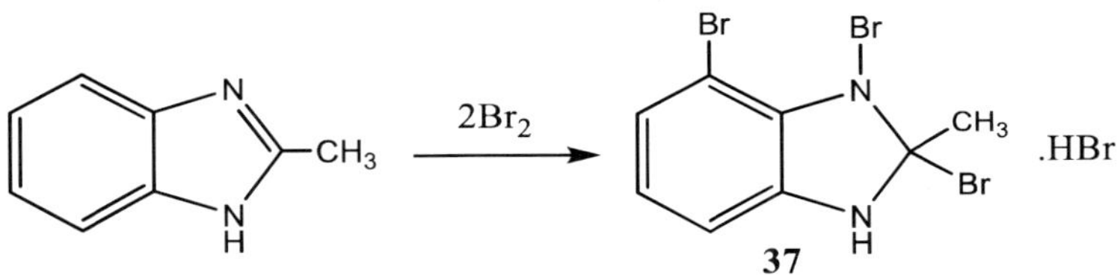

Scheme 13. Benzimidazole bromination reaction.

7.3. Nitration Reaction

In most cases nitration appears to take place preferentially at the 5- or 6-position. However the nitro group may also enter the 4-or 7-position especially if the 5-or 6-position is blocked.

8. Benzimidazole Substituent Reactions

A number of reactions of benzimidazoles take place transformations such as the conversion of an acid to its ester etc.

8.1. 2-Methyl or Methylene Reactions

The methyl group of 2-methylbenzimidazoles is comparable in its activity to the methyl group of picoline, quinaldine or methyl ketones and shows most of the same reactions of these compounds (Scheme 14). The benzimidazole ring like the pyridine and quinoline ring because of its electron-attracting nature imparts a positive character to the carbon atom of the 2-methyl group. 2-methyl benzimidazoles (38). For example react with aromatic aldehydes in aldol-type condensations in a manner analogous to form picoline and quinaldine [58].

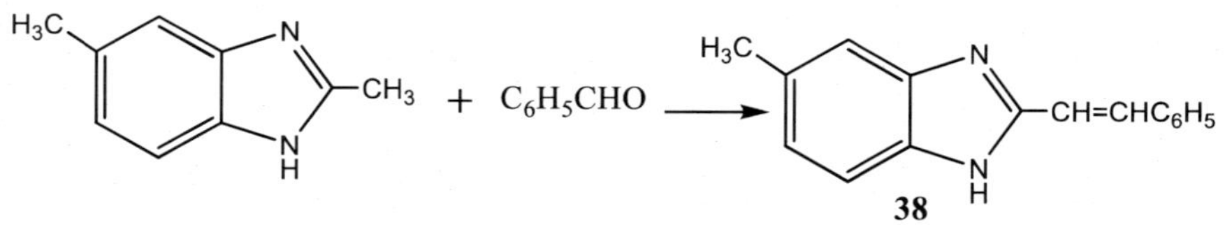

38

Scheme 14. Benzimidazole methylation.

8.2. Carboxylic Acid Reactions

Benzimidazoles containing a carboxyl group in the 2-position readily undergo decarboxylation on Heating [59]. 2-Benzimidazolecarboxylicacid on heating above its melting point for example yields benzimidazole (39). 5 or 6-Methyl-2-benzimidazolecarboxylicacid is decarboxylated in a similar manner by melting or boiling the acid in acetic acid [60]. l-Phenyl-5-nitro-2- benzimidazole carboxylic acid and the corresponding ethyl ester may be decarboxylated by heating with concentrated hydrochloric acid in a sealed tube at150°C. 2-Benzimidazolecarboxylicacid may not be converted to its acid chloride with phosphorus chlorides or thionyl chlorides (Scheme 15).

With excess thionyl chloride a yellow cyclic diamide containing no chlorine is obtained in 86 percent yield [61].

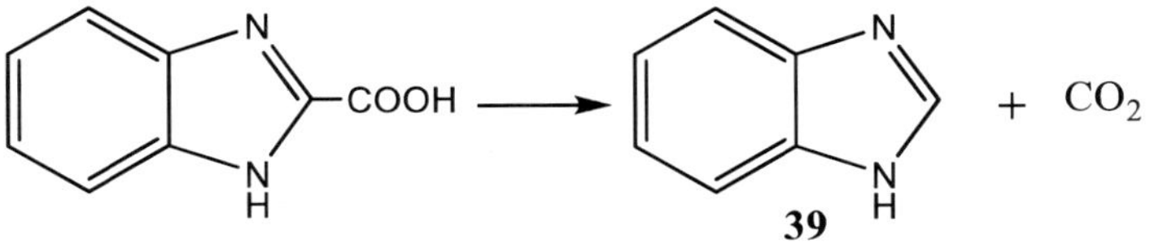

Scheme 15. Benzimidazole decarboxylation reaction.

8.3. 2-Haloalkyl Reactions

2-Chloromethylbenzimidazole (40) and other 2-(a-halo alkyl) beneimidazoles are highly reactive. Structurally they are related to allyl and benzyl halides since they possess the same =CCH and l grouping [62]. The reactivity of these benzimidazoles is probably due to the existence of such resonating structures as (41). 2-Chloromethylbenzimidazoleis quantitatively hydrolyzed with boiling water in 30-60 min. Other 2 or chloroalkylbenzimidazoles that were studied are hydrolyzed in a somewhat shorter time and the 2-(chloroisopropyl) derivative is completely hydrolyzed by water at room temperature (Scheme 16).

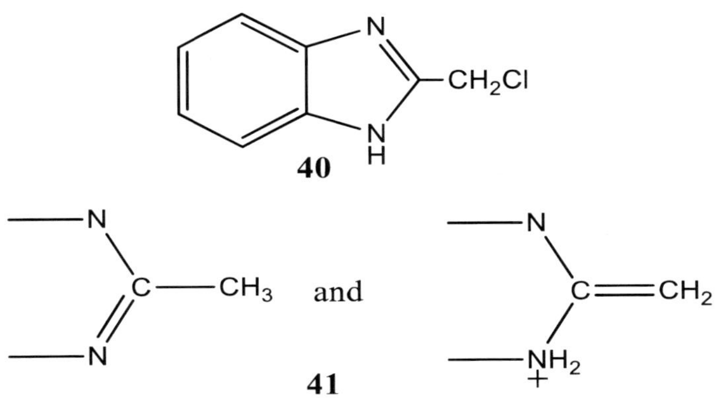

Scheme 16. Benzimidazole 2-haloalkyl reaction.

8.4. 2-(*3H*)-Benzimidazole Reactions

2(*3H*)-Benzimidazolones or 2-hydroxybenzimidazolesare extremely stable substances. 2-(*3H*)-Benzimidazolone is not split by treatment with benzoyl chloride in alkaline solution [63]. 2(*3H*)-benzimidazolones show many of the reactions of 2-hydroxypyridines and 2-

hydroxyquinolines for example treatment of 2-(*3H*)-benzimidazolone with phosphorus oxychloride or phosphorus pentachloride yields 2-chloro derivative (42). 2-Hydroxybenzimidazolea have been alkylated on the oxygen in good yields (43) by treatment with chloroacetic acid in sodium carbonate solution [64] (Scheme 17).

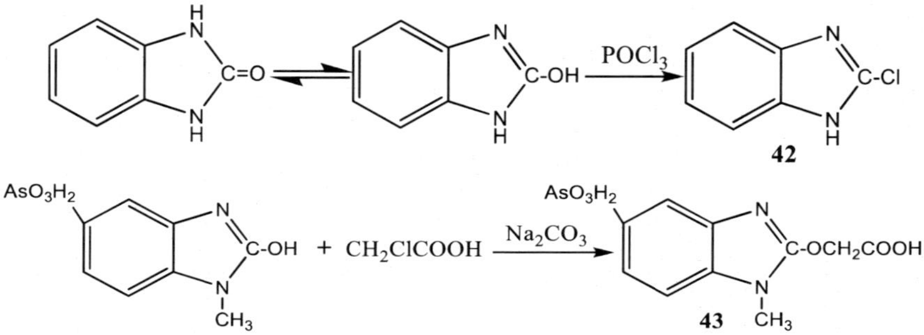

Scheme 17. Benzimidazole reaction with phosphorus oxychloride or phosphorus pentachloride.

2(*3H*)-benzimidazolethiones or 2-mercaptobenzimidazolesare generally stable substances and are soluble in dilute alkali. Alkylation occurs readily with replacement of the mercapto hydrogen to yield S-alkylated derivatives and a number of these derivatives have been prepared. 2(*3H*)-benzimidazolethione for example may be alkylated (44) with chloroacetic acid by refluxing in 2N sodium hydroxide solution [65-66] (Scheme 18).

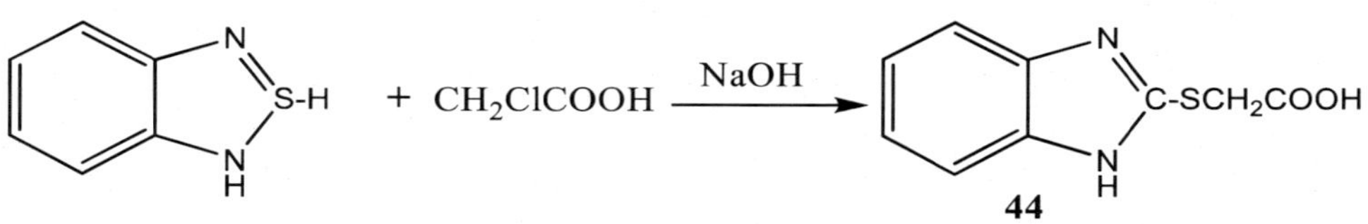

Scheme 18. Benzimidazole reaction with chloroacetic acid.

8.5. Aminobenzimidazoles Reaction

2-Aminobenzimidazole is soluble in water and alkali.2-Aminobenzimidazole is converted to 2-(*3H*)-benzimidazolone upon heating with barium hydroxide and a little water to 180-190°C or by treatment with nitrous acid [67]. Hypochlorous or hypobromous acid causes a series of color changes to take place [68]. Treatment with two moles of cyanogen bromide in the presence of water and potassium bicarbonate or treatment with biuret at 180°C yields a tricyclic compound (45) (Scheme 19).

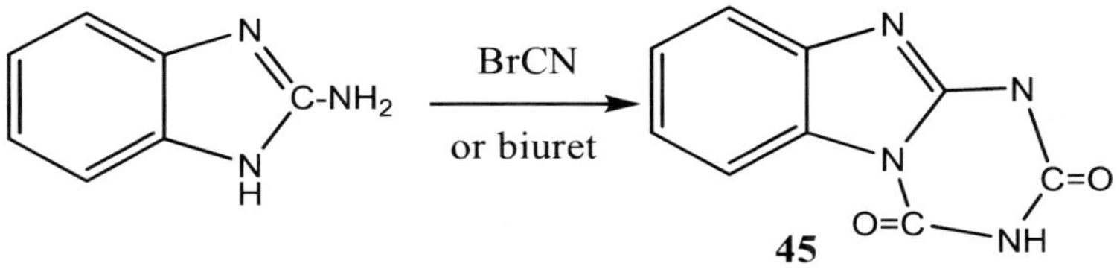

Scheme 19. Aminobenzimidazoles reaction with barium hydroxide.

8.6. Oxidation Reaction

Benzimidazoles are stable to oxidation [69]. Under highly vigorous conditions of oxidation (potassium permanganate in hot alkaline solution), it is possible to partially oxidize benzimidazole to obtain a small amount of imidazole dicarboxylic acid (46). Because of stability of the benzimidazole ring to oxidation it is possible to oxidize substituent groups without affecting the ring [70]. By the oxidation of substituent groups a variety of benzimidazole carboxylic acids have been prepared.2-Benzimidazolecarboxylic acid (47) may be conveniently prepared in good yield by the oxidation of 2-hydroxymethylbenzimidazole which may be prepared by the action of glycolic acid on o-phenylenediamine [71] (Scheme 20).

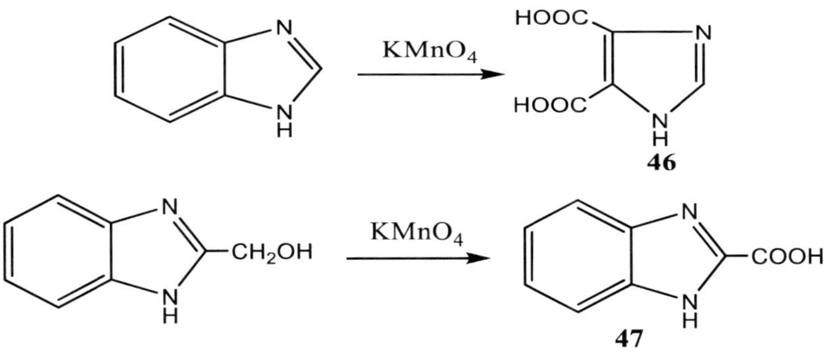

Scheme 20. Benzimidazoles reaction with potassium permanganate.

9. Synthetic Methods for Benzimidazole Molecules

Practically all synthesis of benzimidazole start with benzene derivatives possessing nitrogen containing functions ortho to each other, that is the starting material possess the functional unit designated by the given formula (48) as shown below (Scheme 21). Following are different methods by which we can prepare benzimidazole molecules.

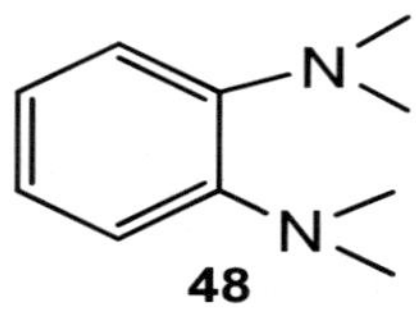

48

Scheme 21. Benzene ortho amine derivative.

9.1. Method 1; Traditional Synthetic Method

Traditionally, benzimidazole (51) have most commonly been prepared from the reaction of 1,2-diaminobenzenes (49) with carboxylic acids (50) under harsh dehydrating reaction conditions, utilizing strong acids such as polyphosphoric acid, hydrochloric acid, boric acid or p-toluene sulfonic acid [72]. However, the use of milder reagents, particularly Lewis acids [73], inorganic clays [74], or mineral acids has improved both the yield and purity of this reaction [75], example as given below (Scheme 22).

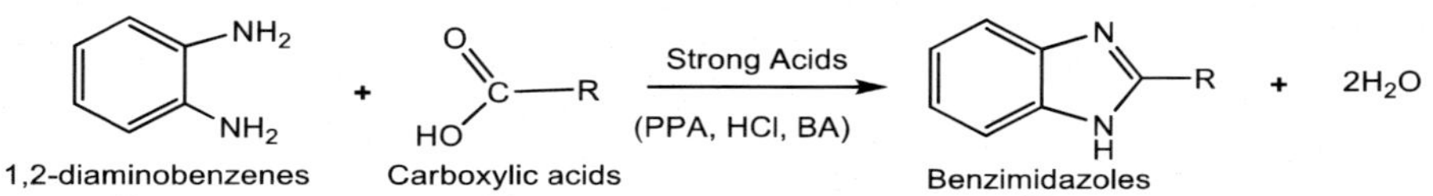

Scheme 22. Benzimidazole traditional synthetic method.

On the other hand, the synthesis of benzimidazole via the condensation of 1, 2-diaminobenzenes with aldehydes requires an oxidative reagent to generate the benzimidazole nucleus. Various oxidative reagents, such as nitrobenzene, benzoquinone, sodium metabisulfite, mercuric oxide, lead tetra acetate, iodine, copper (II) acetate, indium perfluorooctane sulfonates, ytterbium perfluorooctane sulfonates and even air, have been employed for this purpose [76-77].

9.2. Method 2; Benzimidazole Ring Closure Method

Various substituted benzimidazole have been synthesized in very good yields in solvent free conditions from 1, 2-diaminobenzene and aldehydes in the presence of titanium (IV) chloride as a catalyst. The method is applicable to most aromatic, unsaturated and aliphatic aldehydes and to substituted1, 2-diaminobenzenes without significant differences [78] (Scheme 23). A palladium catalyzed N-acylation reaction provided a novel synthesis of benzimidazole (52) from (o-bromophenyl) amidine (53) precursors under microwave irradiation. It has been found that 50% aqueous dimethyl ether (DME) is an optimal solvent for the reaction and that catalyst loading of palladium can be reduced to 1 mol.

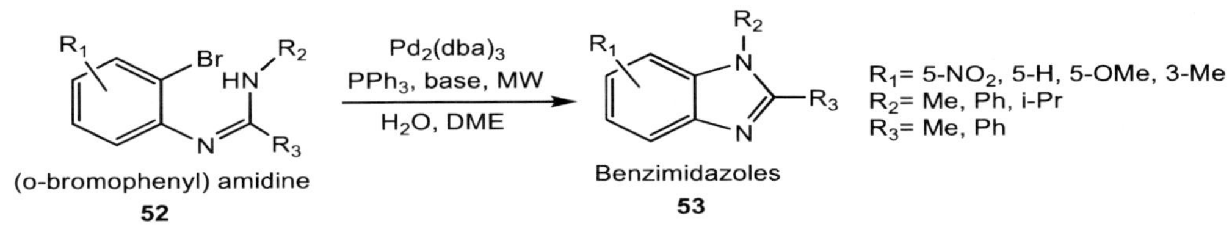

Scheme 23. Benzimidazole ring closure synthetic method.

9.3. Method 3; Diels-Alder Reaction Method

Imidazole o-quinodimethane intermediate (54), synthesized from 2-bromo-4, 5-bis (bromo methyl) imidazole via N-bromosuccinimide mediated bromination of imidazole undergoes a Diels-Alder reaction with several symmetrically and asymmetrically substituted dienophiles to yield the benzimidazole derivatives (55-58) in moderate yields [79] (Scheme 24).

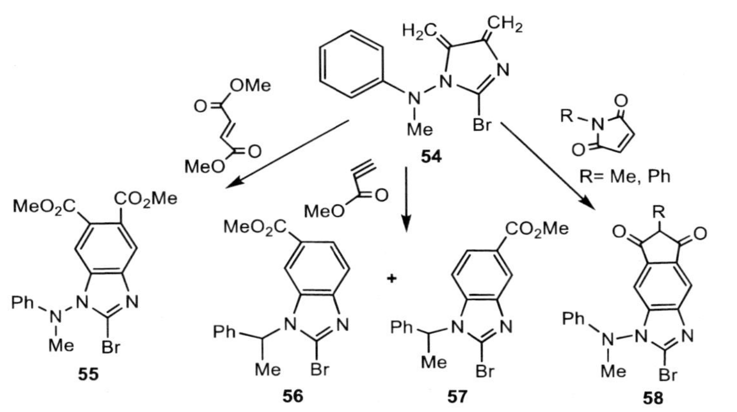

Scheme 24. Benzimidazole Diels-Alder synthetic method.

9.4. Method 4; Microwave Assisted Synthetic Method

The use of microwave irradiation as a source of heat in synthetic chemistry has been heralded as a promising method of increasing productivity, quality and reducing reaction time since its first use in 1986 [80]. Recently, 2-alkyl and 2-aryl-substituted benzimidazole derivatives (61) have been synthesized from 1,2-diaminobenzene dihydrochloride (59) and its corresponding acids (60) in the presence of polyphosphoric acid using microwave assisted methods. The reaction time required for the synthesis of benzimidazole derivatives (61) was reduced to minutes by this method as compared to conventional synthesis, which requires up to four hours of heating to complete the reaction (Scheme 25). Furthermore, it was found that the application of microwave irradiation increased yields by 10-50% (Table 1) [81].

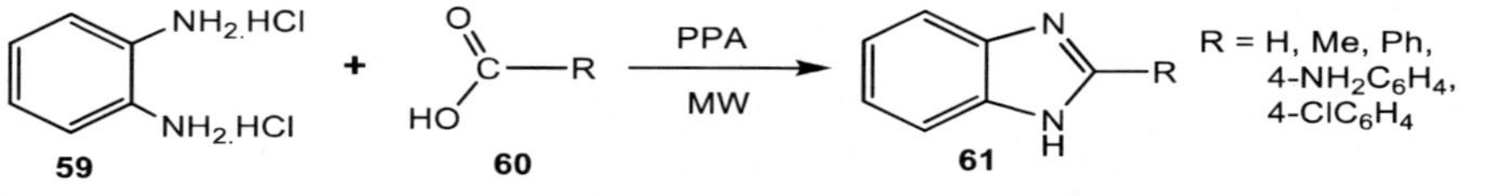

Scheme 25. Benzimidazole microwave assisted synthetic method.

Table 1. Yield and reaction time for benzimidazole synthesis by using microwave irradiation

R	RT	Yield (%)	MW	Yield (%)
H	2h	80	1 min 20 s	92
Me	45 min	48	1 min 20 s	89
Ph.	4h	34	4 min 30 s	84
4-$NH_2C_6H_4$	4h	57	5 min 20 s	95
4-ClC_6H_4	4h	43	4 min 30 s	89

RT- Reaction time, MW-Microwave irradiation.

9.5. Method 5; Benzimidazole Natural Products

Benzimidazole derived alkaloids are rare in nature and only a few examples of these natural products can be found in the literature. On the other hand, the occurrence of the imidazole skeleton in various natural sources is quite common. The benzimidazole alkaloid kealiiquinone (62) has been isolated from a yellow button like Micronesian sponge species of Lauretta [82]. Recently, Nakamura et al. successfully synthesized a regioisomer (63) of kealiiquinone [83] (Scheme 26).

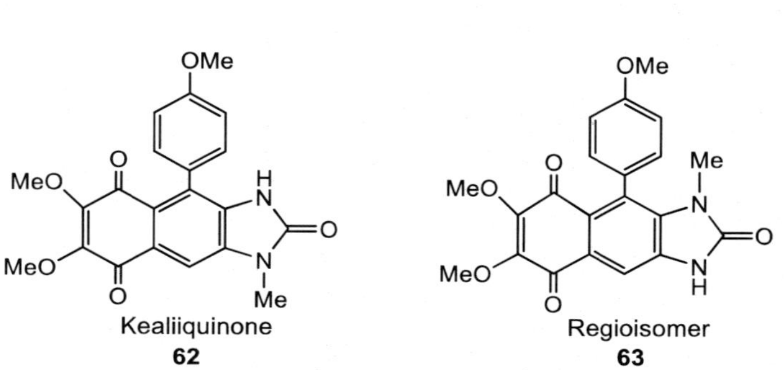

Scheme 26. Benzimidazole ring containing natural products.

10. Benzimidazole Scaffold for Various Biological Applications

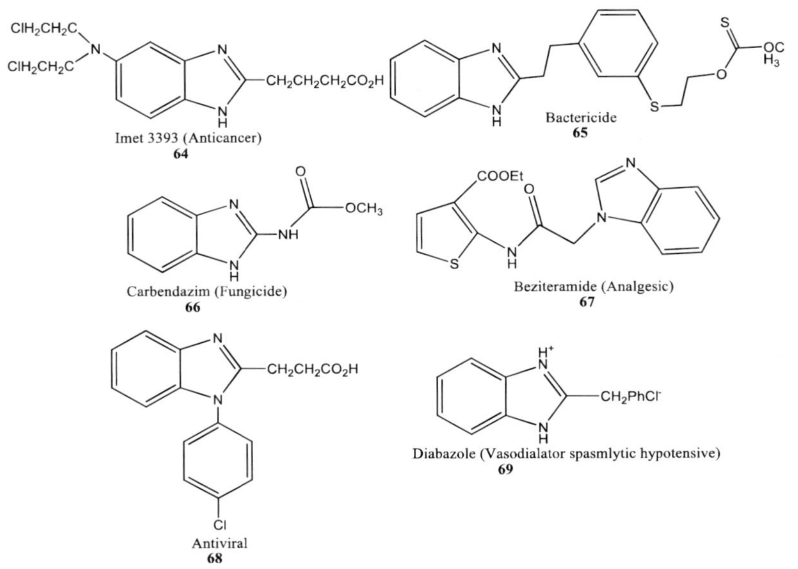

Scheme 27. Benzimidazole scaffold for various biological activity.

Various structural modifications were carried out in benzimidazole ring system to obtain promising compounds. These structural changes resulted in some fruitful biological activities of the compounds; some of these activities are described here such as anticancer (64) [84] bactericidal (65) [85], fungicidal (66) [86-87], analgesic (67) [88] and antiviral (68) [89], cardiovascular selective (69) [90] and some derivatives evaluated for HIV-1 infectivity [91] (Scheme 27).

10.1. Anticancer/Antitumor/Antiproliferative Activity

In cancer chemotherapy there is currently much interest in the design of small molecules that bind to DNA with sequence selectivity and noncovalent interactions. A possible lead for this new class of compounds is Hoechst 33258 (70), which recognize adenine/thymine sequences in human DNA and is also an effective inhibitor of mammalian DNA topoisomerase [92]. Several structure-activity relationship studies have been performed on the Hoechst motif. The replacement of the terminal piperazine ring with an amidinium, an imidazoline or a tetrahydropyridinium group significantly reinforces the affinity of the drug for the adenine/thymine stretches [93] (Scheme 28).

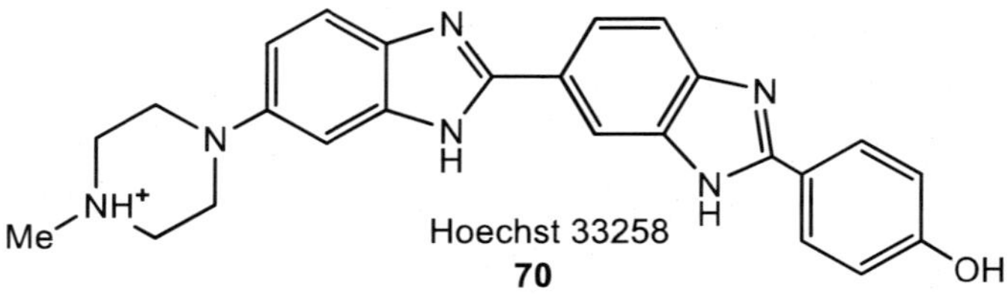

Scheme 28. Hoechst 33258 chemical structure.

Novel bisbenzimidazoles with general formula (71-44) incorporating benzimidazole, pyridoimidazole and imidazoquinone moieties as one of the units of bisbenzimidazole with a piperazinyl functional group have been synthesized [94]. The series of bisbenzimidazoles contains different leaving groups along with p-methoxy substituents. The latter may be expected to have some influence on the nitrogen lone pair and consequently on the binding characteristics of the ligand. These novel bisbenzimidazoles are found to be actively cytotoxic against many human cancer cell lines with GI_{50} values of between 0.01 and 100 μM, especially in the cases of renal cancer, CNS cancer, Colon cancer, Melanoma and Breast cancer cell lines. The pyridoimidazole compounds (71) and (73) are generally more potent (Scheme 29).

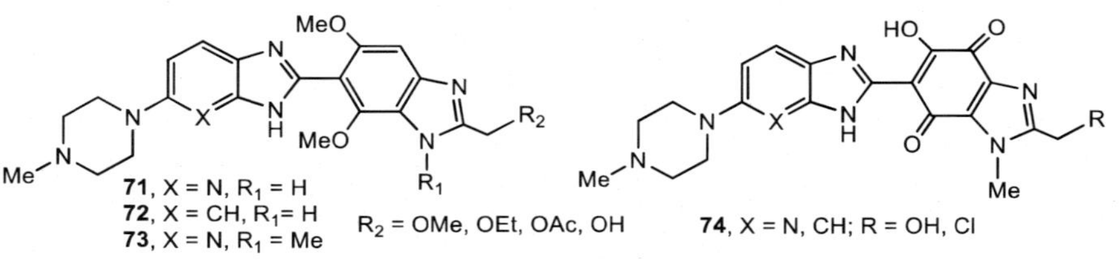

Scheme 29. Novel bisbenzimidazoles compounds.

Furthermore, a novel head to head bisbenzimidazole compound (75) binds with high affinity to the minor groove of double stranded B-DNA with a strong preference for adenine/thymine rich regions. The bisbenzimidazole (75) showed potent growth inhibition in human ovarian carcinoma cell lines (IC_{50} = 200-300 nM), with no significant cross-resistance in two acquired cisplatin resistant cell lines and a low level of cross-resistance in the p-glycoprotein over expressing doxorubicin resistant cell line [95] (Scheme 30).

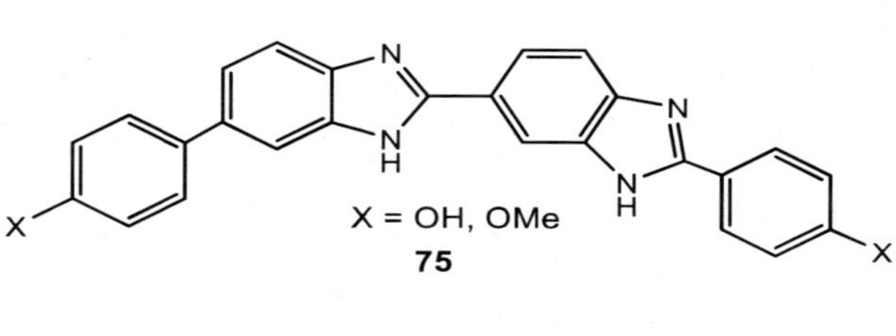

Scheme 30. Bisbenzimidazoles compounds.

The benzimidazole-6, 9-dione (76) has been found to be 300 times more cytotoxic towards the human skin fibroblast cell line in the MTT assay than the clinically used bio reductive drug, Mitomycin C. Attaching methyl substituents onto the quinone moiety increased reductive potential, decreased cytotoxicity and selectivity towards hypoxia [96]. In addition, the alkyl-linked bisbenzimidazole (77) [97] and thiazolylbenzimidazole-4, 7-diones (78) [98] exhibited cytotoxic activity against tumor cell lines (Scheme 31).

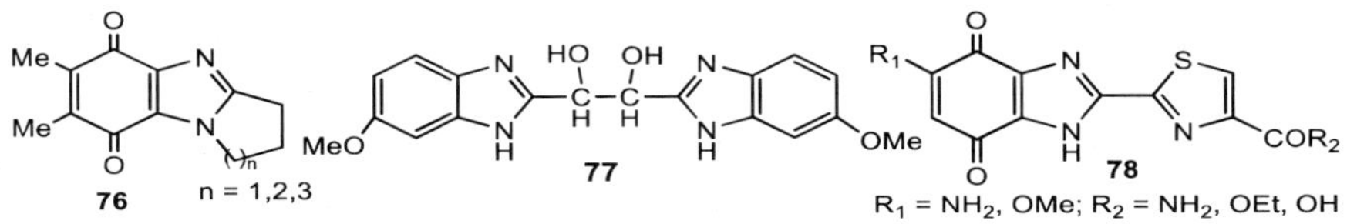

Scheme 31. Bisbenzimidazoles alkyl-linked compounds.

P. Xiang et al., synthesized five novel benzothiazole-2-thiol, benzimidazole-2-thiol and benzoxazole-2-thiol derivatives for lead optimization and their *in-vitro* antitumor activity were evaluated. To our delight, compounds (80-82) showed comparable antitumor activities and better solubility compared with the lead compound (79) [99] (Scheme 32).

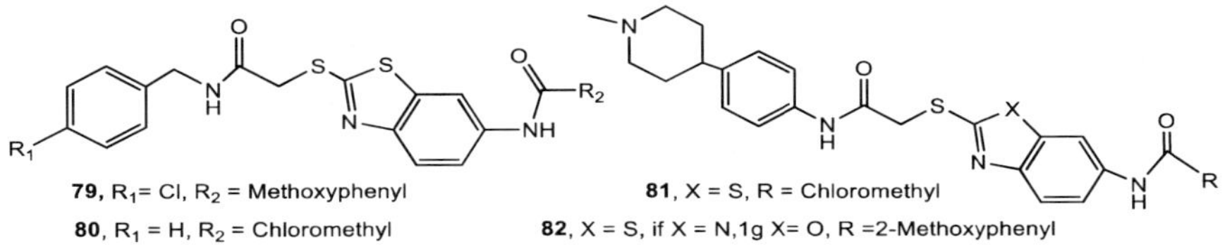

Scheme 32. Benzimidazole-2-thiol, benzothiazole-2-thiol and benzoxazole-2-thiol derivatives.

Tilomisole (WY-18, 83) has been widely studied. It showed several potent activities, such as antinflammatory activity. It has been reported to possess remarkable anticancer activity since it could be considered an analog of Levamisole, a well-known immunomodulator which is used for the adjuvant treatment of the Colon cancer (Scheme 33). However, Tilomisole (Wy-18, 83) has favorable biological response effects *in-vivo* and it is a suitable alternative to Levamisole in cancer treatments [100].

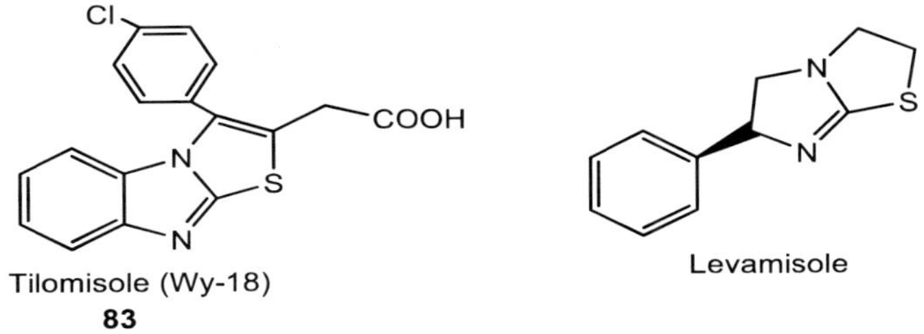

Scheme 33. Tilomisole and levamisole chemical structure.

Metal complexes of 2-methyl-1H-benzimidazole-5-carboxylic acid hydrazide (84, 85) and its schiff base 2-methyl-N-(propan-2-ylidene)-1H-benzimidazole-5-carbohydrazide with transition metal ions, for example copper, silver, nickel, iron and manganese were prepared (Scheme 34). The antitumor activity of the synthesized compounds has been studied. The silver complex (86) was found to display cytotoxicity (IC_{50}¼ 2 mM) against both human Lung cancer cell line A549 and human breast cancer cell line MCF-7 [101].

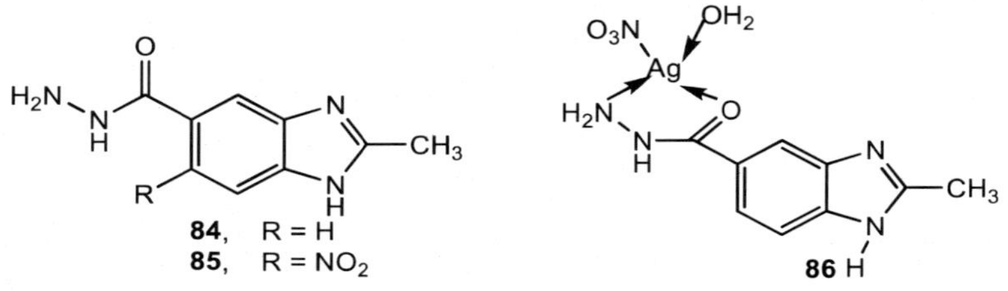

Scheme 34. 2-Methyl-1H-benzimidazole-5-carboxylic acid hydrazide metal complexes and its Schiff base.

K. Starcevic et al., prepared a set of heterocyclic benzimidazole derivatives bearing amidino substituents at C-5 of benzimidazole ring by introducing various heterocyclic nuclei (pyridine, N-methyl-pyrrole or imidazole) at C-2 and evaluated their antitumor and antiviral activities (87-95). The most pronounced antitumor activity was shown with compounds (92) and (95) having imidazolinylamidino substituent (Scheme 35). The most distinct and selective antiviral activity toward coxsackie viruses and echoviruses was observed with compounds having pyridine ring at C-2 [102].

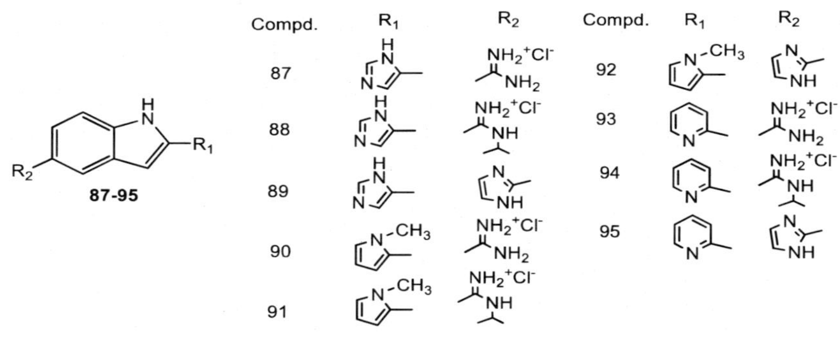

Scheme 35. Benzimidazole derivatives bearing amidino.

Inhibitors of the PI3-kinase/AKT (protein kinase B) pathway are under investigation as anticancer and antiviral agents. The benzimidazole derivative AKT inhibitor-IV (Chem. Bridge 5233705) affects this pathway and exhibits potent anticancer and antiviral activity. To probe its biological activity, Q. Sun et al., synthesized AKT inhibitor-IV and (96) analogues using a novel six-step route based on $ZrCl_4$ catalyzed cyclization of 1, 2-arylenediamines with R, β-unsaturated aldehydes and examined their effects on viability of HeLa carcinoma cells, viability of normal human cells (NHBE), replication of recombinant parainfluenza virus-5 (PIV5) in HeLa cells and replication of the intracellular bacterium M. fortuitism in HeLa cells [103] (Scheme 36).

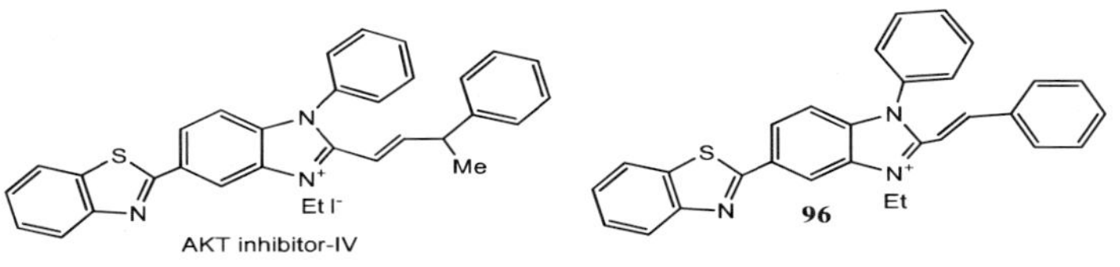

Scheme 36. Benzimidazole derivative and AKT inhibitor.

Gold (I) complexes such as auranofin have been used for decades to treat symptoms of rheumatoid arthritis and have also demonstrated a considerable potential as new anticancer drugs. The enzyme thioredoxin reductase (TrxR) is considered as the most relevant molecular target for these species. The investigated gold (I) complexes with benzimidazole derived N-heterocyclic carbine (NHC) ligands (97-100) represent a promising class of gold coordination compounds with a good stability against the thiol glutathione. TrxR was selectively inhibited by (97-100) in comparison to the closely related enzyme glutathione reductase and all complexes triggered significant antiproliferative effects in cultured tumor cells [104] (Scheme 37).

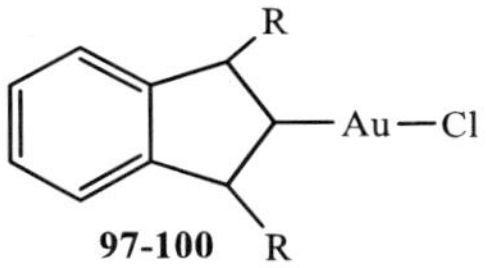

97, R = Methyl
98, R = Ethyl
99, R = Benzyl
100, R = Diphenylmethyl

Scheme 37. Auranofin and its derivatives.

T. D. Penning et al., synthesized a series of cyclic amine containing benzimidazole carboxamide PARP inhibitors with a methyl substituted quaternary center at the point of attachment to the benzimidazole ring system. These compounds exhibit excellent PARP enzyme potency as well as single digit nanomolar cellular potency. These efforts led to the identification of (101) (2-[(R)-2-methylpyrrolidin-2-yl]-1H-benzimidazole-4-carboxamide, ABT-888), currently in human phase I clinical trials [105] (Scheme 38).

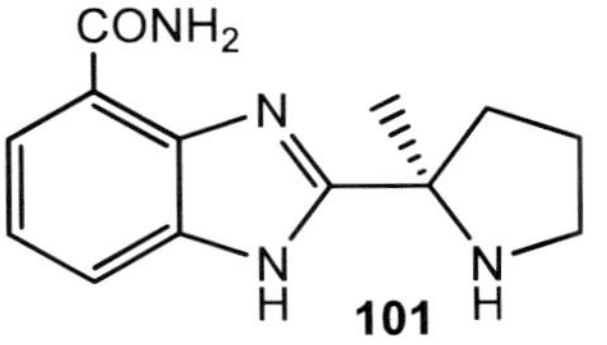

Scheme 38. Benzimidazole carboxamide containing PARP inhibitors.

Methyl-2-benzimidazolecarbamate (Carbendazim, FB642) (102) is an anticancer agent that induces apoptosis of cancer cells. In-vitro, FB642 demonstrated potent antitumor activity against both the murine B16 melanoma (IC_{50} = 8.5μm) and human HT-29 colon carcinoma (IC_{50} = 9.5μm) cell lines (Scheme 39). FB642 was also highly active against both murine tumor models and human tumor xenografts at varying doses and schedules. Toxicity of FB642 in-vivo appeared to be dose-dependent. Lower doses in the range of 2000-3000 mg/kg were better tolerated, while still preserving antitumor activity. Evaluation of FB642 in phase I clinical trials of adult patients with advanced malignancies is currently ongoing [106].

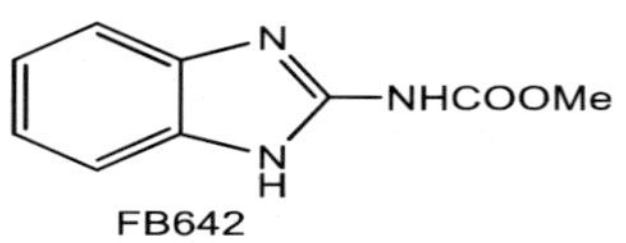

FB642
102

Scheme 39. Structure of methyl-2-benzimidazolecarbamate.

Two scaffolds based on 5, 6-fused heterocyclic backbones were designed and synthesized as Raf kinase inhibitors (103-105). The scaffolds were assessed for *in-vitro* pan-Raf inhibition, activity in cell proliferation, target modulation assays and pharmacokinetic parameters [107] (Scheme 40).

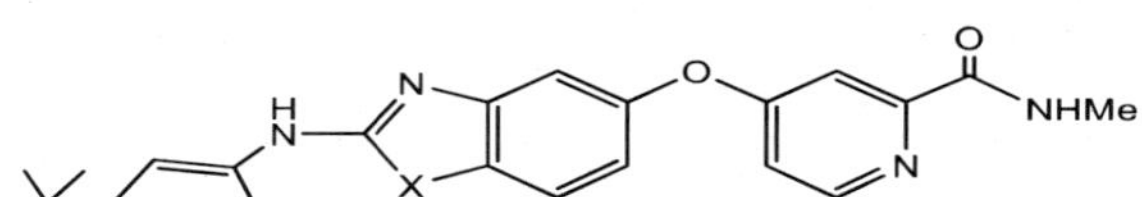

103-105
69, X = NMe, 70, X = O, 71, X = S

Scheme 40. 5,6-Fused heterocyclic compounds as Raf kinase inhibitors.

R. Abonia et al., synthesized some novel methyl 1-(5-tert-butyl-1H-pyrazol-3-yl)-2-(aryl)-1H-benzo[d]imidazole-5-carboxylates (106-119) (Scheme 41). Several of the obtained compounds were screened by the US National Cancer Institute (NCI) for their ability to inhibit 60 different human tumor cell lines. Products (107) and (119) exhibited the highest activity against a range of cancer cell lines with remarkable values in panels of non-small cell Lung cancer, Melanoma and Leukemia, with GI_{50} range of 1.15-7.33 μM and 0.167-7.59 μM, respectively and suitable LC_{50} with values greater than 100 μM [108].

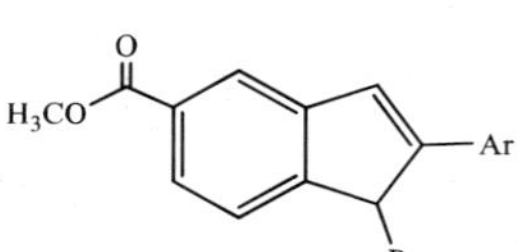

106-119

Compd.	Ar	Compd.	Ar
106	4-Br-C_6H_4	**113**	2-OH-4-$OCH_3C_6H_3$
107	4-Cl-C_6H_4	**114**	4-$(CH_3)_2NC_6H_4$
108	4-CH_3-C_6H_4	**115**	3,4-$(CH_3O)_2C_6H_3$
109	4-CH_3O-C_6H_4	**116**	C_6H_5
110	4-F-C_6H_4	**117**	4-CF_3-C_6H_4
111	3,4,5-$(CH_3O)_3C_6H_2$	**118**	1,3-Benzodioxol-5-yl
112	2-OH-C_6H_4	**119**	2-Oxo-1,2-dihydroquinolin-3-yl

Scheme 41. Novel methyl 1-(5-tert-butyl-1H-pyrazol-3-yl)-2-(aryl)-1H-benzo[d] imidazole-5-carboxylates.

Two series of benzimidazole clubbed with triazolo-thiadiazoles (121-144) and triazolo-thiadiazines (145-149) were synthesized from 3-(4-amino-5-mercapto-4H-1, 2, 4-triazol-3-yl)-1-(1H-benzo[d]imidazol-2-yl) propan-1-one (120) with an aim to produce promising anticancer agents [109] (Scheme 42). In-vitro anticancer activities of synthesized compounds were investigated at the National Cancer Institute (NCI) against NCI 60 cell line panel; results showed good to remarkable broad-spectrum anticancer activity. Among them, one compound 129 (1-(1H-benzo[d]imidazol-2-yl)-3-(6-(2,4-dichlorophenyl)-[1,2,4]triazolo[3,4-b][1,3,4]thiadiazol-3-yl) pro pan-1-oneexhibited significant growth inhibition with GI_{50} values ranging from 0.20 to 2.58μM and found superior selectivity for the leukemia cell lines and further screened at 10-fold dilutions of five different concentrations (0.01, 0.1, 1, 10 and 100 μM).

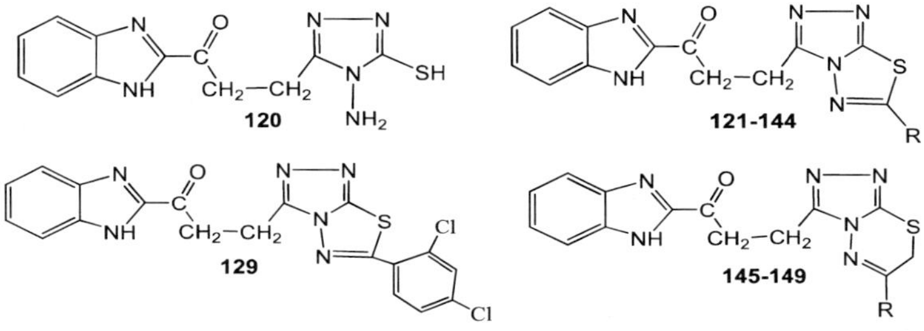

Scheme 42. Benzimidazole with triazolo-thiadiazoles (121-144) and triazolo-thiadiazines.

A. Husain et al., synthesized a series of benzimidazole bearing oxadiazole (150-160) and triazolothiadiazoles derivatives (161-165) from 4-(1H-benzo[d]imidazol-2-yl)-4-oxobutane hydrazide (153) and screened for their *in-vitro* anticancer activities at the National Cancer Institute (NCI), USA, against full NCI 60 human cell lines on nine cancer subpanel with significant results (Scheme 43). Among them, compound 158 (3-(5-(4-amino-2,6-dibromo phenyl)-1,3,4-oxadiazol-2-yl)-1-(1H-benzo[d] imidazole-2-yl)propan-1-one) exhibited significant growth inhibition and further screened at 10-fold dilutions of five different concentrations (0.01, 0.1, 1, 10 and 100 μM) with GI_{50} values ranging from 0.49 to 48.0 μM and found superior for the non-small cell lung cancer cell lines like HOP-92 (GI_{50} 0.49, TGI 19.9, LC_{50} >100 and $Log_{10}GI_{50}$ -6.30, $Log_{10}TGI$ -4.70, $Log_{10}LC_{50}$> -4.00) [110].

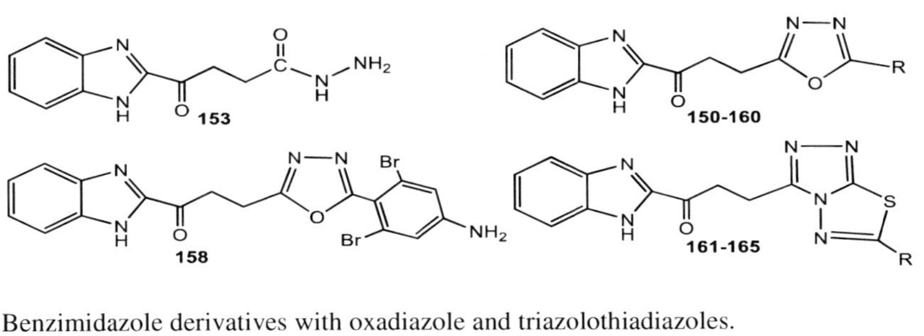

Scheme 43. Benzimidazole derivatives with oxadiazole and triazolothiadiazoles.

M. Rashid et al., synthesized twenty new 1-(1H-benzo[d]imidazol-2-yl)-3-(1, 3, 4-oxadiazol-5-substitute dderivatives-2-yl) propan-1-ones (167-185) under microwave irradiation in good yields. Further, compound 166 was reacted with different secondary amines under microwave irradiation to produce novel 1-(1H-benzo[d]imidazol-2-yl)-3-(5-(methyl substituted)-1, 3, 4-oxadiazol-2-yl) propan-1-ones (186-190) (Scheme 44). The title compounds were screened for their *in-vitro* anticancer activity at National Cancer Institute (NCI), USA; at a single dose (10 μM) in NCI 60 cell line panel and results showed significant to good anticancer activity. One compound 180 (1-(1H-benzo[d]imidazol-2-yl)-3-(5-(2,4-dichlorophenyl)-1,3,4-oxadiazol-2-yl) propan-1-one) emerged as lead compound; it was selected for five-dose level screening and found to have significant growth inhibition activity [111].

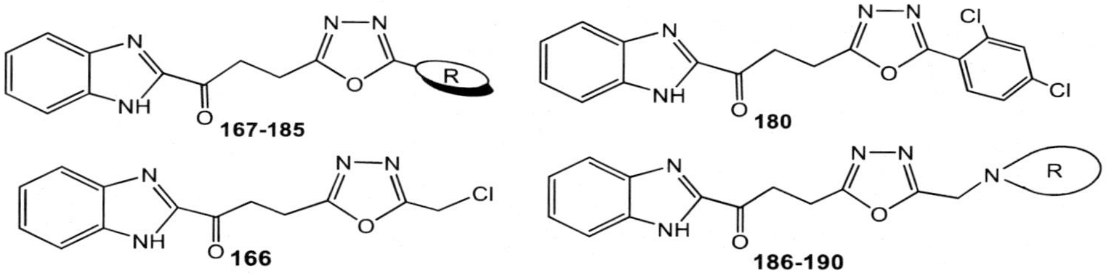

Scheme 44. Benzimidazole derivatives with oxadiazole ring and seconday amine.

M. Rashid et al., synthesized, four series (190-196, 197-202, 203-204 and 205-206) of benzimidazole derivatives based on the heterocyclic core of bendamustine were designed and synthesized starting from 4-(1H-benzo[d]imidazol-2-yl)-4-oxobutanehydrazide. In the rational design of target molecules, the benzimidazole ring of bendamustine was retained and the bis-(chloroethyl) amine group (mechlorethamine) was substituted with several biologically active scaffolds such as oxadiazole, thiadiazole, and triazolothiadiazines, in the hope of obtaining novel cytotoxic agents with improved efficacy and safety (Scheme 45). Cytotoxic activities of the designed analogues were carried out at the National Cancer Institute (NCI), USA, against full NCI 60 human cell lines. Among all the tested compounds, 195 (761982/1) exhibited significant antiproliferative activity and was further screened at 10-fold dilutions of five different concentrations (0.01, 0.1, 1, 10 and 100 lM) with GI50 values ranging from 0.09 to 16.2 lM and found superior for CNS cancer cell line SNB-75 (GI50 0.09, TGI 1.39, LC50 >100 and log10GI50 _7.0, log10TGI-5.86, log10LC50 >_4.00). Docking study was also performed to provide an insight about the binding mode into binding sites of topoisomerase enzyme. Hopefully in future, compound 4f could be used as a lead compound for developing new anticancer agents [112].

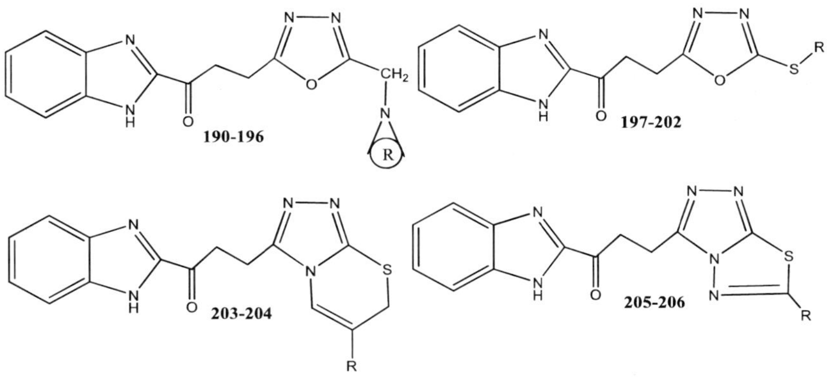

Scheme 45. Benzimidazole derivatives based on heterocyclic core of bendamustine.

10.2. Antimicrobial Activity

Recently, an efficient and rapid synthesis of novel benzimidazole azetidin-2-ones (207-216) has been established and antibacterial screening revealed that all newly synthesized azetidin-2-ones exhibited potent antibacterial activity against *B. subtilis, S. aureus* and *E. coli*. Among all of the compounds investigated (215) and (216) exhibited the greatest antibacterial activity against Gram-negative *E. coli* as compared to the antibiotic streptomycin [113] (Scheme 46).

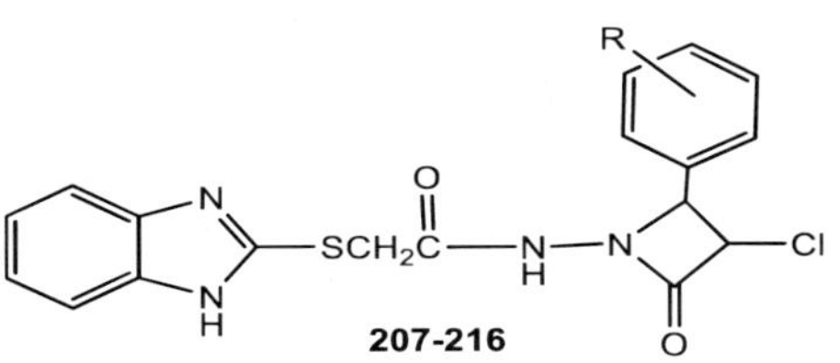

Compd.	R	Compd.	R
207	4-NO	212	2-OMe
208	3,4,5-(OMe)$_3$	213	4-OMe
209	2-OH	214	2-Cl
210	3-OH	215	3-Cl
211	4-OH	216	4-Cl

Scheme 46. Benzimidazole derivatives with azetidin-2-ones.

Benzimidazole benzyl ethers (217-224) have exhibited good antibacterial activity against *S. aureus* and antifungal activity against *C. albicans* and *C. krusei*. In general the dichlorophenyl substituted benzimidazoles (221, 222 & 224) showed the best antibacterial (MIC 3.12 μg/mL) and antifungal (MIC 12.5 μg/mL) activity [114] (Scheme 47).

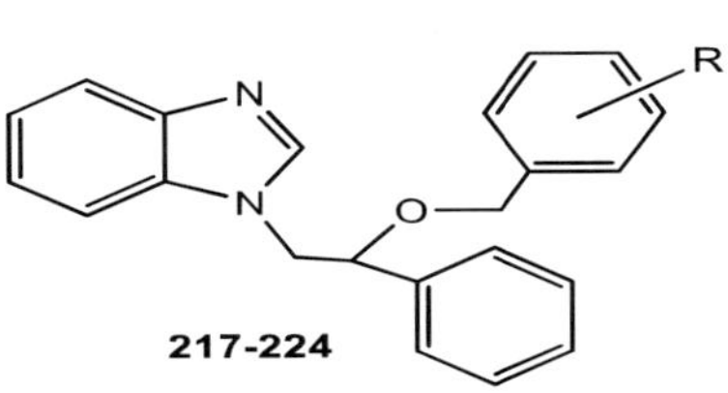

Compd.	R	Compd.	R
217	4-H	221	2,4-di-Cl
218	4-F	222	2,6-di-Cl
219	4-Cl	223	2,5-di-Cl
220	4-Br	224	3,4-di-Cl

Scheme 47. Benzimidazole derivatives with benzyl ethers.

Benzimidazole containing oxadiazole nucleus is beneficial to design and synthesize some new benzimidazole derivatives bearing oxadiazole moiety and screen them for potential biological activities. Some benzimidazole derivative containing oxadiazole like, 1-{[5-(alkyl/aryl)-1, 3, 4-oxadiazol-2-yl] methyl}-2-alkyl-1H-benzimidazoles (225) are synthesized for their antimicrobial activities [115] (Scheme 48).

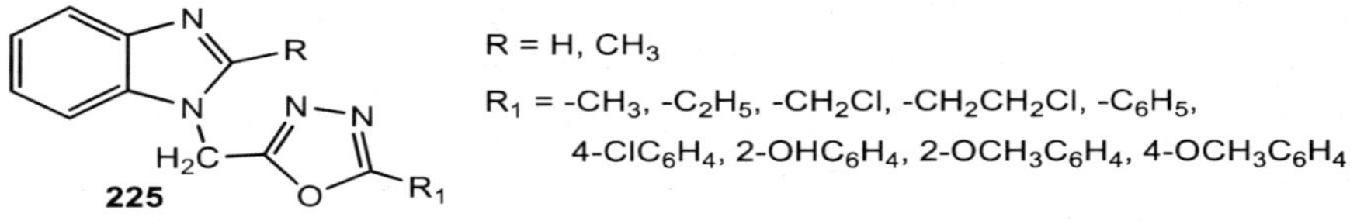

Scheme 48. Benzimidazole derivatives with oxadiazole ring.

Some fluoroquinolones substituted benzimidazole derivatives have been reported by microwave assisted method. The synthesized compounds are reported to be the derivatives of Ciprofloxacin (226) and Norfloxacin (227) and evaluated their antimicrobial activity [116] (Scheme 49).

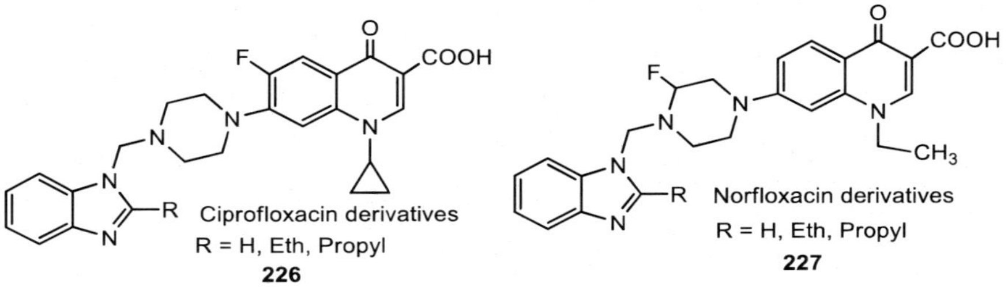

Scheme 49. Benzimidazole derivatives with fluoroquinolones.

A series of 1, 2-disubstituted benzimidazole-5(6)-carboxamides was prepared and evaluated *in-vitro* for antimicrobial activity against *S. aureus*, *E. coli* and *C. albicans* (Scheme 50). Among the investigated compounds (228-230) showed the best activity. Methyl benzimidazole-5-carboxylate derivatives (228) had potent activity comparable to that of fluconazole against *C. albicans*. Among the synthetic derivatives (228-230), the compound (229) showed best result against *S. aureus* [117].

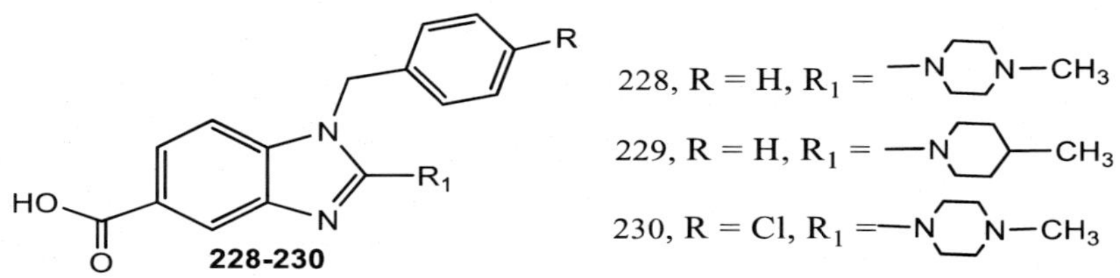

Scheme 50. 1, 2-disubstituted benzimidazole-5(6)-carboxamides.

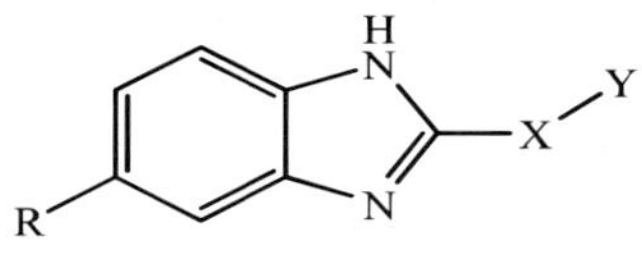

231-237

Compd.	R	X	Y
231	H	-	cyclopentyl
232	Cl	-	cyclopentyl
233	H	CH_2	cyclopentyl
234	Cl	CH_2	cyclopentyl
235	H	C_2H_4	cyclopentyl
236	Cl	C_2H_4	cyclopentyl
237	H	C_2H_4	cyclohexyl

Scheme 51. Benzimidazole with cyclohexyl or cyclopentyl moieties.

A new series of 5-substituted benzimidazoles holding cyclohexyl or cyclopentyl moieties at position-2 was synthesized in order to determine their antimicrobial activities and feasible structure-activity relationships. The synthesized compounds (231-237) are able to inhibit in -vitro growth of screened microorganisms, provided antibacterial activity against *S. aureus, S. faecalis* and *B. subtilis* possessing MIC values between 25-100 mg/mL, except the derivative 5-chloro-2-(2-cyclohexylethyl) benzimidazole (237), which was found active at a MIC value of 12.5 mg/mL against each of the screened Gram-positive bacteria strains [118] (Scheme 51).

10.3. Anthelmintic Activity

Bearing in mind previous benzimidazole anthelmintic (e.g., Albendazole, Mebendazole), the search for new anthelmintic drugs is being actively pursued. Synthetic benzimidazole piperazine derivatives exhibited 50% anthelmintic activity in mice infected with *S. obvelata* [119]. Furthermore, piperazine derivatives of 5(6)-substituted-(1H-benzimidazol-2-ylthio) acetic acids (238-241) [120] and benzimidazolyl crotonic acid anilide (241) have shown good anthelmintic activity [121] (Scheme 52).

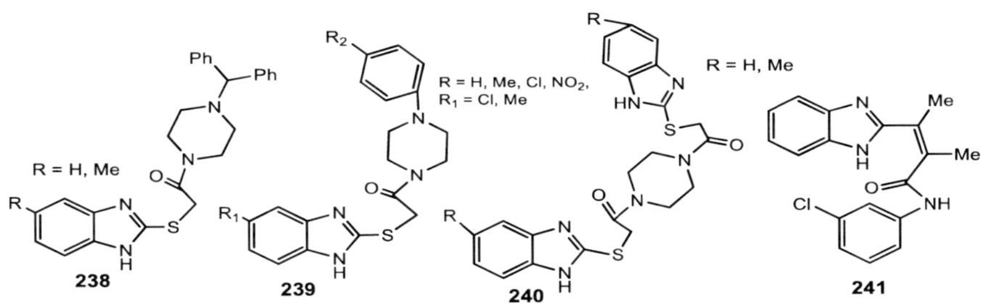

Scheme 52. Benzimidazole with piperazine and crotonic acid anilide.

R. Sharma et al., synthesized certain methyl 5-(α-hydroxyl-α-substituted-methyl)-1H-benzimidazole-2-carbomates (242) and screened them for anthelmintic activity [122] (Scheme 53).

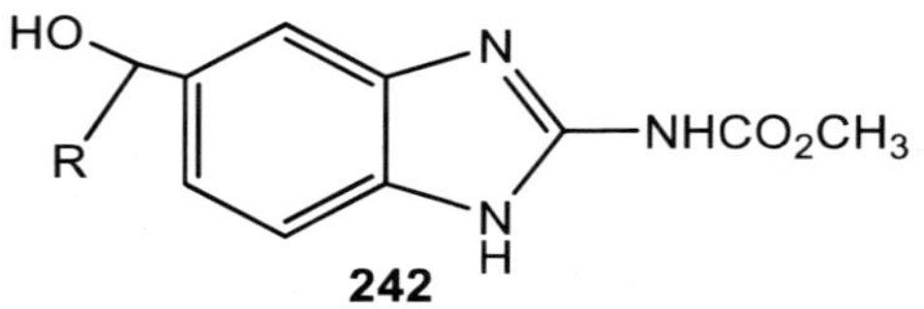

242

R = C_2H_5 (Mebendazole)

R = C_6H_4-NH_2, CH_3, C_2H_5, C_6H_4-NH_2

Scheme 53. Methyl 5-(α-hydroxyl-α-substituted-methyl)-1H-benzimidazole-2-carbomates.

M. Himaja et al., synthesized 6-nitrobenzimidazol-1-acetyl amino acid (243) and evaluated their anthelmintic activity and they found it as potent anthelmintic agent [123] (Scheme 54).

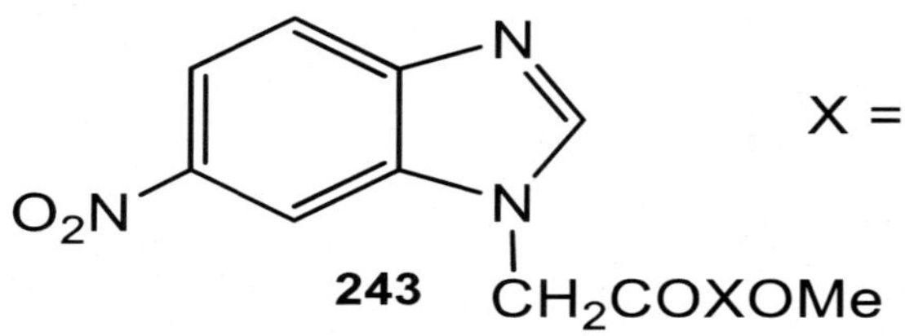

243

X = L-Tyr, L-Gly

Scheme 54. 6-Nitrobenzimidazol-1-acetyl amino acid.

H. Luis et al., synthesized some new 2-(trifluoromethyl)-1H-benzimidazole derivatives (244-246) and showed the most desirable in-vitro antiparasitic profile against *G. intestinalis, E. histolytica, T. vaginalis* and *T. spiralis* [124] (Scheme 55).

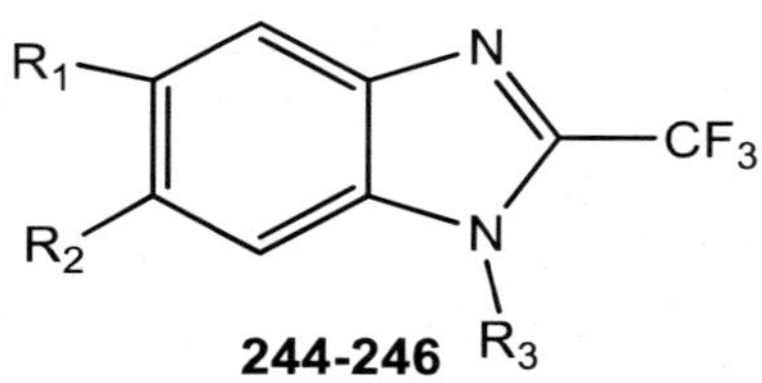

244-246

244: R_1= 2,3-$Cl_2C_6H_3O$, R_2 = H, R_3 = CH_3

245: R_1= H, R_2 = 2,3-$Cl_2C_6H_3O$, R_3 = CH_3

246: R_1 = $C_{10}H_7O$, R_2 = Cl, R_3 = H

Scheme 55. 2-(Trifluoromethyl)-1H-benzimidazole derivatives.

10.4. Antihypertensive Activity

Toshiro et al., observed a supporting evidence by studying the pharmacokinetic properties of various serotonin-3(5-HT3) receptor antagonists, including ondansetron (247) and ramosetron (248) especially their ability to induce or inhibit CYP450 isoenzymes in human liver microsomes (Scheme 56). Ramosetron, a 1H-benzimidazole derivative, neither inhibited nor induced any of the metabolic activities, whereas ondansetron, an aliphatic substituted imidazole derivative, competitively inhibited CYP1A2 and CYP2D6 activities [125].

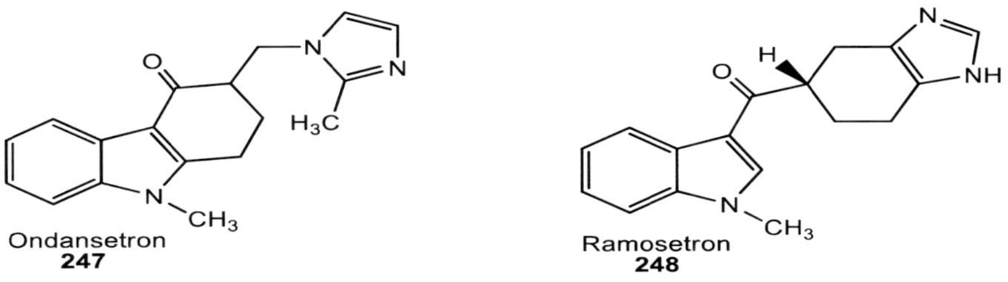

Scheme 56. Ondansetron and ramosetron chemical structure.

Wienen et al., studied the pharmacokinetics of angiotensin II-receptor antagonist class used widely as highly potent and selective antihypertensive, especially losartan (249) and Telmisartan (250) (Scheme 57). Based on their data under in-vitro and in-vivo conditions, telmisartan, a 1H-benzimidazole derivative, did not induce or significantly inhibit CYP450 isoenzyme activity and was not metabolized by phase I reactions. Wienen et al. pointed out that the medical use of Telmisartan did not include the potential risk of drug interactions with compounds subject to CYP450-dependant biotransformation is low [126].

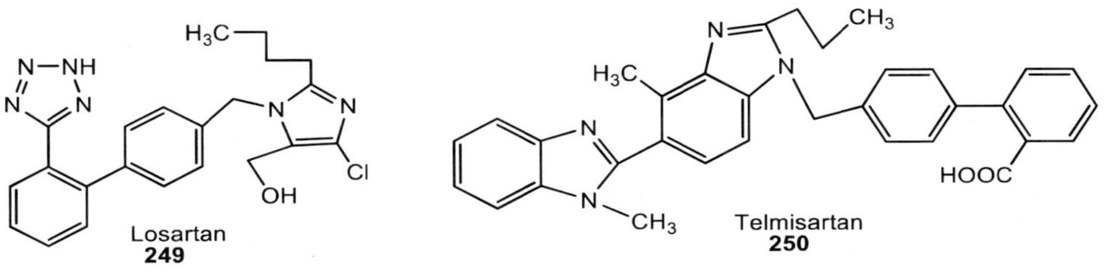

Scheme 57. Losartan and Telmisartan chemical structure.

10.5. Antiretroviral Activity

Reverse transcriptase is a key enzyme which plays an essential and multifunctional role in the replication of HIV-1 and thus constitutes an attractive target for the development of new drugs that could be used in AIDS therapy. Thiazolo-benzimidazoles (251) proved to be a highly potent inhibitor of HIV-1 induced cytopathic effects. Structure activity relationship studies showed that the C-1 substituents in benzimidazole (252-255) greatly influence the interaction of the active compound with the receptor (Scheme 58). Substitution on the benzene fused ring influences the inhibitory potency depending on the nature and position of the substituent, the presence of a methyl group at C-3 is favorable to the pharmacological profile [127].

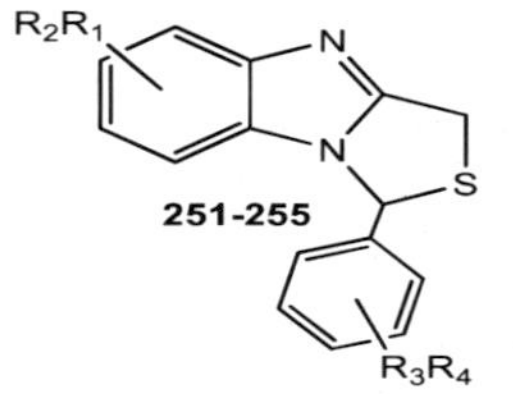

Compd.	R_1	R_2	R_3	R_4
251	H	H	2-F	6-F
252	H	H	3-NO_2	H
253	6-Me	7-Me	2-F	6-F
254	5-Me	H	2-F	6-F
255	8-Me	H	2-F	6-F

Scheme 58. Thiazolo-benzimidazoles derivatives.

V. K. Pandey et al., synthesized 1, 2-disubstituted benzimidazoles (**256**) and screened them for antiviral activity [128] (Scheme 59).

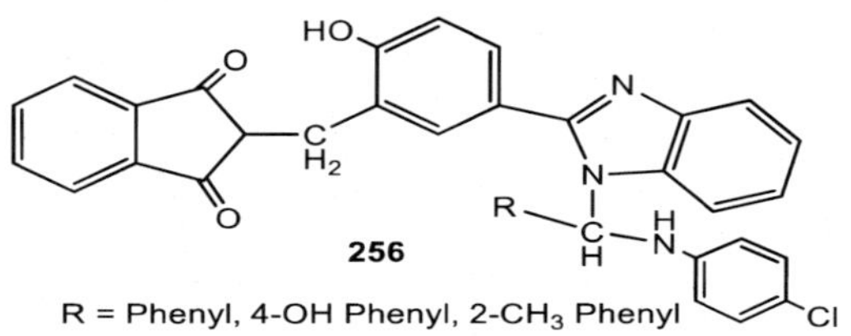

Scheme 59. 1, 2-Disubstituted benzimidazoles.

A series of 4-oxo-4H-quinolizine-3-carboxylic acid derivatives bearing sulfamido, carboxylamido, and benzimidazole and benzothiazole substituents have been designed and synthesized by Y. S. Xu et al., and the synthetic compounds (257-259) were screened for possible HIV integrase inhibitory activity [129] (Scheme 60).

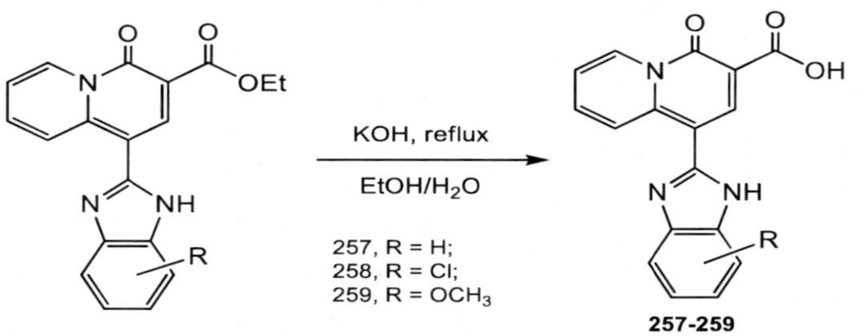

Scheme 60. 4-Oxo-4H-quinolizine-3-carboxylic acid derivatives of benzimidazole.

A. Chimirri et al., synthesized a new derivative of 1H,3H-oxazolo[3,4-a] benzimidazoles (OBZs) as HIV-1 non-nucleoside reverse transcriptase inhibitors (NNRTI) to extend the structure activity relationships observed for an early series of related 1H,3H-thiazolo[3,4-a] benzimidazole derivatives (TBZs) (Scheme 61). The new compounds (260-262) showed inhibitory activity against the replication of various HIV-1 strains, including NNRTI-resistant strains. Testing of a representative OBZ derivative in an HPLC assay on biological fluids, indicated that the sulphur substitution appreciably improved the metabolic stability of the TBZ compound [130].

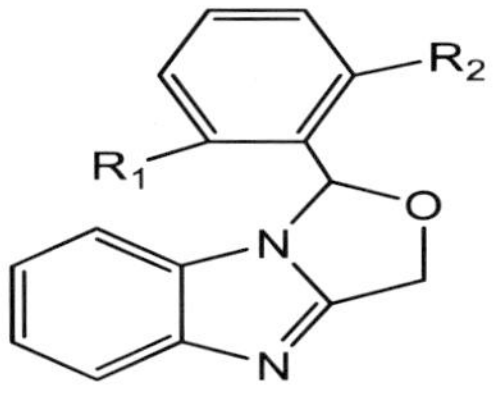

260-262

Compd.	R_1	R_2
260	H	H
261	Cl	F
262	F	F

Scheme 61. 1H,3H-oxazolo[3,4-a] derivatives of benzimidazole.

R. Zou et al., synthesized a series of disubstituted benzimidazole ribonucleosides (263-264) and evaluated their antiviral activity [131] (Scheme 62).

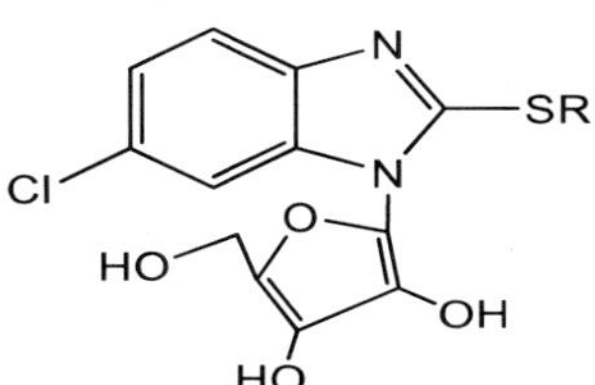

263-264

263: R = Me

264: R = Bn

Scheme 62. Disubstituted benzimidazole with ribonucleosides.

10.6. Anti-Inflammatory, Analgesic and Antiulcer Activity

Structure activity relationship studies of the 5, 6-dialkoxy-2-thiobenzimidazole derivatives have revealed that compounds (265-275) possess pronounced anti-inflammatory properties. Using the carrageenan model, the most significant anti-inflammatory effects were observed for compounds (265, 267, 270, 271 & 272) (Scheme 63). While using the bentonite model, the maximum activities were observed for compounds 268 and 272. These results indicated that benzimidazoles are promising leads for the development of new anti-inflammatory agents [132].

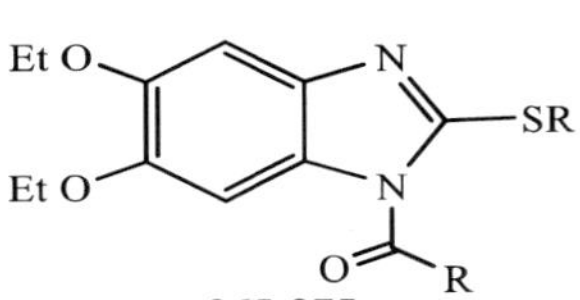

265-275

265	Ph	271	2-Pyridyl
266	4-MeC_6H_4	272	3-Pyridyl
267	4-$MeOC_6H_4$	273	4-Pyridyl
268	4-FC_6H_4	274	2-Thienyl
269	4-ClC_6H_4	275	2-Furyl
270	3-ClC_6H_4		

Scheme 63. 5, 6-Dialkoxy-2-thiobenzimidazole derivatives.

Pyrimidobenzimidazole [133] (276) and dioxinobenzimidazothiazol-9-ones [134] (277) exhibited anti-inflammatory and analgesic activity, as evaluated by carrageenan induced rat paw edema and phenylquinone induced writhing tests (Scheme 64). In addition, N-benzoyl and N-tosylbenzimidazole compounds (278) showed significant anti-inflammatory activity, as indicated by ear swelling induced by xylene in mice, and their ulcer indices were all lower than those of Aspirin [135].

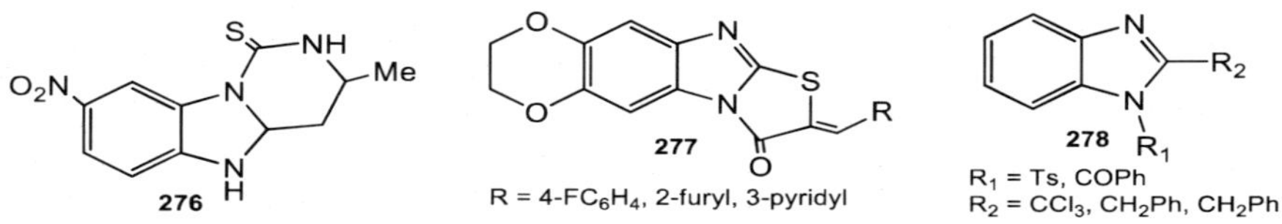

Scheme 64. Pyrimidobenzimidazole and dioxinobenzimidazothiazol-9-ones.

Despite the success of several commercial benzimidazole proton pump inhibitors for the treatment of ulcer disease, work is still in progress to discover new benzimidazole derived antiulcer drugs. Cinitapride (279) related benzimidazole derivatives (280) have been prepared and studied for their antiulcerative activity [136] (Scheme 65).

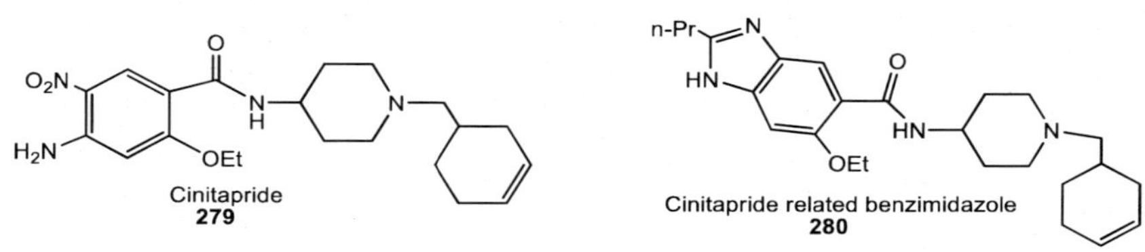

Scheme 65. Cinitapride related benzimidazole derivatives.

S. N. Sawhney et al., synthesized certain 2-(5-substituted-2-yl) (281) and 2-(3-mercapto-4-substituted-4H-1(282) of 2, 4-triazol-5-yl)-benzimidazole and screened them for potential anti-inflammatory activity [137] (Scheme 66).

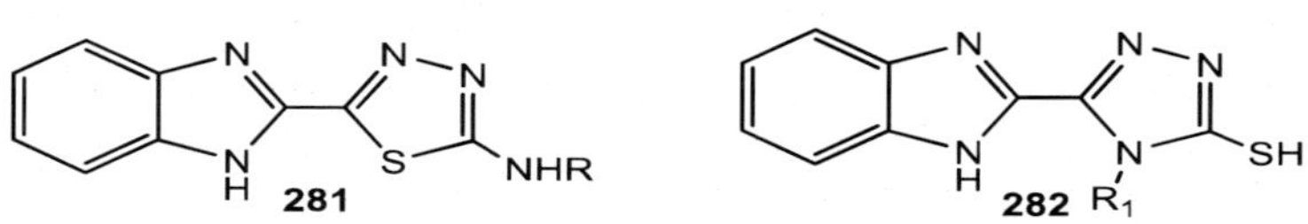

R= CH_3, C_2H_5, C_6H_5, 4-$CH_3OC_6H_5$, 4-ClC_6H_4, 4-$NO_2C_6H_4$

R_1 = CH_3, C_2H_5, CH_2=CH-CH_2, C_6H_5, 4-$CH_3OC_6H_4$

Scheme 66. Triazole and thiadiazole derivatives of benzimidazole.

C. S. Kavitha et al., prepared a series of 2-methylaminobenzimidazole derivatives (283) by the reaction of 2-(chloromethyl)-1H-benzimidazole derivatives with primary aromatic amines (Scheme 67). The newly synthesized compounds were screened for analgesic and anti-inflammatory activities by the author on acetic acid induced writhing in mice and carrageenan induced paw oedema in rats. Some compounds showed a potent analgesic (89% at 100 mg/kg b.w.) and anti-inflammatory (100% at 100 mg/kg b.w.) activities compared with standard drug Nimesulide (100% at 50 mg/kg b.w.), respectively [138].

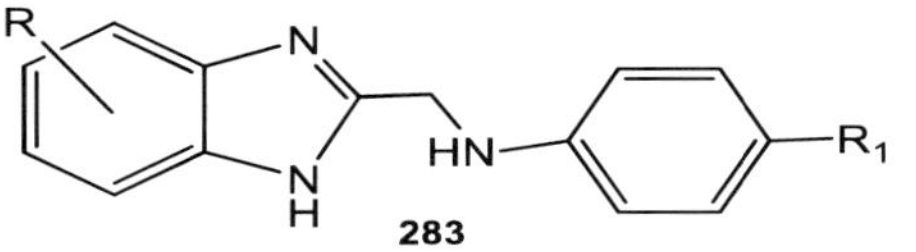

283

$R = H, Br, NO_2$

$R_1 = H, Cl, Br, CH_3, OCH_3$

Scheme 67. 2-Methylaminobenzimidazole derivatives.

10.7. Opportunistic Infections

R. L. Lombardy et al., synthesized and assessed the DNA interactions of bis-benzimidazoles (284) which showed appreciable activity against opportunistic infection [139] (Scheme 68).

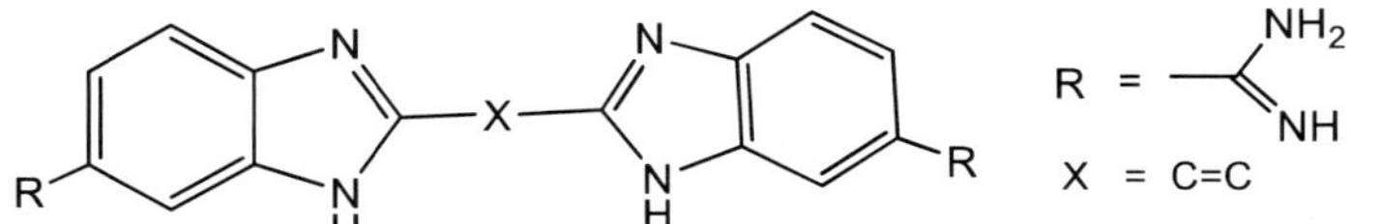

284

X = C=C

Scheme 68. Bis-benzimidazoles derivatives.

10.8. Antioxidant/Radio-Protective Activity

Some novel 5-[(2-(substitutedphenyl)-1H-benzimidazole-1-yl)methyl]methyl-1,3,4-thiadia zole-2-amines (285) were synthesized and tested for antioxidant properties by Kus et al., using various in-vitro systems [140] (Scheme 69).

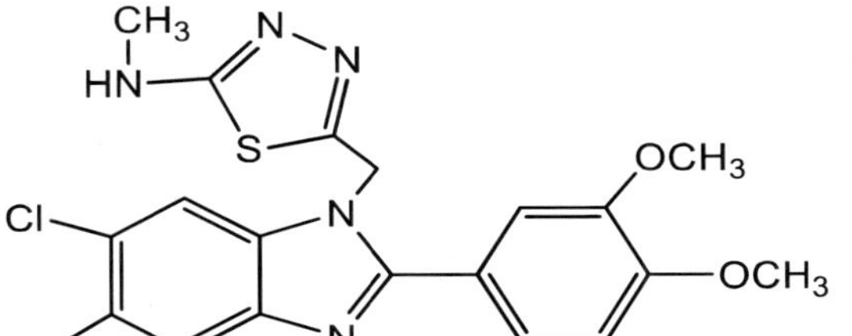

285

Scheme 69. 5-[(2-(substitutedphenyl)-1H-benzimidazole-1-yl)methyl] methyl-1,3,4-thiadia zole-2-amines.

C. Kus et al., reported the synthesis and antioxidant properties of novel N-methyl-1, 3, 4-thiadiazol-2-amine and 4-methyl-2H-1, 2, 4-triazole-3(4H)-thione derivatives (286) of benzimidazole class [141] (Scheme 70).

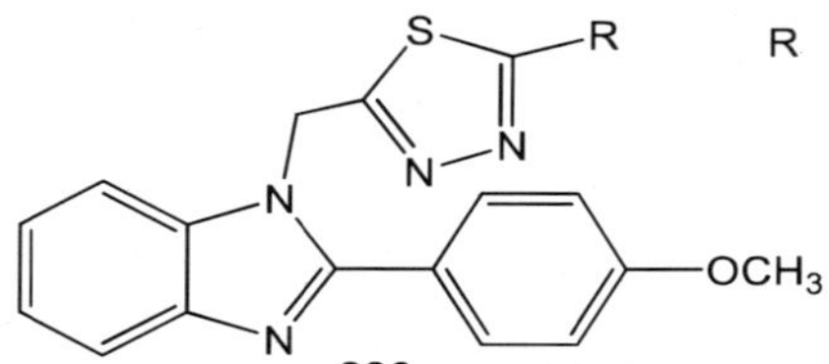

286

R -NHCH3
-CH3

Scheme 70. Thiadiazole derivatives of benzimidazole.

10.9. Carbonic Anhydrase Inhibitors

A series of benzimidazo[1,2-c][1,2,3]thiadiazole-7-sulfonamides (287-289) were synthesized by Daumantas et al., as inhibitors of two isozymes of carbonic anhydrase, human carbonic anhydrase I (hCAI) and bovine carbonic anhydrase II (bCAII). The strongest binder to both isozymes of carbonic anhydrase was compound (287) with the observed K_d of about 0.04 lM (Scheme 71). The most specific binder of hCAI was compound (288) that bound about fourfold stronger to hCAI than to bCAII. The (289) compound bound three fold tighter to bCAII than to hCAI [142].

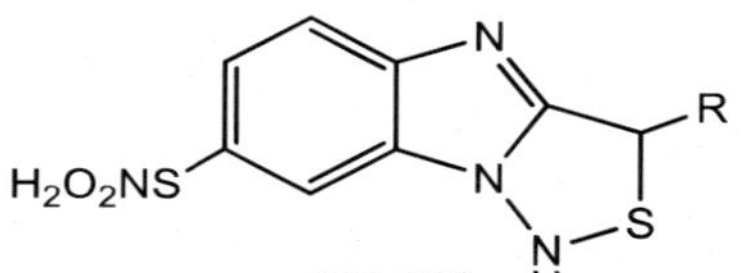

287-289

287: R= Cl

288: R = SO_2CH_3

289: R = $N(CH_2CH_3)_2$

Scheme 71. Benzimidazo [1,2-c][1,2,3]thiadiazole-7-sulfonamides derivatives.

10.10. Anticonvulsant and Antidepressant Activity

A series of 1, 2, 5-trisubstituted benzimidazoles (290-292) derivatives have been synthesized and reported as potential anticonvulsant. The results of QSAR investigation and the study of various physicochemical properties indicates that the change in linker at position one (R_1) does not change the activity of the synthesized compounds and optimum chain length at position two (R_2) is responsible for the anticonvulsant activity [143] (Scheme 72).

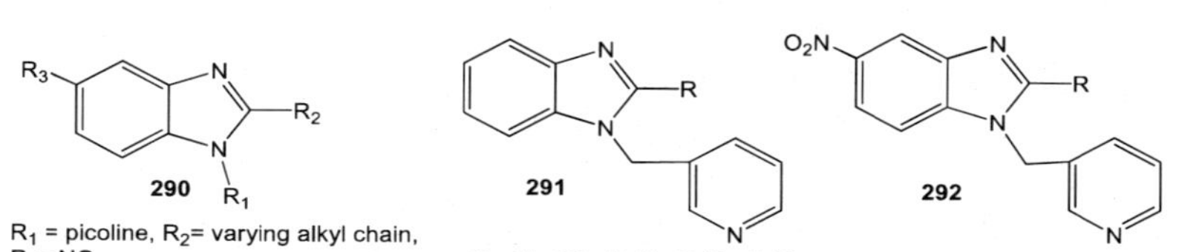

Scheme 72. 1, 2, 5-trisubstituted derivatives of benzimidazoles.

H. Rajak et al., synthesized novel 2,5-disubstituted-1,3,4-thiadiazoles (293) as potential anticonvulsant activity using standard drug Carbamazepine and Phenytoin at concentration 30 and 100 mg/kg [144] (Scheme 73).

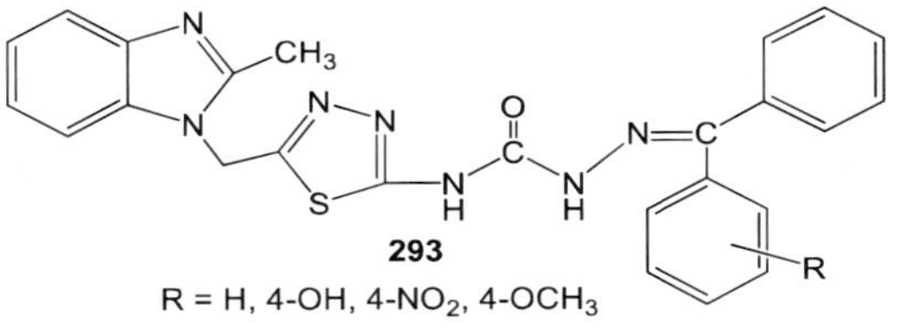

293

Scheme 73. 2,5-disubstituted-1,3,4-thiadiazoles derivative of benzimidazoles.

Pattanayak et al., synthesized the new benzimidazole derivatives of 2-amino-5-sulfanyl-1, 3, 4-thiadiazole (294) and evaluated their antidepressant activity (Scheme 74). The synthetic compounds exhibited good to remarkable antidepressant activity in comparison to reference drugs [145].

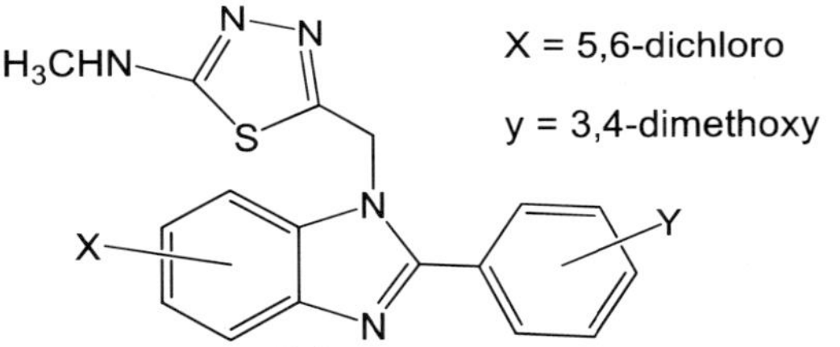

294

Scheme 74. New benzimidazole derivatives of 2-amino-5-sulfanyl-1, 3, 4-thiadiazole.

10.11. Antihistaminic

A number of benzimidazoles having several substituents on the azole and benzene nuclei and C-2 (methyl amino ethylene diamine, morpholine, piperazine, piperidines) were prepared regioselective synthesis was designed for the N-alkyl substituted benzimidazoles (Scheme 75). X-ray structure analysis of (14) was revealed. Compounds were evaluated for their in vitro H_1-antihistaminic activity in the isolated guinea-pig ileum method [146]. The benzimidazole core of the selective non-brain-penetrating H_1-antihistamine mizolastine was used to identify a series of brain-penetrating H_1-antihistamines for the potential treatment of insomnia. Using cassette PK studies, brain-penetrating H_1-antihistamines were (295) identified and in vivo efficacy was demonstrated in a rat EEG/EMG model. Further optimization focused on strategies to attenuate an identified hERG liability [147].

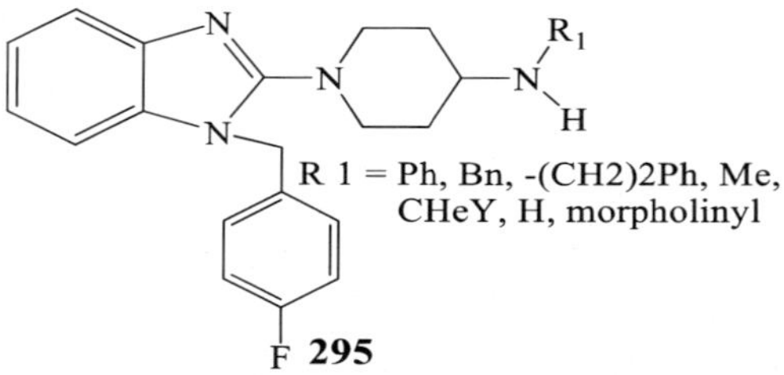

295

Scheme 75. N-alkyl substituted benzimidazoles.

10.12. Antiasthmatic and Antidiabetic

Synthesis of some novel 2-(6-fluorochroman-2-yl)-1-alkyl/acyl/aroyl-1H-benzimidazoles (296) by the condensation of o-phenylenediamine with 6-fluoro-3, 4-dihydro-2H-chroman-2-carboxylic acid and subsequent reactions with different types of electrophiles, have been reported (Scheme 76). Some compounds exhibited promising anti-bacterial activity against *salmonella typhimurium*; however, they showed poor activity against *S. aureus*. The biological activity against PDE IV for potential anti-asthmatic effect, and against DP-IV and PTB IB for potential antidiabetic effects was also determined [148].

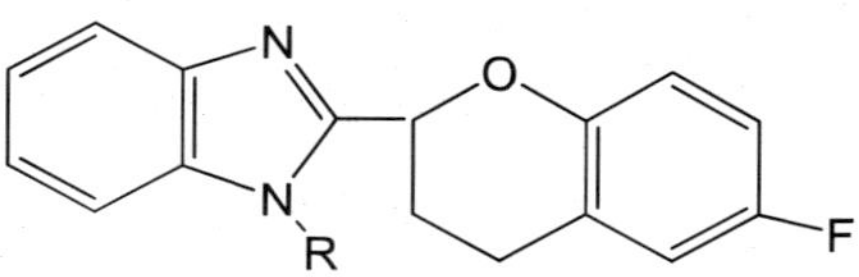

R = methyl, ethyl, propyl, n-Butyl
-coo-ethyl, -coo-tert-Butyl

296

Scheme 76. 2-(6-fluorochroman-2-yl)-1-alkyl/acyl/aroyl-1H-benzimidazoles.

Synthesis of a series of novel and functionalized benzimidazole derivatives by the condensation of OPDA with 4-bromobenzoic acid and subsequent reactions of the product obtained with phenyl acetylene and 6-ethynyl-4, 4-dimethylthiochroman utilizing sonogashira coupling has been reported. The Sonogashira coupling products were then alkylated at the benzimidazole –NH with different electrophilic reagents leading to functionalized derivatives. All the compounds synthesized were screened for their potential anti-bacterial, anti-asthmatic and anti-diabetic properties, which exhibited moderate activities in screening studies in vitro. A novel series of non-covalent, benzimidazole-based inhibitors of DPP-4 has been developed from a small fragment hit using structure-based drug design. A highly versatile synthetic route was created for the development of SAR, which led to the discovery of potent and selective inhibitors with excellent pharmaceutical properties [149].

11. Recently Discovered Anticancer Drugs Containing Benzimidazole Moieties

Recently discovered anticancer agents having benzimidazole and other related heterocyclic moiety are: Nocodazole (NSC-238189) (297) [150], FB642 (298) [151], A-620223 (299) [152], Hoechst-33258 (300)[153], ABT-888 (301)[154], Phortress (302) [155], SNS-032 (303) [156], Proxazole (304)[157], CYC116 (305)[158], Thiadiazole derivative (306)[159], Levamisole (307)[160], Imidazo[2,1-b] [1, 3,4]thiadiazole analogues (308)[161], Triazolo[1,3,4]thiadiazole derivative (309) [162], ATEK 10934 (310)[162] and IMC-094332 (311) [163] are shown in Figure 7. Benzimidazole clubbed with other heterocyclic ring systems; oxadiazole, thiadiazole, triazolo-thiadiazoles and triazolo-thiadiazinesmoieties, also showed highly promising results and may soon be available in the market.

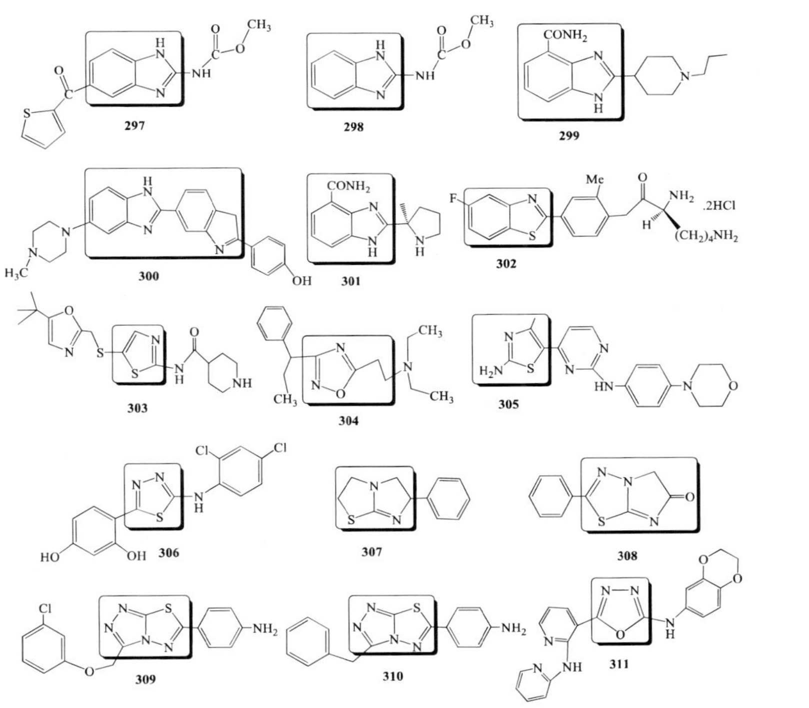

Figure 7. Structure of some biologically active antitumor agents containing the benzimidazole and other related heterocyclic scaffold.

Rashid et al. to shed some rational design target compounds from market approved anticancer drugs like Bendamustine and Chlorambucil having benzimidazole nucleus and other related heterocyclic moiety. Treanda a trade name of Bendamustine hydrochloride made up by Cephalon Inc. USA, chemically as 1H-benzimidazole-2-butanoicacid-5-[bis(2-chloro ethyl)amino]-1-methyl monohydrochloride. It comprises of mechlorethamine and benzimidazole heterocyclic ring with a butyric acid substituent, a rationally design purine analog and alkylation hybrid of Chlorambucil (Figure 8). On 20 March 2008, it was approved by the FDA (U.S.) for the treatment of chronic lymphocytic leukemia (CLL) and about 6 months later (31 October 2008), it also approved for patients with indolent B-cell non-Hodgkin's lymphoma (NHL) [164-165].

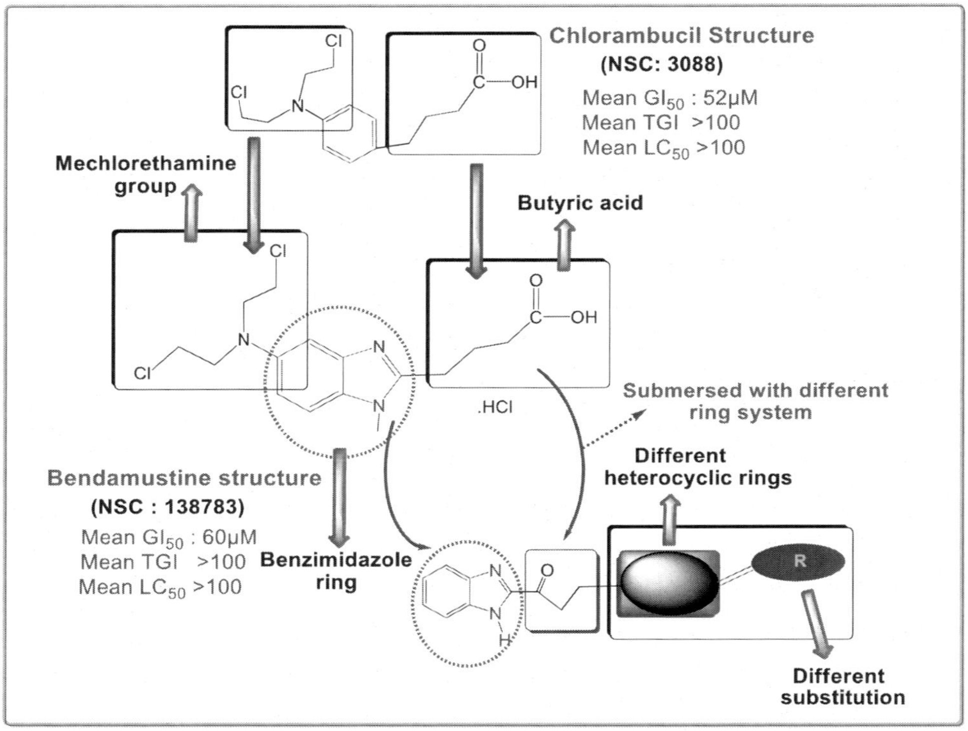

Figure 8. Retionally design and prepared a template for synthetic Schemes from marketed anticancer drug like Bendamustine and Chlorambucil.

A rational template has been designed having benzimidazole residue as pharmacophoric group with butyric acid substituent and further clubbed with other heterocyclic ring systems; oxadiazole, thiadiazole, triazolo-thiadiazoles and triazolo-thiadiazines with a view to produce potential anticancer agents (Figure 8). This rational template was further used for preparation a number of compounds through different synthetics schemes along with other heterocyclic agents having antitumor activity like Proxazole [166], IMC-094332(having oxadiazole ring) [167], SNS-032(having thiazole ring) [168], Levamisole [169], ATEK 10934 (having diazole-thiazole & trizolo-thiadiazole ring) [170] and Hoechst-33258 (having imidazole and azole ring) [171] represented in Figure 9.

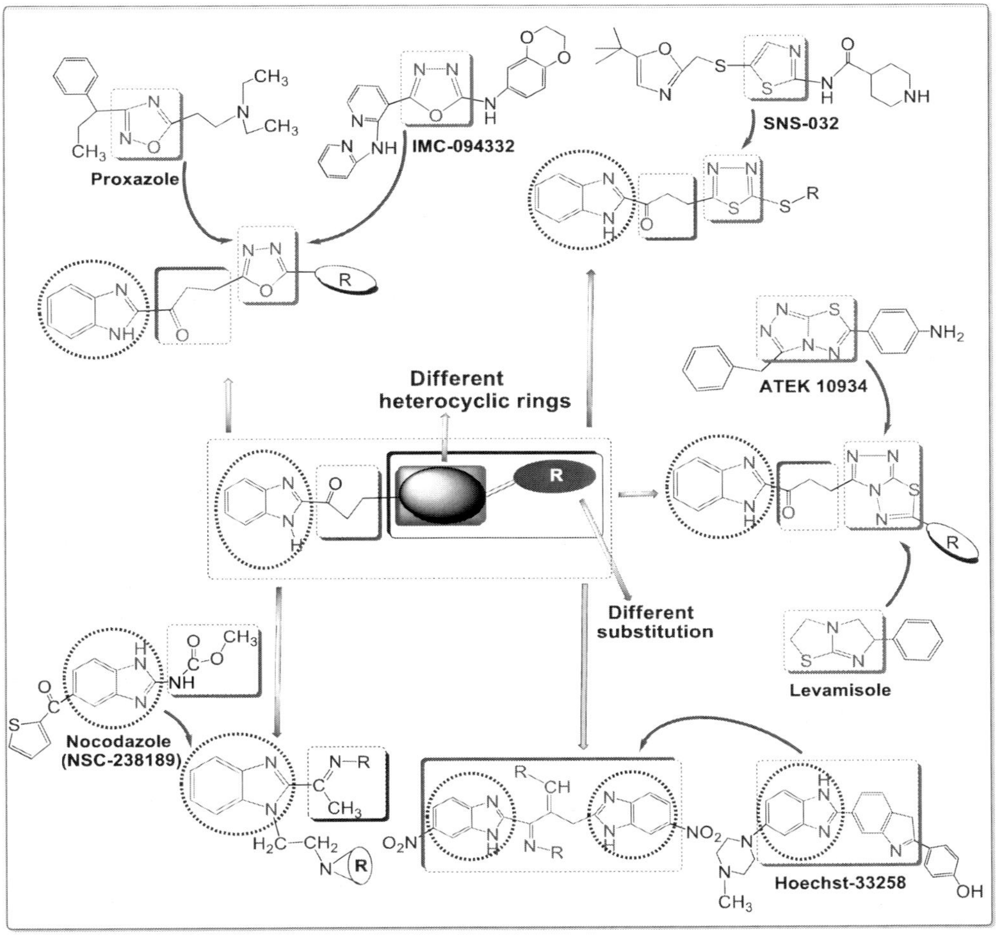

Figure 9. Rational template used for preparing different synthetic Schemes.

12. General Hypothetical Prediction of Different Synthetic Compounds of Benzimidazole

The present work describes the synthesis and anticancer activity of 'Targeted hybrid compounds' of benzimidazole clubbed with other heterocyclic ring systems like oxadiazole, thiadiazole, triazole, triazolo-thiadiazole and triazolo-thiadiazine. Some of these compounds are water soluble, crystalline powder with amphoteric properties due to heterocyclic ring and butyric substituted acid chain. The benzimidazole ring and heterocyclic substitution may be responsible for their cytotoxic activity. The benzimidazole may act as purine analog (nitrogen base of DNA), and could be responsible for cross-linking between DNA bases (intra and inter strand cross linking). General hypothetical representation in regard to water solubility, mechanism of anticancer action, receptor binding etc. is illustrated in Figure 10.

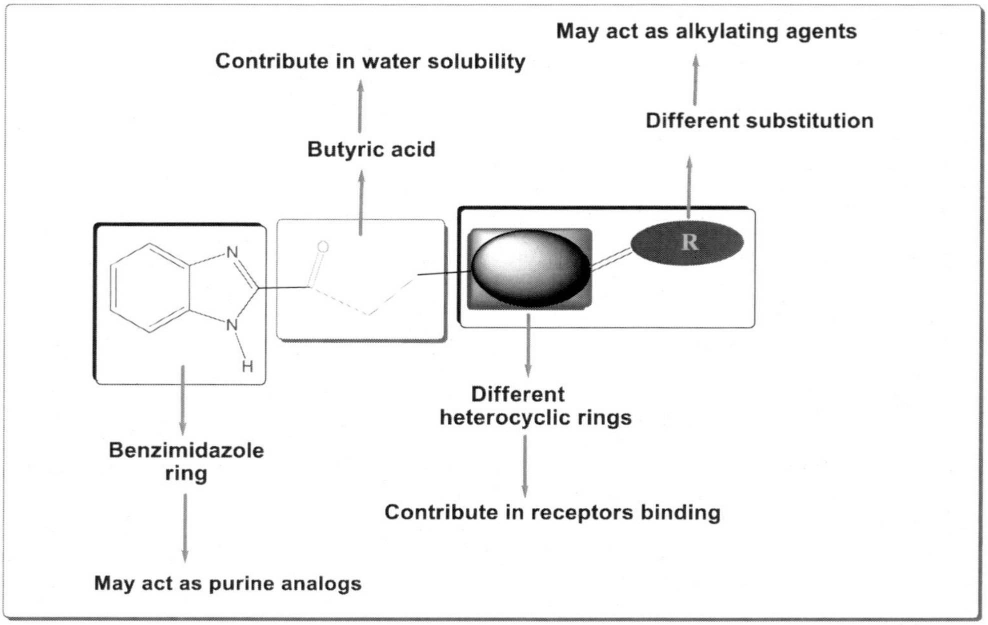

Figure 10. General hypothetical representation of synthetic "Targeted Hybrid Compounds."

References

[1] Agrawal, O. P. (2008). *Organic chemistry reactions and reagents*, Goal publishing house, New Delhi, India, 627, 686.

[2] Eicher, T., Hauptmann, S. (2003). *The chemistry of heterocycles: Structure, reactions, synthesis and applications*, John Wiley and Sons, 2nd edition, USA.

[3] Vita, V. T., Hellman, S., Rosenberg, S. A. (1992). *Cancer principle practical Oncology*. 4th ed. Philadelphia; J. B. Lippincott, Co. USA.

[4] Hartwell, L. H., Kashan. M. B. (1994). *Science* 266, 1821.

[5] Li, Y., Tan, C., Gao, C., Zhang, C., Luan, X., Chen, X., Liu, H., Chen, Y., Jiang Y. (2011). *Bioorganic Medicinal Chemistry*. 19, 4529.

[6] Demirayak, S., Kayagil, I., Yurttas L. (2011). *European Journal of Medicinal Chemistry*. 46, 411.

[7] Sondhi, S. M., Rani, R., Singh, J., Roy, P., Agrawal, S. K., Saxena, A. K. (2010). *Bioorganic Medicinal Chemistry Letter*. 20, 2306.

[8] Penning, T. D., Zhu, G. D., Gandhi, V. B., Gong, J., Liu, X., Shi, Y., Klinghoffer, V., Johnson, E. F., Donawho, C. K., Frost, D. J., Diaz, V. B., Bouska, J. J., Osterling,

D. J., Olson, A. M., Marsh K. C., Luo, Y., Giranda, V. L. (2009). *Journal of Medicinal Chemistry*. 52, 514.

[9] Lio, S. C., Johnson, J., Chatterjee, A., Ludwig, J. W., Millis, D., Banie, H., Sircar, J. C., Sinha, A., Richards M. L. (2008). *Cancer Chemotherapy Pharmacology*. 61, 1045.

[10] Sharma, D., Narasimhan, B., Kumar, P., Judge, V., Narang, R., Clercq, E. D., Balzarini, J. (2009). *Journal of Enzyme Inhibition and Medicinal Chemistry*. 24, 1161.

[11] Salahuddin, Shaharyar, M., Mazumder, A. (2017). Benzimidazoles: A biologically active compounds. *Arabian Journal of Chemistry*. 10, S157–S173.

[12] Kumar, B. V. S., Vaidya, S. D., Kumar, R. V., Bhirud, S. B., Mane R. B. (2006). *European Journal of Medicinal Chemistry*. 41, 599.

[13] Goker, H., Kus, C., Boykin, D. W., Yldz, S., Altanlar N. (2002). *Bioorganic Medicinal Chemistry*. 10, 2589.

[14] Kerimov, I., Kilcigil, G. A., Eke, B. C., Altanlar N. (2007). *Journal of Enzyme Inhibition and Medicinal Chemistry*. 17, 696.

[15] Francisca, P., Helgi, J. C., Jaime, P. V., Juan, C. P., Sergio, R. M., Guadalupe, P. H., Nayeli, L. B., Alicia, H. C., Rafael, C., Francisco H. L. (2009). *European Journal of Medicinal Chemistry*. 44, 1794.

[16] Padmavathi, V., Reddy, S. G., Padmaja, A. Kondaiah, P., Shazia A. (2009). *European Journal of Medicinal Chemistry*. 44, 2106.

[17] El-Sayed, N. S., El-Bendary, E. R., El-Ashry, S. M., El-Kerdawy M. M. (2011). *European Journal of Medicinal Chemistry*. 46, 3714.

[18] Formagio, A. S. N., Tonin, L. T. D., Foglio, M. A., Madjarof, C., Carvalho, J. E., Costa, W. F. D., Cardoso, F. P., Sarragiotto, M. H. (2008). *Bioorganic Medicinal Chemistry*. 16, 9660.

[19] Matysiak, J. (2007). *European Journal of Medicinal Chemistry*. 42, 940.

[20] Elderfield, R. C. (1957). *Heterocyclic compounds,* Wiley, New York, 5.

[21] Acheson, M., King, F. E., Spensleyp, C. (1947). *Nature*. 53, 160.

[22] Grimmest, M. R. (1970). *Advanced Heterocyclic Chemistry*. 12, 103.

[23] Sharif, N. A., Su, S. X., Yanni, J. M. (1994). *Journal of Ocular Pharmacology*. 10, 653.

[24] Day, A. R. (1983). *Electronic Mechanisms of Organic Reactions*. American Book, 242.

[25] Kartaraski, (1984). *Compound Heterocyclic Chemistry*. 5, 345

[26] Gilchrist, T. L. 2005. In *Heterocyclic Chemistry*. Published by Pearson Education Pvt. Ltd, vol III, 315.

[27] Remick, A. E. (1949). *Electronic Interpretations of Organic Chemistry,* John Wiley and Sons, Inc. New York, USA.

[28] Jensen, K. A. Fridioehr, K. D. V. (1943). *Selskab Math. fys. Med.* 20, 1 Chem. Abstracts (1945), 39, 2068.

[29] Smith, W., Rasmussen, S., Ballard, A. (1949). *Journal of the American Chemical Society* 7(1), 1082.

[30] Fischer, O. (1889). *Berichte [Reports]*. 23, 645.

[31] Vanagsg, Dombrowsak, I. (1942). *Chemische Berichte [Chemical Reports]*. 76B, 82.

[32] Hobreckerf, (1872). *Berichte [Reports],* 6,920.

[33] Acheson, R. M., King, F. E., Spensley, P. C. (1947). *Nature* 160, 53.

[34] Shieh, W. C., Lozanov, M., Repic, O. (2003). *Tetrahedron Letter.* 44, 6943.

[35] Combs, A. P., Zhu, W., Crawley, M. L., Glass, B., Polam, P., Sparks, R. B., Modi, D., Takvorian, A., Mc-Laughlin, E., Yue, E. W., Wasserman, Z., Bower, M., Wei, M., Rupar, M., Ala, P. J., Reid, B. M., Ellis, D., Gonneville, L., Emm, T., Taylor, N., Yeleswaram, S., Li, Y., Wynn, R., Burn, T. C., Hollis, G., Liu, P. C. C., Metcalf, B. J. (2006). *Medicinal Chemistry.* 49, 3774.

[36] Ng, R. A., Guan, J., Alford, V. C., Lanter, J. C., Allan, G. F., Sbriscia, T., Linton, O., Linden, S. G., Sui, Z. (2007). *Bioorganic Medicinal Chemistry Letter.* 17, 784.

[37] Bellini, F., Calandri, C., Cauteruccio, S., Rossi, R. (2007). *Tetrahedron.* 63, 1970.

[38] Fouchard, D. M. D., Tillekeratne, L. M. V., Hudson, R. A. (2004). *Journal of Organic Chemistry.* 69, 2626.

[39] Dalla, V. L., Gia, O., Magno, S. M., Settimo, A. D., Marini, A. M., Primofiore, G., Settimo, F. D., Salerno, S. (2001). *II Farmaco [II Medication]*. 56, 159.

[40] Pandey, V. K., Raj, N., Srivastava, U. K. (1986). *Acta Pharma Jugoslavica [Pharma Act Yugoslavia]*. 36, 281.

[41] Vostrova, L. N., Voronina, T. A., Karaseva, T. L., Gernega, S. A., Ivanov, E. I., Kirichenko, A. M., Totrova, M. Y. (1986). *Khim-Farm Zh.* 20, 690.

[42] Frere, S., Thiery, V., Bailly, C., Besson, T. (2003). *Tetrahedron.* 59, 773.

[43] Hao, P., Zhang, S., Sun, W. H., Shi, Q., Adewuyi, S., Lu, X., Li, P. (2007). *Organometallics*, 26, 2439.

[44] Arjmand, F., Mohani, B., Ahmad, S. (2005). *European Journal of Medicinal Chemistry.* 40, 1103.

[45] Gudasi, K. B., Shenoy, R. V., Vadavi, R. S., Patil, M. S., Patil, S. A., Hanchinal, R. R., Desai, S. A., Lohithaswa, H. (2006). *Bioinorganic Chemistry Application.* 9, 1.

[46] Cirulisa, Straumanmis. (1943). *Journal fur Praktische Chemie [Journal of Practical Chemistry]*. 162, 307-28.

[47] Bambergeerand, Berlb, B. (1893). *Acronyms.* 273, 360.

[48] Gerngros, O. S., German patent 282(1915). *Chemical Zentra.* 491, 88.

[49] Hellerg. (1904). *Chemische Berichte [Chemical Reports]*. 37, 3116.

[50] Bambergeer, Lorenzejn. (1893). *Acronyms.* 273, 287, 289.

[51] Fischer, O. (1924). *Journal fur Praktische Chemie [Journal of Practical Chemistry]*. 107(2), 1649.

[52] Blocli, J. J. (1919). *Journal of the Society of Chemical Industry*. 38, 118.

[53] Hartmanmn, Panizzoln. (1938). *Helvetica Chimica Acta [Helvetica Chemical Act]*. 21, 1692.

[54] Wasere, Gratsosa. (1928). *Helvetica Chimica Acta [Helvetica Chemical Act]*. 11, 952.

[55] Weidenhagern, Wegnerh. (1938). *Chemische Berichte [Chemical Reports]*, 71B, 2129.

[56] Gerngros, O. S. (1913). *Berichte [Reports]*. 46, 1913.

[57] Wolff, L. (1913). *Acronyms*. 399, 297.

[58] Baczy, K. I., Niementowsks, I., Bull, Accad, S. C. (1902), 421. *Chem. Zentr*. 73, 940.

[59] Bistrzyckai, Lrzeworsk, G. (1912). *Berichte [Reports]*, 46, 3489.

[60] Bambergeerand, Berle, B. (1893). *Ann*. 273, 333.

[61] Reisserta, Goll, G. (1905). *Berichte [Reports]*. 38, 102.

[62] Copelandr, A. B., Daya, R. (1943). *Journal of the American Chemical Society*. 66, 1072.

[63] Manuellci, Recchiv. (1900). *Chemical Zentra*. 71(11), 1207.

[64] Streitwolkf, F., Herrmanwn. (1933). *U. S. patent 1*, 919, 579. Chemical Abstracts 27, 4817.

[65] Fichterf. (1906). *Journal fur Praktische Chemie [Journal of Practical Chemistry]*. 74(2), 326.

[66] Everettj, G. (1930). *Journal Chemical Society*. 240, 2.

[67] Pierronp, (1908). *Ann. Chim. phys*. 16 (8), 194. *Chemical Zentra*. 79(11), 1586.

[68] Pellizzagri, Gaitera, *Gaza chim ital*. (1918).48(11), 151. Chemical Abstracts. (1919).13, 1584.

[69] Bambergeer, berlb, B. (1893). *Ann*. 273, 323.

[70] Copelandr, A. B., Daya, R., (1943). *Journal of the American Chemical Society*. 66, 1072.

[71] Andrew, T., Athanasia, V., (1997). *Arzeim-Forsch/Drug Research*. 47(1), 307.

[72] Jing, X., Zhu, Q., Xu, F., Ren, X., Li, D., Yan, C. (2006). *Synthetic Communications*, 36, 2597.

[73] Tandon, V. K., Kumar, M. (2004). *Tetrahedron Letter*. 45, 4185.

[74] Bougrin, K., Loupy, A., Petit, A., Daou, B., Soufiaoui, M. (2001). *Tetrahedron*. 57, 163.

[75] Vanvliet, D. S., Gillespie, P., Scicinski, J. J. (2005). *Tetrahedron Letter*. 46, 6741.

[76] Hegedus, A., Hell, Z., Potor, A. (2006). *Synthetic Communications*. 36, 3625.

[77] Shatha, I. A. (2017). Synthesis approaches to benzimidazoles from o-phenylenediamine: A literature review, *Journal of Saudi Chemical Society*, 21, 229-237.

[78] Nagawade, R. R., Shinde, D. B. (2007). *Indian Journal of Chemistry* (Section B). 46B, 349.

[79] Neochoritis, C., Livadiotou, D., Stephanidou, S. J., Tsoleridis, C. A. (2007). *Tetrahedron Letter.* 48, 2275.

[80] Gedye, R., Smith, F., Westaway, K., Ali, H., Baldisera, L., Laberge, L., Russell, J. (1986). *Tetrahedron Letter.* 27, 279.

[81] Anshul, C., Ramandeep, K., Anju, G. (2011). Importance of Microwave Reactions in the Synthesis of Novel Benzimidazole Derivatives: A Review. *Journal of Chemical and Pharmaceutical Research*. 3(6), 925-944.

[82] Faulkner, D. J. (1992). *Natural Product Reports.* 9, 323.

[83] Nakamura, S., Tsuno, N., Yamashita, M., Kawasaki, I. Ohta, S., Ohishi, Y. (2001). *Journal of the Chemical Society, Perkin Transactions.* 1, 429.

[84] Kurakata, S., Fujiwaa, K. (2001). *Fujita.* 2000-JP4858, 2001005402, 20000719.

[85] Carcanague, D., Shue, Y. K., Wuonola, M. A. T., Urianickelsen, M., Jourbran, C., Abedi, J. K., Jones, J., Kuehler, T. C. (2002). *Journal of Medicinal Chemistry.* 45, 4300-4309.

[86] Lezcano, M., Al-Soufi, W., Novo, M., Rodriguez N. E., Tato, J. V. (2002). *Journal of Agricultural and Food Chemistry.* 50, 108-112.

[87] Aghatabay, N. M. Somer, M. Senel, M. Dulge, B. Gucin, F. (2007). *European Journal of Medicinal Chemistry.* 42, 1069-1075.

[88] Demirayak, S., Karaburun, A. C., Kayagil, I., Ucucu, U., Beis, R. (2005). *Phosphorous Sulfur Silicon*. 180, 489-493.

[89] Tiwari, A. K., Mishra, A. (2006). *Indian Journal of Chemistry (Section B).* 45, 489-493.

[90] Shatha, I. A. (2017) *Journal of Saudi Chemical Society,* 21, 2, 229-237.

[91] Gardiner, J. M., Lyons, C. R., Burke, A., Khan, A., Mahmood, N. (1995). *Bioorganic & Medicinal Chemistry Letters,* 7, 1251-1254.

[92] Baraldi, P. G., Bovero, A., Fruttarolo, F., Preti, D., Tabrizi, M. A., Pavani, M. G., Romagnoli, R. (2004). *Medicinal Research Reviews.* 24, 475.

[93] Reddy, B. S. P., Sharma, S. K., Lawn, J. W. (2001). *Current Medicinal Chemistry*, 8, 475.

[94] Singh, A. K., Lown J. W. (2000). *Anticancer Drug Design.* 15, 265.

[95] Joubert, A., Sun, X. W., Johansson, E., Bailly, C., Mann, J., Neidle, S. (2003). *Biochemistry.* 42, 5984.

[96] Lynch, M., Hehir, S., Kavanagh, P., Leech, D., Shaughnessy, J. O., Carty, M. P., Aldabbagh, F. (2007). *Chemistry- A European Journal.* 13, 3218.

[97] Kubota, Y., Fuji, H., Liu, J., Tani S. (2001). *Nucleic Acids Research.* 12, 101.

[98] Garuti, L., Roberti, M., Pession, A., Leoncini, E., Hrelia S. (2001). *Bioorganic & Medicinal Chemistry Letters.* 11, 3147.

[99] Xiang, P., Zhou, T., Wang, L., Sun, C. Y., Hu, J., Zhao, Y. L., Yang L. (2012). *Molecules*. 17, 873.

[100] Dillman, R. O. Ryan, K. P. Dillman, J. B. Shawler, D. L. Maguire R. (1992). *Molecular Biotherapy*. 4, 10.

[101] Galal, S. A., Hegab, K. H., Kassab, A. S., Rodriguez, M. L., Kerwin, S. M., Khamry, A. M. A. E., Diwani, H. I. E. (2009). *European Journal* of *Medicinal Chemistry*. 44, 1500.

[102] Starcevic, K., Kralj, M., Ester, K., Sabol, I., Grace, M., Pavelic. K., Zamola G. K. (2007). *Bioorganic & Medicinal Chemistry*. 15, 4419.

[103] Sun, Q., Runzhi, W., Sutang, C., Lin, Y., Sellers, L., Sakamoto, K., He, B., Peterson, B. R. (2011). *Journal of Medicinal Chemistry*. 54, 1126.

[104] Rubbiani, R., Kitanovic, I., Alborzinia, H., Can, S., Kitanovic, A., Onambele, L. A. (2010). *Journal of Medicinal Chemistry*. 53, 8608.

[105] Penning, T. D., Zhu, G. D., Gandhi, V. B., Gong, J., Liu, X., Shi, Y., Klinghofer, V., Johnson, E. F., Donawho, C. K., Frost, D. J., Diaz, V. B., Bouska, J. J., Osterling, D. J. (2009). *Journal of Medicinal Chemistry*. 52, 514.

[106] Hao, D., Rizzo, J. D., Stringer, S., Moore, R. V., Marty, J., Dexter, D. L., Mangold, G. L., Camden, J. B., Von-Hoff, D. D., Weitman, S. D. (2002). *Investigational New Drugs*. 20, 261.

[107] Ramurthy, S., Aikawa, M., Amiri, P., Costales, A., Hashash, A., Jansen, J. M., Lin, S., Mab, S., Renhowe, P. A., Shafer, C. M., Subramanian, S., Sung, L., Verhagen, J. (2011). *Bioorganic & Medicinal Chemistry Letters*. 21, 3286.

[108] Abonia, R., Cortés, E., Insuasty, B., Quiroga, J., Nogueras, M., CoboJ. (2011). *European Journal of Medicinal Chemistry*. 46, 4062.

[109] Husain, A., Rashid, M., Siddiqui, A. A., Mishra R. (2013). *European Journal* of *Medicinal Chemistry*. 62, 785.

[110] Husain, A., Rashid, M., Mishra, R., Praveen, S., Shin, D. S., Kumar, D. (2012). *Bioorganic & Medicinal Chemistry Letters*. 22, 5438.

[111] Rashid, M., Husain, A., Mishra R. (2012). *European Journal of Medicinal Chemistry*. 54, 855.

[112] Rashid, M., Husain, A., Mishra, R., Karim, S., Khan, S., Ahmad, M., Al-wabel, N., Husain, A., Ahmad, A., Khan, S. A. (2015). *Arabian Journal of Chemistry*. 3, 2341.

[113] Desai, K. G., Desai, K. R. (2006). *Bioorganic & Medicinal Chemistry*. 14, 8271.

[114] Guven, O. O., Erdogan, T., Goeker, H., Yildiz, S. (2007). *Bioorganic & Medicinal Chemistry Letters*. 17, 2233.

[115] Ansari, K. F., Lal, C. (2009). *European Journal of Medicinal Chemistry*. 44, 2294.

[116] Jubie, S., Rajeshkumar, R., Yella, R. B., Siddhartha, G., Sandeep, M., Surendrareddy, K., Dushyatha, H. S., Elango, K. (2010). *Journal of Pharmaceutical Sciences and Research*. 2(2), 69.

[117] Giiker, H., Tuqbilek, M., Ayhan, G., Altanlar N. (1998). *Il Farmaco [Il Medication]*. 53, 415.

[118] Orena, I. Temiza, O. Yalcýna, I. Senera, E. Altanlarb N. (1998). *European Journal of Pharmaceutical Sciences*. 7, 153.

[119] Anichina, K. K., Vuchev, D. I., Mavrova, A. T. (2006). *Journal of Parasitic Disease*. 34, 19.

[120] Mavrova, A. T., Anichina, K. K., Vuchev, D. I., Tsenov, J. A., Denkova, P. S., Kondeva, M. S., Micheva, M. K. (2006). *European Journal of Medicinal Chemistry*. 41, 1412.

[121] Gaur, N. M., Patil, S. V., Mourya, V. K., Wagh, S. B. (2000). *Indian* Journal *of Heterocyclic Chemistry*. 9, 227.

[122] Rashmi, S., Jujher, U. V. S., Bhanuki, D. S. (1989). *Indian Journal of Chemistry (Section B)*. 28B, 702.

[123] Rajiv. M. H., Ramana, M. V. (2002). *Indian Journal of Heterocyclic Chemistry*. 2, 121.

[124] Hernández, L. F., Hernández, C. A., Castillo, R., Navarrete, V. G., Soria, A. O., Hernández, H. M., Yepez, M. L. (2010). *European Journal of Medicinal Chemistry*. 45(7), 3135.

[125] Toshiro, N., Sachiko, Y., Miho, S., Naoto, O., Kaoru, I., Yasuo, N., Akira T. (2006). *Biological and Pharmaceutical Bulletin*. 29(9), 1931.

[126] Wienen, W., Entzeroth, M., Van-Meel, J. C. A., Stangier, J., Busch, U., Ebner, T., Schmidt, J., Lehmann, H., Matzek K., Rawson, J. K., Gladigau, V., Hauel, N. (2000). *Cardiovascular Drug Reviews*. 18(2), 127.

[127] Barreca, M. L., Chimirri, A., Clercq, E. D., Luca, L. D., Monforte, A. M., Monforte, P., Rao, A., Zappala, M. (2003). *Il Farmaco [Il Medication]*. 58, 259.

[128] Pandey, V. K. Gupta, V. D. Tiwari D. N. (2005). *Indian Journal of Heterocyclic Chemistry*. 14, 217.

[129] Xu, Y. S., Zeng, C. C., Jiao, Z. G., Hu, L. M., Zhong, R. G. (2009). *Molecules*. 14, 868.

[130] Chimirri, A., Monforte, P., Rao, A., Zappalà, M., Monforte, A. M., Sarro, G. D., Pannecouque, C., Witvrouw, M., Balzarini, J., Clercq, E. D., (2001). *Antiviral Chemistry & Chemotherapy*. 12, 169.

[131] Zou, R., Ayres, K. R., Drach, J. C., Townsend, L. B. (1996). *Journal of Medicinal Chemistry*. 39, 3477.

[132] Labanauskas, L. K., Brukstus, A. B., Gaidelis, P. G., Buchinskaite, V. A., Udrenait, E. B., Dauksas, V. K. (2000). *Pharmaceutical Chemistry Journal*. 34, 353.

[133] Sondhi, S. M., Rajvanshi, S., Johar, M., Bharti, N., Azam, A., Singh, A. K. (2002). *European Journal of Medicinal Chemistry*. 37, 835.

[134] Labanauskas, L., Brukstus, A., Udrenaite, E., Gaidelis, P., Bucinskaite, V., Dauksas, V. (2000). *Chemija Journal [Chemistry journal]*. 11, 211.

[135] Puratchikody, A., Sivajothi, V., Jaswanth, A., Ruckmani, K., Nallu M. (2002). *Indian Journal of Heterocyclic Chemistry*. 11, 241.

[136] Srinivasulu, G., Reddy, P. P., Hegde, P., Chakrabart, R. (2005). *Heterocyclic Communication*. 11, 23.

[137] Sawhney, S. N., Asha, G. (1991). *Indian Journal of Chemistry*. 30B, 407.

[138] Kavitha, C. S., Achar, Hosamani, K. M., Seetharamareddy, H. R. (2010). *European Journal of Medicinal Chemistry*. 45, 2048.

[139] Lombardy, R. L., Tanious, F. A. (1996). *Journal of Medicinal Chemistry*. 39, 1452.

[140] Kus, C., Kilcigil, G. A., Ozbey, S., Kaynak, F. B., Kaya, M., Coban, T., Eke, B. C. *Bioorganic & Medicinal Chemistry*. 16, 4294 (2008).

[141] Vikrant, S. P., Amey, J. M., Shukla, S. R. (2009). *European Journal of Medicinal Chemistry*. 44, 5112.

[142] Dudutien, V., Baranauskien, L., Matulis, D. (2007). *Bioorganic & Medicinal Chemistry Letter*. 17, 3335.

[143] Singh, J., Grover, P., Pathak, D. P. (2010). *Acta Pharmaceutica Sciencia*. 52, 511.

[144] Rajak, H. Aggarwal, N. Kashaw, S. Kharya, M. D. Mishra P. (2010). *Journal of the Korean Chemical Society*, 54(1), 158.

[145] Pattanayak, P., Sharma, R., Sahoo, P. K. (2009). *Medicinal Chemistry Research*. 18(5), 351.

[146] Canetea, R., Escobedob, A. A., Almirallc, A., Gonzálezd, M. E., Britoe, K., Cimermanf. S. 2009. *Transaction of the Royal Society of Tropical Medicine and Hygiene*. 103, 437.

[147] Keiji, K., Yoshiyoki, I. (1993). *Journal of Medicinal Chemistry*. 36(12), 1772.

[148] Ivana, S., Katarina, M., Anahi, M., Ariel, M. S., Ivo, P., Miroslav, B., Ljubica, G. O. 2009. *Bioorganic & Medicinal Chemistry*. 17, 2544.

[149] Refaat, Hanan, M., Khalil, Omneya, M., Abuel, M., Suzan, M. 2009. *Journal of Chemical Research*. 7, 448.

[150] Duanmu, C., Shahrik, L. K., Holly, H. H., Hamel, E. (1989). *Cancer Research*. 49, 1344.

[151] Britten, C. D., Delioukina, M., Boulos, L., Reiswig, L., Gicanov, N., Rizzo, J., Hao, D., Tolcher, A., Weitman, S., Rugg, T., Von Hoff. D., Camden, J., Rosen, L. S. (2001). *American Society of Clinical Oncology*. 20, 2129.

[152] Penning, T. D., Zhu, G. D., Gandhi, V. B., Gong, J., Thomas, S., Lubisch, W., Grandel, R., Wernet, W., Park, C. H., Fry, E. H., Luo, Y., Liu, X., Shi, Y., Klinghofer, V., Johnson, E. F., Frost, D. J., Donawho, C. K., Bontcheva, D. V., Bouska, J. J., Olson, A. M., Marsh, K. C., Rosenberg, S. H., Giranda, V. L. (2008). *Bioorganic & Medicinal Chemistry*. 16, 6965.

[153] Tolner, B., Hartley, J. A., Hoch Hauser, D. (2001). *Molecular Pharmacology*. 59, 699.

[154] Albert, J. M., Cao, C., Kim, K. W., Willey, C. D., Geng, L., Xiao, D., Wang, H., Sandler, A., Johnson, H. D., Colevas, A. D., Low, J., Rothenberg, M. L., Lu, B. (2007). *Clinical Cancer Research.* 13, 3033.

[155] Bradshaw, T. D., West well, A. D. (2004). *Current Medicinal Chemistry.* 11, 1241.

[156] Mcclue, S. J., Blake, D., Clarke, R., Cowan, A., Cummings, L., Fisher, P. M., Mackenzie, M., Melville, J., Stewart, K., Wang, S., Zhelev, N., Zheleva, D., Lane, D. P. (2002). *International Journal of Cancer.* 102, 463.

[157] Dalip, K., Gautam, P., Angela, K., Chavers, Kuei, H. C., Kavita, S. (2011). *European Journal of Medicinal Chemistry.* 46, 3085.

[158] Griffiths, G., Scaerou, F., Midgely, C., McClure, S. (2008). *American Society of Clinical Oncology.* 49, 56, 44.

[159] Matysiak, J., Opolski, A. (2006). *Bioorganic & Medicinal Chemistry.* 14, 4483.

[160] Remrs, W. A., Wilson, A., Griswold's, B. (1982). *Text Book of Organic, Medicinal and Pharmaceuticals Chemistry,* Lippincott Company, Philadelphia, USA.

[161] Azza, T. T., Hanan, H. G., Hussein, E. S. (2012). *European Journal of Medicinal Chemistry.* 47, 445.

[162] Ibrahim, D. A. (2009). *European Journal of Medicinal Chemistry.* 44, 2776.

[163] Maria, C. T., Asra, M., Xiaohu, O., David, S., James, F., Stan, M., Margarita, C., Bridget, F., Jacqueline, D., Eugene, L. P., Chekler, P. K., James R. T. (2010). *Home of Clone Systems*, Inc. New York, USA, 3(5), 318.

[164] Lissitchkov, T., Arenado, G., Pitched, D., Merkle, K. (2006). *Journal of Cancer Research and Clinical Oncology.* 132 (2), 99.

[165] Knauf, W. U., Lissichkov, T., Aldaoud A. (2009). *Journal of Clinical Oncology.* 27, 4378.

[166] Dalip, K., Gautam, P., Angela, K., Chavers, H. C., Kuei, S. K. (2011). *European Journal of Medicinal Chemistry.* 46, 3085.

[167] Maria, C. T., Asra, M., Xiao Hu, O., David, S., James, F., Stan, M., Margarita, C., Bridget, F., Jacqueline, D., Eugene, L. P., Chekler, P. K., James, R. T. (2010). *I'm. Clon. Systems.* Inc. New York, USA, 3(5), 318.

[168] Mcclue, S. J., Blake, D., Clarke, R., Cowan, A., Cummings, L., Fisher, P. M., Mackenzie, M., Melville, J., Stewart, K., Wang, S., Zhelev, N., Zheleva, D., Lane, D. P. (2002). *International Journal of Cancer,* 102, 463.

[169] Remrs, W. A., Wilson, A., Griswold's, B. (1982). *Text Book of Organic, Medicinal and Pharmaceuticals Chemistry,* Lippincott Company, Philadelphia.

[170] Ibrahim, D. A. (2009). *European Journal of Medicinal Chemistry.* 44, 2776.

[171] Tolner, B., Hartley, J. A., Hoch Hauser, D. (2001). *Molecular Pharmacology.* 59, 699.

Biographical Sketch

Dr. Mohammad Rashid

Affiliation: Department of Medicinal Chemistry and Pharmacognosy, College of Pharmacy and Dentistry, Buraydah Colleges, Buraydah, Al-Qassim-31717, Kingdom of Saudi Arabia.

Education: Ph.D., M. Pharm, PGDPRA, M.B.A. (IB), B. Pharm.

Business Address: Department of Medicinal Chemistry and Pharmacognosy, College of Pharmacy and Dentistry, Buraydah Colleges, Buraydah, Al-Qassim-31717, Kingdom of Saudi Arabia Research and Professional Experience: 8 Years (Research Experience), 9 Years (Teaching Experience)

Professional Appointments: Senior Assistant Professor

Honors:

1. Young Scientist Award from Science & Engineering Research Board (SERB) under Department of Science & Technology (DST), Government of India, New Delhi.
2. Senior Research Fellowship in Engineering & Technology under the Scheme of University Grant Commission, Government of India, New Delhi (INDIA) F. No. 10-01/2008 (SA)].
3. Junior Research Fellowship in Engineering & Technology under the Scheme of University Grant Commission, Government of India, New Delhi (INDIA). [F. No. 10-01/2008 (SA)].
4. Project Research Fellowship under the scheme of Major Research Project, University Grant Commission, Government of India, New Delhi (INDIA) [F. No. 36-107/2008 (SR)].
5. Granted "Fellowship Travel award" from Eurotox-2013 for attending the 'The 49th Congress of the European Societies of Toxicology' held in Interlaken, Switzerland on September 1-4th, 2013.
6. Got "Junior Fellowship Travel award" from IUTOX for attending the 'The 8th Congress of Toxicology in Developing Countries (8CTDC)' held in Bangkok, Thailand on September 10-13, 2012.
7. Got "International travel awarded" from DST (Department of Science and Technology) for presented a poster in a International Conference, The 8th Congress

of Toxicology in Developing Countries (8CTDC). The meeting held in Bangkok, Thailand on September 10-13, 2012.

8. Got "International travel awarded" from ICMR (INDIAN COUNCIL OF MEDICAL RESEARCH) for presented a poster in a International Symposium for Thin-Layer Chromatography BASEL, 06-08, July, 2011 at Congress Center Basel, Switzerland.
9. University Grant Commission Fellowship for Master of Pharmacy.
10. Dr. Manjeet Singh Fellowship award for PGDPRA.
11. My research topic selected and presented as a speaker in "Invention Convention Programme" on the 12th July as in final 15 list out of 5000 participates from 50 countries, ESOF-2012, at Dublin, Ireland.

Professional Recognitions:

1. Got "Certificate of Recognition" of my research work presented in 'National Science Day' on the 28-Feb-2013, held at Convention Centre of Jamia Hamdard (Hamdard University), New Delhi-62.
2. Editorial Board Member of Cancer Studies: Open Access, Aperito Online Publishing, Bascom Avenue, Suite 700, Campbell, CA, 95008, USA.
3. Appointed as Editorial Board member for Journal Clinical Cancer Drugs-CCAND: Bentham Science, Bentham Publisher, Dubai, UAE.

- **Membership in Professional Body**

1. IPGA (Indian Pharmacy Graduates Association), Membership No. LM-4921.
2. ASPET (American Society for Pharmacology and Experimental Therapeutics) USA, Membership No. 31180.
3. UPC (Uttarakhand Pharmacy Council), Dehradun, Membership No. 3314.
4. IATDMCT (International Association of Therapeutic Drug Monitoring and Clinical Toxicology) USA.

Reviewer Recognition Award: Got Outstanding and recognition reviewer award from the following Journal.

1. Saudi Pharmaceutical Journal (Recognition contribution awarded in 2018)
2. Bioorganic Chemistry (Outstanding contribution in reviewing awarded in 2017)
3. Arabian Journal of Chemistry (Recognition contribution awarded in 2015)
4. Journal of Natural Gas Science & Engineering (Recognition contribution awarded in 2012)

Reviewer: Review the manuscript of the following Journals

1. Letters in Drug Design & Discovery (Benthamscience.com)
2. Bioorganic Chemistry (Elsevier.com)
3. Journal of Enzyme Inhibition and Medicinal Chemistry (Informahealthcare.com)
4. Journal of Natural Gas Science & Engineering (Elsevier.com)
5. Arabian Journal of Chemistry (Elsevier.com)

6. Chinese Journal of Pharmaceutical Science

Non-Teaching Work Experience: In the following program committees and units as follows.

1. The Academic, development and research committee (Head)
2. College website unit (Head-English Web Master)
3. Laboratory Committee (Member)
4. Quality and Academic Accreditation Committee (Member)
5. Dr. Hassan Azzazy Research Group (Member)
6. Developing Learning Resources Project (Member)
7. Guide and Consult the students as Academic Advisor (Member)

Publications from the Last 3 Years:

1. Neelima Shrivastava, Asif Husain, Mohammad Rashid*, Shahid Karim, Nimer Fehaid Alsabeelah, Nasir Ali Siddiqui, Recent advances towards treatment of HIV: Synthesis and SAR studies, *Mini Review Article in Medicinal Chemistry*, Bentham Science, 2019 (Accepted) (Impact Factor: 2.845).
2. Mohammad Rashid*, Babar Ali, Shamshir khan, Makhmur Ahmad, Shahid Karim, Nimer Fehaid Alsabeelah, Asif Husain, Ravinesh Mishra, Plant Derived Semisynthetic Analogues as Anti-cancer Agents, *Journal of Chemical, Biological and Physical Sciences,* February 2018 – April 2018, Vol. 8, No. 2; 396-433. (Impact Factor: 1.30).
3. Mohammad Rashid*, Shahid Karim, Babar Ali, Shamshir Khan, Makhmur Ahmad, Asif Husain, Ravinesh Mishra, PI3K signaling pathway targeting by using different molecular approaches to treat cancer, *Journal of Chinese Pharmaceutical Sciences,* 2017, 26 (9), 621–634, http://www.jcps.ac.cn. (*Impact Factor*: 0.691).
4. Sana Hasmi, Gamal O. El-Hassan, Shamshir Khan, Mohammad Rashid*, A Valuable Insight Impact of Pharmacist in Pharmacogenomics, *Scholars Academic Journal of Pharmacy* (SAJP), Scholars Academic and Scientific Publisher, 2017, 12(6), 527-535. (Impact Factor: 1.4).
5. Ravinesh Mishra, Mohd Rashid ‘Phenothiazines and related drugs as Multi Drug Resistance reversal agents in cancer chemotherapy’ *Current Cancer Therapy Reviews*, Bentham Publication, 2017, 13(1): 28 - 42. (Impact Factor: 0.45).
6. Vikas Kumar, Prakash Chandra Bhatt, Gaurav Kaithwas, Mohd Rashid, F. A. Al-abbasi, Jalaluddin A. J. Khan, Firoz Anwar, Amita Verma, α-Mangostin Mediated Pharmacological Modulation of Hepatic Carbohydrate Metabolism in Diabetes Induced Wistar Rat, *Beni-Suef University Journal of Basic and Applied Sciences,* 2016, Elsevier, http://dx.doi.org/doi: 10.1016/j.bjbas.2016.07.001.

7. Amita Verma, Prakash Chandra Bhatt, Gaurav kaithwas, Nikunj Sethi, Mohd Rashid, Yashwant Singh, Mahfoozur Rahman, Fahad Al-Abbasi, Firoz Anwar, Vikas Kumar, Chemomodulatory effect Melastoma Malabathricum Linn against chemically induced renal carcinogenesis rats via attenuation of inflammation, oxidative stress, and early markers of tumor Expansion, *Inflammopharmacology,* 2016, DOI 10.1007/s10787-016-0276-1. (Impact Factor: 3.35).
8. Asif Husain, Diwakar Madhesia, Mohd Rashid, Aftab Ahmad, Shah Alam Khan, Synthesis and in-vivo diuretic activity of some new benzothiazole sulfonamides containing quinoxaline ring system, *Journal of Enzyme Inhibition And Medicinal Chemistry*, 2016 Taylor & Francis. DOI: 10.3109/14756366.2015.1128425. (Impact Factor: 2.33).
9. Manoj K. Pal, Mohammad Rashid, Manisha Bisht, Multiplexed magnetic nanoparticle-antibody conjugates (MNPs-ABS) based prognostic detection of ovarian Cancer biomarkers, CA-125, β-2M and ApoA1 using fluorescence spectroscopy with comparison of surface Plasmon resonance (SPR) analysis, *Biosensors and Bioelectronics*, 2015, 73 146–152. (doi: http://dx.doi.org/10.1016/j.bios.2015.05.051). (Impact Factor: 8.173).
10. Mohd Rashid, Asif Husain, Ravinesh Mishra, Shahid Karim, Shamshir Khan, Makhmur Ahmad, Naser Al-wabel, Amjad Husain, Aftab Ahmad, Shah Alam Khan, Design and synthesis of benzimidazoles containing substituted oxadiazole, thiadiazole and triazolo-thiadiazines as a source of new anticancer agents. *Arabian Journal of Chemistry-2015* (http://dx.doi.org/10.1016/j.arabjc.2015.08.019). (Impact Factor: 3.72).
11. Mohammad Asif, Asif Husain, Aftab Ahmad, Shah Alam Khan, Mohd Rashid, Karun Arora, Dherya Bahl and Farah Iram 'An overview on antitubercular activities of fluoroquinolones and other related analogues' *European Journal of Experimental Biology*, Pelagia Research Library, 2015, 5(2): 96-109. (Journal Impact: 0.23*).
12. Mohd Rashid, Asif Husain, Mohammad Shaharyar, Ravinesh Mishra, Afzal Hussain, Obaid Afzal 'Design and Synthesis of Pyrimidine Molecules Endowed with Thiazolidin-4-one as New Anticancer agents' *European Journal of Medicinal Chemistry*, 2014, 83, 630-645. (Impact Factor: 4.816).
13. Mohd Rashid, Asif Husain*, Mohammad Shaharyar and Mohd Sarafroz 'Anticancer Activity of New Compounds Using Benzimidazole as a Scaffold' *Anti-Cancer Agents in Medicinal Chemistry*, 2014, 14 (7), 1003-1018. (Impact Factor: 3.37).

Prof. (Dr.) Asif Husain

Affiliation: Department of Pharmaceutical Chemistry, School of Pharmaceutical Education and Research (SPER), Jamia Hamdard, New Delhi-1100 62, India

Education: B. Pharm., M. Pharm., Ph.D. (Pharmaceutical Chemistry)

Business Address: Department of Pharmaceutical Chemistry, School of Pharmaceutical Education and Research (SPER), Jamia Hamdard, New Delhi-1100 62, India

Research and Professional Experience: 23 years (Research Experience), 21 years (Teaching Experience)

Professional Appointments: Professor

Honors:

1. In-Charge, Central Instrumentation Facility (CIF), Jamia Hamdard, New Delhi, India.
2. Albert Nelson Marquis Life Time Achievement Award (2018), by Marquis Who`s Who.
3. Scientist of the Year (2016), VIRA awards (Venus International Research Awards), Chennai, India
4. Foreign Pharmacy Graduate Equivalency Certificate (FPGEC certification-2013), NABP, USA
5. Young scientist award- 2007, Department of Science & Technology (D.S.T.), Govt. of India, under the scheme of fast track proposal for young scientists.
6. Scientist of the Year-2008, National Environmental Science Academy (NESA), New Delhi.
7. Travel grant (U.G.C.), 19th American Peptide Symposium, San Diego, California, U.S.A. (2005).
8. Received more than 75 certificates for attending different conferences, symposiums, workshops, QIP & training programs, resource person, and delivering invited lectures.
9. Expert Reviewer of 10 international journals and 5 national journals.
10. Name selected for 'TOP 100 Scientists' list by the International Biographic Center (IBC), Great Britain, 2008 & 2015.
11. Biography published and regularly updated in `Who is Who`, IBC, Great Britain.
12. Asstt. Dean, Student's Welfare (A.DSW), Jamia Hamdard, New Delhi (2001-2010).

Publications from the Last 3 Years:

1. Asif Husain, Diwakar Madhesia, Mohd Rashid, Aftab Ahmad and Shah Alam Khan (2016), Synthesis and *in vivo* diuretic activity of some new benzothiazole sulfonamides containing quinoxaline ring system. *J. Enzym. Inh. Med. Chem.*, 31(6), 1682-1689. (IF = 3.638)
2. Asif Husain, Priyanka Ahuja, Aftab Ahmad, Shah A. Khan (2016), Synthesis, Biological Evaluation and Pharmacokinetic Studies of Mefenamic Acid-N-Hydroxymethylsuccinimide Ester Prodrug as safer NSAID. *Med. Chem.*, 12(6), 585-591. (IF = 2.631)
3. Sukhbir Lal Khokra, Jyoti, Chetan, Pawan Kaushik, M. M. Alam, M. S. Zaman, Aftab Ahmad, Shah Alam khan, Asif Husain (2016), Quinoline based furanones and their nitrogen analogues: Docking, synthesis and biological evaluation. *Saudi Pharm. J.*, 24(6), 705-717. (IF = 3.346)
4. Asif Husain, Aftab Ahmad, Shah A. Khan (2016), Synthesis and biological evaluation of a mutual prodrug of norfloxacin and fenbufen. *J. Taibah Univ. Med. Sci.*, 11(3), 277-281.
5. S. L. Khokra, S. A. Khan, D. Choudhary, S. M. Hasan, A. Ahmad, Asif Husain (2016), Rational Design and Synthesis of Biologically Active Disubstituted 2(3*H*) Furanones and Pyrrolone Derivatives as Potent and Safer Non Steroidal Anti-inflammatory Agents. *Anti-Inflammatory & Anti-Allergy Agents in Medicinal Chemistry*, 15, 54-71.
6. Sukhbir Lal Khokra, Shah Alam Khan, Pramila Thakur, Deepika Chowdhary, Aftab Ahmad, Asif Husain (2016), Synthesis, Molecular Docking and Potential Antioxidant Activity of Di/Trisubstituted Pyridazinone Derivatives. *J. Chinese Chem. Soc.*, 63, 739-750.
7. Farah Iram, Sadaf Ali, Aftab Ahmad, Sha Alam Khan, Asif Husain (2016), A review on dronedarone: Pharmacological, pharmacodynamic and pharmacokinetic profile. *J. Acute Disease*, 5, 102-108.
8. Yakub Ali, Mohammad Sarwar Alam, Hinna Hamid, Asif Husain, Abhijeet Dhulap, Firasat Hussain, Sameena Bano and Chetna Kharbanda (2016), Molecular modeling and synthesis of some new 2-imino-4-thiazolidinone derivatives with promising TNF-a inhibitory activity. *New J. Chem.*, 40, 711-723. (IF = 3.277)
9. Asif Husain, Aftab Ahmad, Shah Alam Khan, Mohd Asif, Rubina Bhutani, Fahad A. Al-Abbasi (2016), Synthesis, molecular properties, toxicity and biological evaluation of some new substituted imidazolidine derivatives in search of potent anti-inflammatory agents. *Saudi Pharm. J.*, 24, 104-114. (IF = 2.611)
10. Aftab Ahmad, Asif Husain, Shah Alam khan, Mohd Mujeeb, Anil Bhandari (2016), Synthesis, antimicrobial and antitubercular activities of some novel pyrazoline derivatives, *J. Saudi Chem. Soc.*, 20(5), 577-584. (IF = 2.912)

11. Yakub Ali, Mohammad Sarwar Alam, Hinna Hamid, Asif Husain, Abhijeet Dhulap, Sameena Bano, Chetna Kharbanda (2017), Novel 2,4-dichlorophenoxy acetic acid substituted thiazolidin-4-ones as anti-inflammatory agents: Design, synthesis and biological screening. *Bioorg. Med. Chem. Lett.*, 27(4), 1017-1025. (IF = 2.442)
12. Deepak Kumar, G. Mariappan, Asif Husain, J. Monga, S. Kumar (2017), Design, synthesis and cytotoxic evaluation of novel imidazolone fused quinazolinone Derivatives. *Arabian J. Chem.*, 10, 344-350. (IF = 4.008)
13. Mohd Rashid, Asif Husain, Ravinesh Mishra, Shahid Karim, Shamshir Khan, Makhmur Ahmad, Naser Al-wabel, Amjad Husain, Aftab Ahmad, Shah Alam khan (2017), Design and synthesis of benzimidazoles containing substituted oxadiazole, thiadiazole and triazolo-thiadiazines as a sourse of new anticancer agents. *Arabian J. Chem.*, (in Press). (IF = 4.008)
14. Farah Iram, Shah Alam Khan, Asif Husain (2017), Phytochemistry and potential therapeutic actions of Boswellic acids: A mini-review. *Asian Pacf. J. Trop. Biomed.* 7, 513-523.
15. Shah Alam Khan, Priyanka Ahuja, Asif Husain (2017), Oxidative Cyclization of Isoniazid with Fluoroquinolones: Synthesis, Antibacterial and Antitubercular Activity of New 2,5-disubstituted-1,3,4-Oxadiazoles. *J. Chinese Chem. Soc.*, 64, 918-924. (IF = 0.862)
16. Syed Misbahul Hasan, Shah Alam Khan, Kavita Golani, Jaswinder Lamba, Aftab Ahmad, Asif Husain (2017), Chemical modification of acetaminophen: PASS assisted *in vivo* anti-inflammatory activity of chalcones, flavanones and Schiff bases. *J. Chin. Pharm. Sci.*, 26(8), 611-620.
17. Mohammad Rashid, Shahid Karim, Babar Ali, Shamshir Khan, Makhmur Ahmad, Asif Husain (2017), PI3K signaling pathway targeting by using different molecular approaches to treat cancer. *J. Chin. Pharm. Sci.*, 26(9), 621-634.
18. Farah Iram, Shah Alam Khan, Aftab Ahmad, Anees A. Siddiqui, Asif Husain (2017), Eslicarbazepine acetate: A therapeutic agent of paramount importance in acute and chronic anticonvulsant therapy. *J Acute Dis.*, 6(6): 245-254.
19. Shah Alam Khan, S. Monawwar Imam, Aftab Ahmad, Syed Hussain Basha, Asif Husain (2018), Synthesis, molecular docking with COX I & II enzyme, ADMET screening and *in vivo* anti-inflammatory activity of oxadiazole, thiadiazole and triazole analogs of Felbinac. *J. Saudi Chem. Soc.,* 22, 469-484. (IF = 2.912)
20. Asif Husain, Munendra M. Varshney, Versha Parcha, Aftab Ahmad, Shah A. Khan (2018), Nalidixic Acid Schiff bases: Synthesis and Biological Evaluation. *Lett. Drug Design Discov.* 15(1), 103-111. (IF = 0.924)
21. Jokha Al Namani, Esra Baqir, Ajwaa Al Abri, Tamadher Al Hubaish, Asif Husain, Shah Alam Khan (2018), Phytochemical Screening, Phenolic content and

Antioxidant Activity of *Citrus aurantifolia* L. Leaves Grown in Two Regions of Oman. *Iranian J. Pharm. Sci.,* 14(1), 27-34.

22. Rubina Bhutani, Dharam Pal Pathak, Garima Kapoor, Asif Husain, Ravi Kant, Md Azhar Iqbal (2018), Synthesis, molecular modeling studies and ADME prediction of benzothiazole clubbed oxadiazole-Mannich bases, and evaluation of their anti-diabetic activity through in vivo model. *Bioorg. Chem.* 77, 06-15. (IF= 3.929)
23. Aftab Ahmad, Fahad A. Al-Abbasi, Saida Sadath, Soad Shaker Ali, Mohammed F. Abuzinadah, Hani A. Alhadrami, Anwar Ali Mohammad Alghamdi, Ali H. Aseeri, Shah Alam Khan, Asif Husain (2018), Ameliorative effect of camel`s milk and *Nigella sativa* Oil against thioacetamide-induced hepatorenal damage in rats. *Pharmacognosy Magazine*, 14(53), 27-35. (IF= 1.069)
24. Kanika Arora, Sukhbir Lal Khokra, Shah Alam khan and Asif Husain (2018), 3D-QSAR and Pharmacophore Identification of Benzothiazole Derivatives as Potential $p56^{lck}$ Inhibitors. *Chiang Mai J. Sci.,* 45(2), 1062-1072.
25. Farah Iram, Perwez Alam, Nasir A. Siddiqui, Saleh I. Alqasoumi, Anees A. Siddiqui, Shah Alam Khan, Asif Husain (2018), Development of a stress induced validated UPLC-PDA method for the analysis of Eslicarbazepine acetate. *Saudi Pharm. J.,* 26, 286-291. (IF= 3.346)
26. Mohammad Asif, Anita Singh, Shah Alam Khan, Asif Husain (2018), Studies on new substituted pyridazinones: synthesis and biological evaluation. *Braz. J. Pharm. Sci.* 54(3), 01-12.
27. Silky Bedi, Shah A. Khan, Majed M. AbuKhader, Perwez Alam, Nasir A. Siddiqui, Asif Husain (2018), A comprehensive review on Brigatinib – A wonder drug for targeted cancer therapy in non-small cell lung cancer. *Saudi Pharm. J.* 26(6), 755-763. (IF= 3.346)
28. Rubina Bhutani, Dharam Pal Pathak, Garima Kapoor, Asif Husain, Azhar Iqbal (2019), Novel Hybrids of Benzothiazole-1,3,4-oxadiazole-4-thiazolidinone: Synthesis, in silico ADME study, molecular docking and in vivo anti-diabetic assessment. *Bioorg. Chem.* 83, 06-19. (IF= 3.929)
29. Sukhbir L. Khokra, Kanika Arora, Shah A. Khan, Pawan Kaushik, Reetu Saini and Asif Husain (2019), Synthesis, Computational Studies and Anticonvulsant Activity of Novel Benzothiazole Coupled Sulfonamide Derivatives. *Iranian J. Pharm. Res.*, 18(1), 1-15.
30. Husain A, Khan S. A., Iram F., Iqbal M. A., Asif M (2019), Insights into the chemistry and therapeutic potential of furanones: A versatile pharmacophore. *Eur. J. Med. Chem.,* 171, 66-92. (IF = 4.816)
31. Garima Kapoor, Rubina Bhutani, Sandeep Jain, Asif Husain, Md Azhar Iqbal, Dharam Pal Pathak (2019), Synthesis, ADME, Docking studies and in vivo Anti-Hyperglycemic Potential Estimation of Novel Schiff Base Derivatives from Octadec-9-enoic Acid. *Bioorg. Chem.* 84, 478-492. (IF= 3.929)

32. Neelima Shrivastava, Asif Husain, Mohammad Rashid, Babar Ali, Nimer Fehaid Alsabeelah, Shahid Karim and Nasir Ali Siddiqui (2019), Recent Advances Towards Treatment of HIV: Synthesis and SAR Studies. *Mini Rev. Med. Chem.*, 19, 01-29. (IF = 2.903)

In: Benzimidazole
Editor: Aksel A. Vestergaard
ISBN: 978-1-53616-986-7

Chapter 3

CHIRAL BENZIMIDAZOLES IN MEDICINAL CHEMISTRY: SYNTHESES AND APPLICATIONS

Maria Marinescu*
Department of Organic Chemistry, Biochemistry and Catalysis,
University of Bucharest, Bucharest, Romania

ABSTRACT

The chemistry of chiral benzimidazoles is a relatively young branch of stereochemistry which has been required to be more developed over the past decades because of the superior therapeutic properties of the pure isomers as compared to racemic mixtures. Recent studies refer to methods of chiral functionalization of benzimidazoles as well as to the medicinal activities of chiral benzimidazoles. In this chapter benzimidazole syntheses with central, axial and planar chirality, with the specific reaction conditions were presented briefly. A wide range of therapeutic activities for benzimidazole chiral compounds is known, after almost 60 years of research in medicinal chemistry, such as: anticancer, antimicrobial, antiviral, antithrombotic, antihypertensive, anti-inflammatory, analgesic, hypoglycemic, antimalarial, FRX and CRF1 antagonists and renin inhibitors. In this chapter we propose a presentation of the synthetic routes of chiral benzimidazoles and their applications in the field of medicinal chemistry. The literature review presented in this chapter, with the synthesis of chiral compounds and the subsequent application of these as therapeutic compounds, will open new horizons in medicinal chemistry, in order to better evaluate the results so far.

* Corresponding Author's Email: maria.marinescu@chimie.unibuc.ro.

Keywords: chiral benzimidazole, stereochemistry, synthesis, medicinal chemistry

INTRODUCTION

A central position in medicinal chemistry is held by benzimidazole derivatives, key materials for the design of new therapeutics [1, 2]. A series of benzimidazole-containing drugs are marketed, like pantoprazole, lansoprazole, esomeprazole and ilarazole as proton pump inhibitors; albendazole, mebendazole, thiabendazole, flubendazole as antihelmintics; ridinazole as antibacterial, astemizole and bilastine as antihistamines; bezitramide as analgesic; selumetinib, bendamustine, pracinostat and galeterone, as antitumor agents; enviradine, samatasvir and maribavir as antivirals, candesartan and mibefradil as antihypertensive [1, 3-6]. Some studies mention chiral benzimidazole derivatives as NaV1.8 (Voltage-gated sodium channels) blockers, which play a key role in the transmission of pain signals and possess excellent preclinical *in vitro* ADME [7]. Recent research mention benzimidazole derivatives as potential DNA binding ligands [8, 9], EGFR and erbB2 inhibitors [10, 11], antitumor agents [12-14], anti-Alzheimer agents [15, 16], antimicrobial agents [17, 18], antidiabetic agents [19, 20], anti-parasitic agents [8, 21], antiquorum-sensing agents [12] and antimalarial agents [21]. Also, intensive studies found the benzimidazole scaffold as key pharmacophore in clinically approved analgesic and anti-inflammatory agents [22]. Some benzimidazoles have been shown to be anti-HIV-1 agents through the protection of APOBEC3G protein [23]. Also, SAR studies were accomplished for anticancer, antiviral, antimicrobial, antihelmintic, antimycobacterial, antidiabetic, antipsychotic, antiprotozoal, antidepressant and antioxidant benzimidazole derivatives [1, 24-26]. Figure 1 shows the structures of chiral benzimidazole drugs.

This review aims to present synthesis strategies and medicinal applications of the chiral benzimidazoles.

SYNTHESIS OF CHIRAL BENIMIDAZOLES

Researchers synthesized chiral benzimidazoles with different chiral elements in the form of point chirality, planar chirality and axial chirality. Therefore, different routes employed to get the chiral benzimidazoles will be presented.

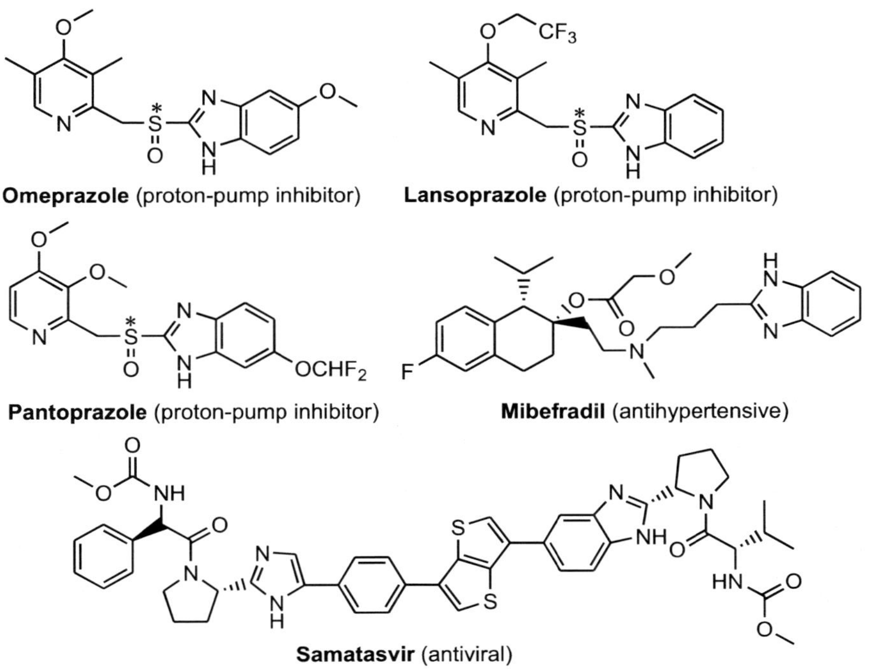

Figure 1. Chiral benzimidazole drugs.

1. Synthesis of Benzimidazoles with Central Chirality

1.1. Synthesis of Benzimidazoles with Chiral Substituents at C-2

1.1.1. Synthesis of Benzimidazoles from Chiral Carboxilic Acids

The reaction between 1,2-phenylenediamine **1** and **2a-b** chiral carboxylic acid (lactic acid or mandelic acid) in the presence of 4N hydrochloric acid, known as Phillips condensation (Scheme 1) is the most common method for benzmidazoles **3a-3b** [27]. This reaction has also been adapted for chiral amino acids (alanine, proline), aldonic acids or gluconic acid [27-30].

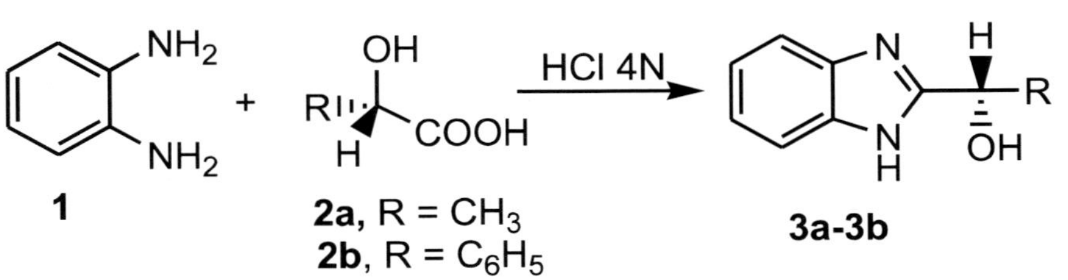

Scheme 1. Synthesis of chiral benzimidazoles by Phillips condensation.

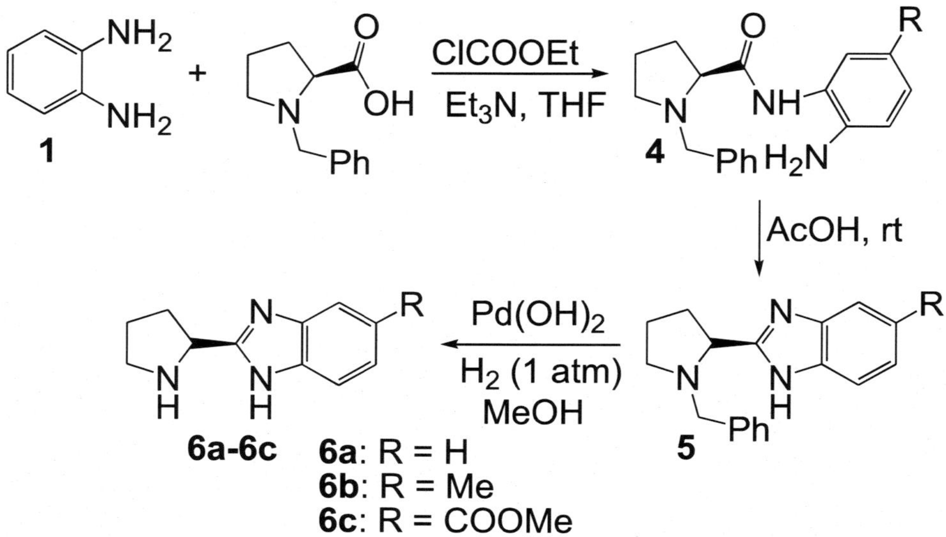

Scheme 2. Synthesis of chiral benzimidazoles **6a-6c**.

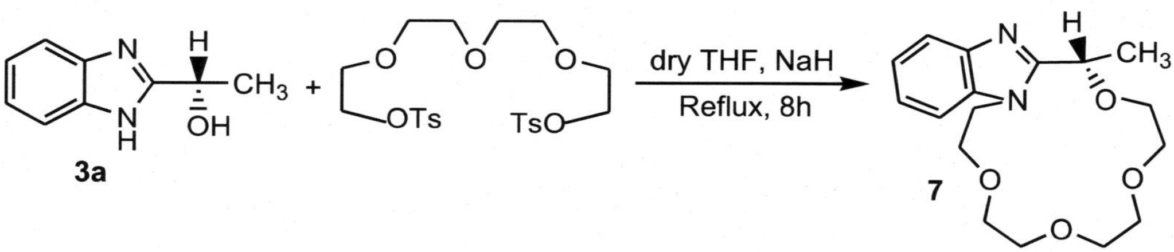

Scheme 3. Synthesis of chiral mono azacrown ether of benzimidazole **7**.

Benzimidazoles **6a-6c** were synthesized from N-benzylproline and **1** in presence of ethyl chloroformate to give amide **4**, which affords benzimidazole **5** in presence of acetic acid. Hydrogenolysis of **5** gave chiral benzimidazoles **6** [31].

Pandey et al. synthesized a chiral mono azacrown ether of benzimidazole in a single step reaction, by treating benzimidazole **3a** with tetraethylene glycol ditosylate in the presence of sodium hydride as the base and anhydrous THF as the solvent at reflux under high dilution conditions under an inert atmosphere, as indicate in Scheme 3 [32].

1.1.2. Synthesis of Chiral Benzimidazoles from Aldehydes

This method assumes the presence of a chiral carbon atom neighboring the aldehyde group, therefore aldoses such as D-arabinose or D-galactose are commonly used. Some synthesis methods are described by Khose et al. [27].

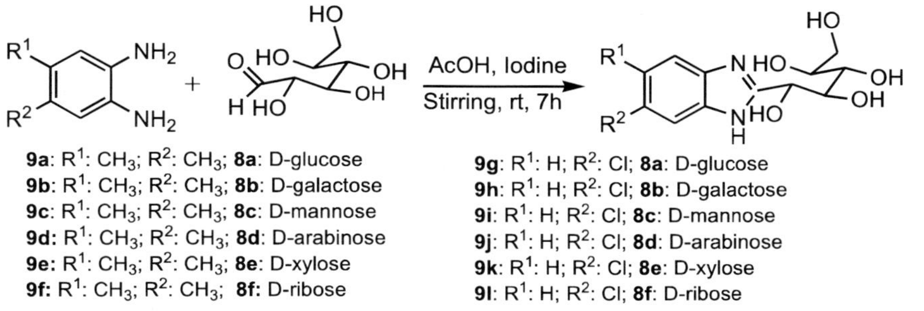

Scheme 4. Synthesis of chiral benzimidazoles **9a-9l**.

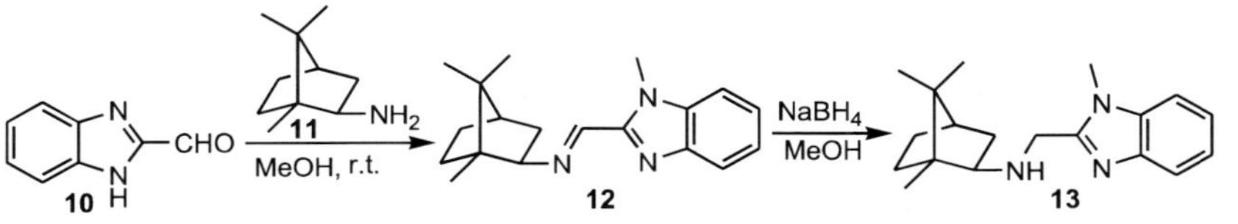

Scheme 5. Synthesis of chiral benzimidazole **13**.

Yadav et al. accomplished a novel synthesis of chiral 2,5,6-trisubstituted-1*H*-benzimidazole nucleoside derivatives **9a-9l** from 1,2-phenylediamines and aldoses **8a-8f** using iodine as oxidant in acetic solution at room temperature [33] (Scheme 4).

Condensation of substituted benzimidazole carbaldehyde **10** with a chiral (-)-exo-bornylamine **11** yield the intermediary Schiff bases **12**, which reduced with sodium borohydride to furnish the chiral benzimidazole ligand **13** (Scheme 5) [34].

1.2. Synthesis of Benzimidazoles with Chiral Substituents at N-1

Wagh et al. reported an efficient protocol to synthesis the racemic heterocyclic bases, thiazolo-benzimidazole **14a-14b** and pyrrolo-benzimidazoles **15a-15b**. Both benzimidazole derivatives **14** and **15** have been prepared by the reduction of benzimidazole based respective ketones followed by intramolecular cyclization (Scheme 6) [35].

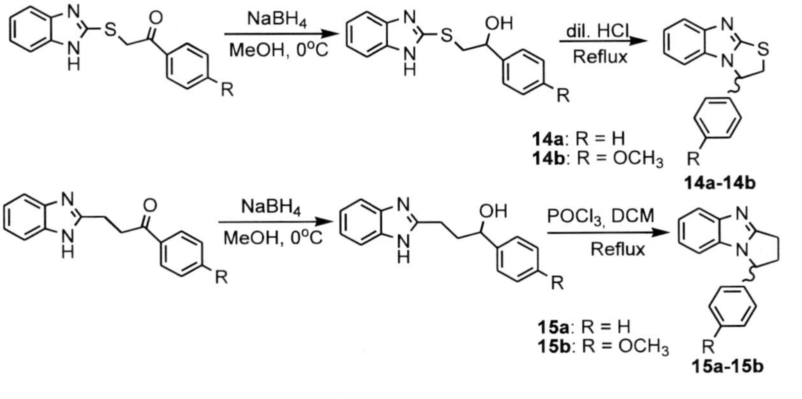

Scheme 6. Synthesis of chiral benzimidazole **14-15**.

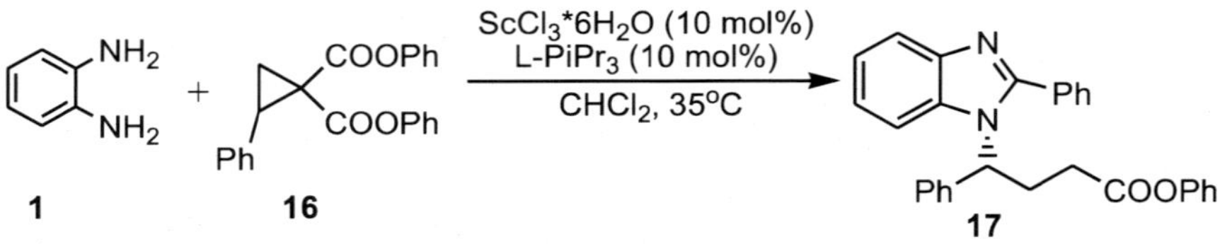

Scheme 7. Synthesis of chiral benzimidazole.

Xia et al. reported a highly efficient method for obtaining enantiopure benzimidazole derivatives **17** through a catalytic asymmetric reaction of benzene-1,2-diamine **1** with cyclopropyl ketones **16** using a chiral N,N'-dioxide/Sc (III) catalyst (Scheme 7) [36].

2. Synthesis of Benzimidazoles with Axial Chirality

Figge et al. developed a new method for synthesis of axially chiral benzimidazoles by refluxing diamine **18** and the aldehyde **19** for 2h in MeOH to give the N,N-acetal 20 in 63% yield. On dehydration of intermediary **20** with MnO_2 in benzene, the compound **21** was precipitated due to its poor solubility in benzene. Intermediate **21** leads to atropisomeric compound **22** in two steps (Scheme 8) [37].

3. Synthesis of Benzimidazoles with Benzimidazoles with Planar Chirality

Cyclophanes, ansa compounds and *trans*-cycloalkanes possess planar chirality. Therefore, benzimidazole scaffold with cyclophane backbone exhibit planar chirality. Rowlands et al. reported synthesis of planar chiral benzimidazoles attached with paracyclophane [38].

Paracyclophane **23** was converted to 4-hydroxy[2.2]paracyclophane *via* formation of the boronic ester and *in situ* oxidation using N-methylmorpholine N-oxide. The carbamate **24** was synthesized under standard acylation conditions. *ortho*-Lithiation with *s*-BuLi and TMEDA followed by reaction with crushed solid CO_2 and subsequent acid hydrolysis gave the acid **25** with 81% yield. The hydroxyl group was converted to triflate 26 in good yield (Scheme 9). Coupling *p*-methoxyaniline, employing the conditions of Buchwald, with Cs_2CO_3 as base and racemic BINAP as ligand, furnished the product **27**, in moderate yield (50%). Both benzimidazole derivatives **14** and **15** have been prepared by the reduction of benzimidazole based respective ketones followed by intramolecular cyclization [38].

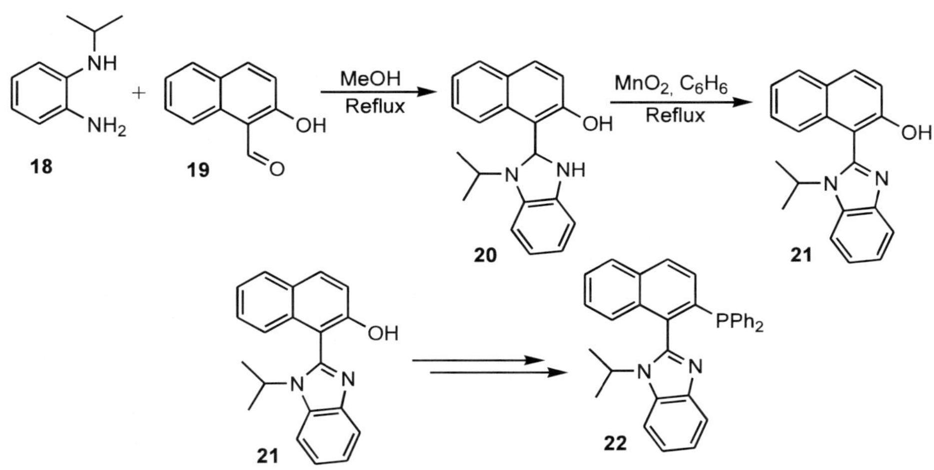

Scheme 8. Synthesis of axially chiral benzimidazoles.

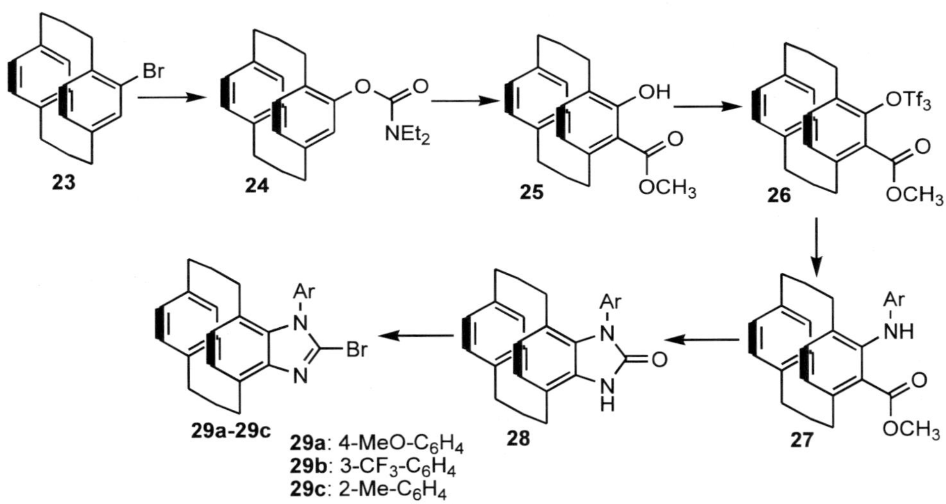

Scheme 9. Synthesis of paracyclophane benzimidazoles **29a-29c**.

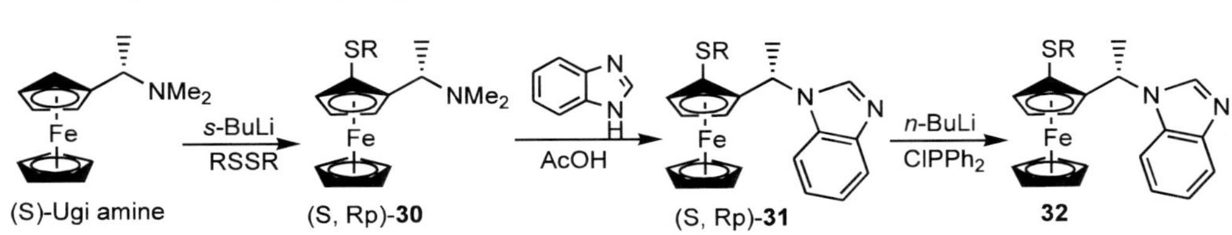

Scheme 10. Synthesis of ferocenyl-benzimidazole **32**.

Ferocenyl-benzimidazole also exhibit planar chirality. Cheung et al. prepared P/S chiral ferrocenyl **32** with benzimidazole moiety as depicted in Scheme 10 [39]. Diastereoselective *ortho*-substitution of (S) enantiomer of Ugi's amine with various disulfides produced aminothioether **30**. By the treatment of benzimidazole with aminothioether in acetic acid enantiomerically pure intermediates (S,Rp) **31** was obtained with high yield. Lithiation of (S,Rp) **31** and subsequent trapping with $ClPPh_2$ gave ligand **32**.

APPLICATIONS OF CHIRAL BENZIMIDAZOLES IN MEDICINAL CHEMISTRY

Benzimidazoles are heterocyclic compound with a broad spectrum of biological activities [1, 13, 40]. In the following we will highlight various activities anticancer, antimicrobial, antiviral, antitubercular, anti-thrombotic, antiinflammatory, anti-hypertensive, hypoglycaemic, etc. of the chiral benzimidazoles and its derivatives.

1. Anticancer Chiral Benzimidazoles

Cancer, which is mostly caused by a mutation of normal cell, has become the disease mostly urgent to be solved, because causing the death of millions of people annually across the globe. Novel effective anticancer heteroaromatic drugs with selectivity and better activity, and also fewer side effects are urgently need to be developed [41-42]

Akhtar et al. reported two new Cu(II)-dipeptide complexes of 2-(4'-thiazolyl)benzimidazole, [Cu(Gly-Gly)(TBZ)(Cl)]*4H_2O **33** and [Cu(Gly-L-Leu)(TBZ)(Cl)]-H_2O **34** (Gly-Gly = glycyl-glycine anion, Gly-L-Leu = glycyl-L-leucine anion and TBZ = 2-(4'-thiazolyl)benzimidazole) which possess superior cytotoxicity (IC_{50} values varying from 33.17 to 100 μM) and selective inhibition on human carcinoma HeLa cell lines [43].

The PARP protein superfamily plays wide-ranging roles in cellular processes such as apoptosis [44, 45], proliferation [46], DNA methylation [47, 48], transcriptional regulation [49, 50] and WNTsignalling [51]. The nuclear protein poly(ADP-ribose) polymerases-1/2 (PARP-1/2) are involved in DNA repair damaged by endogenous or exogenous process and PARP-1/2 inhibitors have been proved to be clinically efficacious for DNA repair deficient tumors in the past decade [52]. Chen et al. reported a series of 4,5,6,7-tetrahydrothienopyridin-2-yl benzimidazole carboxamides as novel and potent PARP-1/2 inhibitors [52]. The S enantiomer (S-**35**) provide a CC_{50} value of 864 nM in the V-C8 cytotoxity assay (Figure 2).

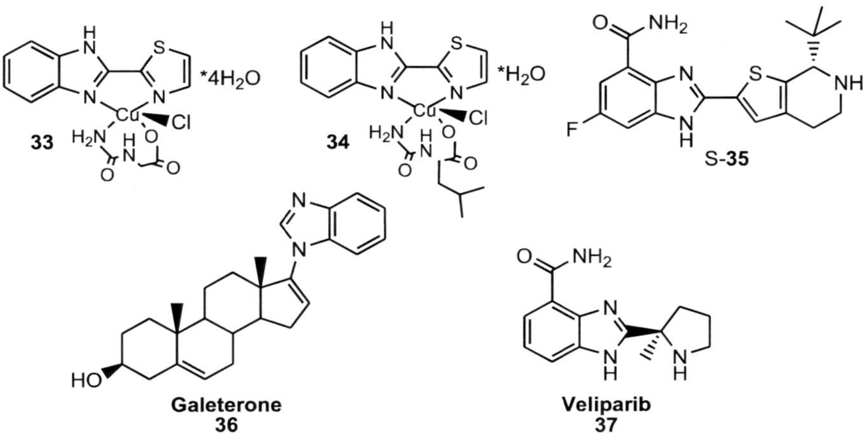

Figure 2. Anticancer chiral benzimidazoles.

Njar and Brodie showed that Galeterone **36** is a first-in-class, multitarget, oral small molecule for the treatment of prostate cancer which acts by disrupting andogen receptor (AR) signaling via three independent points in the pathway [3]. Galeterone was successfully applied phase II clinical development in men with castration resistant (advanced) prostate cancer. Another potential anti-cancer drug is Veliparib (ABT-888), which acting as a PARP inhibitor [53]. Veliparib kills cancer cells by blocking the PARP protein, thereby preventing the repair of DNA or genetic damage in cancer cells and possibly making them more susceptible to anticancer treatments [54].

2. Antimicrobial Chiral Benzimidazoles

The increase in antimicrobial resistance to existing drugs necessitated the search for new molecules for the treatment of bacterial infections [55].

A series of new chiral hydroxamates bearing benzimidazole ring **38a-38g** have been synthesized by Zhang et al. All the compounds were evaluated *in vitro* against *Staphylococcus aureus*, *Klebsiella pneumoniae* and *Sarcina lutea* considering Cefoperazone as standard (Table 1). The compound **38a** showed good bacterial inhibition. The compound with unsubstituted benzimidazole exhibited potent inhibitory activity (Figure 3) [56].

Li et al. synthesized a series of cinnamaldehyde derivatives and tested *in vitro* for their antibacterial activity against a variety of Gram-positive and Gram-negative bacteria [57, 58]. Several compounds exhibited MIC values of 0.25–4 μg/mL against *S. aureus* ATCC25923. Cinnamaldehyde derivatives containing a 2-methylbenzimidazolyl substitution at 1-position and 2,4-dichlorophenyl or 4-fluorophenyl, at the 3-position exhibited the best activity (compounds **39** and **40,** Figure 4). Cong et al. synthesized a series

4-*o*-benzimidazolyl clarithromycin derivatives which had benzimidazoles nucleus and screened for antibacterial activity using the broth microdilution method against *S. aureus* and *Streptococcus pneumoniae* [59]. Compound **41** was the most active [60].

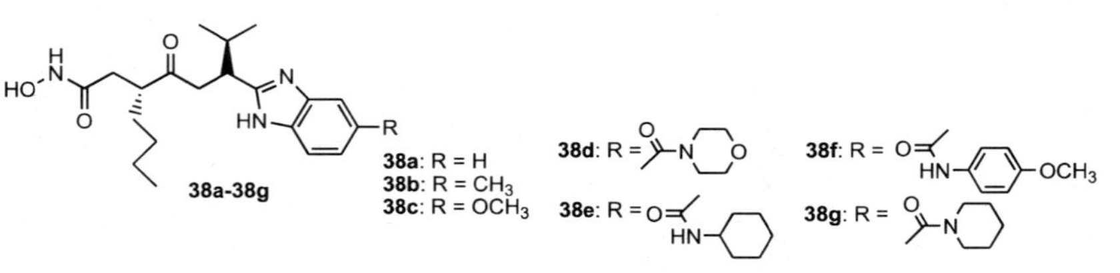

Figure 3. Antimicrobial chiral benzimidazoles.

Table 1. Antibacterial activities of 38a-38g (MIC, µg/mL)

Compound	*Staphylococcus aureus*	*Klebsiella pneumoniae*	*Sarcina lutea*
38a	2	0.5	4
38b	8	4	8
38c	8	8	8
38d	16	8	16
38e	32	16	16
38f	64	32	64
38g	>64	8	> 64
Cefoperazone	0.25	0.25	0.25

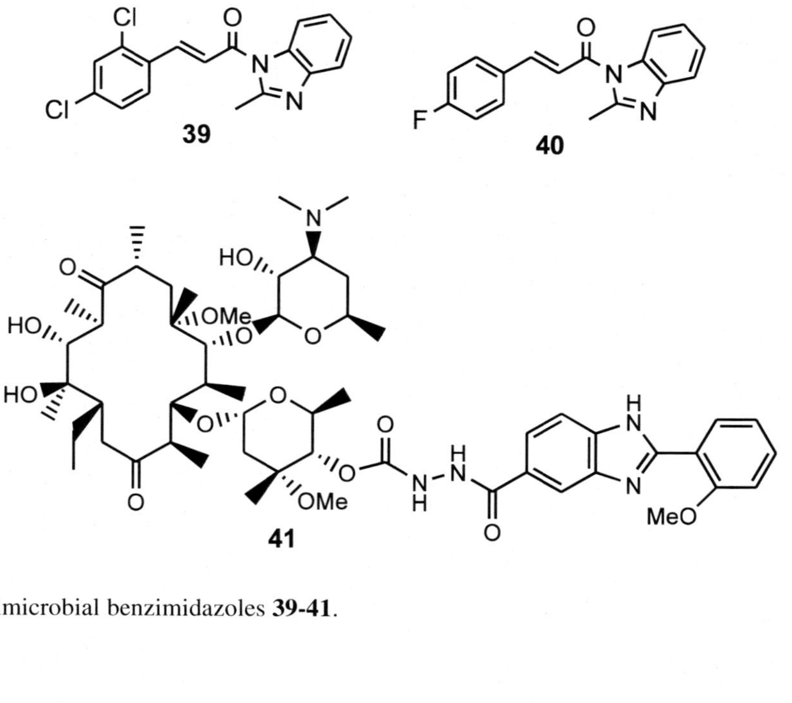

Figure 4. Antimicrobial benzimidazoles **39-41**.

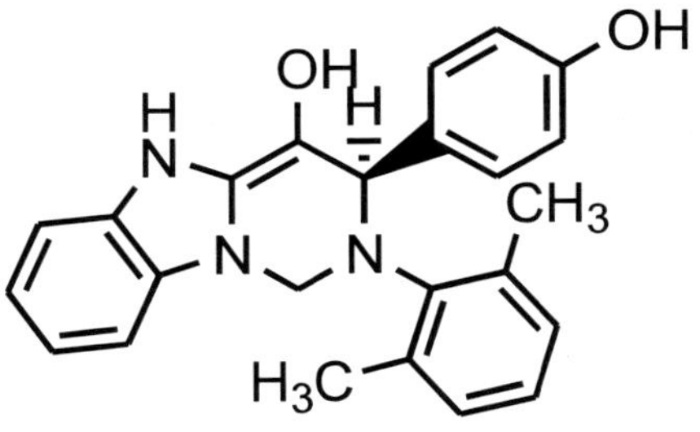

45

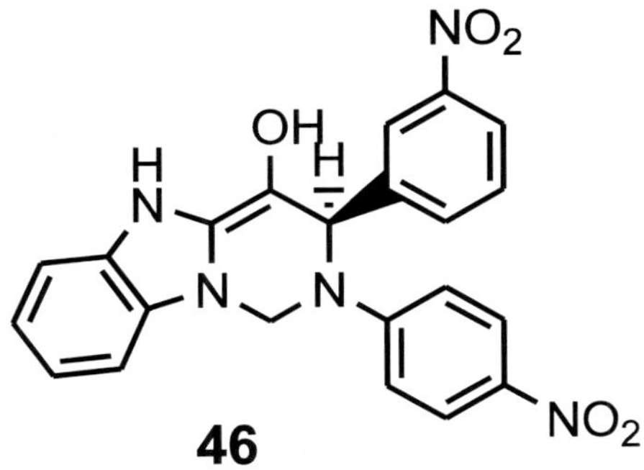

46

Figure 5. Antitubercular chiral benzimidazoles.

Benzimidazoles are well known for their *in vivo* and *in vitro* antitubercular activities and also for their mode of action [1]. A series of (R)-2-(4'-chlorophenyl)-3-(4'-nitrophenyl)-1,2,3,5- tetrahydrobenzo[4,5] imidazo[1,2-c]pyrimidin-4-ol derivatives were designed and docked on the FtsZ protein crystal structure [61]. Compound **45** showed the highest docking score and H-bond interaction with Arg140 and Gly19. Compound **46** exhibited good *in vitro* antitubercular activity 3.13 μg/mL and 4.7 μg/mL whereas compound **45** exhibited *in vitro* antitubercular activity 6.25 μg/mL and 9.4 μg/mL using GAST/Fe medium after week 1 and week 2 respectively against *Mycobacterium tuberculosis* H37Rv (Figure 5).

3. Antiviral Chiral Benzimidazoles

Statistics show that about 3% of the currentworld population are chronically infected by the chronic hepatitis C virus (HCV), which causes a progressive liver disease and can lead to lethal complications, such as cirrhosis and hepatocellular carcinoma [62]. New benzimidazole-coumarin compounds **42a-42d** were screened for anti-hepatitis C virus activity (Figure 6). The compounds showed potent activity with EC_{50} values in the range of 3.1 and 4.3 μM.

Xue et al. prepared a novel series of 2-substituted-1*H*-benzimidazole-4-carboxamide derivatives and screened for activity against four kinds of enteroviruses, i.e., Coxsackie virus A16, B3, B6 and enterovirus 71 in VERO cells. The compound **43** exhibited high antiviral potency (IC_{50} = 1.76 μg/ml) and a remarkable selectivity index [63]. Tallant et al. synthesized new N-substituted 2-phenyl and 2-methyl-2-phenyl-1,4-butanediamine-benzimidazole derivatives and evaluated for CCRS antagonists for the treatment of HIV-1 [64]. Compound **44** was found to be most potent in anti-HIV assays.

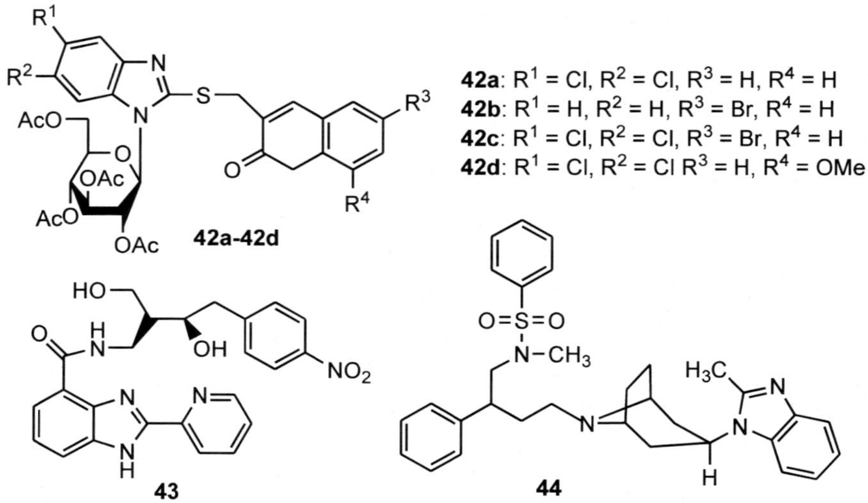

Figure 6. Antiviral benzimidazoles **42-44**.

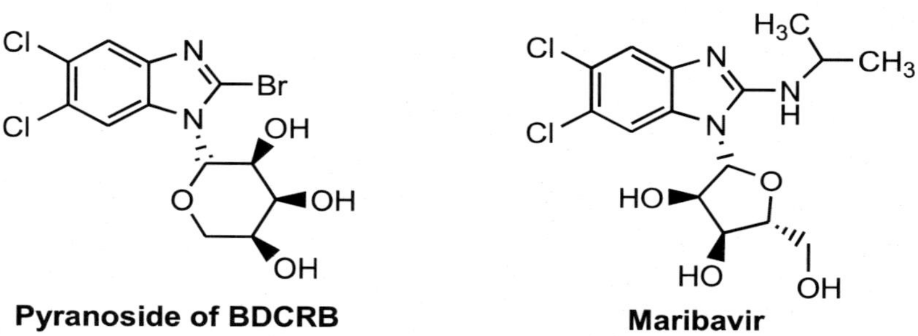

Figure 7. Chiral antiviral benzimidazoles.

Biron et al. conducted structure-activity relationship (SAR) studies on benzimidazole series that led to the emergence of two clinical candidates, the pyranoside of BDCRB and maribavir (Figure 7), each with distinct modes of action. Maribavir is a ribosyl benzimidazole selective, orally bioavailable, which is introduced for the prevention and treatment of HCMV disease in hematopoietic stem cell/bone marrow transplant patients [65]. Thus, a ribosyl moiety at 1-position proved to be very important for the antiviral activity.

4. Antithrombotic Chiral Benzimidazoles

There is a need for antithrombotic or thrombolytic drugs that are safer than established therapies, as various thrombotic diseases are one of the more common causes of death in the Western world [66]. Therapy available today includes thrombolytics, anticoagulants and antiplatelets. Brink et al. describe the design and synthesis of a series of novel, potent and selective 4,5,6,7-tetrahydro-1*H*-benzimidazole-5-carboxylic acid derivatives, small molecule inhibitors of thrombin activatable fibrinolysis inhibitor (TAFIa) using a specific approach, *via* a hydrogenation–alkylation sequence starting from the appropriate benzimidazole. The X-ray structure of compound **47** in complex with a H333Y/H335Q double mutant TAFI indicate that the conformational restriction is responsible for the observed potency increase (Figure 8).

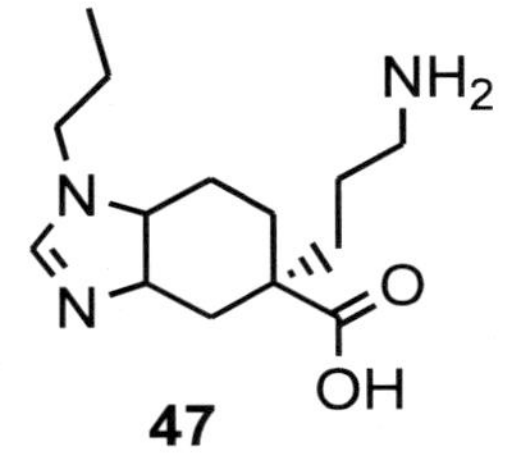

47

Figure 8. Anti-thrombotic chiral benzimidazole.

5. Antihypertensive Chiral Benzimidazoles

Han et al. developed novel angiotensin II receptor type 1 (AT1) blockers bearing 6-substituted carbamoyl benzimidazoles **48-51** with a chiral center as new antihypertensive agents. Benzimidazole **48R** was identified for its excellent efficacy in antihypertension and fairly low toxicity based on plasma analyses, toxicology studies, and chronic oral tests. Also, the screening of decreased blood pressure in hypertensive rats displayed that compound **48S** (IC_{50} = 5.0 nM) was equipotent with Losartan, whereas compounds **49R** (IC_{50} = 7.3 nM), **50R** (IC_{50} = 6.3 nM), and **50S** (IC_{50} = 3.5 nM) were slightly ahead of Losartan, and the most significant activity was demonstrated by compound 8R (IC_{50} = 1.1 nM) [67].

6. Anti-Inflammatory and Analgesic Chiral Benzimidazoles

Leukotrienes are proinflammatory lipid mediators associated with diverse chronic inflammatory diseases such as asthma, arthritis, atherosclerosis, chronic obstructive pulmonary disease, irritable bowel syndrome, dermatitis and cancer. Cellular leukotrienes are produced from arachidonic acid via the 5-lipoxygenase pathway in which the 5-lipoxygenase activating protein, also named as FLAP, plays a critical role by operating as a regulatory protein for efficient transfer of arachidonic acid to 5-lipoxygenase. By blocking leukotriene production, FLAP inhibitors may behave as broad-spectrum leukotriene modulators, which might be of therapeutic use for chronic inflammatory diseases requiring anti-leukotriene therapy [68].

Banoglu et al. characterized compound **51** in *in vitro* experiments as a selective FLAP modulator inhibiting the formation of leukotrienes with IC_{50} values of 0.15 μM and 0.04 μM in neutrophils and monocytes, respectively. It was experimentally demonstrated for the first time that a compound impaired the co-localization of 5-Lipoxygenase with 5-lipoxygenase-activating protein (FLAP) at the nuclear membrane, thereby intervening with the formation of the productive 5-LO/FLAP biosynthetic complex [69]. Janssen group designed an extensive SAR study on 1,2,6-trisubstituted benzimidazoles, which showed excellent potency in both FLAP binding and HWB (human whole blood) assays. Compound **52** is one of the most potent derivatives depicted (K_i for FLAP binding = 1.1 nM and HWB IC_{50} = 29 nM) [70].

Voltage-gated sodium channels (Na_V) are a family of transmembrane (TM) ion channel proteins. Na_V's play a key role in controlling excitability of neurons by regulating the threshold of firing, underlying the upstroke of the action potential and controlling the duration of interspike interval. $Na_V1.8$ is highly (but not exclusively) expressed in nociceptors and its expression and function is modulated by agents that cause pain.

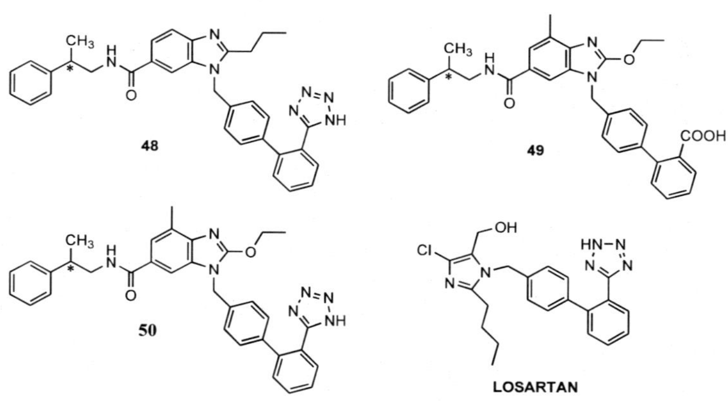

Figure 9. Antihypertensive chiral benzimidazoles and Losartan.

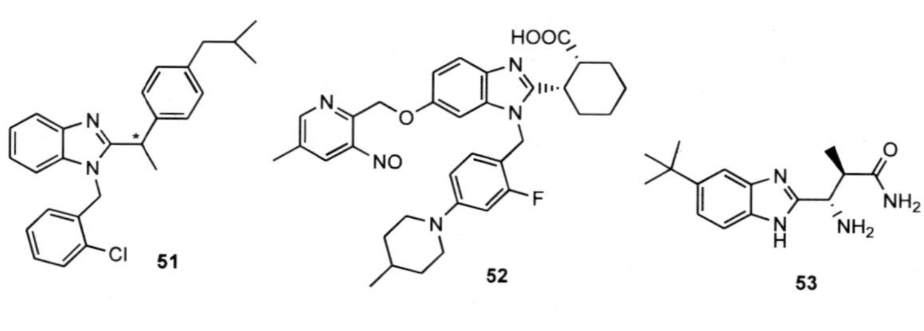

Figure 10. Anti-inflammatory and analgesic chiral benzimidazoles.

Brown et al. synthesized a series of a novel benzimidazole $Na_V1.8$ blockers with high lipophilic efficiency (LipE=-Log(h $Na_V1.8$ IC_{50})-clogP). Compound **53** proved to be the best clinical candidate, with an enhanced solubility profile than previous reported, and therefore a more thorough evaluation of the role of Na*V*1.8 in the treatment of pain [71].

7. Hypoglycemic Chiral Benzimidazoles

Glucokinase (GK), a member of the hexokinase family, is expressed in the liver and in pancreatic β-cells, and catalyzes the first step in glycolysis, the phosphorylation of glucose to glucose 6-phosphate. GK plays an important role as a glucose sensor in maintaining plasma glucose homeostasis by enhancing both glucose uptake in the liver and insulin secretion from pancreatic β-cells. Therefore, a GK activator (GKA) can be expected to function as a hypoglycemic drug in the treatment of type II (insulin-independent) diabetes mellitus [72]. Klapars et al. reported a short and practical synthesis of glucokinase activator **54** [73].

8. Antimalarial Chiral Benzimidazoles

Malaria is one of the most widespread parasitic diseases, largely affecting developing countries of sub-Saharan Africa. According to the World Health Organisation, over 216 million cases and a resultant 445 000 deaths were reported in 2016 [74]. Various benzimidazole derivatives in recent literature possess potent antiplasmodial activities [75-77].

Rylands et al. found that 2-phenylbenzimidazole ligands **55** possess weak antiplasmodial activities (IC_{50} ~ 17.66-22.32 μM) while an enhancement of antiplasmodial activity was observed upon coordination of the ligands with either ruthenium, iridium or rhodium. The new cyclometallated complexes **56-57** were found to be active against parasite *Plasmodium falciparum*, with IC_{50} values in the low to submicromolar range (0.12-5.17 μM) [74].

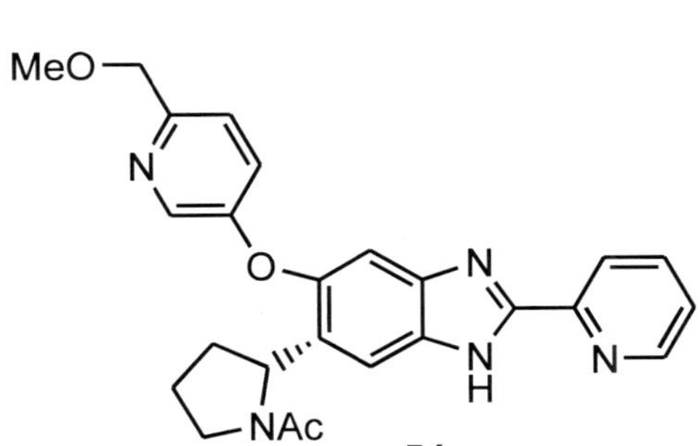

54

Figure 11. Hypoglycemic chiral benzimidazole.

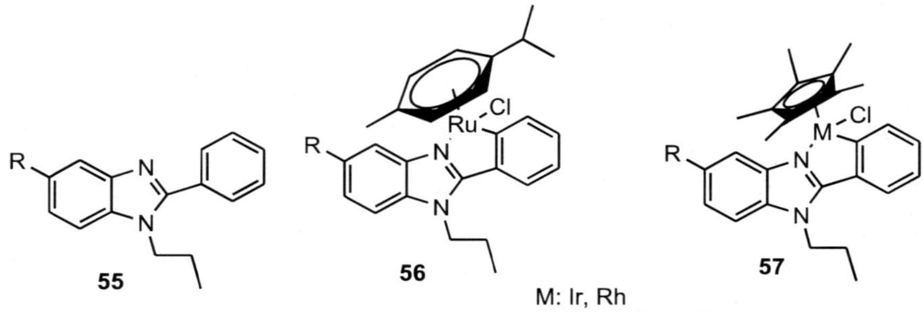

55 56 57

Figure 12. Antimalarial chiral benzimidazoles.

9. Chiral Benzimidazoles as Farnesoid X Receptor (FRX) Antagonists

The farnesoid X receptor (FXR) belongs to the nuclear receptor family and is activated by bile acids [78]. Bile acids are responsible for efective absorption of fats and fat-soluble vitamins, facilitate digestion and are important regulators of cholesterol, triglyceride homeostasis and infammation. Several of these metabolic actions of bile acids involve the activation of the nuclear farnesoid X receptor (FXR) [79]. FXR is used as a target for new drug therapies against metabolic dysregulation associated with obesity, including type 2 diabetes, non-alcoholic fatty liver disease (NAFLD), and atherosclerosis as well as for the cholestatic liver disease primary biliary cholangitis [80, 81]. FXR is primarily expressed in liver, kidney, and intestine and is crucial for maintaining bile acid homeostasis. Teno et al. discovered a new kind of nonsteroidal FXR antagonists. Compound **58** (Figure 13) possess the promising activity (Luciferase reporter assay: IC_{50} = 1.2 nM) [78].

10. Chiral Benzmidazoles as Corticotropin Releasing Factor 1 (CRF_1) Receptor Antagonists

Corticotropin-releasing factor (CRF) is a 41-amino-acid neuropeptide that mediates its actions through two G_s-coupled G protein-coupled receptor subtypes, CRF1 and CRF2. CRF is believed to be the main regulator of the hypothalamus–pituitary–adrenocortical (HPA) axis *via* CRF1 receptors and has an important role as a neurotransmitter in the mediation of stress-related behaviors [82].

Kojima et al. reported a new class of corticotropin releasing factor 1 (CRF1) receptor antagonists characterized by a tricyclic core ring. Compound **59**R, was noted for its good activity (IC_{50} = 110nM). Also, compound **59R** (Figure 14) exhibited dose-dependent inhibition of [^{125}I]-CRF binding in the frontal cortex (5 and 10 mg/kg, p.o.) as well as suppression of locomotor activation induced by intracerebroventricular administration of CRF in rats (10 mg/kg, p.o.) [83].

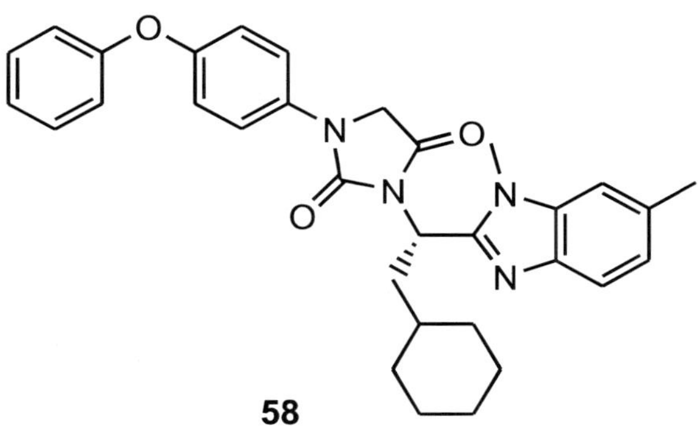

58

Figure 13. Chiral benzimidazole as FRX antagonists.

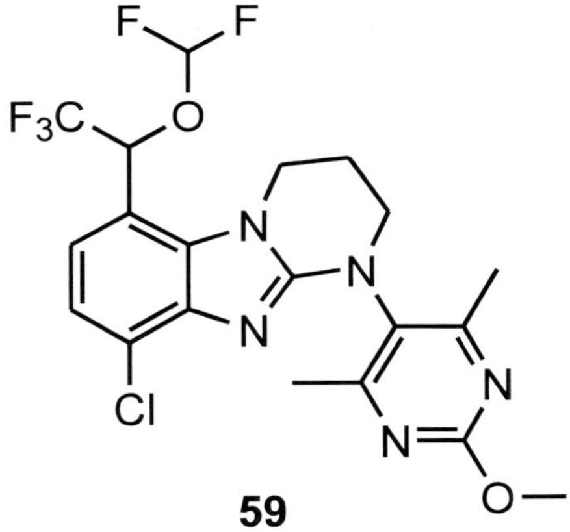

59

Figure 14. Chiral benzimidazole as CRF_1 receptor antagonists.

11. Chiral Benzimidazoles as Renin Inhibitors

Renin is a particularly promising target for blood pressure, progression of cardiovascular and renal diseases because of the action of the aspartyl protease renin which is the rate-limiting initial step of the reninangiotensin-aldosterone system [41].

Tokuhara et al. synthesized a new series of benzimidazoles as renin inhibitors. Compound **60** was identified as a potent renin inhibitor (IC_{50} = 0.29-0.66 nM) with favorable PK profiles in rats and monkeys. It exhibited potent and long-lasting PRA (plasma renin activity) inhibition in cynomolgus monkeys (Figure 15) [84].

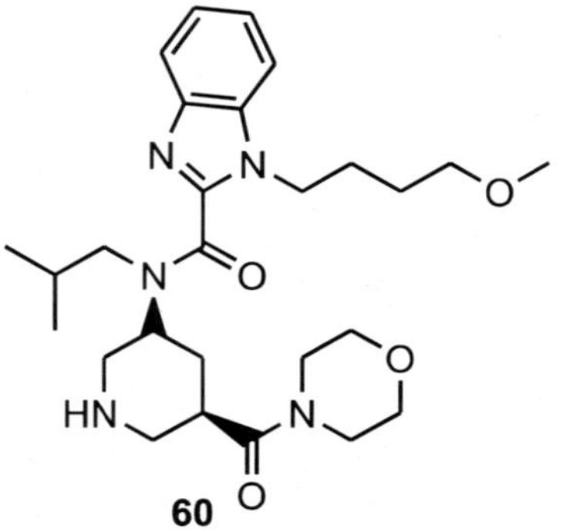

60

Figure 15. Chiral benzimidazole as renin antagonists.

Conclusion

In this chapter we tried to review the methods for the synthesis of chiral benzimidazoles as well as their applications in medicinal chemistry. Benzimidazole syntheses with central, axial and planar chirality, with the specific reaction conditions were presented briefly. Researches on chiral benzimidazoles are fully utilized in their applications as therapeutic compounds. Pantoprazole, Mibefradil, Samatasvir, Omeprazole, Lansoprazole, Galeterone, Veliparib, are drugs that contain chiral benzimidazoles, with very good therapeutic results. We have presented various medicinal applications of benzimidazoles, such as: anticancer, antimicrobial, antiviral, antithrombotic, antihypertensive, anti-inflammatory, analgesic, hypoglycemic, antimalarial, FRX and CRF1 antagonists and renin inhibitors. Current research on chiral benzimidazoles is in continuous development. Also, the knowledge of chiral functionalization methods as well as structure-property relationships reported in the literature for various diseases is required. We hope that the encouraging results presented here will lead to the finding of new stereoselective syntheses, and the further application of these chiral benzimidazoles in medicinal chemistry.

References

[1] Bansal, Y. & Silakar, O. (2012). "The therapeutic journey of benzimidazoles: A review." *European Journal of Medicinal Chemistry*, *20* (21), 6208–6236.

[2] Anand, K. & Wakode, S. (2017). "Development of drugs based on Benzimidazole Heterocycle: Recent advancement and insights." *International Journal of Chemical Studies*, *5* (2), 350-362.

[3] Njar, V. C. & Brodie, A. M. (2015). "Discovery and Development of Galeterone (TOK-001 or VN/124-1) for the Treatment of All Stages of Prostate Cancer." *Journal of Medicinal Chemistry*, *58*(5), 2077-2087.

[4] Moniruzzaman, R. S. & Mahmud, T. (2019). "Quantum chemical and pharmacokinetic studies of some proton pump inhibitor drugs." *American Journal of Biomedical Sciences &Research*, *2*(1), 3-8.

[5] Scholten, W. K., Christensen, A. E., Olesen, A. E. & Drewes, A. M. (2019). "Quantifying the adequacy of opioid analgesic consumption globally: an updated method and early findings." *American Journal of Public Health* (AJPH), *109*(10), 52-57.

[6] Tahlan, S., Kumar, S., Ramasamy, K., Lim, S. M., Shah, S. A. A., Mani, V., Pathania, R. & Narasimhan, B. (2019). "Design, synthesis and biological profile of heterocyclic benzimidazole analogues as prospective antimicrobial and antiproliferative agents." *BMC Chemistry*, *13*(50), 1-15.

[7] Brown, A. D., Bagal, S. K., Blackwell, P., Blakemore, D. C., Brown, B., Bungay, P. J., Corless, M., Crowforth, J., Fengas, D., Fenwick, D. R., Gray, V., Kemp, M., Klute, W., Sanz, L. M., Miller, D., Murata, Y., Payne, C. E., Skerratt, S., Stevens, E. B. & Warmus, J. S. (2019). "The discovery and optimization of benzimidazoles as selective NaV1.8 blockers for the treatment of pain." *European Journal of Medicinal Chemistry*, *27*, 230-239.

[8] Popov, A. B., Stolic, I., Krstulovic, L., Taylor, M. C., Kelly, J. M., Tomic, S., Tumir, L., Bajic, M. & Raić-Malić, S. (2019). "Novel symmetric bis-benzimidazoles: Synthesis, DNA/RNA binding and antitrypanosomal activity." *European Journal of Medicinal Chemistry*, *173*, 63-75.

[9] Ding, Y., Chai, J., Centrella, P. A., Gondo, C., DeLorey, J. L. & Clark, M. A. (2018). "Development and Synthesis of DNA-Encoded Benzimidazole Library." *ACS Combinatorial Science*, *20*, 251-255.

[10] Celik, I., Ayhan-Kilcigil, G., Guven, B., Kara, Z., Gurkan-Alp, A. S., Karayel, A. & Onay-Besikci, A. (2019). "Design, synthesis and docking studies of benzimidazole derivatives as potential EGFR inhibitors." *European Journal of Medicinal Chemistry*, *173*, 240-249.

[11] Akhtar, M. J., Siddiqui, A. A., Khan, A. A., Ali, Z., Dewangan, R. P., Pasha, S. & Yar, M. S. (2017). "Design, synthesis, docking and QSAR study of substituted benzimidazole linked oxadiazole as cytotoxic agents, EGFR and erbB2 receptor inhibitors." *European Journal of Medicinal Chemistry*, *126*, 853-869.

[12] El-Gohary, N. S. & Shaaban, M. I. (2017). "Synthesis, antimicrobial, antiquorum-sensing and antitumor activities of new benzimidazole analogs." *European Journal of Medicinal Chemistry*, *137*, 439-449.

[13] Kanwal, A., Saddique, F. A., Aslam, S., Ahmad, M., Zahoor, A. F. & Moshin, N. A. (2018). "Benzimidazole ring system as a privileged template for anticancer agents." *Pharmaceutical Chemistry Journal*, *51*(12), 1068-1077.

[14] Yadav, S., Narasimhan, B. & Kaur, H. (2016). "Perspectives of benzimidazole derivatives as anticancer agents in the New Era." *Anti-Cancer Agents in Medicinal Chemistry*, *16*(11), 1403-1425.

[15] Chaves, S., Hiremathad, A., Tomas, D., Keri, R. S., Piemontese, L. & Santos, M. A. (2018). "Exploring the chelating capacity of 2-hydroxyphenyl-benzimidazole based hybrids with multi-target ability as anti-Alzheimer's agent." *New Journal of Chemistry*, *42*(20), 16503-16515.

[16] Fang, Y., Zhou, H. & Xu, J. (2019). "Synthesis and evaluation of tetrahydroisoquinoline-benzimidazole hybrids as multifunctional agents for the treatment of Alzheimer's disease." *European Journal of Medicinal Chemistry*, *167*, 133-145.

[17] Marinescu, M., Tudorache, G. D., Marton, G. I., Zalaru, C. M., Popa, M., Chifiriuc, M. C., Stavarache, C. E. & Constantinescu, C. (2017). "Density functional theory molecular modeling, chemical synthesis, and antimicrobial behaviour of selected benzimidazole derivatives." *Journal of Molecular Structure*, *1130*, 463-471.

[18] Bansal, Y., Kaur, M. & Bansal, G. (2019). "Antimicrobial potential of benzimidazole derived molecules." *Mini-Reviews in Medicinal Chemistry*, *19*(8), 624-646.

[19] Aboul-Enein, H. Y. & Rashedy, A. A. E. (2015). "Benzimidazole derivatives as antidiabetic agents." *Medicinal Chemistry*, *5*(7), 318-325.

[20] Adegboye, A. A., Khan, K. M., Salar, U., Aboaba, S. A., Kanwal, C. S., Fatima, I., Taha, M., Wadood, A., Mohammad, J. I., Khan, H. & Parveen, S. (2018). "2-Aryl benzimidazoles: Synthesis, *In vitro* α-amylase inhibitory activity, and molecular docking study." *European Journal of Medicinal Chemistry*, *150*, 248-260.

[21] Farahat, A. A., Ismail, M. A., Kumar, A., Wenzler, T., Brun, R., Paul, A., Wilson, W. D. & Boykin, D. W. (2018). "Indole and benzimidazole bichalcophenes: synthesis, DNA binding and antiparasitic activity." *European Journal of Medicinal Chemistry*, *143*, 1590-1596.

[22] Gaba, M., Singh, S. & Mohan, C. (2014). "Benzimidazole: An emerging scaffold for analgesic and anti-inflammatory agents." *European Journal of Medicinal Chemistry*, *76*, 494-505.

[23] Pan T., He, X., Chen, B., Chen, H., Gheng, G., Luo, H., Zhang, H. & Bai, C. (2015). "Development of benzimidazole derivatives to inhibit HIV-1 replication through protecting APOBEC3G protein." *European Journal of Medicinal Chemistry*, *95*, 500-513.

[24] Siddiqui, M., Alam, M. S., Sahu, M., Yar, M. S., Alam, O. & Siddiqui, M. J. A. (2016). "Antidepressant, analgesic activity and SAR Studies of substituted benzimidazoles." *Asian Journal of Pharmaceutical Research*, *6*(3), 170-174.

[25] Tahlan, S., Ramasamy, K., Lim, S. M., Shah, S. A. A., Mani, V. & Narasimhan, B. (2018). "Design, synthesis and therapeutic potential of 3-(2-(1H-benzo[d]imidazol-2-ylthio) acetamido)-N-(substituted phenyl)benzamide analogues." *Chemistry Central Journal*, *12*(139), 1-12.

[26] Xu, M., Wang, S. L., Zhu, L., Wu, P. Y., Dai, W. B. & Rakesh, K. P. (2019). "Structure-activity relationship (SAR) studies of synthetic glycogen synthase kinase-3β inhibitors: A critical review." *European Journal of Medicinal Chemistry*, *164*, 448-470.

[27] Khose, N. V., John, M. E., Pandey, A. D. & Karnik, A. V. (2017). "Chiral benzimidazoles and their applications in stereodiscrimination bprocesses." *Tetrahedron: Asymmetry*, *28*, 1233–1289.

[28] Lacoste, E., Landais, Y., Schenk, K., Verlhac, J. & Vincent, J. (2004). "Benzoimidazole–pyrrolidine (BIP), a highly reactive chiral organocatalyst for aldol process." *Tetrahedron Letters*, *45*, 8035–8038.

[29] Lacoste, E., Vaique, E., Berlande, M., Pianet, I., Vincent, J. M. & Landais, Y. (2007). "Benzimidazole-pyrrolidine/H^+ (BIP/H^+), a highly reactive organocatalyst for asymmetric processes." *European Journal of Organic Chemistry*, 167–177.

[30] Xu, D., Sun, Q., Quan, Z. & Sun, W. (2017). "The synthesis of chiral tridentate ligands from l-proline and their application in the copper(II)-catalyzed enantioselective Henry reaction." *Tetrahedron: Asymmetry*, *28*, 954–963.

[31] Reddy, K. R., Khrishna, G. G. & Rajasekhar, C. V. (2007). "Synthesis of chiral benzimidazole-pyrrolidine derivatives and their application in organocatalytic aldol and Michael addition reactions." *Synthetic Communications*, *37*, 4289–4299.

[32] Pandey, A., Mohammed, H. & Karnik, A. (2013). *Chiral benzimidazole-derived mono azacrowns: synthesis and enantiomer recognition studies with chiral amines and their ammonium salt.*

[33] Yadav, T. B. & Singh, V. (2013). "Iodine promoted simple synthesis of benzimidazole acyclonucleosides." *Indian Journal of Chemistry*, *52B*, 1536-1539.

[34] Zhou, Y. & Gong, Y. (2011). "Asymmetric Copper (II)-Catalysed Nitroaldol (Henry) Reactions Utilizing a Chiral C_1-Symmetric Dinitrogen Ligand". *European Journal of Organic Chemistry*, 6092–6099.

[35] Wagh, S. J., Tawde, T. S., Sapre, J. V., Khose, V. N. & Karnik, A. V. (2016). A convenient rout to benzimidazole fused chiral heterocyclic bases. *Indian Journal of Chemistry*, *55B*, 707–712.

[36] Xia, Y., Lin, L., Chang, F., Liao, Y., Liu, X. & Feng, X. (2016). "Asymmetric Ring Opening/Cyclization/Retro-Mannich Reaction of Cyclopropyl Ketones with Aryl 1,2-Diamines for the Synthesis of Benzimidazole Derivatives." *Angewandte Chemie International Edition*, *55*, 12228–12232.

[37] Figge, A., Altenbach, H. J., Brauer, D. J. & Tielmann, P. (2002). "Synthesis and resolution of 2-(2-diphenylphosphinyl-naphthalen-1-yl)-1-isopropyl-

1Hbenzoimidazole; a new atropisomeric P,N-chelating ligand for asymmetric catalysis." *Tetrahedron: Asymmetry*, *13*, 137-144.

[38] Hitchcock, P. B., Hodgson, A. C. & Rowlands, G. J. (2006). "The First examples of planar chiral organic benzimidazole derivatives." *Synlett*, 2625–2628.

[39] Cheung, H. Y., Yu, W., Lam, F. L., Au-Yeung, T., Zhou, Z., Chan, T. H. & Chan, A. S. C. (2007). "Enantioselective Pd-Catalyzed Allylic Alkylation of Indoles by a New Class of Chiral Ferrocenyl P/S Ligands." *Organic Letters*, *21*, 4295–4298.

[40] Prasad, P. M. K., Avdhut Kanvinde, S. & Raja, S. (2016). Potent bioogical agent benzimidazole. A review. *International Journal of Pharmacy and Pharmaceutical Sciences*, *8*(12), 22-33.

[41] Marinescu, M. (2017). "2-Aminopyridine - a classic and trendy pharmacophore". *International Journal of Pharma and Bio Sciences*, *8*(2), 338-355.

[42] Gao, C., Li, B., Zhang, B., Sun, Q., Li, L., Li, X., Chen, C., Tan, C., Liu, H. & Jiang, Y. (2015). "Synthesis and biological evaluation of benzimidazole acridine derivatives as potential DNA-binding and apoptosis-inducing agents." *Bioorganic & Medicinal Chemistry*, *23*, 1800-1807.

[43] Akhtar, J., Khan, A. A., Ali, Z., Haider, R. & Shahar Yar, M. (2017). Structure-activity relationship (SAR) study and design strategies of nitrogen-containing heterocyclic moieties for their anticancer activities. *European Journal of Medicinal Chemistry*, *125*, 143-189.

[44] Chou, H. Y. E., Chou, H. T. & Lee, S. C. (2006). "CDK-dependent activation of poly (ADPribose) polymerase member 10 (PARP10)." *Journal of Biological Chemistry*, *281*, 15201-15207.

[45] Herzog, N., Hartkamp, J. D., Verheugd, P., Treude, F., Forst, A. H., Feijs, K. L., Lippok, B. E., Kremmer, E., Kleine, H. & Lüscher, B. (2013). "Caspase-dependent cleavage of the mono-ADP-ribosyltransferase ARTD10 interferes with its pro-apoptotic function." *FEBS Journal*, *280*, 1330-1343.

[46] Yu, M., Schreek, S., Cerni, C., Schamberger, C., Lesniewicz, K., Poreba, E., Vervoorts, J., Walsemann, G., Grotzinger, J. & Kremmer, E. (2005). "PARP-10, a novel Myc-interacting protein with poly (ADP-ribose) polymerase activity, inhibits transformation." *Oncogene*, *24*, 1982-1993.

[47] Krishnakumar, R. & Kraus, W. L. (2010). "The PARP side of the nucleus: molecular actions, physiological outcomes, and clinical targets." *Molecular Cell*, *39*, 8-24.

[48] Luo, X. & Kraus, W. L. (2012). "On PAR with PARP: cellular stress signaling through poly(ADP-ribose) and PARP-1." *Gene & Development*, *26*, 417-432.

[49] Kraus, W. L. (2008). "Transcriptional control by PARP-1: chromatin modulation, enhancer-binding, coregulation, and insulation." *Current Opinion in Cell Biology*, *20*, 294-302.

[50] Ji, Y. & Tulin, A. V. "The roles of PARP1 in gene control and cell differentiation." *Current Opinion in Genetics & Development*, *20*, 512-518.

[51] Huang, S. M. A., Mishina, Y. M., Liu, S., Cheung, A., Stegmeier, F., Michaud, G. A., Charlat, O., Wiellette, E., Zhang, Y., Wiessner, S., Hild, M., Shi, X., Wilson, C. J., Mickanin, C., Myer, V., Fazal, A., Tomlinson, R., Serluca, F., Shao, W., Cheng, H., Shultz, M., Rau, C., Schirle, M., Schlegl, J., Ghidelli, S., Fawell, S., Lu, C., Curtis, D., Kirschner, M. W., Lengauer, C., Finan, P. M., Tallarico, J. A., Bouwmeester, T., Porter, J. A., Bauer, A. & Cong, F. (2009). "Tankyrase inhibition stabilizes axin and antagonizes Wnt signalling." *Nature*, *461*, 614-620.

[52] Chen, X., Huan, X., Liu, Q., Wang, Y., He, Q., Tan, C., Chen, Y., Ding, J., Xu, Y., Miao, Z. & Yang, C. (2018). Design and synthesis of 2-(4,5,6,7-tetrahydrothienopyridin-2-yl)-benzoimidazole carboxamides as novel orally efficacious Poly(ADPribose) polymerase (PARP) inhibitors. *European Journal of Medicinal Chemistry*, *145*, 389-403.

[53] Donawho, C. K., Luo, Y., Luo, Y., Penning, T. D., Bauch, J. L., Bouska, J. J., Bontcheva-Diaz, V. D., Cox, B. F., DeWeese, T. L., Dillehay, L. E., Ferguson, D. C., Ghoreishi-Haack, N. S., Grimm, D. R., Guan, R., Han, E. K., Holley-Shanks, R. R., Hristov, B., Idler, K. B., Jarvis, K., Johnson, E. F., Kleinberg, L. R., Klinghofer, V., Lasko, L. M., Liu, X., Marsh, K . C., McGonigal, T. P., Meulbroek, J. A., Olson, A. M., Palma, J. P., Rodriguez, L. E., Shi, Y., Stavropoulos, J. A., Tsurutani, A. C., Zhu, G. D., Rosenberg, S. H., Giranda, V. L. & Frost, D. J. (2007). "ABT-888, an orally active Poly(ADP-Ribose) Polymerase Inhibitor that potentiates DNA-damaging agents in preclinical tumor models." *Clinical Cancer Research*, *13*, 2728-2737.

[54] Gaba, M. & Mohan, C. (2016). "Development of drugs based on imidazole and benzimidazole bioactive heterocycles: recent advances and future directions." *Medicinal Chemistry Research*, *25*, 173-210.

[55] Tahlan, S., Kumar, S. & Narasimhan, B. (2019). "Antimicrobial potential of 1H-benzo[d]imidazole scafold: a review." *BMC Chemistry*, *13*, 1-27.

[56] Zhang, D., Wang, Z., Xu, W., Sun, F., Tang, L. & Wang, J. (2009). "Design, synthesis and antibacterial activity of novel actinonin derivatives containing benzimidazole heterocycles." *European Journal of Medicinal Chemistry*, *44*, 2202-2210.

[57] Li, X., Sheng, J., Huang, G., Ma, R., Yin, F., Song, D., Zhao, C. & Ma, S. (2015). "Design, synthesis and antibacterial activity of cinnamaldehyde derivatives as inhibitors of the bacterial cell division protein FtsZ." *European Journal of Medicinal Chemistry*, *97*, 32-41.

[58] Haranahalli, K., Tong, S. & Ojima, I. (2016). "Recent advances in the discovery and development of antibacterial agents targeting the cell-division protein FtsZ." *Bioorganic & Medicinal Chemistry*, *24*, 6354-6369.

[59] Cong, C., Wang, H., Hu, Y., Liu, C., Ma, S., Li, X., Cao, J. & Ma, S. (2011). "Synthesis and antibacterial activity of novel 4"-O-benzimidazolyl clarithromycin derivatives." *European Journal of Medicinal Chemistry*, *46*, 3105-3111.

[60] Shah, K., Chhabra, S., Shrivastava, S. K. & Mishra, P. (2013). "Benzimidazole: a promising pharmacophore." *Medicinal Chemistry Research*, *22*(11), 507-5104.

[61] Barot, K. P., Jain, S. V., Gupta, N, Kremer, L., Singh, S., Takale, V. B., Joshi, K. & Ghate, M. D. (2014). " Design, synthesis and docking studies of some novel (R)-2-(4'-chlorophenyl)-3-(4'-nitrophenyl)-1,2,3,5-tetrahydrobenzo[4,5]imidazo [1,2-c] pyrimidin-4-ol derivatives as antitubercular agents." *European Journal of Medicinal Chemistry*, *83*, 245-255.

[62] Hwu, J. R., Singha, R., Hong, S. C., Chang, Y. H., Das, A. R., Vliegen, I., Clercq, E. D. & Neyts, J. (2008). "Synthesis of new benzimidazole-coumarin conjugates as anti-hepatitis C virus agents." *Antiviral Resarch*, *77*, 157–162.

[63] Xue, F., Luo, X., Ye, C., Ye, W. & Wang, Y. (2011). "Inhibitory properties of 2-substituent-1*H*-benzimidazole-4-carboxamide derivatives against enteroviruses." *Bioorganic & Medicinal Chemistry Letters*, *19*, 2641–2649.

[64] Tallant, M. D., Duan, M., Freeman, G. A., Ferris, R. G., Edelstein, M. P., Kazmierski, W. M. & Whellam, P. J. (2011). "Synthesis and evaluation of 2-phenyl-1,4-butanediamine-based CCR5 antagonists for the treatment of HIV-1." *Bioorganic & Medicinal Chemistry Letters*, *21*, 1394–1398.

[65] Biron, K. K., Harvey, R. J., Chamberlain, S. C., Good, S. S., Smith, A. A., Davis, M. G., Talarico, C. L., Miller, W. H., Ferris, R., Dornsife, R. E., Stanat, S. C., Drach, J. C., Townsend, L. B. & Koszalka, G. W. (2002). "Potent and selective inhibition of human cytomegalovirus replication by 1263W94: a benzimidazole L-riboside with a unique mode of action." *Antimicrobial Agents and Chemotherapy*, *46*, 2365–2372.

[66] Brink M., Dahlen, A., Olsson, T., Polla, M. & Svensson, T. (2014). "Design and synthesis of conformationally restricted inhibitors of active thrombin activatable fibrinolysis inhibitor (TAFIa)." *Bioorganic & Medicinal Chemistry Letters*, *22*, 2261-2268.

[67] Han, X. F., He, X., Wang, M., Xu, D., Hao, L. P., Liang, A. H., Zhang, J. & Zhou, Z. M. (2015). "Discovery of novel, potent and low-toxicity angiotensin II receptor type 1 (AT_1) blockers: Design, synthesis and biological evaluation of 6-substituted aminocarbonyl benzimidazoles with a chiral center." *European Journal of Medicinal Chemistry*, *103*, 473-487.

[68] Tugce, Z., Caliscan, B. & anoglu, E. (2018). "Drug discovery approaches targeting 5-lipoxygenase-activating protein (FLAP) for inhibition of cellular leukotriene biosynthesis." *European Journal of Medicinal Chemistry*, *153*, 34-48.

[69] Pergola, C., Gerstmeier, J., Monch, B., Caliskan, B., Luderer, S., Weinigel, C., Barz, D., Maczewsky, J., Pace, S., Rossi, A., Sautebin, L., Banoglu, E. & Werz, O. (2014). "The novel benzimidazole derivative BRP-7 inhibits leukotriene biosynthesis *in vitro* and *in vivo* by targeting 5- lipoxygenase-activating protein (FLAP)." *British Journal of Pharmacology*, *171*, 3051-3064.

[70] Chai, W., Dvorak, C. A., Eccles, W., Edwards, J. P., Goldberg, S. D., Krawczuk, P. J., Lebsack, A. D., Liu, J., Pippel, D. J. & Sales, Z. S. (2014). *1,2,6-substituted Benzimidazoles as Flap Modulators,* September 18, 2014. Patent 20140275029.

[71] Brown, A. D., Bagal, S. K., Blackwell, P., Blakemore, D. C., Brown, B., Bungay, P. J., Corless, M., Crawforth, J., Fengas, D., Fenwick, D. R., Gray, V., Kemp, M., Klute, W., Sanz, L. M., Miller, D., Murata, Y., Payne, C. E., Skerratt, S., Stevens, E. B. & Warmus, J. S. "The discovery and optimization of benzimidazoles as selective $Na_V1.8$ blockers for the treatment of pain." *Bioorganic & Medicinal Chemistry Letters*, *27*, 230-239.

[72] Takahashi, K., Hashimoto, N., Nakama, C., Kamata, K., Sasaki, K., Yoshimoto, R., Ohyama, S., Hosaka, H., Maruki, H., Nagata, Y., Eiki, J. I. & Nishimura, T. (2009). "The design and optimization of a series of 2-(pyridin-2-yl)-1H-benzimidazole compounds as allosteric glucokinase activators." *Bioorganic & Medicinal Chemistry Letters*, *17*, 7042-7051.

[73] Klapars, A., Campos, K. R., Waldman, J. H., Zewge, D., Dormer, P. G. & Chen, C. Y. (2009). "Enantioselective Pd-catalyzed α-arylation of N-Boc-Pyrrolidine: the key to an efficient and practical synthesis of a Glucokinase activator." *Journal of Organic Chemistry*, *73*, 4986-4993.

[74] Rylands, L. i., Welsh, A., Maepa, K., Stringer, T., Taylor, D., Chibale, K. & Smith, G. S. (2019). "Structure-activity relationship studies of antiplasmodial cyclometallated ruthenium(II), rhodium(III) and iridium(III) complexes of 2-phenylbenzimidazoles." *European Journal of Medicinal Chemistry*, *161*, 11-21.

[75] Farahat, A. A., Ismail, M. A., Kumar, A., Wenzler, T., Brun, R., Paul, A., Wilson, W. D. & Boykin, D. W. (2018). "Indole- and benzimidazole bichalcophenes: synthesis, DNA binding and antiparasitic activity." *European Journal of Medicinal Chemistry*, *143*, 1590-1596.

[76] Singh, K., Okombo, J., Brunschwig, C., Ndubi, F., Barnard, L., Wilkinson, C., Njogu, P. M., Njoroge, M., Laing, L., Machado, M., Prudencio, M., Reader, J., Botha, M., Nondaba, S., Birkholtz, L., Lauterbach, S., Churchyard, A., Coetzer, T. L., Burrows, J. N., Yeates, C., Denti, P., Wiesner, L., Egan, T. J., Wittlin, S. & Chibale, K. (2017). "Antimalarial pyrido[1,2-a]benzimidazoles: lead optimization, parasite life cycle stage profile, mechanistic evaluation, killing kinetics, and *in vivo* oral efficacy in a mouse model." *Journal of Medicinal Chemistry*, *60*, 1432-1448.

[77] Roman, G., Crandall, I. E. & Szarek, W. A. (2013). "Synthesis and anti-Plasmodium activity of benzimidazole analogues structurally related to astemizole." *Chem Med Chem*, *8*, 1795-1804.

[78] Teno, N., Iguchi, Y., Yamashita, Y., Mori, N., Une, M., Nishimaki-Mogami, T. & Gohda, K. (2017). "Discovery and optimization of benzimidazole derivatives as a novel chemotype of farnesoid X receptor (FXR) antagonists." *Bioorganic & Medicinal Chemistry Letters*, *25*, 1787-1794.

[79] van de Wiel, S. M. W., Bijsmans, I. T. G. W., van Mil, S. W. C. & van de Graaf, S. F. J. (2019). "Identification of FDA-approved drugs targeting the Farnesoid X Receptor." *Scientific Reports*, *9*, 2193.

[80] Fang, S., Suh, J. M., Reilly, S. M., Yu, E., Osborn, O., Lackey, D., Yoshihara, E., Perno, A., Jacinto, S., Yoshihara, Y., Atkins, A. R., Khvat, A., Schnabl, B., Yu, R. T., Brenner, D. A., Coulter, S., Liddle, C., Schoonjans, K., Olefsky, J. M., Saltiel, A. R., Downes, M. & Evans, R. M. (2015). "Intestinal FXR agonism promotes adipose tissue browning and reduces obesity and insulin resistance." *Nature medicine*, *21*, 159–165.

[81] Yang, Z. X., Shen, W. & Sun, H. (2010). Effects of nuclear receptor FXR on the regulation of liver lipid metabolism in patients with nonalcoholic fatty liver disease. *Hepatology International*, *4*, 741–748.

[82] Mochizuki, M., Kori, M., Kobayashi, K., Yano, Y., Sako, Y., Tanaka, M., Kanzaki, N., Gyorkos, A. C., Corrette, C. P., Cho, S. Y., Pratt, S. A. & Aso, K. (2016). "Design and Synthesis of Benzimidazoles As Novel Corticotropin-Releasing Factor 1 Receptor Antagonists." *Journal of Medicinal Chemistry*, *59*, 2551-2566.

[83] Kojima, T., Mochizuki, M., Takai T., Hoashi, Y., Morimoto, S., Seto, M., Nakamura, M., Kobayashi, K., Sako, Y., Tanaka, M., Kanzaki, N., Kozugi, Y., Yano, T. & Aso, K. (2018). "Discovery of 1,2,3,4-tetrahydropyrimido[1,2-a]benzimidazoles as novel class of corticotropin releasing factor 1 receptor antagonists." *Bioorganic & Medicinal Chemistry Letters*, *26*, 2229-2250.

[84] Tokuhara, H., Imaeda, Y., Fukase, Y., Iwanaga, K., Taya, N., Watanabe, K., Kanagawa, R., Matsuda, K., Kajimoto, Y., Kusumoto, K., Kondo, M., Snell, G., Behnke, C. C. & Kuroita, T. (2018). " Discovery of benzimidazole derivatives as orally active renin inhibitors: Optimization of 3,5-disubstituted piperidine to improve pharmacokinetic profile." *Bioorganic & Medicinal Chemistry Letters*, *26*, 3261-3286.

In: Benzimidazole
Editor: Aksel A. Vestergaard
ISBN: 978-1-53616-986-7

Chapter 4

BENZIMIDAZOLE: PREPARATION AND APPLICATIONS: BENZIMIDAZOLE DERIVATIVES IN ORGANIC LIGHT EMITTING DEVICE

J. Jayabharathi, PhD and V. Thanikachalam*, PhD
Department of Chemistry, Material Science Lab, Annamalai University, Annamalainagar, Tamilnadu, India

ABSTRACT

Benzimidazole derivatives (BI) play a vital-role in organic light emitting diodes (OLEDs) as emissive materials, host materials, electron transport materials, hole blocking materials, etc., and have pharmaceutical applications such as antimicrobials, antihypertensive, anti-HIVs, anticancer, etc. The non-doped blue emitters with BI derivatives, N,N-diphenyl-4′-(1-phenyl-1H-benzo[d]imidazol-2-yl)-[1,1′-biphenyl]-4-amine (TPABBI) / N,N-diphenyl-4′-(1-phenyl-1H-benzo[d] imidazol-2-yl)-[1,1′-biphenyl]-3-amine (*m*-TPABBBI) [1] show low efficiency roll-off with high luminance, current efficiency (η_c) -3.56/5.1 cdA^{-1}; power efficiency (η_c)- 3.11/5.77 lmW^{-1}; external quantum efficiency (η_{ex}) -4.23/5.66%. The BI used as green emitters shows η_c -7.8 cd/A [2] and 3,7-bis(1-(biphenyl-4-yl)-benzimidazol-2-yl)-10-methylphenothiazine (PhBBMP) exhibits η_c -4.5 cd/A, η_{ex}-0.9% and luminance (L)-12000 cd/m^2 with CIE (0.20, 0.55) [3]. The Pt(II) complex based 2-(2'-pyridyl)benzimidazolyl functionalized molecular stars afford red emission [3b]. The white-emitting device with BI derivative show L-260500 cdm^{-2} and η_p-12.8 lmW^{-1} [4]. 2, 2', 2''-(1,3,5-benzinetriyl)-tris(1-phenyl-1- H-benzimidazole) (TPBi) used as hole blocking layer and electron transport layer in blue OLEDs with CIE (0.15, 0.09) and high η_{ex} - 4.3% due to effective carrier recombination [5, 6]. 2,2'-bis(-phenyl-1H-benzo[d]imidazol-2-yl)biphenyl (BImBP) used as host material exhibit η_c- 57.2 cdA^{-1} with low efficiency roll-off, η_p -50.4 lmW^{-1} and η_{ex} -25.7% with CIE (0.15, 0.38) and BImBP basedPHOLED show high η_c-53 cd/A [7]. Among the green host

* Corresponding Author: jtchalam2005@yahoo.co.in.

materials, 9-phenyl-3-(4-(1-phenyl-1H-benzo[d]imidazol-2-yl)phenyl)-9H-carbazole (PBIC) and 9-(naphthyl-2-yl)-3-(4-(1-phenyl-1H-benzo[d]imidazol-2-yl)phenyl)-9H-carbazole (NBIC), PBIC host exhibit higher efficiency with η_c and η_p enhanced by 1.6 and 2.0 times compared with CBP host device [8]. Among the BI based hosts, *o*-mCPBI, *m*-mCPBI and *p*-mCPBI, *o*-mCPBI PHOLEDs exhibit, η_{ex} of 20% [9]. The blue PHOLEDs with BI based Ir(III) emitter demonstrated longer lifetime than 300 h and show η_{ex} of 20.1% [10]. Among the two bipolar molecules, tris(3′-(1-phenyl-1*H*-benzimidazol-2-yl)biphenyl-4-yl)amine(TBBI) and tris(2-methyl-3′-(1-phenyl-1*H*-benzimidazol-2-yl)biphenyl-4-yl)amine (Me-TBBI), Me-TBBI device exhibit excellent device efficiency (L-16400 cd m^{-2}, η_c- 27.4 cdA^{-1}, η_p- 4.5 lmW^{-1}) than TBBI [11]. The two novel and stable red luminescent radicals namely, bis[4-(1H-benzimidazolyl)-2,6-dichlorophenyl](2,4,6-trichlorophenyl) methyl radical (TTM-2Bi) and tris[4-(1H-benzimidazolyl)-2,6-dichlorophenyl]methyl radical (TTM-3Bi) reported, TTM-2Bi exhibits better performance [12]. Jayabharathi et al., reported synthesis, characterization and electroluminescent studies of BI derivatives along with their Ir(III) complexes [13]. Herein at first, the BI based emitters with multi-functional abilities have been reviewed comprehensively. The molecular design- strategy and synthetic routes for BI materials and BI based host materials for RGB PHOLEDs are discussed.

Keywords: Benzimidazole, synthesis, applications, OLEDs, emissive materials, host, HLCT, maximum efficiencies

1. INTRODUCTION

OLEDs have drawn considerable attention due to their potential applications in solid-state lighting, electronics industry and low power consumption flat-panel displays [14]. In fluorescent OLEDs, internal quantum efficiency (IQE) is restricted by deactivation of triplet excitons to only 25% (Figure 1). To overcome this disadvantage, intensive research was focused on both phosphorescent and fluorescent OLEDs [15]. When electron-withdrawing BI is coupled to the para carbon of tris(2,4,6-trichlorophenyl)methyl (TTM), the photostability has been enhanced compared to TTM core, delocalization of benzimidazole is likely to be the reason. Among the doped red OLEDs based on bis[4-(1H-benzimidazolyl)-2,6-dichlorophenyl](2,4,6-trichlorophenyl)methyl radical (TTM-2Bi) and tris[4-(1H-benzimidazolyl)-2,6-dichlorophenyl]methyl radical (TTM-3Bi), TTM-2Bi shows maximum η_{ex} of 5.4% and η_c of 9.8 cdA^{-1} [16]. Green-emitting D-A molecule with non-planar geomentry, 3, 7-bis(1-(biphenyl-4-yl)benzimidazol-2-yl)10-methylphenothiazine (PhBBMP) consists of electron donor (electron-rich sulfur phenothiazine core) and electron acceptor(biphenylbenzimidazole substituent) [12, 16] exist in amorphous form with decreased crystallinity leads to enhanced efficiency [17-20]. The device configuration of ITO/NPB (40 nm)/PhBBMP (30 nm)/BCP (10 nm)/TPBi (30 nm)/LiF (1 nm)/Al (100 nm) exhibits η_c of 4.5 cdA^{-1}, η_{ex} of 0.9% and luminance of 12000 cd/m^2 with CIE (0.20, 0.55). The electron transport materials (ETM) TPBI confined holes and excitons in emissive layer (EML) and enhanced the emission efficiency.

The non-doped blue device with BI derivatives namely, 4-(1-phenyl-1H-benzo[d]imidazol-2-yl)-[1,1-biphenyl]-4-amine (TPABBI) and N,N-diphenyl-4-(1-phenyl-1H-benzo [d]imidazol-2-yl)-[1,1′-biphenyl]-3-amine (m-TPABBI) consist of BI and triphenylamine with biphenyl or phenyl center at *para*-position of BI exhibit excellent performances. The blue device based on TPABBI exhibit maximum η_c (3.56/ cd/A), η_p (3.11 lm/W) and η_{ex} (4.23%) [21].

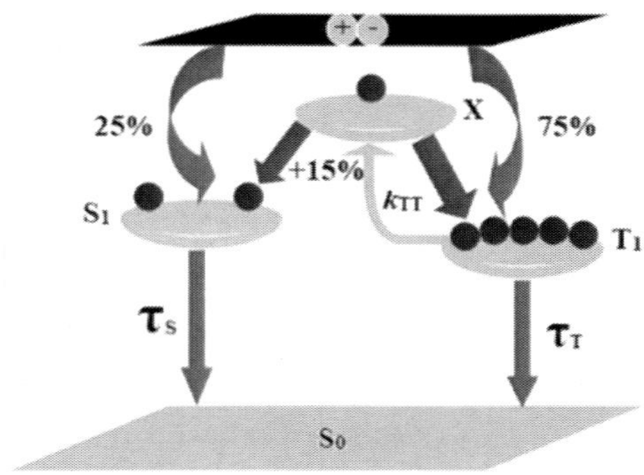

Figure 1. Spin-statistics in electrically driven OLEDs.

From phosphorescent OLEDs (PhOLEDs), 100% internal quantum efficiency can be obtained when triplet emitters are incorporated. To minimise concentration quenching (triplet–triplet annihilation) of PhOLEDs, phosphorescent organo-transition metal complex emitters are doped into a suitable host material. For efficient PhOLEDs, host materials require large triplet (E_T) energy which effectively confine the triplet exciton on phosphorescent dopant and to prevent back transfer of triplet exciton from metal complexes to host and adjacent functional layer [22]. Carbazoles are widely used as a key building block for hole-transporting host materials because they possess high triplet energies and excellent hole transport properties [20]: carbazole-based material, 4,4’-bis(9-carbazolyl)-2,2’-biphenyl (CBP) with E_T of 2.56 eV and 1,3-bis(9-carbazolyl)benzene (mCP) with E_T (2.9 eV) have been prominently used as hosts for green/red phosphorescent emitters [23-28]. However, glass-transition temperature (T_g), 62 °C (CBP) and 60°C (mCP) and low electron mobilities of CBP and mCP hindered their application as host materials in OLEDs [24, 25]. The host materials with effective carrier transporting properties are required to increase the opportunity of electron and hole recombination within the emissive layer [26]. Therefore, steps were made for the development of host materials with bipolar characteristics by linking hole-transporting (donor) and electron-transporting (acceptor) moieties [27]. Novel bipolar host materials were reported with different combination of hole and electron transporting moieties. Since carbazoles and arylamines have attractive hole mobilities (μ_h-10^{-3} to 10^{-5} cm^2 V^{-1} s^{-1}) [28, 29], they used as hole/electron-transporting moieties in the materials with oxadiazole, triazine, phenanthroline and benzimidazole units [28]; benzimidazole groups are widely used to promote electron injection/transportation

[29]. Ma et al., reported host materials comprising carbazole/benzimidazole hybrid for green PHOLEDs and maximum η_{ex}, η_c and η_p obtained from these devices are of 18.7%, 70.2 cd/A, and 73.4 lmW^{-1}, respectively [30]. Anzenbacher et al., employed 1,3,5-tris(N-phenylbenzimidazol-2-yl) benzene/carbazole and diphenylamine moieties as host for green PhOLEDs and harvested maximum η_{ex}, η_c and η_p of 14%, 48.2 cdA^{-1} and 46.0 lmW^{-1}, respectively [31]. However, the detailed study on benzimidazole based fluorophores/phosphores for OLEDs have not been reported yet, this review, focuses molecular design strategy, synthesis of BI based fluorophores/phosphores and their EL characteristics. This review affords valuable insight in terms of design criteria for BI materials toward efficient OLEDs.

2. Synthesis of Benzimidazole Derivatives (1-6)

A mixture of aldehyde (2 mmol) in ethanol, ammonium acetate (2.5 mmol) and *o*-phenylenediamine (1 mmol) was refluxed at 80 °C. Purification was made by column chromatography (petroleum ether: ethyl acetate (9:1)).

2.1. Synthesis of Benzimidazole Derivatives (7-12)

A mixture of N-phenyl-o-phenylenediamine (1 mmol), aldehyde (1 mmol), ammonium acetate (2.5 mmol) and cobalt(II) hydroxide (10 or 15 mol %) was refluxed in ethanol at 80 °C for different time intervals. Purification was made by column with petroleum ether (60-80 °C) as eluent (Figure 2) [32-38].

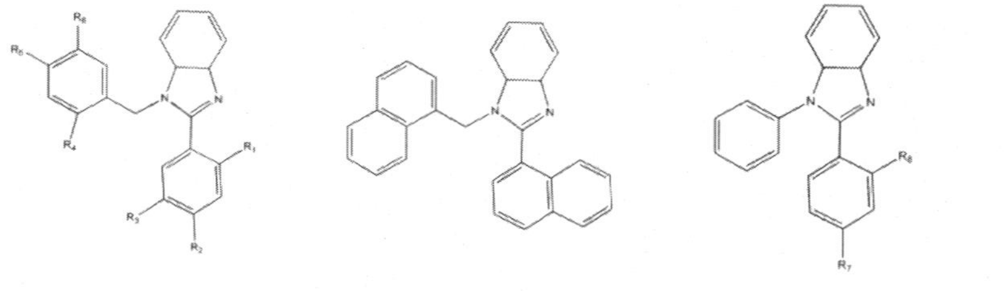

1, 2, 4-6 **3** **7-12**

Ligand	R_1	R_2	R_3	R_4	R_5	R_6
1	H	H	H	H	H	H
2	H	F	H	H	F	H
4	OH	H	Br	OH	H	Br
5	H	CH_3	H	H	CH_3	H
6	H	OCH_3	H	H	OCH_3	H

Ligand	R_7	R_8
7	H	H
8	F	H
9	CH_3	H
10	OCH_3	H
11	CF_3	H
12	H	OH

Figure 2. Various BI derivatives.

2.2. Synthesis of Phosphorescent Emitters

2.2.1. Synthesis of Dimer

The dimer has been prepared by Nonoyama route [39]. The BI ligand (2.2 mmol) and iridium(III) chloride trihydrate (1 mmol) in 2-ethoxyethanol:H_2O (3:1) (15 mL) was refluxed at 120 °C (N_2 steam). The obtained dimer was washed with hexane.

2.2.2. Synthesis of Iridium(III) Complexes

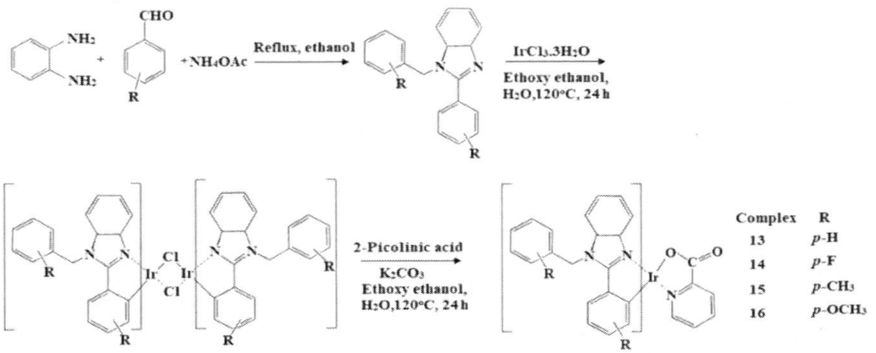

Scheme 1. Synthetic route of iridium(III) complexes 13-16.

The corresponding dimer (1 mmol), potassium carbonate (2.5 mmol) and 2-picolinic acid (2.2 mmol) or acetylacetone (2.2 mmol) in 2-ethoxyethanol was refluxed at 120 °C (N_2 steam). The iridium(III) complex was washed with hexane (Scheme 1).

3. Spectral Characteristics

Jayabharathi et al., [33-38] reported mass, NMR spectral studies and XRD of BI derivatives and iridium(III) complexes. Their photophysical studies were discussed along with detailed DFT studies.

3.1. 1-(4-Methybenzyl)-2-*p*-tolyl-1H-benzo[d]imidazole (5)

Yield: 50%. Mp–114 °C, ^{1}H NMR (400 MHz, $CDCl_3$): δ 2.32 (s, 3H), 2.39 (s, 3H), 5.39 (s, 2H),6.99 (d, 2H, J = 8.0 Hz), 7.12 (d, 2H, J = 8.0 Hz), 7.17-7.30 (m, 6H), 7.58 (d, 2H, J = 8.0 Hz), 7.85 (d, 1H, J = 8.0 Hz). ^{13}C NMR (100 MHz, $CDCl_3$): δ 21.47, 21.13 (methyl carbons), 47.71 (-CH_2 carbon), 110.54, 119.86, 122.57, 122.86. 125.93, 127.23, 129.20, 129.47, 123.73, 133.50, 136.15, 137.46, 140.05, 143.23, 154.38 (Aromatic carbons). $C_{22}H_{20}N_2$, MS: m/e 313.35, calcd 313.16 [M^{+1}].

3.2. 1-(4-Methoxybenzyl)-2-(4-methoxyphenyl)-1H-benzo[d]imidazole (6)

Yield: 60%. Mp–162 °C, ^{1}H NMR (400 MHz, $CDCl_3$): δ 3.79 (s, 3H), 3.85 (s, 3H), 5.39 (s, 2H), 6.85 (d, 2H, J = 8.0 Hz), 6.97 (d, 2H, J = 8.8 Hz), 7.03 (d, 2H, J = 8.0 Hz), 7.21-7.32 (m, 4H), 7.63 (d,2H, J = 8.0 Hz), 7.84 (s, 1H, J = 8.0 Hz). ^{13}C NMR (100 MHz, $CDCl_3$): δ 47.91 (-CH_2 carbon), 55.32, 55.40 (methoxy carbons), 110.43, 114.44, 119.73, 122.47, 122.75, 127.23, 130.73, 136.11, 143.19, 159.12, 160.91 (Aromatic carbons). $C_{22}H_{20}N_2O_2$, MS: m/e 345.29, calcd 345.15[M^{+1}].

3.3. 1,2-Diphenyl-1H-benzo[d]imidazole (7)

Yield: 55%. Mp–107 °C, ^{1}H NMR (400 MHz, $CDCl_3$): δ 7.24-7.33 (q, 7H), 7.46 (d, 3H, J = 7.2 Hz), 7.56 (d, 2H, J = 7.2 Hz), 7.90 (d, 1H, J = 7.6 Hz). ^{13}C NMR (100 MHz, $CDCl_3$): δ 110.52, 119.89, 123.04, 123.39, 127.44, 128.35, 128.59, 129.26, 129.50, 129.90, 129.98, 137.00, 137.24, 143.03, 152.43. $C_{19}H_{14}N_2$, MS: m/e 271.26, calcd m/e 271.12 [M^{+1}].

3.4. 2-(4-Fluorophenyl)-1-phenyl-1H-benzo[d]imidazole (8)

Yield: 62%. Mp–96 °C, ^{1}H NMR (400 MHz, $CDCl_3$): δ 6.98-7.02 (t, 2H, J = 8.0 Hz), 7.26-7.36 (m, 4H), 7.50-7.55 (m, 6H), 7.87 (d, 1H, J = 8.0 Hz). ^{13}C NMR (100 MHz, $CDCl_3$): δ 110.50, 115.44, 115.64, 119.83, 123.13, 123.47, 126.14, 126.18, 127.43, 128.75, 130.01, 131.41, 131.49, 136.84, 137.19, 142.90, 151.47, 162.16, 164.65. $C_{19}H_{13}FN_2$, MS: m/e 289.17, calcd m/e 289.11 [M^{+1}].

3.5. 2-(4-Methylphenyl)-1-phenyl-1H-benzo[d]imidazole (9)

Yield: 68%. Mp–111 °C, ^{1}H NMR (400 MHz, $CDCl_3$): δ 2.33 (s, 3H), 7.10 (d, 2H, J = 8.0 Hz), 7.24-7.32 (m, 5H), 7.44-7.50 (m, 5H), 7.88 (d, 1H, J = 8.0 Hz). ^{13}C NMR (100 MHz, $CDCl_3$): δ 21.42, 110.41, 119.73, 122.93, 123.18, 127.05, 127.48, 128.53, 129.07, 129.37, 129.88, 137.14, 137.24, 139.62, 143.02, 152.60. $C_{20}H_{16}N_2$, MS: m/e 285.30, calcd m/e 285.13 [M^{+1}].

3.6. 2-(4-Methoxyphenyl)-1-phenyl-1H-benzo[d]imidazole (10)

Yield: 69%. Mp-118 °C, ^{1}H NMR (400 MHz, $CDCl_3$): δ 3.80 (s, 3H), 6.82 (d, 2H, *J* = 9.2 Hz), 7.23-7.26 (m, 2H), 7.31-7.33 (m, 3H), 7.47-7.52 (m, 5H), 7.86 (d, 1H, *J* = 8.0 Hz). ^{13}C NMR (100 MHz, $CDCl_3$): δ 55.29, 110.32, 113.77, 119.55, 122.32, 122.89, 123.02, 127.50, 128.53, 129.90, 130.92, 137.21, 143.00, 152.41, 160.51. $C_{20}H_{16}N_2O$, MS: m/e 301.21, calcd m/e 301.13 [M^{+1}].

3.7. 2-(4-(Trifluoromethyl)phenyl)-1-phenyl-1H-benzo[d]imidazole (11)

Yield: 70%. Mp-108 °C, ^{1}H NMR (400 MHz, $CDCl_3$): δ 7.26-7.39 (m, 5H), 7.51-7.58 (m, 5H), 7.70 (d, 2H, *J* = 8.0 Hz), 7.91 (d, 1H, *J* = 8.0 Hz). ^{13}C NMR (100 MHz, $CDCl_3$): δ 110.68, 120.15, 122.51, 123.40, 123.99, 125.25, 125.29, 125.33, 127.40, 129.01, 129.67, 130.16, 131.00, 131.32, 133.45, 136.61, 137.35, 142.90, 150.68. $C_{20}H_{13}F_3N_2O$, MS: m/e 339.38, calcd m/e 339.11 [M^{+1}].

3.8. 2-(1-Phenyl-1H-benzo[d]imidazol-2-yl)phenol (12)

Yield: 69%. Mp-114 °C, ^{1}H NMR (400 MHz, $CDCl_3$): δ 6.51-6.55 (t, 1H, *J* = 8.0 Hz), 6.85 (d, 1H, *J* = 8.0 Hz), 7.08-7.12 (q, 2H), 7.21-7.28 (m, 3H), 7.33-7.43 (m, 3H), 7.60-7.65 (m, 3H), 7.81 (d, 1H, *J* = 8.0 Hz), 13.56 (s, 1H). ^{13}C NMR (100 MHz, $CDCl_3$): δ 110.42, 112.37, 118.05, 118.06, 118.71, 123.48, 123.78, 127.33, 127.95, 129.58, 130.43, 131.45, 136.56, 137.21, 140.04, 150.82, 159.65. $C_{19}H_{14}N_2O$, MS: m/e 287.26, calcd m/e 287.11 [M^{+1}].

3.9. Iridium(III)bis(1,2-diphenyl-1H-benzimidazolato-N,C$^{2'}$)(acetylacetonate) (17)

Yield: 97%, ^{1}H NMR ($CDCl_3$, 400 MHz): δ 1.85 (s, 6H), 5.24 (s, 1H), 6.44-6.56 (m, 8H), 7.12-7.13 (m, 1H), 7.27-7.30 (m, 4H), 7.57-7.73 (m, 12H). ^{13}C NMR (100 MHz, $CDCl_3$): δ 9.54, 11.36, 110.03, 111.36, 114.15, 117.51, 120.15, 120.82, 122.94, 123.51, 123.81, 124.88, 125.17, 125.38, 127.51, 127.73, 127.92, 128.26, 128.51, 128.66, 129.55, 129.93, 130.15, 130.24, 130.32, 130.59, 133.93, 134.13, 134.83, 135.38, 135.44, 136.48, 136.95, 137.31, 139.61, 139.68, 149.54, 151.09, 152.47, 152.93. $C_{44}H_{33}IrN_4O_2$, MS: m/z 830.1, calcd. 830.20.

4. Crystal Structure

Crystal structure of compounds **5, 7-12** and **17** is displayed in Figure 3.

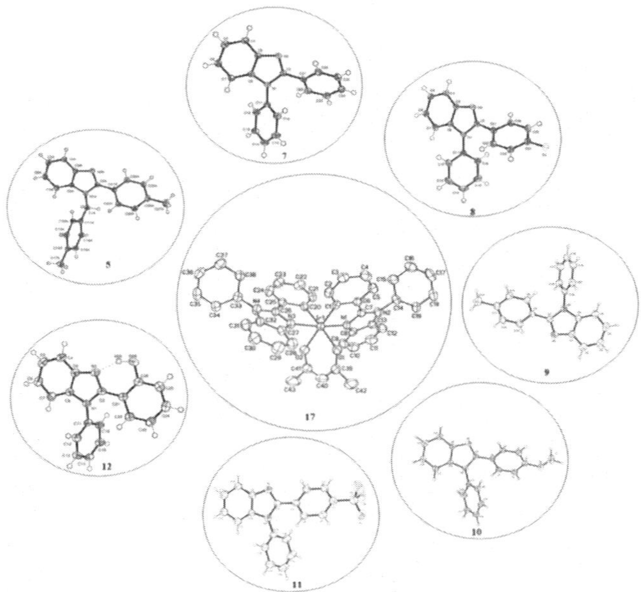

Figure 3. ORTEP of BI derivatives (5, 7-12) and Ir(III) complex (17).

4.1. Crystal Structure of 1-(4-methybenzyl)-2-*p*-tolyl-1H-benzo[d] imidazole (5)

Crystalline 1-(4-methybenzyl)-2-p-tolyl-1H-benzo[d] imidazole (**5**) is triclinic [33], crystallizes with P-1 space group and cell dimensions:a = 9.6610 Å, b = 10.2900Å, c = 17.7271 Å. ORTEP diagram of **5** show that benzimidazole core is planar (Figure 3). In one asymmetric unit two molecules (A&B) crystallize out. The dihedral angle (Φ) between benzimidazole plane and phenyl ring of *p*-methylbenzyl and *p*-tolyl group in molecule A are 76.64° (3) and 46.87° (4) and B are 86.31° (2) and 39.14° (4), respectively. The dihedral angle (Φ) between two benzene ring is 73.73° (3) and 80.69° (4) in both A as well as B, respectively (Table 1).

The HOMO and LUMO orbital picture and the DOS spectrum is displayed in Figure 4 which supports 1-(4-methybenzyl)-2-*p*-tolyl-1H-benzo[d] imidazole is a bipolar material.

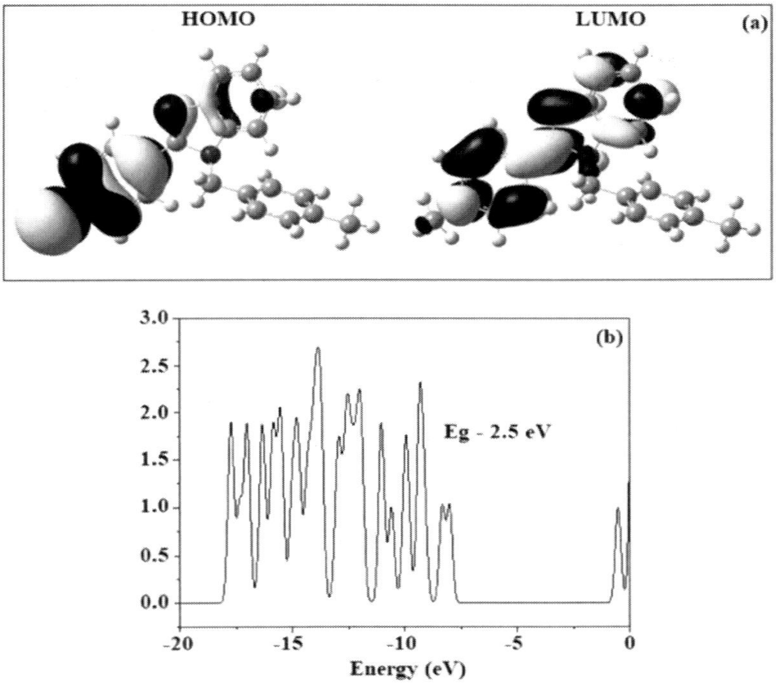

Figure 4. HOMO -LUMO contour map (a) and DOS spectra (b) of BI **5**.

4.2. Crystal Structure of Iridium(III)bis(1,2-diphenyl-1H-benzimidazolato - N,C$^{2\prime}$) (acetylacetonate) (17)

Iridium(III) bis(1,2-diphenyl-1H-benzimidazolato-N,C$^{2\prime}$) (acetylace-tonate) is orthorhombic and crystallizes in space group Pbca. Cell dimensions: a = 13.7242 (8) Å, b = 16.1154 (9) Å and c = 31.1268 (18) Å. This complex shows octahedral geometry around iridium and prefers *cis*-C,C as well as *trans*-N,N chelate disposition rather than *trans*-C,C/*trans*-N,N chelate. Electron rich aryl group show strong influence and *trans* effect: *trans*-C, C configuration will be thermodynamically higher energy and kinetically more labile called *transphobia* which is confirmed by Ir–C bond length (Ir–C_{av}-1.998 Å) < Ir–N bond length (Ir–N_{av}-2.029 Å) (Table 1). Optimization have been made by DFT B3LYP/6-31G (d,p) using Gaussian-03. The XRD parameters are in close with theoretical values, however, the optimized bond angle, bond length and dihedral angles are higher than XRD parameters: theoretical calculations were carried out for isolated molecule in gaseous phase whereas XRD measurement in solid state is likely to be the reason.

Table 1. Bond length (Å)/bond angle (°) of 16 by DFT/ B3LYP/6-31 (d,p)

Bond length (Å)	Crystal (DFT) (Å)	Bond angle (°)	Crystal (DFT) (°)
Ir(1)–O(1)	2.137(2.139)	O(1)–Ir(1)–N(1)	86.50(87.10)
Ir(1)–O(2)	2.138(2.139)	O(1)–Ir(1)–N(2)	96.88(97.02)
Ir(1)–N(1))	2.028(2.030)	O(1)–Ir(1)–C(26)	177.48(178.26)
Ir(1)–N(3)	2.031(2.032)	O(1)–Ir(1)–C(46)	93.26(93.87)
Ir(1)–C(20)	2.002(2.004)	N(1)–Ir(1)–N(2)	87.59(88.01)
Ir(1)–C(1)	1.994(1.995)	N(1)–Ir(1)–C(26)	95.45(96.29)
Bond length (Å)/bond angle (°) of 17 by DFT/B3LYP/6-31 (d,p).			
Ir(1)–O(1)	2.137	O(1)–Ir(1)–N(1)	86.50
Ir(1)–C(1)	2.143	O(1)–Ir(1)–N(2)	96.88
Ir(1)–N(1))	2.053	O(1)–Ir(1)–C(26)	177.48
Ir(1)–N(2)	2.080	O(1)–Ir(1)–C(46)	93.26
Ir(1)–C(26)	2.005	N(1)–Ir(1)–N(2)	87.59
Ir(1)–C(46)	2.071	N(1)–Ir(1)–C(26)	95.45

4.3. Optical and Electrochemical Studies

The absorption bands of iridium(III) complexes show three kinds of bands. The strong absorption ~ 307 nm in UV region attributed to ligand centered (π-π*) transition. The extinction coefficient, band position and size of the bands (449-484 nm) suggest that these are MLCT transitions (^{1}MLCT and ^{3}MLCT). The lower energy tail peaks are attributed to ^{1}MLCT ← S_0 (461 nm) and ^{3}MLCT ← S_0 (483 nm) transitions: closure in intensity of these transitions is due to spin-orbit coupling (S-T mixing). Absorption around 460 nm corresponds to ^{1}MLCT transition (extinction coefficient of the order 10^3).

4.3.1. Mixing of Excited States (LC and MLCT)

The lowest excited state is assigned to ^{3}MLCT for Ir(dpb)$_2$(pic). An analogous argument was used to assign the excited states for complexes. The broad spectra and short lifetime indicates lowest triplet states are dominantly having ^{3}MLCT. The excited triplet state is mixture of Φ_T (π-π*) and Φ_T (MLCT) according to: $\Phi_T = a\ \Phi_T\ (\pi\text{-}\pi^*) + b\ \Phi_T$ (MLCT), a and b are normalized co-efficients, Φ_T (π-π*) and Φ_T (MLCT) are the wave function of ^{3}(π-π*) and ^{3}(MLCT) excited states, respectively. The triplet state is attributed to ^{3}π-π* excited state when a > b and dominantly ^{3}MLCT excited state when b > a. The emission spectra with vibronic progression is attributed fromligand centered ^{3}π-π* state whereas broad shape is from ^{3}MLCT contribution. Complexes Ir(dpb)$_2$(pic), Ir(fppb)$_2$(pic) and Ir(mppb)$_2$(pic) have excited state with large contribution of ^{3}MLCT whereas Ir(mbtb)$_2$ (pic) has excited state with large contribution of ^{3}π-π*. These complexes show green emission at 504, 498, 524 and 522 nm in dichloromethane, respectively. The emission spectra of $Ir(mppb)_2(pic)$ and $Ir(mbtb)_2(pic)$ are red shifted when compared to $Ir(dpb)_2(pic)$, this may be due to the electronic effect of substituent. The emission spectrum of $Ir(fppb)_2(pic)$ is blue shifted compared to $Ir(dpb)_2(pic)$ which is due to electron withdrawing fluoro substituent into phenyl ring attached to C(2) carbon of imidazole ring.

Table 2. Absorption (λ_{abs}), emission (λ_{emi}), fluorescence quantum yield (Φ), lifetime (τ, μs), radiative rate constant (k_r, s^{-1}), nonradiative rate constant (k_{nr}, s^{-1}) and electrochemical behaviour

Complex	λ_{abs} (nm)	λ_{emi}(nm)	Φ	τ	k_r	k_{nr}	E_{onset} (V)	$E^{1/2}$ (V)	HOMO (eV)	LUMO (eV)	E_g
$Ir(dpb)_2(pic)$	307, 461, 483	504	0.45	1.54	0.29	0.36	461	0.44	5.20	2.58	2.62
$Ir(fppb)_2(pic)$	307, 452, 477	498	0.30	1.56	0.19	0.45	458	0.55	5.34	2.67	2.67
$Ir(mbtb)_2(pic)$	308, 461, 484	524	0.38	1.71	0.22	0.36	478	0.34	5.11	2.49	2.62
$Ir(mppb)_2(pic)$	307, 449, 485	522	0.36	1.81	0.20	0.35	472	0.29	5.06	2.48	2.58

4.3.2. Depopulation Kinetics and Frontier Orbital Energies

The radiative and non-radiative decay rate constants (k_r and k_{nr}) are calculated from quantum yield and lifetime (Table 2). The radiative lifetime lies in range of 1.54–1.81 μs. The formulae employed to calculate the radiative (k_r) and non-radiative (k_{nr}) rate constants are $\Phi_p = \Phi_{ISC}\ \{k_r/(k_r + k_{nr})\}$, $k_r = \Phi_p/\tau$, $k_{nr} = (1/\tau) - (\Phi_p/\tau)$ and $\tau = (k_r + k_{nr})^{-1}$, where Φ_{ISC} is the intersystem-crossing yield. For iridium complexes, Φ_{ISC} is assumed to be 1.0 because of strong spin-orbit interaction caused by heavy atom effects of iridium. The HOMO/LUMO energies of complexes were calculated to be -5.15/-2.58, -5.26/2.67, -5.02/-2.49 and -4.96/-2.48 eV for $Ir(dpb)_2(pic)$, $Ir(fppb)_2(pic)$, $Ir(mbtb)_2$ (pic) and $Ir(mppb)_2(pic)$, respectively. The metal site in the complex undergo oxidation (one-electron: HOMO) whereas reduction occurs at heterocyclic portion of the complex (LUMO) (Table 2).

5. ELECTROLUMINESCENCE

5.1. BI as Host Materials

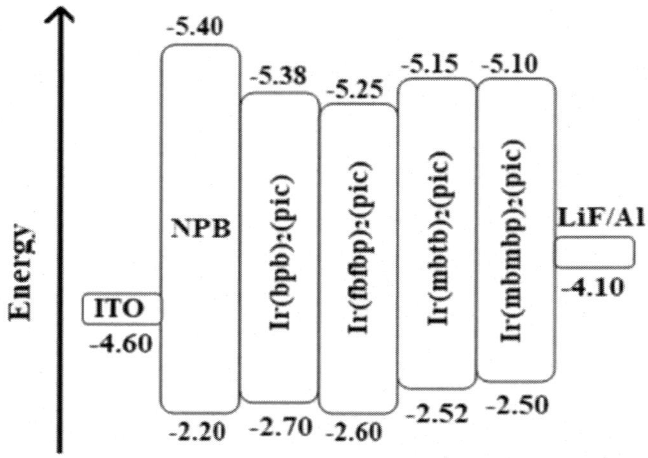

Figure 5. Device with molecular structures of functional materials.

Devices with benzimidazole based Ir(III) complexes, $Ir(bpb)_2(pic)$ and $Ir(fbfpb)_2(pic)$ show high brightness of 130301 cd/m^2 (14.5 V) and 103156 cd/m^2 at 18 V, respectively. Device with $Ir(bpb)_2(pic)$ as dopant shows high η_p of 26.2 Im/w (5.5 V). Device with $Ir(dpb)_2(pic)$ exhibit high brightness of 110421 cd/m^2 at 14.5 V, high η_p of 24.1 lm/W (5.5 V) and η_c of 34.6 cd/A at 6V. The reduction in low driving voltage from $Ir(bpb)_2(pic)$ is due to narrower $E_{HOMO–LUMO}$ energy gap (Figure 5).

Five new BI derivatives, 1,4-bis[2-(2'-pyridyl)benzimidazolyl]benzene (1,4-bmb), 4,4'- bis[2-(2'-pyridyl)benzimidazolyl] biphenyl (bmbp), 1-bromo-4-[2-(2'-pyridyl) benzimidazolyl] benzene (Brmb), 1,3-bis-[2-(2'-pyridyl)benzimidazolyl]benzene (1,3-bmb) and 1,3,5-tris[2-(2'-pyridyl) benzimidazolyl]benzene (tmb) have been synthesized by Ullmann condensation. The 2-(2'-pyridyl)benzimidazole based Pt complexes, Pt_2(1,4-bmb)Ph_4, Pt_2(bmbp)Ph_4, Pt(Brmb)Ph_2, Pt_2(1,3-bmb)Ph_4, and Pt_3(tmb)Ph_6 display orange/red emission at 77 K in solution and also in solid state due to metal-to-ligand charge transfer (MLCT) [40]. Temperature dependent ^{1}H NMR of complexes shows that they exist in isomeric forms due to restricted rotation. The novel BI based Ir(III) complexes (**18a-c**) employed as emissive phosphors in PHOLEDs display excellent device performance to meet the requirement of commercial applications (Figure 6) [10].

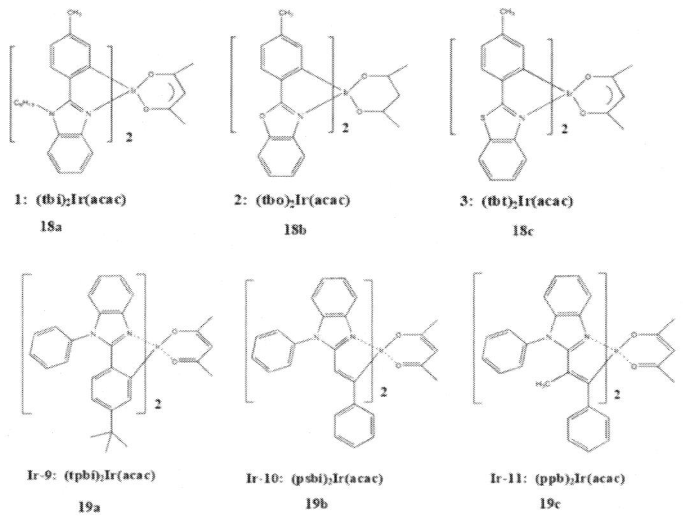

Figure 6. BI based iridium (III) acetylacetonate complexes.

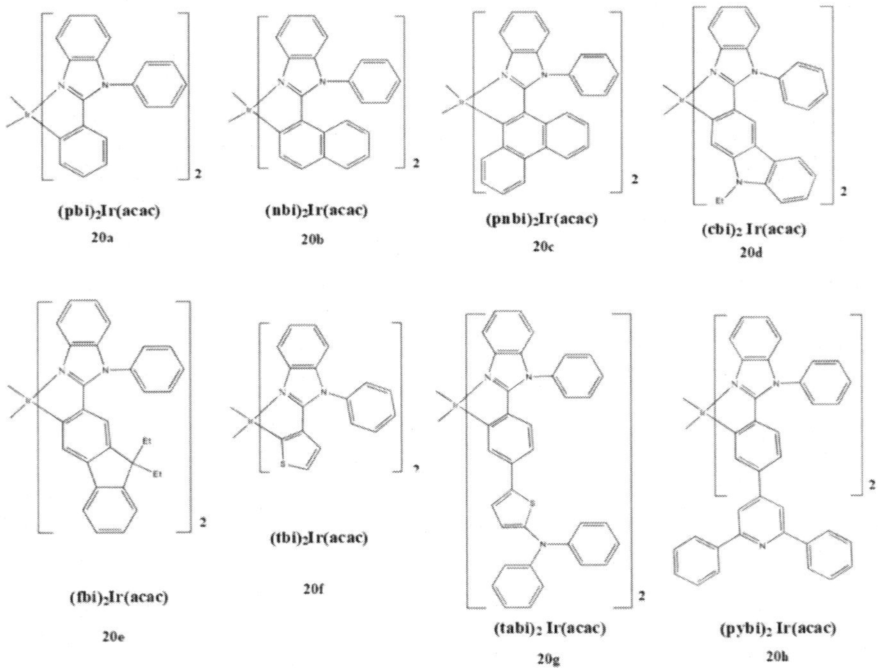

Figure 7. BI based iridium (III) complexes.

The benzimidazole, oxazole or thiazole-based heterocyclic ligands based iridium (III) complexes $(tpbi)_2Ir(acac)$ (**19a**), $(psbi)_2Ir(acac)$ (**19b**) and $(ppbi)_2Ir(acac)$ (**19c**) are used as dopant in PHOLEDs. An electrophosphorescent device with $(tpbi)_2Ir(acac)$ (**19a**) as a dopant show yellow emission with maximum η_c of 9.8 cd/A at 24.2 mA/cm^2 and maximum brightness of 7985 cd/m^2 at 19.6 V [41]. New BI based bis-cyclometalated iridium complex, $(bi)_2Ir(acac)$ used as dopant: emission color varied from green to red. Green- and yellow-OLEDs show high efficiencies (**20a-h**) (Figure 7) [42].

The single-layer device with blue-emitting fluorophores show high device performances. In view of good electron mobility ($10^{-5}cm^2V^{-1}s^{-1}$) of 1,3,5-tris(phenyl-2-benzimidazolyl) benzene (TPBI) and good hole mobility of arylamines (10^{-3}–$10^{-5}cm^2V^{-1}s^{-1}$), the arylamine unit was linked to a phenyl-2-benzimidazolylbenzene unit with fluorene spacer which prevent strong charge transfer (CT) from amine to benzimidazole (**21a-f**) (Figure 8) [29, 43].

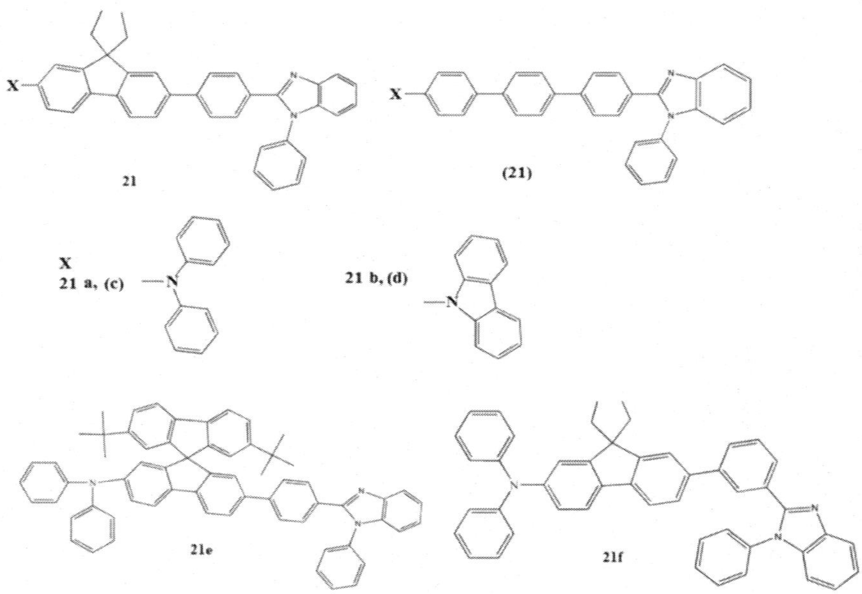

Figure 8. BI derivatives as host materials.

The benzimidazole unit based benzimidazole/triphenylamine hybrids (**22a & 22b**) used as an electron accepting component owing to its good electron mobility. The higher triplet energy of the hybrids employed as host for green and red PHOLEDs. The blue emitter **22a** exhibit maximum η_c of 2.2 cdA^{-1} and η_{ex} of 2.9%, with CIE (0.17, 0.07) [44].

The novel ambipolar hosts of *o*-CbzBz (**23a**) and *o*-DiCbzBz (**23b**) with orthogonal conformation show high η_c and η_p with low efficiency roll-off. *o*-DiCbzBz (**23b**) employed as host for green, sky-blue and orange-red PhOLEDs and exhibit maximum η_c of 78.4, 57.5 and 60.3 cd/A, respectively [45]. The two BI based bipolar materials tris(3′-(1-phenyl-1*H*-benzimidazol-2-yl) biphenyl-4-yl)amine (TBBI) (**24**) and tris(2-methyl-3′-(1-phenyl-1*H*-benzimidazol-2-yl)biphenyl-4-yl)amine (Me-TBBI) (**24**) exhibit maximum electroluminescent efficiencies. TBBI (**24**) (T_g- 148; T_d-552 °C) and Me-TBBI (**24**) (T_g -

144; T_d-515 °C) shows high thermal stability. The Me-TBBI (**24**) and TBBI (**24**) were employed as host material for green dopant Ir(ppy)$_3$. The Me-TBBI (**24**) device show higher efficiency than TBBI (**24a**) (16400 cdm^{-2}, 27.4 cdA^{-1} and 4.5 lmW^{-1}) due to localization of frontier orbitals with an increase of singlet-triplet (ΔE_{S-T}) energy gap [46].

5.1.1. BI as triphenylamine–azole Type Bipolar Host Materials

The 1,3,5-tris(phenyl-2-benzimidazolyl)benzene (TPBI) (**25a**) with high electron mobility (3.3–8 105 cm^{2}V^{1}s^1) used as electron transport (ET) material in OLEDs. Replacing the phenyl attached to phenylbenzimidazole with triphenylamine group (TPBI-Da) (**25b**), the triplet energy (E_T) of TPBI-Da (2.70 eV) is comparable to that of TPBI (2.67 eV). The device with TPBI-Da shows improved device performances with maximum η_{ex}, η_c and η_p of 15.0%, 59.2 cd/A and 70.0 lm/W, respectively, as compared to the device based on TPBI (η_{ex}, η_c and η_p of 4.7%, 18.8 cd/A and 21.0 lm/W). The BI based bipolar hosts TPAPBA1 (**22a**) and TPAPBA2 (**22b**) employed as hosts for green phosphorescent emitters. Owing to more twisted configuration, TPAPBA1 shows higher (E_T) of 2.55 eV than that of TPAPBA2 (2.41 eV). The TPAPBA2: Ir(ppy)$_3$ based device exhibit inferior device efficiency than TPAPBA1:Ir(ppy)$_3$ based device (Figure 9).

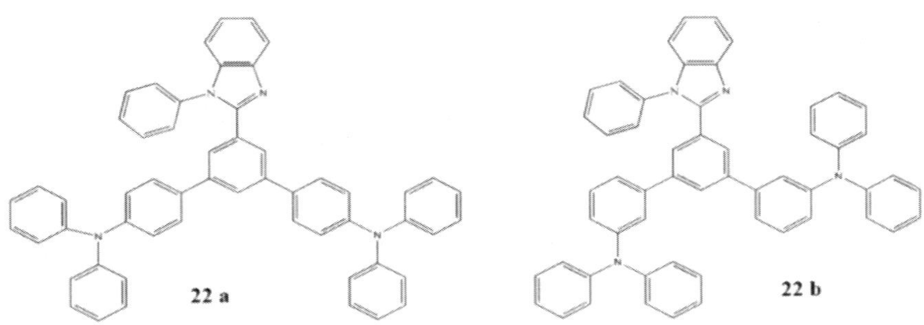

Figure 9. BI as bipolar host materials.

The benzimidazole /triphenylamine bipolar host *p*-BISiTPA with high E_T (2.69 eV) shows good thermal stability with T_g of 102 °C and employed as host for FIrpic-doped blue PHOLED exhibit η_{ex}, η_c and η_p of 16.1%, 35.1 cdA^{-1} and 26.1 lmW^{-1}, respectively. The orange PHOLED with (fbi)$_2$Ir(acac) as an emitter shows even better performance with η_{ex} above 20.0%. The high efficiency for both blue/orange PHOLEDs with *p*-BISiTPA: two-color based WOLED was fabricated using FIrpic (**26**) /(fbi)$_2$Ir(acac) : *p*-BISiTPA in single EML exhibits high η_{ex}, η_c and η_p of 19.1%, 51.8 cdA^{-1} and 42.7 lmW^{-1}, respectively (Figure 10) [47].

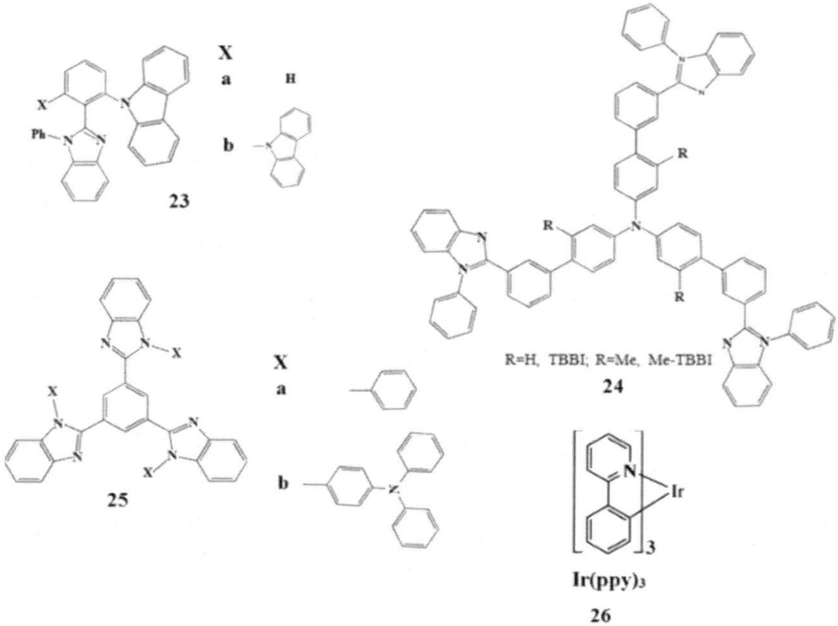

Figure 10. BI derivatives as emissive materials.

5.1.2. BI as carbazole–azole Type Bipolar Host Materials

BI based bipolar host with phenyl spacer, mBICP, mPhBINCP and PhBIDmpCP exhibits high thermal stability: mBICP (T_g–84 °C) and mPhBINCP (T_g-113 °C). The large bulky fluorene-based spacer compounds CzFCBI and CzFNBI show even higher T_g around 170 °C. Owing to extended conjugation, mPhBINCP and PhBIDmpCP possess relatively lower E_T of 2.61 eV and 2.59 eV, respectively. Benefiting from high E_T (3.0 eV) for mBICP, the blue PHOLED with FIrpic show high η_{ex} of 18.7% and η_p of 33.6 lm/W. The mPhBINCP based green PHOLED doped with Ir(ppy)$_3$ exhibit high η_c of 77.6 cd/A and η_p of 80.3 lm/W. Using CzFCBI and CzFNBI as hosts for red emitter Os(bpftz)$_2$(PPhMe$_2$)$_2$, both devices show comparable efficiency with η_{ex} over 18.0%. These two bipolar materials used as hosts for green to red PHOLEDs with high efficiencies. The yellow PHOLEDs with PhCzNBI/ PhCzCBI:(bt)$_2$Ir(acac) show good device performance with η_p of 40 lm/W.

The BI based bipolar host 1,8-mBICz show high E_T of 2.70 eV comparable to that of FIrpic: 1,8-mBICz:FIrpic-doped PHOLED show high EL performance (η_c of 27.6 cd/A and η_p of 27.0 lm/W) while 1,8-mBICz:Ir(ppy)$_3$-doped green PHOLED exhibits better performances: η_c of 64.0 cd/A and η_p of 70.0 lm/W. The asymmetric fluorine bridged carbazole/ benzimidazole hybrids CzFCBI and CzFNBI with high T_g above 170°C used as host for red PHOLEDs. Lai et al., reported blue emitting materials composed of hole transporting arylmine and electron transporting phenyl-2-benzimidazolylbenzene units linked with fluorene spacer exhibit an impressive η_{ex} of 2.8% with CIE (0.14, 0.11) [43].

Single layer orange light emitting devices using such fluorescent blue light emitting material as the host for an iridium complex show η_{ex} of 6.9%.

5.2. BI as Electron Transporting Materials

The benzimidazole (**27**) based py1, py2, mpy1 and mpy2 show high quantum yield of 99.89–100% in dichloromethane and 54.56–86.78% in film. The meta-linked mpy1 and mpy2 compounds show higher quantum yield in solution/film than para-connected py1 and py2 compounds. These compounds show high thermal stability form stable amorphous film with high T_g (182-215 °C). The meta- connected mpy2 as electron-transporting and or emissive layer exhibit blue color (0.16, 0.10) with high η_c and η_p of 2.70 cdA^{-1} and 1.97 lmW^{-1}, respectively [48]. Among the benzimidazole–phosphine oxide hybrid electron transporters BIPO (**28**) and *p*PBIPO (**29**), *p*PBIPO with lowest LUMO energy show highest electron mobility (10^{-4} $cm^2V^{-1}s^{-1}$) (Figure 11). In the multi-layer CBP and $Ir(ppy)_3$ based green PhOLEDs, BIPO used as ETL with efficient performances as well as reduced roll-off. *p*PBIPO based device show excellent device performances (η_{ex} and η_p of 26.7% and 103.8 lmW^{-1}) than BIPO [49].

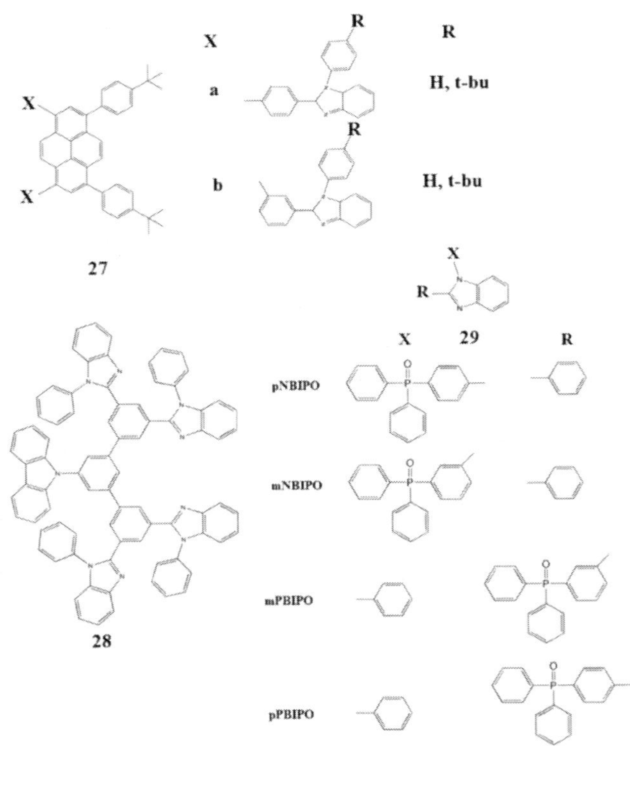

Figure 11. BI as electron transport materials.

BIPO used as electron transport material (ETM) in the green PHOLEDs: ITO/NPB (30 nm)/TCTA(10 nm)/CBP: $Ir(ppy)_3$ (30 nm)/ BIPO (40 nm)/LiF (1 nm)/Al (100 nm). The BIPO (ETL) effectively block the holes and enhanced the effective recombination in the EML results in enhanced device efficiency (22.19%, 68.37 cdA^{-1} and 24.44 lmW^{-1}) with luminance of 72 080 cd m^{-2} compared to TmPyPB (**30**) (η_{ex} -17.03%, η_{c-} 52.02 cdA^{-1} and η_p -20.97 lmW^{-1}) with maximum luminance of 33 490 cdm^{-2} (Figure 12) [50].

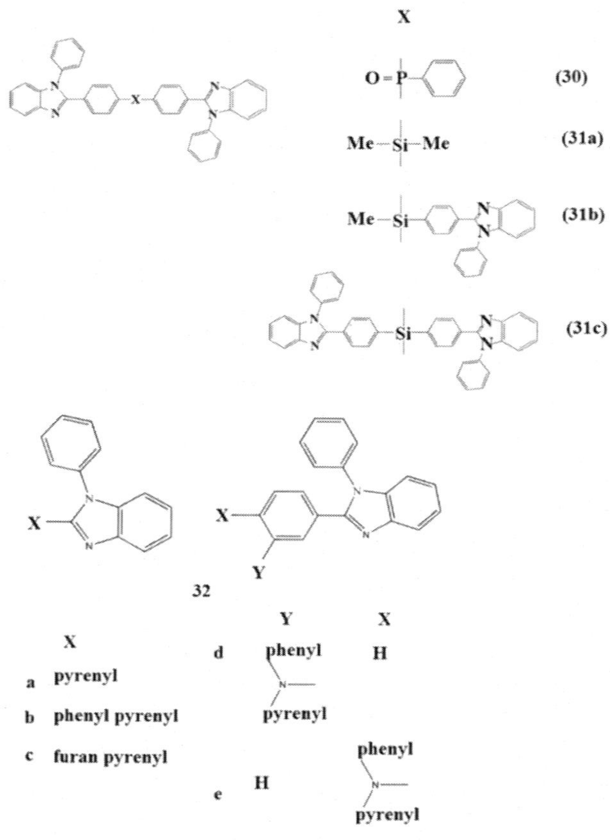

Figure 12. BI for OLEDs.

In green PHOLEDs, iTPBI-CN (**31**) used as an electron transporting layer exhibit maximum η_c and η_p of 37.7 cdA^{-1} and 29.0 lmW^{-1}, respectively than CBP (η_c-31.3 cdA^{-1} and η_p-23.9 lmW^{-1}. These efficiencies are lower than TPBI based device 26.1 cdA^{-1}, 12.2 lmW^{-1} and 20.6 cdA^{-1}, 7.4 lmW^{-1}. Superior device efficiency have been harvested for iTPBI-CN (**31**) over TPBI [51]. The silicon atom annulated with 1,2-diphenylbenzoimidazole (DBI) unit (**32**) used as ETM in green PHOLEDs. Compound **32** transport electrons effectively resulting in high efficiencies: η_{ex} of 18.0%, η_c of 62.8 cd/A and η_p of 61.8 lmW^{-1} [52]. Two bis-cyclometalated iridium complexes using tetraphenylimidodiphosphinate (tpip) as ancillary ligand and TPBi as ETL employed in the device structure of ITO/TAPC (40 nm)/ tpip: mCP (20 nm, *x* wt %)/TPBi (40 nm)/LiF (1 nm)/Al (100 nm): tpip based devices (6 wt %) exhibit maximum luminance of 64 351 cdm^{-2}, η_c and η_p of 67.95 cdA^{-1} and 69.90 lmW^{-1}, respectively; tpip based device with 10 wt % exhibit maximum luminance of 38 963 cdm^{-2} (11.5 V) [53]. The D–A molecules based on naphtho[1,2-*d*]imidazole isomers used as blue emitters for OLEDs: 1*H*-naphtho[1,2-*d*]imidazole derivatives exhibit slightly blue-shifted absorption (λ_{abs})/emission (λ_{emi}) relative to 3*H*-naphtho[1,2-*d*]imidazole compounds. Single-layer device with 1*H*-naphtho[1,2-*d*]imidazole (NI-2-PhTPA) as a non-doped emitting layer show high η_{ex} of 4% compared to 3*H*-naphtho[1,2-*d*]imidazole [54]. The D-A molecule CPhBzIm exhibit high morphological and thermal stability [T_g-170 °C], possess bipolar carrier transport properties ($\mu_e \approx \mu_h \approx 10^{-6}$–$10^{-5}$ $cm^2V^{-1}s^{-1}$) with triplet energy (E_T-2.48 eV) making it appropriate host for green PHOLEDs [4]. The benzimidazole based two bipolar *m*CP-BPBI and CP-QPBI (**28**) compounds used as host for FIrpic and $Ir(mppy)_3$ based blue and green PHOLEDs. The operating voltage of CP-QPBI as host was reduced (~3 V) relative to *m*CP-BPBI. The η_p was enhanced by 2.5 times at the corresponding voltage [55].

5.3. BI as Blue Emitters

Two benzimidazole derivatives, SBFBI and SAFBI used as blue emitters in nondoped OLEDs [ITO/MoO_3 (10 nm)/NPB (80 nm)/TCTA (5 nm)/SBFBI or SAFBI (20 nm)/TPBi (40 nm)/LiF (1 nm)/Al (100 nm)]. Device based on SAFBI show η_c and η_p 1.96 Cd/A and 1.34 lm/W, respectively with CIE (0.15, 0.10) [56]. The blue emissive dopant consist of pyrene and benzimidazole moieties (**33a-e**) bonded directly to or with different π-linkers show deep blue photoluminescence emission whereas thiophene or arylamine based compounds **33c-e** show red-shifted emission to cyan. Dye **33** show deep blue EL emission with high efficiencies: η_{ext} of 1.50%, η_c of 0.5 cdA^{-1}, η_p of 0.2 lmW^{-1}, brightness of 714 cdm^{-2} and CIE (0.16, 0.05) at 100 cd/m^2. Pyrene-benzimidazole conjugates are promising target molecules for blue emitting solution processed OLEDs (Figure 13) [57].

Among the doped OLEDs based on TTM-2Bi (**34**) and TTM-3Bi (**34**), TTM-2Bi emitter show maximum η_{ex} of 5.4% and maximum η_c of 9.8 cdA^{-1} (exciton formation ratio of 69%) [12]. A series of blue to deep blue emitters consist of triphenylamine, thiophene, fluorene, pyrene and carbazole at C2/C4/C7 position of BI nucleus and show high Φ_F in DCM solution (0.33-0.95) and in solid state (0.48-0.62) [58, 59]. Among the two triphenylamine based BI blue emitters T1B (**35**) and T2B (**35**) in non-doped FOLEDs, the device with T2B exhibits high η_{ex} of 4.67%: 1.4 times higher than theoretical maximum because the emitter suppresses the exciton quenching thereby increasing the η_{ex} in the non-doped FOLEDs [60].

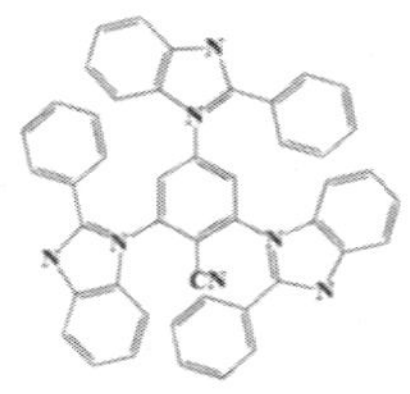

33

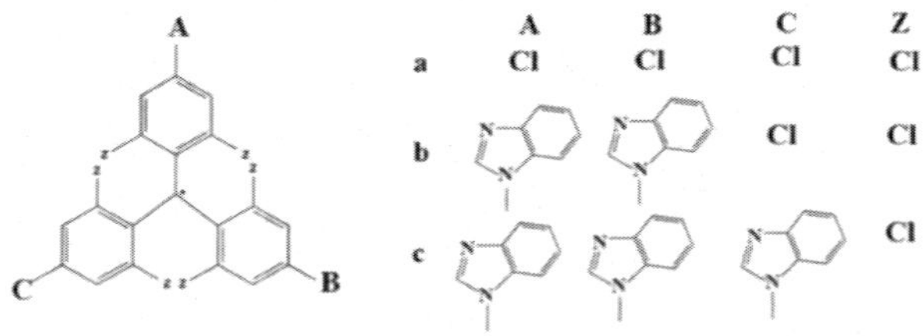

34

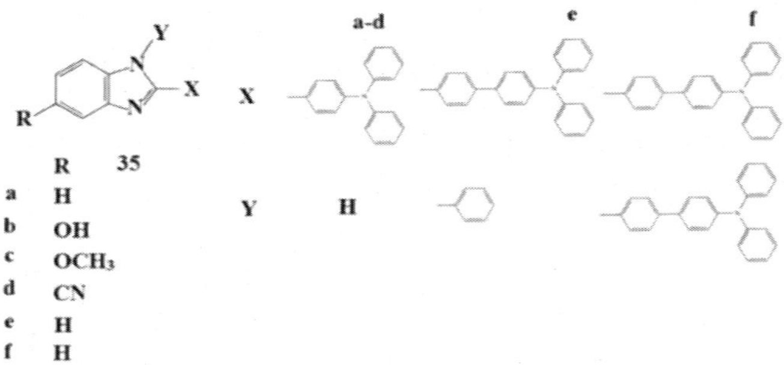

35

Figure 13. BI as blue emitters.

Blue OLEDs with Cz-2pbb (**36**) show η_{ex} of 4.1% with CIE (0.16, 0.05) [61]. The blue-emitting highly emissive compounds **31a** and **37** comprising benzimidazole with arylamine units exhibit ambipolar carrier-transport properties (Figure 14).

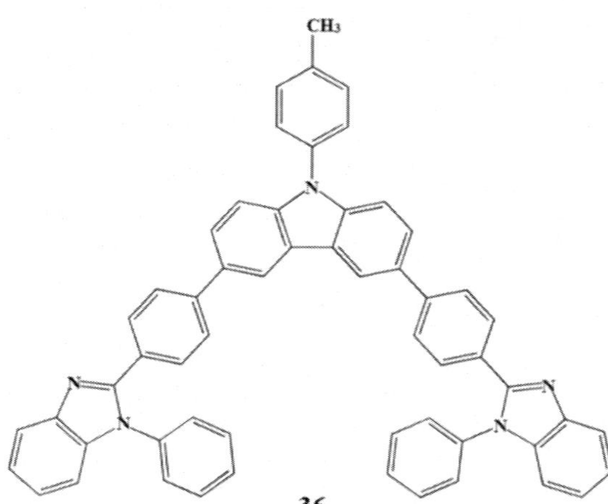

36

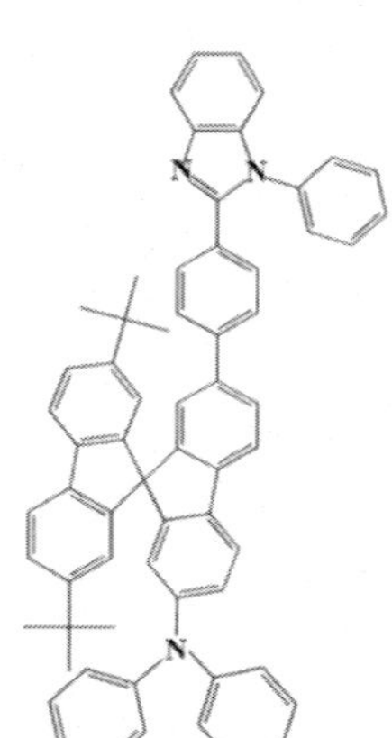

37

Figure 14. Efficient blue emitting BI derivatives.

Efficient single-layer device with phosphorescent dopant with **39a** and **39b** as hosts exhibit excellent efficiency [62-64]. Because of excellent electron-injection/transporting capacities, BI widely used as a building block in OLEDs. Lin et al., fabricated efficient single-layer OLEDs with blue-emitting fluorescent materials [65]. These molecules are hybrid of BI and arylamine linked with spacer (fluorene) to suppress CT. Ambipolar charge-transporting properties are confirmed by comparable μ_h and μ_e (10^{-5} $cm^{-2}V^{-1}s^{-1}$). In a single layer device (ITO/EML/LiF/Al), the molecule with 9,9-diethyl-9H-fluorene linker shows η_{ex} of 2.4% with CIE (0.15, 0.12) at 100 $mAcm^{-2}$ [29].

5.4. BI as White Emitters

A blue emitting bipolar material having electron transporting phenyl-2-benzimidazolyl benzene moieties attached on 3,6-positions of hole transporting carbazole unit employed as blue emissive material show η_{ex} of 4.1% with CIE (0.16, 0.05). Moreover, hybrid WOLEDs having a single emissive layer composed of Cz-2pbb, fac-tris(2-phenylpyridyl)iridium ($Ir(ppy)_3$) and iridium(III) bis(2-phenylquinolyl-N,C-2')acetylacetonate ($PQ_2Ir(acac)$) used as blue, green, and red emitters, respectively show η_{ex}>10% [64].

5.5. BI as Green Emitters

The two BI based emitters with D–π–A structure, TPE-TPAPBI (**38a**) and TPE-DPBI (**38b**), exhibit excellent aggregation-induced-emission (AIE) properties [65]. The BI moieties endow the two emitters with high μ_{es} (~10^{-5} and 10^{-4} $cm^2V^{-1}s^{-1}$ for TPETPAPBI and TPE-DPBI, respectively). A nondoped green device with TPE-TPAPBI show green emission (520 nm) with η_{ex} of 5.8% and high brightness (125 300 cdm^{-2}). The AIEgen, silole coupled with BI derivatives, $(PBI)_2DMTPS$ (**39a**) and $(PBI)_2MPPS$) (**39b**) [66] show higher μ_{es} of 1.24×10^{-4} and 3.42×10^{-4} $cm^2V^{-1}s^{-1}$, respectively than TPBI. A bilayer device: ITO/N,N′-bis(naphthalen-1-yl)-N,N′-bis(phenyl)-benzidine(NPB)/$(PBI)_2DMTPS$ /LiF/Al show comparable performances (η_p -14.51 lmW^{-1} and η_{ex} -4.25%) to those of trilayer OLED (η_p- 12.58 lmW^{-1} and η_{ex}- 4.22%) (Figure 15). Huang et al. used anthracene emitters (**40a-d**) with BI and TPA as electron and hole-transporting groups [67]: a single-layer device of ITO/**40d**/LiF/Al demonstrates η_c of 3.33 cdA^{-1} with CIE (0.16, 0.16) at 20 $mAcm^{-2}$ [68].

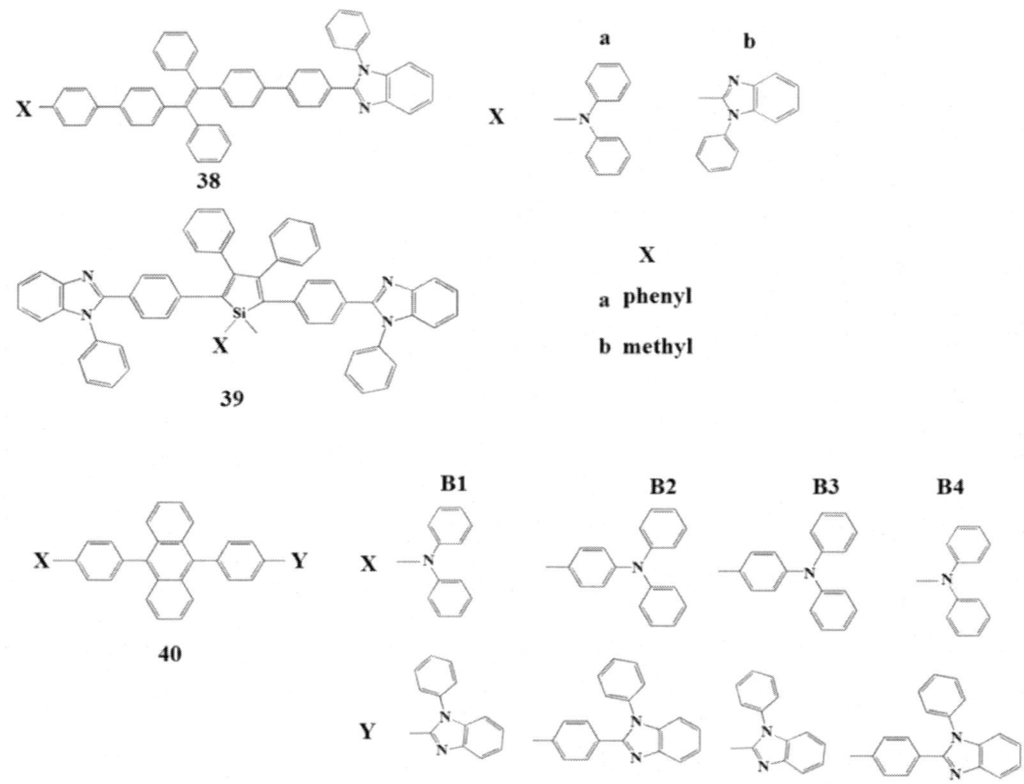

Figure 15. BI as green emitters.

The OLED with M2 (**41**) shows higher luminance (2 times) at same current density than M3 (**42**) based OLED which indicates the former OLED exhibit high efficiency. M2 based OLED show high current efficiency and power efficiency than M3 based OLED (Figure 16). The better device efficiency of M2 agrees with higher PLQY and shows maximum η_c of 7.8 cd/A corresponding to an EQE of 2.7%. The maximum η_c for M3 is 3.6 cd/A which is about two times lower than M2 based OLEDs. The peak position is identical whereas the EL of M3 shows slightly broader bandwidth due to residue emission from the host: energy transfer from CBP to guest is incomplete [68].

A new green-emitting material, 3,7-bis(1-(biphenyl-4-yl)-benzimi-dazol-2-yl)-10-methylphenothiazine (PhBBMP) containing electron-donating phenothiazine core and electron-accepting biphenyl benzimidazole substituent. The optimized device with configuration of ITO/NPB (40 nm)/PhBBMP (30 nm)/BCP (10 nm)/TPBI (30 nm)/LiF (1 nm)/Al (100 nm) exhibits η_c of 4.5 cd/A and η_{ex} of 0.9% at 100 mA/cm^2 and high luminance (12000 cd/m^2) with CIE (0.20, 0.55) [69].

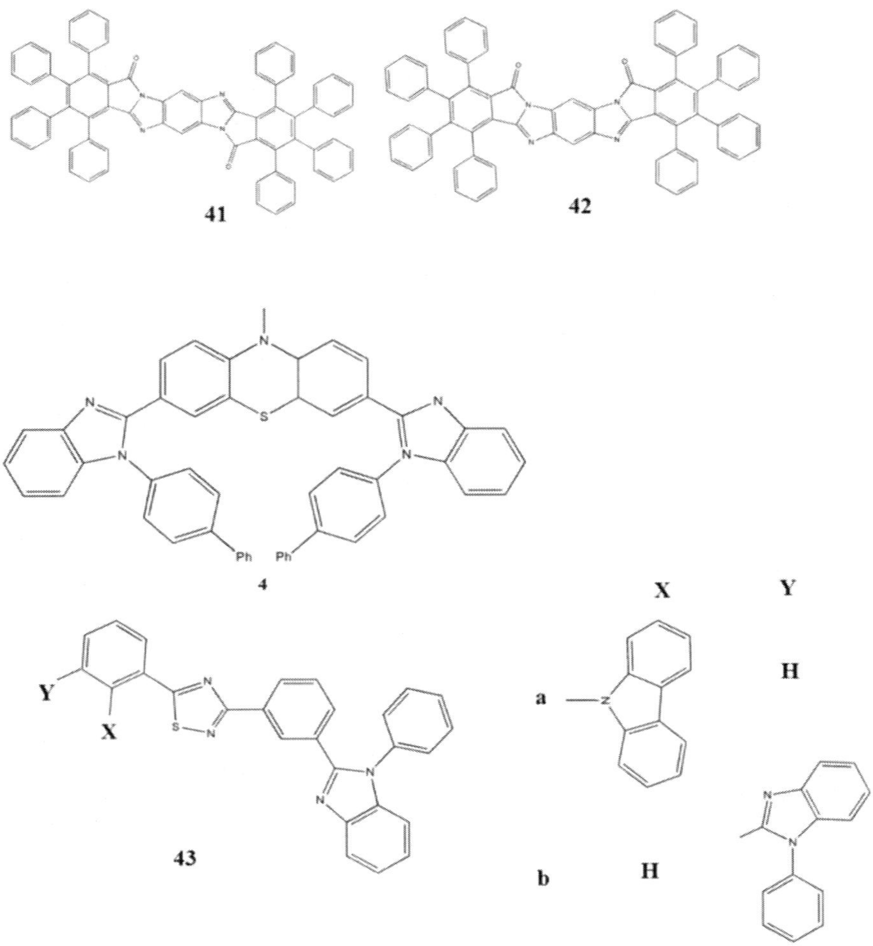

Figure 16. Potential BI based green emissive materials.

Two novel host materials: bipolar type-BzCzTHZ (**43a**) (5-(2-(9H-carbazol-9-yl)phenyl)-3-(3-(1-phenyl-1H-benzo[d]imidazol-2-yl)phenyl)-1,2,4-thiadiazole) and electron type-DBzTHZ (**43b**) (3,5-bis(3-(1-phenyl-1H-benzo[d]imidazol-2-yl)phenyl)-1,2,4-thiadiazole) used as host in the device ITO/MoO_3 (10 nm)/TAPC (50 nm)/TCTA (5 nm)/host: $Ir(ppy)_3$ (3 wt %) (20 nm)/TmPyPB (45 nm)/LiF (1 nm)/Al (100 nm). Devices hosted by DBzTHZ (**43b**) and BzCzTHZ (**43a**) exhibit maximum η_p of 95.2 and 86.2 lm/W corresponding to maximum η_{ex} of 23.4 and 22.2%, respectively. Moreover, device with electron type host DBzTHZ (**43b**) show low turn-on voltage (2.5 V at 1.0 cd/m^2), electron-type hosts are more competitive with bipolar-type hosts [70].

Table 3. Summary of device efficiencies with reported non-doped emitters

Emitters	Host	E_T (eV)	T_g (°C)	Dopant	η_{ex} (%)	η_c (cd/A)	η_p (lm/W)	Ref
TPA +PBI	TPBI-Da	2.70	_	$(ppy)_2Ir(acac)$	15.0	59.2	70	31
TPA + PBI	TPAPBA1	2.55	109	$Ir(ppy)_3$	17.0	64.3	68.3	44
TPA + PBI	TPAPBA2	2.41	129	$Ir(ppy)_3$	15.5	48.7	48.7	44
TPA + aromatic silane + PBI	*p*-BISiTPA	2.69	102	FIrpic	16.1	26.1	26.1	72
TPA + aromatic silane + PBI	*p*-BISiTPA	2.69	102	$(fbi)_{2Ir}(acac)$	20.5	51.9	51.9	72
Cz + dimesitylborane	CzB_1	2.88	105	$Ir(ppy)_3$	6.50	-	-	73
Cz+ dimesitylborane	CzB_2	2.72	89	$Ir(ppy)_3$	23.8	-	-	73
Cz+ dimesitylborane	BDDPC	2.83	110	$(ppy)_2Ir(acac)$	-	38.6	-	74
Cz + BI	*m*BICP	3.0	84	FIrpic	18.7	36.4	33.6	75
Cz + BI	*m*PhBINCP	2.61	113	$Ir(ppy)_3$	21.0	77.6	80.3	76
Cz + BI	PhBIDmPCP	2.59	147	$Ir(ppy)_3$	20.2	74.3	74.4	77
Cz + BI	CzFNBI	2.72	181	$Os(bpftz)_2(PPhMe_2)_2$	18.6	18.6	13.0	78
Cz + BI	CzFCBI	2.52	175	$Os(bpftz)_2(PPhMe_2)_2$	18.1	18.2	17.0	78
Cz + BI	PhCzNBI	2.71	124	$(bt)_2Ir(acac)$	18.7	47	42.2	79
Cz + BI	PhCzCBI	2.55	130	$(bt)_2Ir(acac)$	19.6	51	43.1	79
Cz + BI	1,8-mBICz	2.70	154	$Ir(ppy)_3$	17.5	64.0	70.0	80
Cz + BI	TICCBI	2.61	174	$Os(bpftz)_2(PPhMe_2)_2$	22.0	28.0	22.1	81
Cz + BI	TICNBI	2.61	168	$Os(bpftz)_2(PPhMe_2)_2$	21.0	26.6	25.0	81
Imidazole	Pt-Az8	-	-	Pt	18.1	-	23.3	82
BI	Pt-Az9	-	-	Pt	11.5	38.9	27.2	83
BI	Pt-Az10	-	-	Pt	11.8	40.3	28.1	84
BI	Pt-Az11	-	-	Pt	10.0	34.7	26.4	84
Imidazole	Pt-Az12	-	-	Pt	23.1	-	48.8	85
Imidazole	Pt-Az14	-	-	Pt	26.5	69.8	61.4	86
Imidazole + pyrazole	Pt-Az15	-	-	Pt	23.1	64.5	53.5	86
Imidazole + pyrazole	Ir-Az3	-	-	Ir	26.7	80.3	68.3	86
Imidazole	Ir-Az5	-	-	Ir	23.1	61.5	43.7	87
Imidazole	Ir-Az6	-	-	Ir	33.2	73.6	71.9	88
BI	Ir-Az7	-	-	Ir	16.7	61.0	20.0	88
BI	Ir-NN1	-	-	Ir	10.2	10.2	-	88
Cz + BI	Ir-NN2	-	-	Ir	10.3	24.7	-	89
Cz + BI	Ir-NN3	-	-	Ir	5.00	17.2	-	90
Cz + BI + oxadiazole	Ir-NN5	-	-	Ir	7.40	25.5	-	91
Cz + BI	Ir-NN5	-	-	Ir	6.12	18.3	-	92

*TPA-Triphenylamime, BI- Benzimidazole, PBI-Phenlybenzimidazole and Cz- Carbazole.

5.6. BI as Red Emitters

Benzimidazole based Schiff bases have been shown as fluorophores, different substituents on π-conjugated backbone induced solid-state emission from blue to red. Fluorescence emitter **44** undergo ESIPT (excited state proton transfer) together with AIE results in red fluorescence [71]. Table 3 summarizes the luminous efficiencies of benzimidazole derivatives [31, 44, 72-92] which reveal that benzimidazole derivatives are the potential candidates for OLEDs (Figure 17).

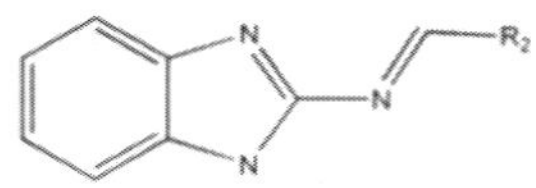

44

R_1	R_2
CH_3	$-C_6H_4N(CH_3)_2$
CN	$-C_6H_4N(CH_3)_2$
NO_2	$-C_6H_4N(CH_3)_2$
CN	$-o-(OH)C_6H_4N(CH_2CH_3)_2$
NO_2	$-o-(OH)C_6H_4N(CH_2CH_3)_2$

Figure 17. BI based Schiff bases.

5.7. Other Applications

5.7.1. BI as Sensor

Benzimidazole based fluorescent sensors shows preferential binding for Zn^{2+} over Pb^{2+}, Fe^{2+}, Co^{2+}, Ni^{2+}, Ca^{2+} and Hg^{2+} metal ions. The high selectivity of BI towards Zn^{2+} is noticeable by fluorescence enhancement due to photoinduced electron transfer. Similarly BI derivatives are selective fluorescent probe for Cu^{2+} metal ion by quenching [93, 94].

5.7.2. BI as Photoinduced Electron-Transferer

Benzimidazole derivative adsorbed on nanoparticles surface via azomethine nitrogen. Using Forester's non-radiation energy transfer theory (FRET), the distance of BI with nanoparticles is deduced as ~ 3.12 nm with energy transfer distance as 2.30 nm. The $-\Delta G_{et}$ reveal that electron transfer is favourable by thermodynamically. Quenching of fluorescence by nano alumina is due to energy transfer [95]. BI derivative gets ducked in the void rutile TiO_2 (Figure 18) enabling binding of titanium of rutile with two azomethine nitrogen, however, in TiO_2 (A) the ducking is sterically hindered (Figure 18). The ducking of BPBI lowered E_{LUMO} and E_{HOMO} of BPBI enabling emission due to electron transfer from LUMO BPBI to CB of rutile. In TiO_2 (A) surface adsorption may not influence frontier energies of BPBI and electron transfer results in quenching. The emission behaviour of TiO_2 (P25) is found by anatase which present as ~80% (TiO_2P25) and ~60% (Hombikat) [96, 97]. The photoelectron transfer studies of BI derivatives with other nanomaterials CuO, Fe_2O_3, and WO_3 were also reported by our research group.

Electron transfer from fluorophore 2-(4-fluorophenyl)-1-phenyl-1H-benzo[d]imidazole (FPPBI) to ZnO nanoparticles is confirmed by logarithmic relationship between the binding constant *Vs* $\Delta E_{LUMO-CB}$ and binding constant *Vs* $\Delta E_{HOMO-CB}$ confirms energy transfer [98]. The electron transfer from FPPBI* to ZnO nanoparticle has been determined by CB energy position (Figure 19). The apparent binding constant was determined by photoinduced electron transfer mechanism [99].

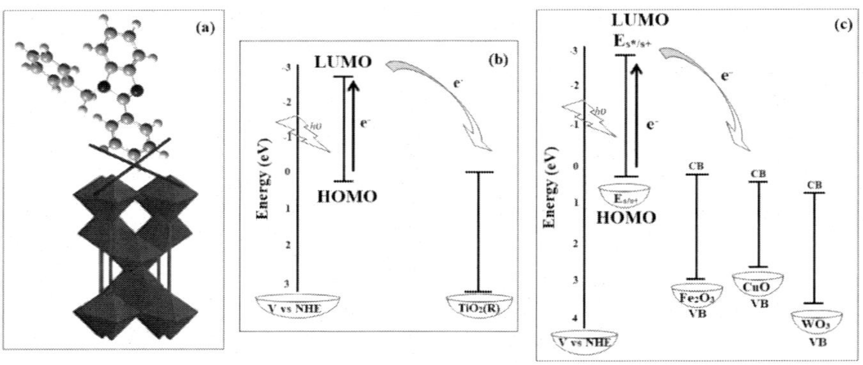

Figure 18. (a) Steric restriction of BI into TiO_2 (A); Schematic diagram describing the electron-donating energy level of BI (b) and (c).

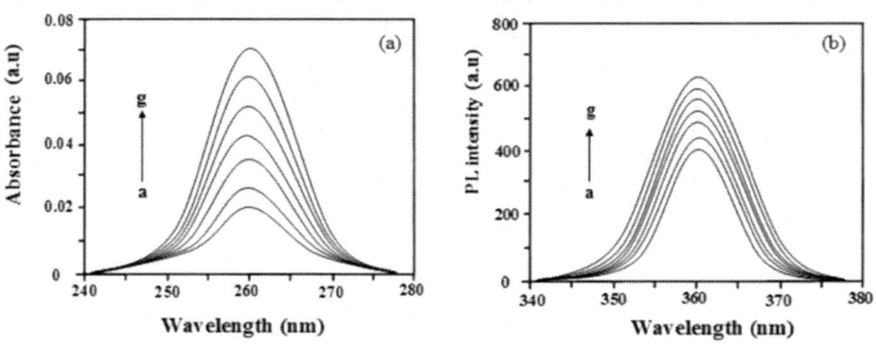

Figure 19. (a)Absorption (bare $1*10^{-8}$ M, (b) to (g) $1*10^{-8}$ M and ZnO $1*10^{-5}$ M to $4.5*10^{-5}$ M) and (b) emission (Bare $1*10^{-8}$ M, (b) to (g) $1*10^{-8}$ M and ZnO $1*10^{-5}$ M to $4.5*10^{-5}$ M) spectra in presence and absence of ZnO.

5.7.3. BI as Biocide

The anticancer screening of some BI derivatives have been carried out against human liver carcinoma cell line (HEPG2) and pheochromocytoma of the rat adrenal medulla cells (PC12). Benzimidazole-2-isoxazole exhibits potency over HEPG2 as well as PC12. BI chalcones, BI mercaptoacetohydrazide and BI thiosemicarbazide (**45 & 46**) derivatives show high activity against PC12 cells (Figure 20) [100].

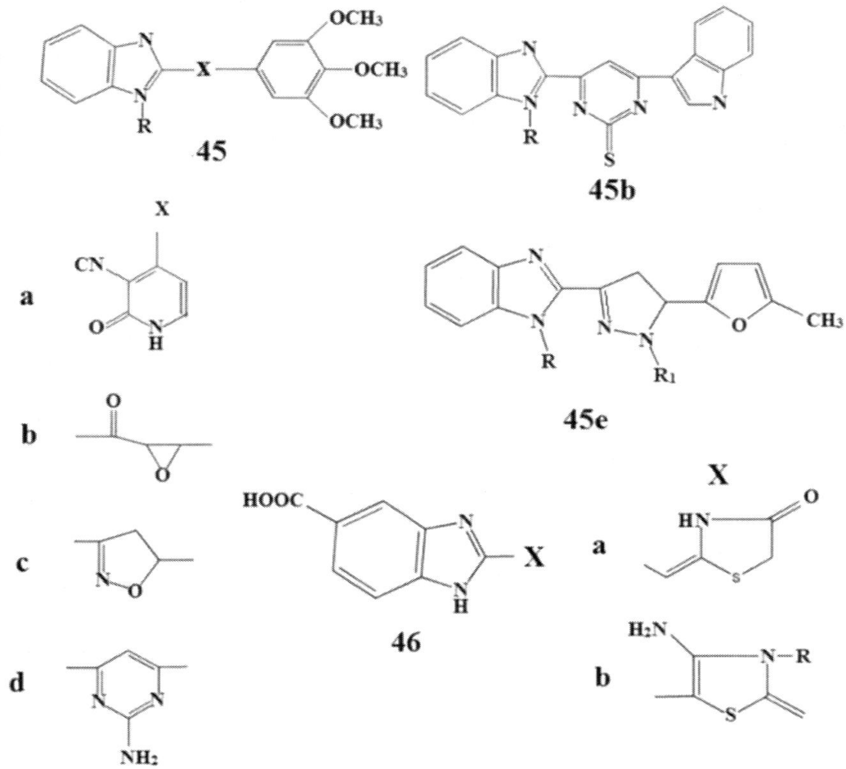

Figure 20. BI based thiosemicarbazide derivatives.

The 2- substituted BI (**47**) with chloro substituent show antitumor activity over MCF7, HEPG2 and HCT 116 [101]. 2-substituted-1-[{(5-substituted-alkyl/aryl)-1,3,4-oxadziazolyl-2-yl}] 3-cyano-2,3-dihydropyro-lo[1,2-a] benzimidazole-1(H)-one exhibit good antibacterial and antifungal activities. The antiviral/anticancer activities are remarkably shown by 7-arylamidoalkyl-3,4-diphenyl-isoquinolinyl-[1,5-c]-benzimidazole and benzimidazole substituted Schiff bases (**48**) (Figure 21) [102].

1H-benzimidazol-2-ylmethyl)-phenylamine derivatives have been tested for their *in-vivo* analgesic and anti-inflammatory activities and show significant analgesic and anti-inflammatory activities [103]. A series of sulphur containing BI derivatives (**49**) are screened for *in vitro* and *in vivo* antitubercular activity against M. tuberculosis H37Rv. These compounds found to be active in mice models infected with M. tuberculosis and also inhibit mycobacterial enzymes *viz.,* pantothenate synthetase, isocitrate lyase and chorismate mutase.

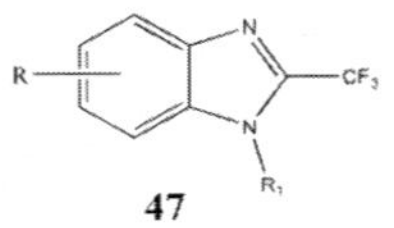

47

R_1 = 5(6)-H, 5(6)-Cl, 5(6)-F, 5(6)-CF_3, 5(6)-CN, 5-CF_3, 6- CF_3

R_2 = H, CH_3

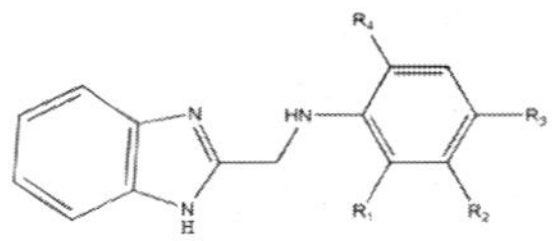

48

Ligand	R_1	R_2	R_3	R_4	Ligand	R_1	R_2	R_3	R_4
a	H	Cl	Cl	H	f	H	H	F	H
b	Cl	Cl	H	H	g	Br	H	Br	Br
c	NO_2	H	H	H	h	Cl	H	NO_2	H
d	H	H	NO_2	H	i	H	H	SO_2NH_2	H
e	H	NO_2	H	H	j	H	H	I	H

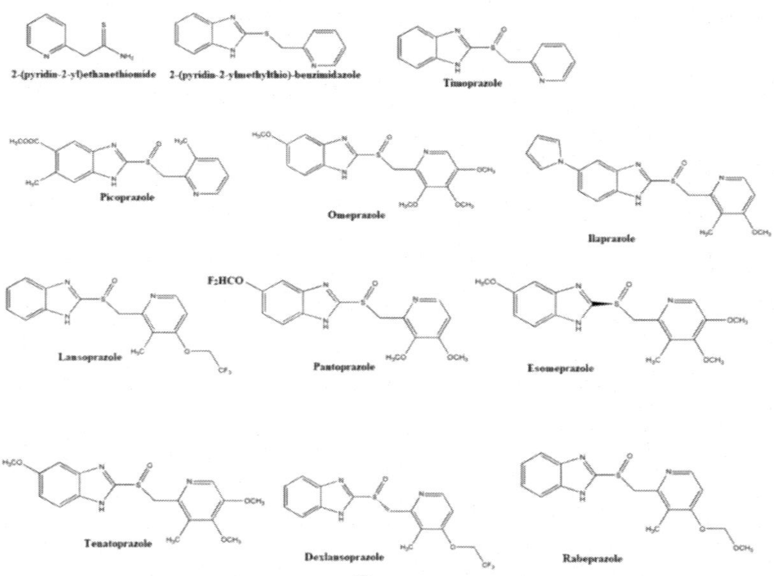

49

Figure 21. BI derivatives as biocide materials.

BI with hydroxylnaphthalene is the potent BI exhibits inhibition of 67.56%, 47.56% and 53.45% over isocitrate lyase, chorismate mutase and pantothenate synthetase, respectively which is comparable with that of reference streptomycin sulphate (Figure 22). BI derivatives show excellent antimicrobial activity relative to standard cefadroxil (antibacterial) and fluconazole (antifungal) drugs. [104]. Among **50a-e**, **50h-j**, **50m**, **50o**, **50a-f**, **51j** and **51l**, methoxy and halo compounds exhibit higher *in vitro* over 60 human-tumor than that of **49a** [105].

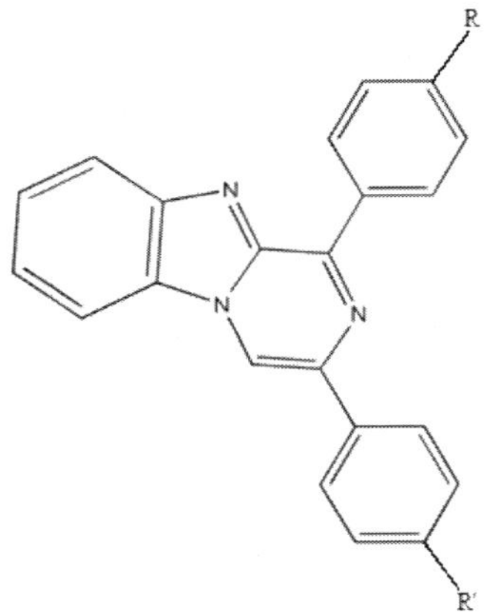

50	R	R'	**51**	R	R'
a	H	H	i	OCH_3	F
b	H	CH_3	j	OCH_3	Cl
c	H	OCH_3	k	Cl	H
d	H	F	l	Cl	CH_3
e	H	Cl	m	Cl	O
f	OCH_3	H	n	Cl	F
g	OCH_3	CH_3	o	Cl	Cl
h	OCH_3	OCH_3			

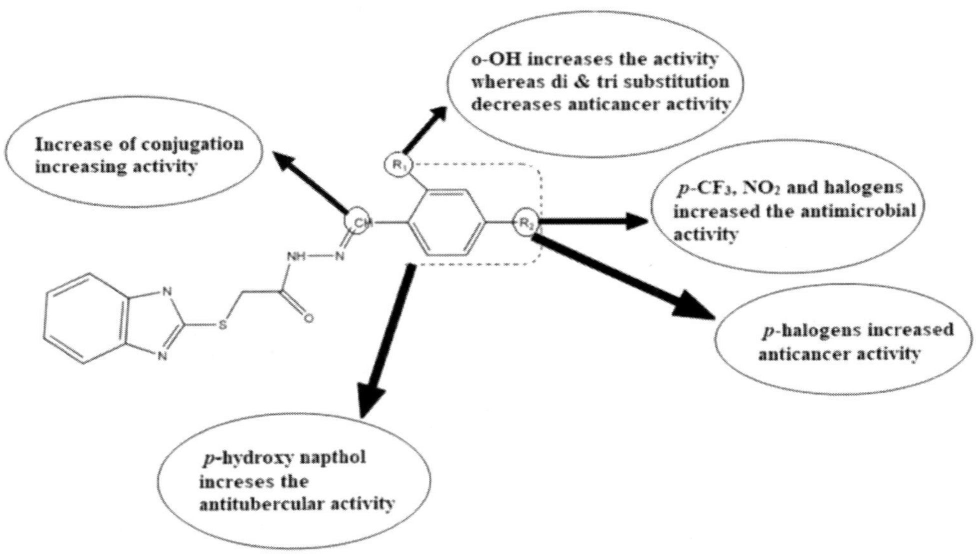

Figure 22. Schematic diagram of structure-activity relationship.

5.7.4. BI as Drug

5,6-dimethyl-1-(α-D-ribofuranosyl)benzimidazole was found in the structure of vitamin B_{12} [106]. The benzimidazole containing drugs (**52**), astemizole, bilastine, clemizole, mizolastine and emedastine play a key role as H1-receptor antagonists. Bilastine is a potent antagonist at H1 receptor for rhinoconjunctivitis and urticaria [107]. Astemizole and mizolastine was used as H1-receptor antagonist and non-sedating antihistaminic drug, respectively. Emedastine and clemizole used to treat allergic conjunctivitis [108], itching and allergic reactions, respectively (Figure 23).

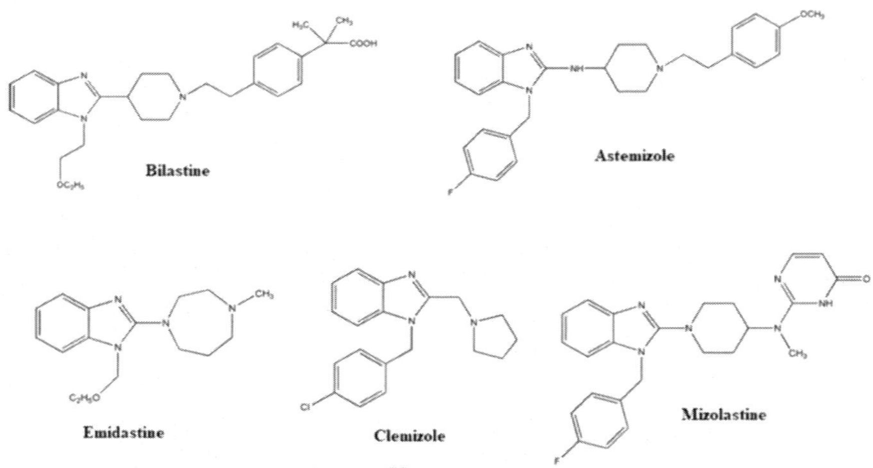

52

Figure 23. BI as drug.

Benzimidazole derivative show pharmacological activities namely, antiparasitic, anticancer, antiulcer, antihistaminic and antihypertensive. Many benzimidazole containing drugs such as norastemizole (antihistaminic), omeprazole (antiulcer) thiabendazole (antiparasitic) and telmisartan (antihypertensive) are used widely to treat the said ailments [109, 110].

Conclusion

The progress of benzimidazole based OLED emitters, host materials for PhOLEDs, RGB emissive materials, ETM and hole blocking materials with thorough attention paid to design, synthesis, crystal structure, performances of fluorescent and phosphorescent OLEDs, pharmaceutical and sensing applications have been reviewed. BI derivatives are potential candidates in OLEDs to enhance the device efficiency. Among all the reported BI based fluorophores, bipolar fluorophores are high potent materials in improving the device performances with deep color coordinates. The bipolar fluorophores broaden the exciton-formation zone as well as lower the roll-off efficiency. BI fluorophores used as electron-transporting (TPBI) building block in fluorescent OLEDs and PhOLEDs. Compared with other emitters, BI fluorophores have some advantages, rigid skeleton, high thermal stability as well as good film-forming ability, ease of fluorophore modification at N1 and C2 positions, high carrier mobilities: leading to promising applications in high performance fluorescent OLEDs and PhOLEDs. Some BI based fluorophores exhibit high efficiency and exceeds the theoretical limited EQE (5%) by translating triplet exciton to singlet state through RISC. Therefore, organic fluorophores with BI functional groups are promising materials for OLEDs with improved device efficiency.

REFERENCES

[1] Ouyang, X. H., Zhang, X. Y. and Ge, Z. Y. (2014). A facile and template-free method for preparation of polythiophene microspheres and their dispersion for waterborne corrosion protection coatings. *Syn. Met.,* 191 (7): 41-46.

[2] Shen, Y., Li, J., Chen, S., Wan, J., Zhao, Y., Lu, Z. H., Li, M., Yang, L., Yu, X., Chen, D. and Pan, Z. (2017). Novel benzimidazole-containing heterocyclic compounds: synthesis, physical properties and OLED application. *Chem. Select*, 2 (34): 11206-11210.

[3] Lee, S. H., Kim, M. S., Cha, Y. B., Hyunahn, K. and Chulkim, Y. (2010). Electroluminescence characteristics of a new green-emitting phenothiazine derivative with biphenyl benzimidazole substituent. *Mol. Cryst. Liq. Cryst.*, 36 (43): 312 -319.

[4] Hung, W. Y., Chi, L. C., Chen, W. J., Chen, Y. M., Choub, S. H. and Wong, K. T. (2010). A new benzimidazole/carbazole hybrid bipolar material for highly efficient deep-blue electrofluorescence, yellow-green electrophosphorescence, and two-color-based white OLEDs. *J. Mater. Chem.*, 20 (45): 10113-10119.

[5] (a) Wang, Z. B., Helander, M. G., Qiu, J., Puzzo, D. P., Greiner, M. T., Liu, Z. W. and Lu, Z. H., (2011). Highly simplified phosphorescent organic light emitting diode with> 20% external quantum efficiency at> 10, 000 cd/m^2. *Appl. Phys. Lett.,* 98 (7): 073310-073313. (b) Lin, S. L., Chan, L. H., Lee, R. H., Yen, M. Y., Kuo, W. J., Chen, C. T. and Jeng, R. J. (2008). Highly efficient carbazole-π-dimesitylborane bipolar fluorophores for nondoped blue organic light-emitting diodes. *Adv. Mater.*, 20 (20): 3947-3952.

[6] Jou, J. H., Weng, J. W., Chavhan, S. D., Kumar Yadav, R. A. and Liang, T.W. (2018). Investigation of charge-transporting layers for high-efficiency organic light-emitting diode. *J. Phys. D: Appl. Phys.,* 51 (45): 454002.

[7] Huang, J. J., Leung, M. K., Chiu, T. L., Chuang, Y. T., Chou, P. T. and Hung, Y. H. (2014). Novel benzimidazole derivatives as electron-transporting type host to achieve highly efficient sky-blue phosphorescent organic light-emitting diode (PhOLED) device. *Org. Lett.*, 16 (20): 5398-5401.

[8] Park, J. H., Kim, E. K., El-Deeb, I. M., Jung,S. J., Choi, D. H., Kim, D. H., Yoo, K. H., Jang Kwon, H. and Lee, S. H. (2011). New bipolar green host materials containing benzimidazole-carbazole moiety in phosphorescent OLEDs. *Bull. Korean Chem. Soc.*, 32, (3): 841-846.[9] Zhao, Y., Wu, C., Qiu, P., Li, X., Wang, Q., Chen, J. and Ma, D. (2016). New benzimidazole-based bipolar hosts: highly efficient phosphorescent and thermally activated delayed fluorescent organic light-emitting diodes employing the same device structure. *ACS Appl. Mater. Interfaces*, 8 (4): 2635-2643.

[10] Zhao, J. H., Hu, Y. X., Lu, H. Y., Lü, Y. L. and Li, X. (2016). Progress on benzimidazole based iridium(III) complexes for application in phosphorescent OLEDs, *Org. Electron.,* 41 (8): 56-72.

[11] Ge, Z., Hayakawa, T., Ando, S., Ueda, M., Akiike, T., Miyamoto, H., Kajita, T. and Kakimoto, M. (2008). Solution-processible bipolar triphenylamine-benzimidazole derivatives for highly efficient single-layer organic light-emitting diodes. *Chem. Mater.* 20 (7): 2532-2537.

[12] Gao, Y., Xu, W., Ma, H., Obolda, A., Yan, W., Dong, S., Zhang, M. and Li, F. (2017). Novel luminescent benzimidazole-substituent tris (2, 4, 6-trichlorophenyl) methyl radicals: photophysics, stability, and highly efficient red-orange electroluminescence. *Chem. Mater.*, 29 (16): 6733-6739.

[13] (a) Jayabharathi, J., Thanikachalam, V. and Jayamoorthy, K. (2014). Synthesis, photophysical and electroluminescent properties of green organic light emitting devices based on novel iridium complexes containing benzimidazole ligands. *J. Organomet. Chem.*, 761 (3): 74-83. (b) Jayabharathi, J., Thanikachalam, V. and Jayamoorthy, K. (2013).Synthesis of some fluorescent benzimidazole derivatives using cobalt (ii) hydroxide as highly efficient catalyst-spectral and physico-chemical studies and ESIPT process. *Photochem. Photobiol. Sci.*, 12 (10): 1761-1773. (c) J. Jayabharathi, J., Thanikachalam, V. and Jayamoorthy, K. (2013). Fluorescence quenching of bovine serum albumin by NNMB. *Spectrochim. ActaPart A.*, 108 (19): 146-150. (d) Jayabharathi, J., Thanikachalam, V., Jayamoorthy, K. and Srinivasan, N. (2013). Synthesis, spectral studies and solvatochromism of some novel benzimidazole derivatives–ESIPT process. *Spectrochim. Acta Part A.*, 105 (32): 223-228. (e) Jayabharathi, J., Thanikachalam, V. and Jayamoorthy, K. (2012). Physicochemical studies of chemosensor imidazole derivatives: DFT based ESIPT process. *Spectrochim. Acta Part A.*, 89 (301): 168-176. (f) Jayabharathi, J., Thanikachalam, V. and Jayamoorthy, K. (2012). Computational studies of 1, 2-disubstituted benzimidazole derivatives. *Spectrochim. Acta Part A.,* 97 (20): 131-136. (g) Jayabharathi, J., Thanikachalam, V. and Jayamoorthy, K. (2012). Antioxidant benzimidazole and bovine serum albumin. *J. Photochem. Photobiol. B*, 115 (12): 85-92. (h) Jayabharathi, J., Thanikachalam, V. and Jayamoorthy, K. (2012). Docking investigation and binding interaction of benzimidazole derivative with bovine serum albumin. *J. Photochem.Photobiol. B*, 117 (4): 27-32. (i) Jayabharathi, J., Karunakaran, C., Jayamoorthy, K. and Vinayagamoorthy, P.(2014). Benzimidazole based Ir (III) picolinate complexes as emitting materials and the fluorescent behavior of benzimidazole bound to Mn–TiO_2@ ZnO core/shell nanospheres. *Mater. Exp.,* 4 (4): 279-292.

[14] Tang, C. W. and VanSlyke, S. A. (1987). Organic electroluminescent diodes. *Appl. Phys. Lett.*, 51 (12): 913-915.

[15] (a) Jayabharathi, J., Nethaji, P., Thanikachalam, V. and Ramya, R. (2019). Interfacial charge-transfer in Cu-TiO_2-HBDPPIN-Ag film and AIEE-active chemosensor. *J. Photochem.Photobiol. A.,* 377 (35): 318-338. (b) Jayabharathi, J., Nethaji, P. and Thanikachalam, V. (2019). Promotional effect of silver nanoparticle embedded Ga-Zr-codoped TiO_2 as an alternative anode for efficient blue, green and red PhOLEDs. *RSC Adv.,* 9, (24): 13664-13676. (c) Jayabharathi, J., Nethaji, P., Thanikachalam, V. and Ramya, R. (2019). Derivatives of cyanonaphthyl-substituted phenanthroimidazole as blue emitters for nondoped organic light-emitting diodes. *ACS Omega.,* 4 (3), 4553-4570. (d) Jayabharathi, J., Thanikachalam, V. and Abirama Sundari, G. (2019). Efficient electroluminescent hybridized local and charge-transfer host materials with small singlet-triplet splitting to enhance exciton utilization efficiency: excited state transition configuration *RSC Adv.,* 9 (12): 6658-6680. (e) Jayabharathi, J., Ramya, R., Thanikachalam, V., Jeeva, P. and Sarojpurani, E. Efficient full-colour organic light-emitting diodes based on donor-acceptor electroluminescent materials with a reduced singlet–triplet splitting energy gap. (2019). *RSC Adv.,* 9 (6): 2948-2966. (f) Jayabharathi, J., Panimozhi, S. and Thanikachalam, V. (2018). A hybrid inorganic–organic light-emitting diode using Ti-doped ZrO_2 as an electron-injection layer *RSC Adv.,* 8 (15): 37324–37338. (g) Jayabharathi, J., Ramya, R., Thanikachalam, V. and Nethaji, *P.* (2018). Tailoring the molecular design of twisted dihydrobenzodioxin phenanthroi-midazole derivatives for non-doped blue organic light-emitting devices. *RSC Adv.,* 8(51): 29031-29043. (h) Jayabharathi, J., Abirami Sundari, G., Thanikachalam, V. and Panimozhi, S. (2018). Enhanced internal quantum efficiency of organic light-emitting diodes: a synergistic effect. *J.Photochem Photobiol. A,* 364, (38): 577-587. (i) Thanikachalam, V., Sarojpurani, E., Jayabharathi, J. and Jeeva, P. (2017). Efficient phenanthroimidazole-styryl-triphenyl-amine derivatives for blue OLEDs: a combined experimental and theoretical study. *New J. Chem.,* 41 (6): 2443-2457. (j) Jayabharathi, J., Sujatha, P., Thanikachalam, V., Jeeva, P. and Panimozhi, S. (2017). Enhancement of electroluminescent green emission by far-field coupling of Au nanoparticles in organic light emitting diodes. *Ind. Eng. Chem. Res.,* 56 (24): 6952-6961. (k) Thanikachalam, V., Jeeva, P. and Jayabharathi, J. (2017). Hybridised-local and charge-transfer excited states in donor-spacer-acceptor molecules for efficient OLEDs: combined experimental and theoretical study. *Chem. Select.,* 2 (5): 1860-1870. (l) Jayabharathi, J., Prabhakaran, A., Thanikachalam V. and Jeeva P. (2016). Efficient non-doped blue emitting devices based on bis (phenanthrimidazolyl) biphenyl derivatives. *New J. Chem.,* 40 (10): 8768-8777. (m) Thanikachalam, V., Jeeva, P., Jayabharathi, J., Ramanathan, P., Prabhakaran, A. and SarojPurani, E. (2016). Nondoped blue organic light emitting devices with donor-π-acceptor derivatives as the emissive material. *Ind. Eng. Chem. Res.,* 55 (36): 9639-9647. (n) Jayabharathi, J., Prabhakaran, A., Thanikachalam, V., Jeeva, P. and Sundharesan, M.

(2016). Air stable HyLEDs using efficient electron injection and emitting materials. *Ind. Eng. Chem. Res.,* 55 (29): 8087-8095. (o) Jayabharathi, J., Prabhakaran, A., Thanikachalam, V. and Sundharesan, M. (2016). Improved efficiency of hybrid inorganic-organic light emitting diodes. *Chem. Select.,* 1 (11): 2642-2651.

[17] Bauer, R., Finkenzeller, W. J., Bogner, U., Thompson, M. E. and Yersin, H., (2008). Matrix influence on the OLED emitter $Ir(btp)_2(acac)$ in polymeric host materials-Studies by persistent spectral hole burning. *Org. Electron.,* 9 (5): 641-648.

[18] Tokito, S., Iijima, T., Suzuri, Y., Kita H., Tsuzuki, T. and Sato, F. (2003). Confinement of triplet energy on phosphorescent molecules for highly-efficient organic blue-light-emitting devices. *Appl. Phys.Lett.,* 83 (3): 569-571.

[19] (a) Adamovich, V., Brooks, J., Tamayo, A., Alexander, A. M., Djurovich, P. I., D'Andrade, B. W., Adachi, C., Forrest, S. R. and Thompson, M. E. (2002). High efficiency single dopant white electrophosphorescent light emitting diodes. *New J. Chem.,* 26 (9): 1171-1178. (b) Holmes, R. J., Forrest, S. R., Tung, Y. J., Kwong, R. C., Brown, J. J., Garon, S. and Thompson, M. E. (2003). Blue organic electrophosphorescence using exothermic host-guest energy transfer *Appl. Phys. Lett.,* 82 (15): 2422-2424.

[20] Chou, H. H., Shih, H. H. and Cheng, C. H. (2010). Triptycene derivatives as high-T_g host materials for various electrophosphorescent devices. *J. Mater. Chem.,* 20 (4): 798-805.

[20] Hu, D., Lu, P., Wang, C., Liu, H., Wang, H., Wang, Z., Fei, T., Gu, X. and Ma, Y. (2009). Silane coupling di-carbazoles with high triplet energy as host materials for highly efficient blue phosphorescent devices. *J. Mater. Chem.,* 19 (34): 6143-6148.

[21] (a) Gao, Y., Xu, W., Ma, H., Obolda, A., Yan, W., Dong, S., Zhang, M. and Li, F., 2017. Novel luminescent benzimidazole-substituent tris (2, 4, 6-trichlorophenyl) methyl radicals: photophysics, stability, and highly efficient red-orange electroluminescence. *Chem. Mater.,* 29 (16): 6733-6739. (b) Gao, Y. H., Chen, C., Tang, Q., Su, B., Zhang, G., Bo, B. X. and Jiang, W. L., 2017. Comparison study of two isomers of benzimidazole for effective blue OLEDs. *J. Mater. Sci.:Mater. Electron.,* 28 (10): 7204-7211.

[22] Adamovich, V. I., Cordero, S. R., Djurovich, P. I., Tamayo, A., Thompson, M. E., D'Andrade, B. W. and Forrest S. R. (2003). New charge-carrier blocking materials for high efficiency OLEDs. *Org. Electron.,* 4 (2-3): 77-87.

[23] Chopra, N., Lee, J., Zheng, Y., Eom, S. H., Xue, J. and So, F. (2008). High efficiency blue phosphorescent organic light-emitting device. *Appl. Phys.Lett.,* 93 (14): 374.

[24] Tsai, M. H., Hong, Y. H., Chang, C. H., Su, H. C., Wu, C. C., Matoliukstyte, A., Simokaitiene, J., Grigalevicius, S., Grazulevicius, J. V. and Hsu, C. P. (2007). 3-(9-carbazolyl) carbazoles and 3, 6-di (9-carbazolyl) carbazoles as effective host materials for efficient blue organic electrophosphorescence. *Adv. Mater.,* 19 (6): 862-866.

[25] Matsusue, N., Suzuki, Y. and Naito H. (2005). Charge carrier transport in neat thin films of phosphorescent iridium complexes. *Jpn. J. Appl. Phys.,* 44 (6R): 3691.

[26] (a) Hughes, G. and Bryce, M. R. (2005). Electron-transporting materials for organic electroluminescent and electrophosphorescent devices. *J. Mater. Chem.,* 15 (1): 94-107. (b) Chen, C. T., Wei, Y., Lin, J. S., Moturu, M. V. R. K., Chao, W. S., Tao, Y. T. and Chien, C. H. (2006). Doubly ortho-linked quinoxaline/diphenylfluorene hybrids as bipolar, fluorescent chameleons for optoelectronic applications. *J.Am. Chem. Soc.,* 128 (34): 10992-10993. (c) Lundin, N. J., Blackman, A. G., Gordon, K. C. and Officer D. L. (2006). Synthesis and characterization of a multicomponent rhenium (I) complex for application as an OLED dopant. *Angew. Chem., Int. Ed.*, 45 (16): 2582-2584.

[27] (a) Cai, X., Padmaperuma, A. B., Sapochak, L. S., Vecchi, P. A. and Burrows, P. E. (2008). Electron and hole transport in a wide bandgap organic phosphine oxide for blue electrophosphorescence. *Appl. Phys. Lett.,* 92 (8): 66. (b) Tao, Y., Wang, Q., Yang, C., Wang, Q., Zhang, Z., Zou, T., Qin, J. and Ma, D., (2008). A simple carbazole/oxadiazole hybrid molecule: an excellent bipolar host for green and red phosphorescent OLEDs. *Angew. Chem., Int. Ed.,* 47 (42): 8104-8107.

[28] (a) Chen, H. F., Yang, S. J., Tsai, Z. H., Hung, W. Y., Wang, T. C. and Wong, K. T. (2009). 1, 3, 5-Triazine derivatives as new electron transport–type host materials for highly efficient green phosphorescent OLEDs. *J. Mater. Chem.,* 19 (43): 8112-8118. (b) Ge, Z., Hayakawa, T., Ando, S., Ueda, M., Akiike, T., Miyamoto, H., Kajita, T. and Kakimoto, M. (2008). Novel bipolar bathophenanthroline containing hosts for highly efficient phosphorescent OLEDs. *Org. Lett.,* 10 (3): 421-424. (c) Li, Y., Fung, M. K., Xie, Z., Lee, S. T., Hung, L. S. and Shi, J. (2002). An efficient pure blue organic light-emitting device with low driving voltages. *Adv. Mater.,* 14 (18): 1317-1321.

[29] Chen, C. H., Huang, W. S., Lai, M. Y., Tsao, W. C, Lin, J. T., Wu, Y. H., Ke, T. H., Chen, L. Y. and Wu, C. C. (2009). Benzimidazole/amine-based ambipolar compounds for electroluminescent applications: single-layer, blue, fluorescent OLEDs, hosts for single-layer, phosphorescent OLEDs. *Adv. Funct. Mater.,* 19 (16): 2661-2670.

[30] Gong, S., Zhao, Y., Yang, C., Zhong, C., Qin, J. and Ma, D. (2010). Tuning the photophysical properties and energy levels by linking spacer and topology between the benzimidazole and carbazole units: bipolar host for highly efficient phosphorescent OLEDs. *J. Phys. Chem. C,* 114 (11): 5193-5198.

[31] Takizawa, S., Montes, V. A. and Anzenbacher, P. (2009). Phenylbenzimidazole-based new bipolar host materials for efficient phosphorescent organic light-emitting diodes. *Jr, Chem. Mater,* 21 (12): 2452-2458.

[32] Tsai, M. H., Lin, H. W., Su, H. C., Ke, T. H., Wu, C., Fang, F. C., Liao, Y. L., Wong, K. T. and Wu, C. I. (2006). Highly efficient organic blue electrophosphorescent

devices based on 3, 6-bis (triphenylsilyl) carbazole as the host material. *Adv. Mater.*, 18 (9): 1216-1220.

[33] Rosepriya, S., Thiruvalluvar, A., Jayamoorthy, K., Jayabharathi, J. and Linden, A. (2011). 1-(4-Methylbenzyl)-2-(4-methylphenyl)-1H-benzimidazole, *Acta Cryst. E*, 67 (12): o3519-o3519.

[34] Thiruvalluvar, A. Rosepriya, S. Jayamoorthy, K. Jayabharathi, J. Yildirim, S. O. Butcher, R. J. 2013. 2-(1-Phenyl-1H-benzimidazol-2-yl)phenol, *Acta Cryst. E*, 69 (1): o62-o62.

[35] Jayamoorthy, K. Rosepriya, S. Thiruvalluvar, A. Jayabharathi, J. Butcher, R. J. 2012. 2-(4-Fluorophenyl)-1-phenyl-1H-benzimidazole, *Acta Cryst. E*, 68 (9): o2708-o2708.

[36] Mohandas, T. Jayamoorthy, K. Sakthivel, P. Jayabharathi, J. 2013. 1-Phenyl-2-p-tolyl-1H-benzimidazole, *Acta Cryst. E*, 69 (3): o334-o334.

[37] Mohandas, T. Jayamoorthy, K. Jayabharathi, J. Sakthivel, P. 2013. 2-(4-Methoxyphenyl)-1-phenyl-1H benzimidazole, *Acta Cryst. E*, 69 (2): o269-o270.

[38] Jayamoorthy, K. Mohandas, T. Sakthivel, P. Jayabharathi, J. 2013. 1-Phenyl-2-[4-(trifluoromethyl)phenyl]-1H-benzimidazole, *Acta Cryst. E*, 69 (2): o244-o244.

[39] Nonoyama, M. 1974. Benzo [h] quinolin-10-yl-N Iridium (III) Complexes. *Bull. Chem. Soc. Jpn.,* 47 (3): 767-768.

[40] Liu, Q. D., Jia, W. L. and Wang, S. (2005). Blue luminescent 2-(2-Pyridyl)benzimidazole derivative ligands and their orange luminescent mononuclear and polynuclear organoplatinum(II) complexes. *Inorg. Chem.,* 44 (5):1332-1343.

[41] Chen, L., Yang, C., Qin, J., Gao, J., You, H. and Ma, D. (2006). Synthesis, structure, electrochemistry, photophysics and electroluminescence of 1, 3, 4-oxadiazole-based ortho-metalated iridium (III) complexes. *J. Organomet. Chem.*, 691 (16): 3519-3530.

[42] Huang, W. S., Lin, J. T., Chien, C. H., Tao, Y. T., Sun, S. S. and Wen, Y. S. (2004). Highly phosphorescent bis-cyclometalated iridium complexes containing benzoimidazole-based ligands. *Chem. Mater.,* 16 (12):2480-2488.

[43] Lai, M. Y., Chen, C. H., Huang, W. S., Lin, J. T., Ke, T. H., Chen, L. Y., Tsai, M. H. and Wu, C. C. (2008). Benzimidazole/amine-based compounds capable of ambipolar transport for application in single-layer blue-emitting OLEDs and as hosts for phosphorescent emitters. *Angew. Chem. Int. Ed.,* 47 (3): 581–585.

[44] Gong, S., Zhao, Y., Wang, M., Yang, C., Zhong, C., Qin, J. and Ma, D. (2010). Versatile benzimidazole/triphenylamine hybrids: efficient nondoped deep-blue electroluminescence and good host materials for phosphorescent emitters. *Chem. Asian J.,* 5 (9): 2093-2099.

[45] Huang, J. J., Hung, Y. H., Ting, P. L., Tsai, Y. N., Gao, H. J., Chiu, T. L., Lee, J. H., Chen, C. L., Chou, P. T. and Leung, M. K. (2016). Orthogonally substituted benzimidazole-carbazole benzene as universal hosts for phosphorescent organic light-emitting diodes. *Org. lett.,* 18 (4): 672-675.

[46] Hayakawa, Ge, Z., Ando, T., Ueda, S., Akiike, M., Miyamoto, T., Kajita, H. T. and Kakimoto, M. A. 2008. Spin-coated highly efficient phosphorescent organic light-emitting diodes based on bipolar triphenylamine-benzimidazole derivatives. *Adv. Funct. Mater.,* 18 (4): 584-590.

[47] Yang, X., Zhou, G. and Wong, W. Y., (2015). Functionalization of phosphorescent emitters and their host materials by main-group elements for phosphorescent organic light-emitting devices. *Chem. Soc. Rev.,* 44 (23): 8484-8575.

[48] Zhang, R., Sun, H., Zhao, Y., Tang, X. and Ni, Z. (2018). Dipolar 1,3,6,8-tetrasubstituted pyrene-based blue emitters containing electro-transporting benzimidazole moieties: syntheses, structures, optical properties, electrochemistry and electroluminescence. *Dyes and Pigments,* 152 (1): 1-13.

[49] Wang, B., Lv, X., Pan, B., Tan, J., Jin, J. and Wang, L. (2015). Benzimidazole–phosphine oxide hybrid electron transporters for unilateral homogeneous phosphorescent organic light-emitting diodes with enhanced power efficiency. *J. Mater. Chem. C,* 3 (42): 11192-11201.

[50] Yoo, S. J., Yun,H. J., Kang, I., Thangaraju, K., Kwon, S. K. and Kim, Y. H. (2013). A new electron transporting material for effective hole-blocking and improved charge balance in highly efficient phosphorescent organic light emitting diodes, *J. Mater. Chem. C,* 1 (11): 2217-2223.

[51] Wang, F., Hu, J., Cao, X., Yang, T., Tao, Y., Mei, L., Zhang, X. and Huang, W. (2015). A low-cost phenylbenzoimidazole containing electron transport material for efficient green phosphorescent and thermally activated delayed fluorescent OLEDs. *J. Mater. Chem. C,* 3 (21):5533-5540.

[52] Yi, S., Kim, J. H., Bae, W. R., Lee, J., Han, W. S., Son, H. J. and Kang, S. O. (2015) Silicon-based electron-transport materials with high thermal stability and triplet energy for efficient phosphorescent OLEDs, *Org. Electron.,* 27 (20): 126-132.

[53] Zhu, Y. C., Zhou, L., Li, H. Y., Xu, Q. L., Teng, M. Y., Zheng, Y. X., Zuo, J. L., Zhang, H. J. and You, X. Z. 2011. Highly efficient green and blue-green phosphorescent OLEDs based on iridium complexes with the tetraphenylimidodiphosphinate ligand. *Adv. Mater.,* 23 (35): 4041-4046.

[54] Liu, M., Li, X. L., Chen, D. C., Xie, Z., Cai, X., Xie, G., Liu, K., Tang, J., Su, S. J. and Cao, Y. 2015. Study of configuration differentia and highly efficient, deep-blue, organic light-emitting diodes based on novel naphtho [1, 2-d] imidazole derivatives. *Adv. Funct. Mater.,* 25 (32): 5190–5198.

[55] Ban, X., Jiang, W., Sun, K., Xie, X., Peng, L., Dong, H., Sun, Y., Huang, B., Duan, L. and Qiu, Y. (2015). Bipolar host with multielectron transport benzimidazole units for low operating voltage and high power efficiency solution-processed phosphorescent OLEDs. *ACS Appl. Mater. Interfaces.,* 7 (13): 7303-7314.

[56] Duan, L., Qiao, J., Sun, Y. and Qiu, Y., (2011). Strategies to design bipolar small molecules for OLEDs: donor-acceptor structure and non-donor-acceptor structure. *Adv. Mater.,* 23 (9): 1137–1144.

[57] Karthik, D., Thomas, K. J., Jou, J. H., Kumar, S., Chen, Y. L. and Jou, Y. C., (2015). Deep-blue emitting pyrene-benzimidazole conjugates for solution processed organic light-emitting diodes. *RSC Adv.,* 5 (12): 8727-8738.

[58] Thomas, K. J. and Bodedla, G. B. (2018). T-shaped benzimidazole derivatives as blue-emitting materials: the role of C2 substituents on photophysical properties. *Asian J. Org. Chem.,* 7 (4): 729-738.

[59] Pina, J., Seixas de Melo, J. S., Batista, R. M., Costa, S. P. and Raposo, M. M. M. 2013. Triphenylamine-benzimidazole derivatives: synthesis, excited-state characterization, and DFT studies. *J. Org. Chem.,* 78 (22): 11389−11395.

[60] Jeong, S., Kim, M. K., Kim, S. H. and Hong, J. I. (2013). Efficient deep-blue emitters based on triphenylamine-linked benzimidazole derivatives for nondoped fluorescent organic light-emitting diodes. *Org. Electron.,* 14 (10): 2497–2504.

[61] Yang, X., Zheng, S., Bottger, R., Chae, H. S., Tanaka, T., Li, S. and Jabbour, G. E. (2011). Efficient fluorescent deep-blue and hybrid white emitting devices based on carbazole/benzimidazole compound. *J. Phys. Chem. C,* 115 (29): 14347–14352.

[62] Lai, M. Y., Chen, C. H., Huang, W. S., Lin, J. T., Ke, T. H., Chen, L. Y., and Wu, C. C. (2008). Benzimidazole/amine-based compounds capable of ambipolar transport for application in single-layer blue-emitting OLEDs and as hosts for phosphorescent emitters, *Angew. Chem. Int. Ed.,* 47 (3): 581–585.

[63] Lai, M. Y., Chen, C. H., Huang, W. S., Lin, J. T., Ke, T. H., Chen, L. Y., Tsai, M. H. and Wu, C. C., (2008). Benzimidazole/amine-based compounds capable of ambipolar transport for application in single-layer blue-emitting OLEDs and as hosts for phosphorescent emitters. *Angewandte Chemie International Edition,* 47 (3): 581-585.

[64] Yang, X., Zheng, S., Bottger, R., Chae, H. S., Tanaka, T., Li, S., Mochizuki, A. and Jabbour, G. E. (2011). White emitting devices efficient fluorescent deep-blue and hybrid white emitting devices based on carbazole/benzimidazole compound. *J. Phys. Chem. C,* 115 (29): 14347-14352.

[65] Lin, G., Peng, H., Chen, L., Nie, H., Luo, W., Li, Y., Chen, S., Hu, R., Qin, A., Zhao, Z. and Tang, B. Z., (2016). Improving electron mobility of tetraphenylethene-based AIEgens to fabricate nondoped organic light-emitting diodes with remarkably high luminance and efficiency. *ACS Appl. Mater. Interfaces,* 8 (26): 16799-16808.

[66] Nie, H., Chen, B., Zeng, J., Xiong, Y., Zhao, Z. and Tang, B. Z., (2018). Excellent n-type light emitters based on AIE-active silole derivatives for efficient simplified organic light-emitting diodes. *J. Mater. Chem. C,* 6 (14): 3690-3698.

[67] Huang, J., Su, J. H., Li, X., Lam, M. K., Fung, K. M., Fan, H. H., Cheah, K. W., Chen, C. H. and Tian, H., (2011). Bipolar anthracene derivatives containing hole-

and electron-transporting moieties for highly efficient blue electroluminescence devices. *J. Mater. Chem.*, 21 (9): 2957-2964.

[68] Zhang, R., Sun, H., Zhao, Y., Tang, X. and Ni, Z., (2018). Dipolar 1, 3, 6, 8-tetrasubstituted pyrene-based blue emitters containing electro-transporting benzimidazole moieties: syntheses, structures, optical properties, electrochemistry and electroluminescence. *Dyes and Pigments,* 152 (1): 1-13.

[69] Lee, S. H., Kim, M. S., Cha, Y. B., Ahn, K. H. and Kim, Y. C., (2010). Electroluminescence characteristics of a new green-emitting phenothiazine derivative with biphenyl benzimidazole substituent. *Mol. Cryst. Liq. Cryst.,* 520 (1): 36-312.

[70] Wang, L., Jin, J., Zhang, W., Pan, B., Huang, Z. and Zhuang, S., (2015). Toward high efficiency green phosphorescent organic light-emitting diodes by fine tuning the charge transporting properties of 1, 2, 4-thiadiazole based hosts. *Org. Electron.,* 16 (24): 177-185.

[71] Horak, E., Robić, M., Šimanović, A., Mandić, V., Vianello, R., Hranjec, M. and Steinberg, I.M., (2019). Tuneable solid-state emitters based on benzimidazole derivatives: Aggregation induced red emission and mechanochromism of D-π-A fluorophores. *Dyes and Pigments,* 162, (77): 688-696.

[72] Gong, S. L., Chen, Y. H., Yang, C. L., Zhong, C., Qin, J. G. and Ma, D. G. (2010). De novo design of silicon-bridged molecule towards a bipolar host: all-phosphor white organic light-emitting devices exhibiting high efficiency and low efficiency roll-off. *Adv.Mater.,* 22 (47):5370–5373.

[73] Jang, H. G., Kim, B. S., Lee, J. Y. and Hwang, S. H. (2014). Synthesis of dimesitylborane-substituted phenylcarbazoles as bipolar host materials and the variation of the green PHOLED performance with the substituent position of the boron atom. *Dalton Trans.,* 43 (21): 7712–7715.

[74] Shi, H. P., Xin, D. H., Dong, X. Q., Dai, J. X., Wu, X. H., Miao, Y. Q., Fang, L., Wang, H. and Choi, M. M. F. J. (2014). A star-shaped bipolar host material based on carbazole and dimesitylboron moieties for fabrication of highly efficient red, green and blue electrophosphorescent devices. *J.Mater. Chem. C,* 2 (12):2160–2168.

[75] Huang, H., Yang, X., Pan, B., Wang, L., Chen, J. S., Ma, D. H. and Yang, C. L. (2012). Benzimidazole–carbazole-based bipolar hosts for high efficiency blue and white electrophosphorescence applications. *J. Mater. Chem.*, 22 (26):13223–13230.

[76] Huang, H., Wang, Y., Zhuang, S., Yang, X., Wang, L. and Yang, C., (2012). Simple phenanthroimidazole/carbazole hybrid bipolar host materials for highly efficient green and yellow phosphorescent organic light-emitting diodes. *J. Phys. Chem. C,* 116 (36): 19458-19466.

[77] Huang, H., Wang, Y., Wang, B., Zhuang, S., Pan, B., Yang, X., Wang, L. and Yang, C., (2013). Controllably tunable phenanthroimidazole-carbazole hybrid bipolar host

materials for efficient green electrophosphorescent devices. *J. Mater. Chem. C,* 1 (37): 5899-5908.

[78] Mondal, E., Hung, W. Y., Chen, Y. H., Cheng, M. H. and Wong, K. T., (2013). Molecular topology tuning of bipolar host materials composed of fluorine-bridged benzimidazole and carbazole for highly efficient electrophosphorescence. *Chem. Eur. J.,* 19 (32): 10563-10572.

[79] Cheng, S.H., Hung, W.Y., Cheng, M.H., Chen, H.F., Chaskar, A., Lee, G.H., Chou, S.H. and Wong, K.T., (2014). Highly twisted biphenyl-linked carbazole-benzimidazole hybrid bipolar host materials for efficient PhOLEDs. *J. Mater. Chem. C,* 2 (40): 8554-8563.

[80] Huang, H., Wang, Y., Pan, B., Yang, X., Wang, L., Chen, J., Ma, D. and Yang, C., (2013). Simple bipolar hosts with high glass transition temperatures based on 1, 8-disubstituted carbazole for efficient blue and green electrophosphorescent devices with "Ideal" Turn-on voltage. *Chem. Eur. J.,* 19 (5): 1828-1834.

[81] Ting, H. C., Chen, Y. M., You, H. W., Hung, W. Y., Lin, S. H., Chaskar, A., Chou, S. H., Chi, Y., Liu, R. H. and Wong, K. T., (2012). Indolo [3, 2-b] carbazole/benzimidazole hybrid bipolar host materials for highly efficient red, yellow, and green phosphorescent organic light emitting diodes. *J. Mater. Chem.,* 22 (17): 8399-8407.

[82] Fleetham, T., Wang, Z. and Li, J., (2012). Efficient deep blue electrophosphorescent devices based on platinum (II) bis (n-methyl-imidazolyl) benzene chloride. *Org. Electron.,* 13 (8): 1430-1435.

[83] Tam, A. Y. Y., Tsang, D. P. K., Chan, M. Y., Zhu, N. and Yam, V. W. W., (2011). A luminescent cyclometalated platinum (II) complex and its green organic light emitting device with high device performance. *Chem. Commun.,* 47 (12): 3383-3385.

[84] Lam, E. S. H., Tsang, D. P. K., Lam, W.H., Tam, A. Y. Y., Chan, M. Y., Wong, W. T. and Yam, V. W. W., (2013). Luminescent platinum (II) complexes of 1, 3-bis (n-alkylbenzimidazol-2′-yl) benzene-type ligands with potential applications in efficient organic light-emitting diodes. *Chem. Eur. J.,* 19 (20): 6385-6397.

[85] Ecton, J., Fleetham, T., Hang, X. and Li, J., (2013), September. Paper No 5.1: Highly efficient blue-green OLEDs from tetradentate cyclometalated platinum complexes. *SID,* 44 (S1): 152-155.

[86] Fleetham, T., Huang, L. and Li, J., (2014). Tetradentate platinum complexes for efficient and stable excimer-based white OLEDs. *Adv. Funct. Mater.,* 24 (38): 6066-6073.

[87] Zhuang, J., Li, W., Su, W., Liu, Y., Shen, Q., Liao, L. and Zhou, M., (2013). Highly efficient phosphorescent organic light-emitting diodes using a homoleptic iridium (III) complex as a sky-blue dopant. *Org. Electron.,* 14 (10): 2596-2601.

[88] Udagawa, K., Sasabe, H., Cai, C. and Kido, J., (2014). Low-driving-voltage blue phosphorescent organic light-emitting devices with external quantum efficiency of 30%. *Adv. Mater.,* 26 (29): 5062-5066.

[89] Ding, J., Gao, J., Cheng, Y., Xie, Z., Wang, L., Ma, D., Jing, X. and Wang, F., (2006). Highly efficient green-emitting phosphorescent iridium dendrimers based on carbazole dendrons. *Adv. Funct. Mater.,* 16 (4): 575-581.

[90] Chen, L., Ding, J., Cheng, Y., Xie, Z., Wang, L., Jing, X. and Wang, F., (2011). Bipolar heteroleptic green iridium dendrimers containing oligocarbazole and oxadiazole dendrons for bright and efficient nondoped electrophosphorescent devices. *Chem. Asian. J.,* 6 (6): 1372-1380.

[91] Cocchi, M., Kalinowski, J., Murphy, L., Williams, J.G. and Fattori, V., (2010). Mixing of molecular exciton and excimer phosphorescence to tune color and efficiency of organic LEDs. *Org. Electron.,* 11 (3): 388-396.

[92] Murphy, L., Brulatti, P., Fattori, V., Cocchi, M. and Williams, J.G., (2012). Blue-shifting the monomer and excimer phosphorescence of tridentate cyclometallated platinum (II) complexes for optimal white-light OLEDs. *Chem. Commun.,* 48 (47): 5817-5819.

[93] Jayabharathi, J., Thanikachalam, V. and Jayamoorthy, K. (2012). Effective fluorescent chemosensors for the detection of Zn^{2+} metal ion. *Spectrochim. Acta Part A: Mol. & Biomol. Spect.,* 95 (20): 143-147.

[94] Jayabharathi, J., Thanikachalam, V., Jayamoorthy, K. and Sathishkumar, R. (2012). Selective quenching of benzimidazole derivatives by Cu^{2+} metal ion. *Spectrochim. Acta Part A: Mol. & Biomol. Spect.,* 97 (51): 384-387.

[95] Karunakaran, C., Jayabharathi, J., Jayamoorthy, K. and Vinayagamoorthy, P. (2013). Photoinduced electron-transfer from benzimidazole to nanocrystals. *J. Mol. Liq.,* 177 (44): 295-300.

[96] Karunakaran, C., Jayabharathi, J. and Jayamoorthy, K. (2013). Benzimidazole derivative vs. different phases of TiO_2-physico-chemical approach. *Spectrochim. Acta Part A: Mol. & Biomol. Spect.,* 114 (41): 303-308.

[97] Karunakaran, C., Jayabharathi, J. and Jayamoorthy, K. (2013). Fluorescence enhancing and quenching of TiO_2 by benzimidazole. *Sensors and Actuators B.,* 188 (27): 207-211.

[98] Karunakaran, C., Jayabharathi, J. and Jayamoorthy, K. (2013). Photoinduced electron transfer from benzimidazole to nano WO_3, CuO and Fe_2O_3. A new approach on LUMO–CB energy-binding efficiency relationship. *Sensors and Actuators B.,* 182 (70): 514-520.

[99] Karunakaran, C., Jayabharathi, J., Jayamoorthy, K. and Sathishkumar, R. (2013). Benzimidazole: dramatic luminescence turn-on by ZnO nanocrystals. *Measurement,* 46 (10): 3883-3886.

[100] Nofal, Z. M., Soliman, E. A., Abd El-Karim, S. S., El Zahar, M. I., Srour, A. M., Sethumadhavan, S. and Maher, T. J. 2011. Novel benzimidazole derivatives as expected anticancer agents. *Acta Poloniae Pharmaceutica-Drug Research,* 68 (4): 519-534.

[101] Refaat, H. M. (2010). Synthesis and anticancer activity of some novel 2-substituted benzimidazole derivatives, *Eur. J. Med. Chem.,* 45 (7): 2949-2956.

[102] Shaharyar, M. and Mazumder, A. (2017). Benzimidazoles: a biologically active compounds. *Arab. J. Chem.,* 10 (25): S157–S173.

[103] Mariappan, G., Hazarika, R., Alam, F., Karki, R., Patangia, U. and Nath, S. (2015). Synthesis and biological evaluation of 2-substituted benzimidazole derivatives. *Arab. J. Chem.,* 8 (5): 715–719.

[104] Yadav, S.,Narasimhan, B., Lim, S. M.,Ramasamy, K.,Vasudevan, M.,Ali Shah, S. A. and Mathur, A. (2018). Synthesis and evaluation of antimicrobial, antitubercular and anticancer activities of benzimidazole derivatives. *Egyptian Journal of Basic and Applied Sciences,* 12 (1): 100–109.

[105] Demirayak, S.,Kayagil, I. and Yurttas, L. (2011). Microwave supported synthesis of some novel 1,3-Diarylpyrazino[1,2-a]benzimidazole derivatives and investigation of their anticancer activities. *Eur. J. Med. Chem.,* 46 (1): 411-416.

[106] Arrang, J. M., Garbarg, M., Lancelot, J. C., Lecomte, J. M., Pollard, H., Robba, M., Schunack, W. and Schwartz, J. C. (1988). Highly potent and selective ligands for a new class H3 of histamine receptor. *Invest. Radiol.,* 23 (1): S130- S132.

[107] (a) Clarke, R. W. and Harris, J., (2002). RX 821002 as a tool for physiological investigation of α2-adrenoceptors. *CNS drug rev.,* 8 (2):177-192. (b) Cooper, D. S. (2005). Antithyroid drugs. *N. Engl. J. Med.,* 352 (9): 905-917]. (c) Corcóstegui, R., Labeaga, L., Innerárity, A., Berisa, A. and Orjales, A. (2005). Preclinical pharmacology of bilastine, a new selective histamine H^1 receptor antagonist. *Drugs in R & D,* 6 (6): 371-384.

[108] Bielory, L., Lien, K.W. and Bigelsen, S. (2005). Efficieacy and tolerability of newer antihistamines in the treatment of allergic conjunctivitis. *Drugs,* 65 (2): 215-228.

[109] Li, C. H., Tao, X., Jiang, J. H., Li, X., Xiao, S. X., Tao, L. M., Zhou, J. F., Zhang, H., Xie, M. A., Zhu, Y. and Xia, Z. (2016). Synthesis, crystal structure and spectroscopic studies of bismuth (III) complex with 2-substituted benzimidazole ligands. *Spectrochim. Acta Part A: Mol. Biomol.Spect.,* 166 (10): 56-61.

[110] Dinparast, L., Valizadeh, H., Bahadori, M. B., Soltani, S., Asghari, B. and Rashidi, M. R. (2016). Design, synthesis, α-glucosidase inhibitory activity, molecular docking and QSAR studies of benzimidazole derivatives. *J. Mol. Struct.,* 1114 (10): 84-94.

List of reviewers:

1. Dr. A.K. Mishra
   Professor
   IIT Madras
   Email: mishra@iitm.ac.in
2. Dr. R. Ramesh
   Professor and Coordinator
   School of Chemistry
   Bharathidasan University
   Email: rramesh@bdu.ac.in
3. Dr. A. Pandurangan
   Professor
   Department of Chemistry
   Institute of Catalysis and Petroleum Technology
   Anna University
   Email: pandurangan_a@yahoo.com, pandu@annauniv.edu

In: Benzimidazole
Editor: Aksel A. Vestergaard
ISBN: 978-1-53616-986-7

Chapter 5

CHARGED GUEST STIMULATED BENZIMIDAZOLE BASED CHEMOSENSORS

Gargi Dhaka, PhD and Navneet Kaur*, PhD
Department of Chemistry, Panjab University, Chandigarh, India

ABSTRACT

Considering the significance of natural receptors in biology, chemistry, medicine and environmental studies, numerous synthetic receptors like cryptands, spherands, crown ethers, calixarenes, thiacalixarenes, porphyrins, cyclodextrins and benzo-fused heterocyclic ring systems etc. have been reported in the literature. Using design strategies similar to nature, synthetic assemblies have been developed to encapsulate a variety of guest molecules, stabilize reactive intermediates and even facilitate chemical reactions. These synthetic receptors play an imperative role in ion recognition. In a related advanced *supramolecular concept*, recognition sites can be coupled to particular groups that are capable of "signaling" the ion coordination process. In this case, the binding process is transduced into a signaling event. These receptor molecules exhibit discriminatory response to specific ions or neutral species to be used as chemosensors. This information is used to analyze the stimulant, both qualitatively and quantitatively. This concept has long been an interest of study in the research field because through the integration of different functionalities into the molecular structures, these systems would respond differently upon the introduction of external stimuli (or target analyte) and organize into interesting supramolecular structures. Such distinctive ingenuity endows these systems with versatile potential to be used in the field of analyte detection. Another chief aspect for molecular recognition is the signaling mechanism involved in the detection process. Chemical sensors utilize specific transduction techniques to yield analyte information. The

* Corresponding Author's Email: neet_chem@yahoo.co.in; neet_chem@pu.ac.in.

most extensively used techniques employed in chemical sensors are optical absorption, luminescence and redox potential etc.

In the present chapter, class of potential chemosensors possessing benzimidazole moieties have been discussed which are capable of both visual and optical estimation of target analyte over a broad concentration range and without interference of contemporary analytes.

Keywords: benzimidazole, cation chemosensor, anion chemosensor, INHIBIT-ESIPT, counter-ion displacement approach

ABBREVATIONS

LOD	limit of detection
UV-vis.	ultraviolet visible, nm, nanometer
ESIPT	excited state intramolecular proton transfer
INHIBIT	inhibition
ESI-MS	Electron spray ionization-mass spectrum
MTCC	microbial type culture collection
μM	micromolar
^{1}H NMR	proton nuclear magnetic resonance
DMSO	Dimethylsulfoxide
CH_3CN	Acetonitrile
CH_3OH	Methanol
DFT	Density functional theory

INTRODUCTION

Inspired by nature's fundamental biological processes involving enzyme-substrate or receptor-guest interactions, which are regulated via light, ions or small molecule concentrations and with the aim of advancing chemical sensor technology, considerable recent attention has been focused on a new generation of abiotic host molecules (chemical sensors) that contain signaling or responsive functional groups as an integral part of a host framework [1-3]. Chemical sensors can be appropriately defined by the "Cambridge definition," which says that *Chemical sensors are miniaturized devices that can deliver real time and online information in the presence of specific compounds or ions in even complex samples* [4]. Now a days, chemosensors find many applications because the role of ions in different areas such as medical diagnostics, physiology, catalysis and environmental chemistry, is being better understood now [5-6]. Moreover, detection of ions is vital as many industrial and agricultural processes can lead to the release of ions to the environment and if remained unchecked, these can have devastating effects. A foremost research effort is being focused towards finding inexpensive, reliable and simple means of detecting ions in solution. Therefore finding new selective ion receptor systems is an important goal which involves chemosensor development, environmental remediation, selective separation and extraction of chemical species [7]. The future envisages that larger interest and legal requirements for food, water and environmental monitoring will boost the need of detection of ionic species, selectively at more and more lower concentrations.

Design of the chemosensors consist of three components as shown in Figure 1; a binding site capable of recognizing the guest of interest generally with high selectivity; a transducer or signaling unit which converts that binding event into a measureable physical change and finally a method of measuring this change and converting it to useful information. Binding sites and signaling subunits can be covalently linked (binding site-signaling subunit approach) or non-covalently linked (displacement approach) [8-10]. In fact, association through above said forces is typical reversible chemical reaction in which changes in the concentration of guest determine the relative amount of associated and free guest. In addition to these systems, analyte can also be observed using irreversible analyte induced chemical reactions. Here the term chemo-dosimeter is used [11].

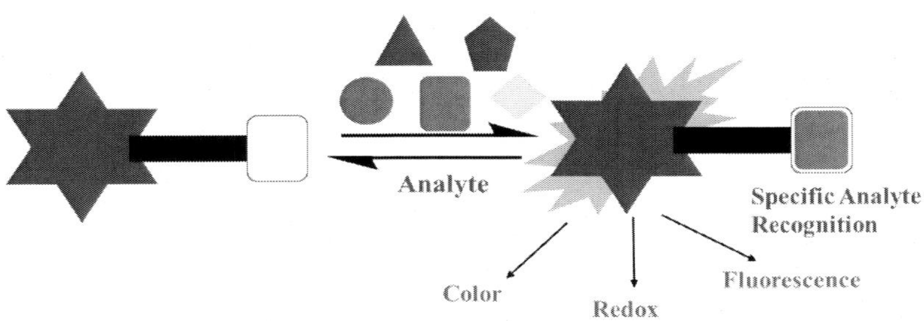

Figure 1. Schematic diagram showing binding of an analyte (guest) by a chemosensor (host), producing a complex with altered optical properties.

Depending on the type of signals produced on the binding event, optical sensors can be classified into two categories; (i) chromogenic and (ii) fluorogenic chemosensors [12]. The development of chormogenic chemosensors is increasingly appreciated since naked eye recognition can offer qualitative and quantitative information without resort to any spectroscopic instrumentation [13]. The fluorescence measurement on the other hand is usually versatile, very sensitive and offer sub-micromolar estimation of guest species. With the swift development of supramolecular chemistry, several new emerging design principles such as excited-state intramolecular proton transfer (ESIPT), fluorescence resonance energy transfer (FRET) have been applied in fluorescent sensing recently [14]. Amongst different processes, ESIPT process is easy to identify in steady-state spectra as the absorbance is normally similar to that of the parent chromophore but the fluorescence is considerably different, resulting in nearly complete lack of spectral overlap between absorption and emission. The ESIPT process commonly involves the transfer of a hydroxyl (or amino) proton to a carbonyl oxygen (or imine nitrogen) less than 2Å away through a pre-existing 6- or 5- membered ring hydrogen bonding configuration [15]. ESIPT molecules such as 2-(2'-aryl) benzimidazole have received substantial attention because of their separated dual fluorescence in a four-stage cycle incorporating the ground and excited states of two different tautomers. It has been observed that the basicity or acidity of aromatic amines and alcohols changes by several orders of magnitude upon electronic excitation.

As heterocyclic rings are known to provide binding sites which can serve as selective and effective analyte receptor systems, many efforts have been made towards designing such kind of synthetic receptors. Amongst different heterocyclic units, the benzimidazole ring acts as an excellent hydrogen bond donor moiety in synthetic anion receptor systems, and the acidity of the NH ptoton of the imidazole can be tuned by changing the electronic properties of the imidazole substituents. A class of benzimidazole derivatives having 2'-substitutions by simple functional groups like –OH, -NH_2 and extended conjugation with anthracene and N-heterocyclic rings have been developed, whose photophysical properties were remarkably changed by this derivatization. Furthermore, the fluorescent conjugated devices incorporating molecular recognition moieties (e.g., N, O and S heteroatoms) can show the high sensitivity of chemosensors to external structural perturbation and to electron density changes within the main chain backbone like pyridine, naphthalene, anthracene, benzothiazole and benzimidazole rings of the conjugated aromatic system, where they can interact and form complexes with metal ions [16-20].

Benzimidazole based chemosensors involve some characteristic features such as [21-23]:

- Conjugated aromatic system.
- Benzo-fused heterocyclic ring.
- Heteroatoms as binding sites.
- Can be used either as chromophore or fluorophore or both.
- Suitable tools for detection of metal ions and anions.
- Can be used in both organic and aqueous media.

In this chapter, design, synthesis and the photo-responsive behavior towards anions and metal ions of various chemosensors possessing benzimidazole moieties have been discussed. Based on the nature of the signaling moiety and analyte, chemosensors have been classified into two different categories viz. Benzimidazole based Cation Chemosensors and Benzimidazole based Anion Chemosensors (A-B).

BENZIMIDAZOLE BASED CATION CHEMOSENSORS

Yi. He and coworkers synthesized two almost identical benzimidazole and pyridine derivatives **1a** and **1b** which bound via ester linkage on the counter and adjoining positions of the benzimidazole benzene ring, respectively [24]. Only receptor **1a** exhibited selectivity towards Cu^{2+} ion via fluorescence quenching response in CH_3CN solution. Whereas receptor **1b** showed no selectivity towards any particular metal ions and found not suitable to be used as a chemosensor. Receptor **1a** had two emission bands centered at 341 and 356 nm. When different cations like K^+, Ag^+, Cd^{2+}, Co^{2+}, Mg^{2+}, Hg^{2+}, Ni^{2+}, Mn^{2+}, Ba^{2+}, Pb^{2+}, Ca^{2+}, Na^+, Fe^{2+} and Cr^{3+} were added into solutions of receptor **1a**, no impact were observed in fluorescence spectra. Only addition of Cu^{2+} metal ion altered the fluorescence emission spectra of **1a** and hugely quenched emission intensity at 340 (95.00%) and 356 nm (92.6%) and the emissive intensity was changed from fluorescence "*on*" to fluorescence "*off*." This observation was ascribed to formation of the complex **1a** -Cu^{2+} and then the paramagnetic characteristic of Cu^{2+} caused the fluorescence quenching. Moreover, the binding affinity (6.02 x 10^4 M^{-1}) of receptor **1a** towards Cu^{2+}, stoichiometric ratio (1:1) and LOD (3.05 x 10^6 M) were also evaluated. The potential application in living cells (HePG2 cells) fluorescence imaging was also explored. Jobs plot, spectral analysis, ^{1}H NMR titration, DFT calculations were also in accord with the results.

Li et al. prepared a ratiometric fluorogenic receptor 1-substituted-2-carbazoleyl-benzoimidazole **2** using benzimidazole and carbazole to investigate Cr^{3+} ion in CH_3CN [25]. To appraise the chemosensing capability of **2** for cations, the spectral changes of **2** after addition of various metal ions such as Na^+, K^+, Ag^+, Ca^{2+}, Mg^{2+}, Mn^{2+}, Zn^{2+}, Cu^{2+}, Fe^{2+}, Fe^{3+}, Cd^{2+}, Ni^{2+}, Co^{2+}, Hg^{2+} and Cr^{3+} were monitored by UV-Vis. and fluorescence spectral data. Initially the UV-Vis. absorption spectrum of **2** showed two absorption maxima at 267 and 297 nm, after successive addition of Cr^{3+} a new absorption band centered at 336 nm arose progressively having an isosbestic point at 306 nm. Similarly, in fluorescence emission spectra of **2** displayed an emission band at 380 nm along with a shoulder peak at 364 nm and a weak broad band between 400 to 550 nm. When Cr^{3+} ion were added successively, fluorescence enhancement at 442 nm and diminishing at 364 and 380 nm was observed indicating ratiometric sensing potential of receptor **2**. This fluorescence enhancement was ascribed to the inhibition of PET process caused by the lone pair electron on the nitrogen atom due to the coordination of Cr^{3+} ions with the nitrogen atoms of the benzimidazole moiety and also increased planarity and rigidity could reduce the non-radiative decay. The fluorescent spectral changes also accompanied by the color change from colorless to fluorescent blue under the UV lamp could be observed by naked eye. The stoichiometric ratio of L: Cr^{3+}, 1:1 was determined by Jobs plot and MALDI-TOF MS analysis. These observations further supported by DFT theoretical studies.

Liu and colleagues designed and synthesized fluorescent colorimetric sensor **3** employing bis-benzimidazole and tri anthraquinone scaffoldings for Co^{2+} ions in CH_3CN

solvent [26]. The UV-Vis. absorption band of free **3** recorded at 396 nm which upon addition of Co^{2+} metal ions decrease along with a blue shift (~ 27 nm) with simultaneous appearance of a new band at 487 nm upon successive addition of Co^{2+} ion, this new absorption peak slowly increased exhibiting two iso-absorption points at 369 and 440 nm indicating good interactions between **3** and Co^{2+}. Other cations such as Li^+, NH_4^+, K^+, Na^+, Ag^+, Zn^{2+}, Ni^{2+}, Cd^{2+}, Fe^{2+}, Ca^{2+}, Cu^{2+}, Hg^{2+}, Pb^{2+}, Cr^{3+}, Al^{3+} and Fe^{3+} had very less effect on the absorption spectra of **3**. Similarly, the fluorescence intensity of emission band at 580 nm (λ_{ex} 410 nm) decreased considerably when Co^{2+} ions were added successively accompanied with color change from orange to red under visible light. The high fluorescence of **3** may be credited to the hindered PET process from benzimidazole to anthraquinone. When Co^{2+} was coordinated to **3**, the PET process from benzimidazole to anthraquinone got turn-on and the fluorescence emission was quenched consequently. The detailed analysis of coordination interaction between **3** and Co^{2+} done by structural analysis of **3**, ^{1}H NMR titration, HRMS and IR spectra and concluded that binding force between two of them mainly operated from Co----O interactions. Additionally, **3** was also utilized for the tracking of Co^{2+} by using a smartphone with a measured limit of detection low to 0.47 μM.

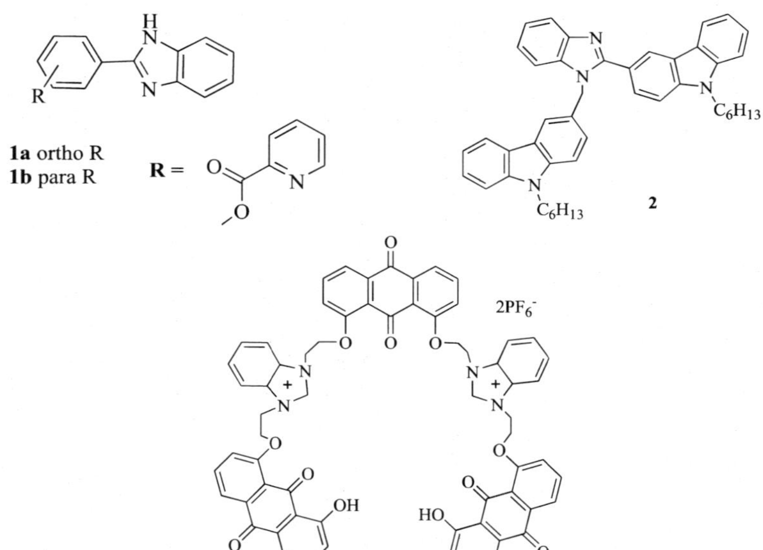

Benzimidazole based fluorescent chemosensor **4** was developed for the detection of Cu^{2+} with the ability to exclude the interference of some other metal ions such as Fe^{3+}, Mg^{2+}, Ag^{2+}, K^+ and Na^+ by Jayabharthi and her co-workers [27]. The absorption spectra of chemosensor **4** showed λ_{max} at 290 nm which was increased remarkably without shifting its absorption maximum with addition of Cu^{2+} ions. With the addition of Cu^{2+} ions to the chemosensor **4**, quenching of fluorescence was observed. The Cu^{2+} metal ion quenched the fluorescence intensity of **4** more than ten times when compared to Fe^{3+}, Mg^{2+}, Ag^{+}, K^{+} and Na^{+} metal ions. The decrease in fluorescence intensity was attributed to electron-transfer between benzimidazole derivative and Cu^{2+} metal ion.

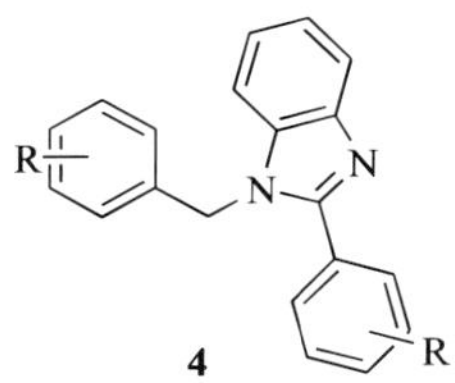
4

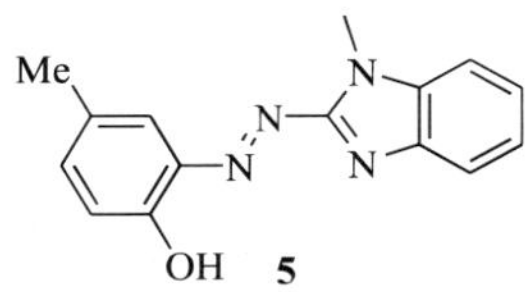
5

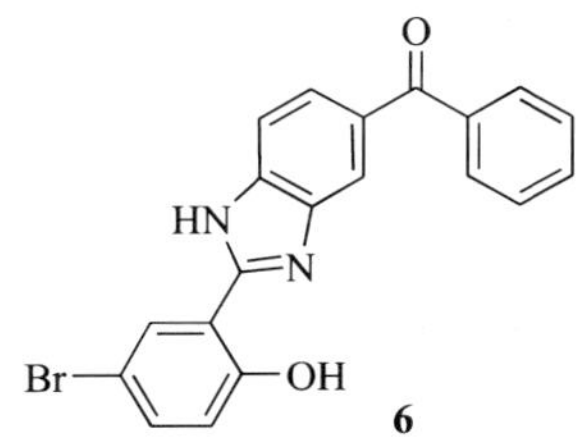
6

Mondal et al. developed a highly selective and sensitive benzimidazole based chemosensor **5** which underwent significant changes in absorption and color with addition of Ni^{2+} ions [28]. On the addition of different metal ions to the solution of **5** in DMSO-HEPES buffer (1:1, v/v, pH =7.4), only Ni^{2+} caused decrease in intensity of absorption band at 403 nm and a new absorption band appeared at 600 nm with color change from orange yellow to blue. The Job's plot confirmed the stoichiometric ratio 1:1 between **5** and Ni^{2+} and limit of detection was 0.14 μM with value of association constant found to be 2.074×10^5 M^{-1}.

Luxami and coworkers developed ESIPT based benzimidazole derivative **6** which was used to recognise Cu^{2+}, Hg^{2+} and CN^- ions in protic/ aqueous system [29. The addition of CN^-/Cu^{2+} ions to MeOH solution of **6** caused decrease in absorption intensity at 340 nm with the appearance of new absorption band at 380 nm. Addition of Hg^{2+} ions caused decrease in intensity at 268 nm and 340 nm with color of solution bleached out. The emission spectra of **6** showed weak fluorescence at 430 nm, which showed enhancement with addition of CN^- ions and then **6**. CN^- complex was used for recognition of Cu^{2+} which further quenched the emission intensity. The probe **6** was also used as 'XOR' logic function in the presence of chemical inputs of Cu^{2+} and CN^- ions.

Krishnamoorthy et al. developed a set of ESIPT based ratiometric fluorescence sensors **7a-7c**, which underwent significant changes in fluorescence and UV-vis. upon addition of different metal ions [30]. The changes exhibited by alkali and alkaline earth metals were small; while transition metals gave significant changes and the changes appeared higher in **7a** than **7b** or **7c**. The UV-vis. spectra of **7a-7c** were perturbed by presence of Zn^{2+} and Fe^{3+} ions and maximum sensitivity was for **7c**. However, there were also changes appeared for **7b** and **7c** upon addition of Ni^{2+}, Co^{2+} and Pd^{2+} ions. In the fluorescence spectra, Zn^{2+} and Fe^{3+} caused increase in tautomer fluorescence intensity ratio of **7a** by 143 and 9 fold, respectively. The Co^{2+} increased the fluorescence ratio of **7b** by 4.7 fold; while Pd^{2+} ion enhanced the fluorescence ratio of **7c** by 11.6 fold.

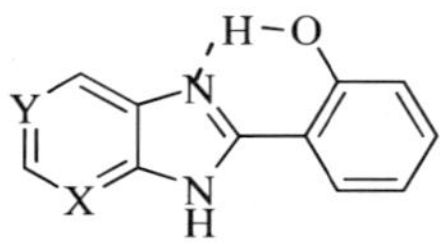

7a. X = Y = CH
7b. X = N, Y = CH
7c. X = CH, Y = N

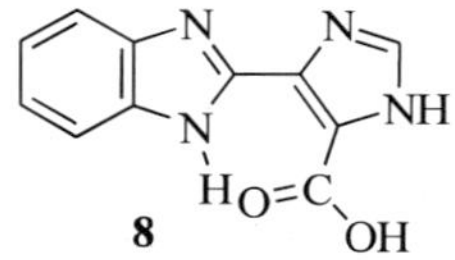

8

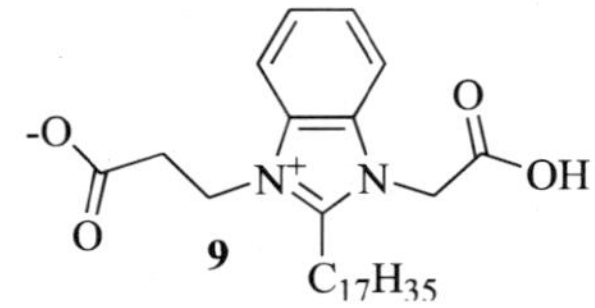

9

Gupta et al. developed a benzimidazole based optical probe **8** and its photophysical properties were studied in CH_3CN/DMF solution [31]. On the addition of different metal ions like Na^+, Ni^{2+}, Zn^{2+}, Pb^{2+}, Mn^{2+}, Mg^{2+}, Ag^+, K^+, Fe^{3+}, Cu^{2+}, Co^{2+}, Cd^{2+}, Ca^{2+}, Li^{2+}, Hg^{2+}, Ce^{4+} or Fe^{2+} to solution of **8**, only Hg^{2+} ions resulted in bathochromic shift ($\Delta\lambda$=12 nm) of $\pi \rightarrow \pi^*$ transition present at 312 nm. The values of association constant for **8.**Hg^{2+} (1:1) complex and detection limit were found to be 1.26 $\times 10^3$ M^{-1} and 1.54$\times 10^6$ M, respectively. In the fluorescence spectrum, there was moderate quenching for Ca^{2+} (10%), Mg^{2+} (20%) and Na^+ (100%); while complete quenching was observed for other metal cations.

Wei et al. synthesized a biscarboxyl funtionalized benzimidazole derivative **9** which formed MOG (metal-organic gels) by inducing with Ag^+ ions [32]. MOG formed thermoreversible opaque gel within few minutes in some polar solvents like DMF, MeOH, iPrOH, MeOH-H_2O etc and thermal stability of MOG was researched by measuring the gel- sol transition temperature, $T_{gel.}$. The nanofibrillar network structure of MOG confirmed by FESEM. There was self association of **9** in the gel state due to the metol ion chelation, intermolecular H-bonding and π-π stacking which was also confirmed by XRD. The cross linking of gel fibres was confirmed by FT-IR which suggested that Ag^+ ion coordinated with carboxyl group of **9** induced gel formation. The MOG underwent sol-gel transformation by addition of reasonable chemicals like NH_4OH, Na_2S and EDTA by change of temperature, pH. MOG also showed good ability in dye sorption which was detected by UV-vis. spectra. Finally, this MOG of **9** was widely applied in material science.

Yao et al. synthesized supramolecular metallogelator chemosensor (**10**-Fe) based on benzimidazole gelator functionalized with Naphthol moieties (**10**) to detect and separate Cr^{3+} ion in water as well as gel state [33]. Benzimidazole and Naphthol moieties were employed into a single gelator (**10**) molecule to provide fluorescent signalling and binding sites. Acylhydrazone groups were introduced as hydrogen bonding self-assembly sites. To strengthen Vander Waals force groups, long alkyl chains were used. Thus obtained self-assembled supramolecular organogel (**10**) showed strong aggregation induced emission (AIE). The dispersion of right amount of Fe^{3+} into **10** formed the desired metallogel (**10**-Fe) which showed high selectivity for Cr^{3+}. To evaluate the ion detection behavior of **10**-Fe various metal ions such as Hg^{2+}, Ag^+, Ca^{2+}, Cu^{2+}, Co^{2+}, Ni^{2+}, Cd^{2+}, Pb^{2+}, Zn^{2+}, Cr^{3+}, La^{3+}, Eu^{3+}, Tb^{3+} and Mg^{2+} were added in aqueous media. Addition of Cr^{3+} ion only perturbed the fluorescence spectra of **10**-Fe by quenching of emission band at 552 nm. The LOD value of **10**-Fe for Cr^{3+} 2.62 X 10^{-8} M and absorbing percentage (96.36%) were determined. This spectral alteration was attributed to the synergistic coordination interaction occurred between supramolecular gelator and metal ion.

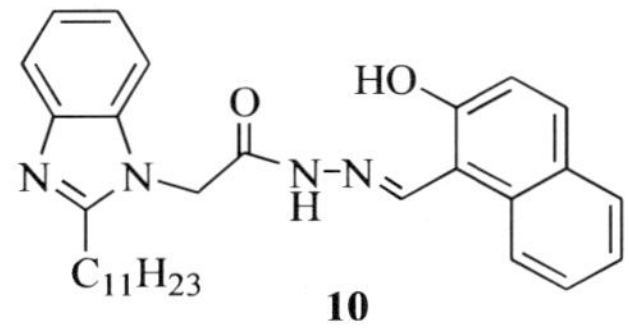

10

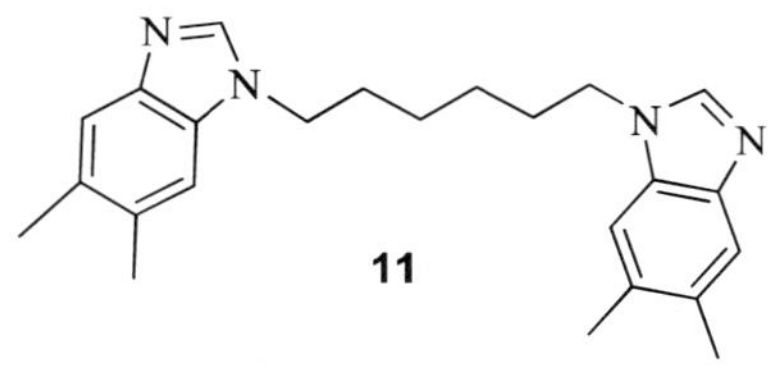

11

Cui and coworkers designed and synthesized two luminescent metal-organic frameworks (**Cd(II)-MOFs**) based on bis-benzimidazole (**11**) and dicarboxylate ligands, which detected in aqueous solution [34]. Upon excitation, these **Cd(II)-MOFs** showed emission band maxima at 378 nm (λ_{ex} 288 nm, due to the intraligand π-π* charge transition) and 310 nm (λ_{ex} 290 nm, due to Ligand to Metal Charge Transfer), respectively. To identify the detection ability towards different cations like Na^+, Ba^{2+}, Pb^{2+}, Mg^{2+}, Cd^{2+}, Ni^{2+}, Zn^{2+}, Ag^+, Cu^{2+} and Fe^{3+}, luminescence spectra of the **Cd(II)-MOFs** were recorded in the presence of the said metal ions. In the case Fe^{3+} only, the fluorescence emission of both the **Cd(II)-MOFs** quenched completely due to the coordination interactions between scaffoldings and Fe^{3+} ions.

Das and co-workers synthesized styryl-imidazolium base chromogenic receptor **12** to monitor both anion (HSO_4^-) and cations (Cu^{2+} and Al^{3+}) in CH_3OH solvent [35]. The receptor **12** have two absorption maxima at 269 and 394 nm (due to π-π* transition). After adding multiple anions, only HSO_4^- ion caused noticeable changes in the absorption spectrum. Upon incremental addition of HSO_4^- ion, gradual decrease in absorbance at 394 nm and increase in absorbance at 527 nm (huge red shift 133 nm) along with an apparent color change (yellow to dark red) was noticed. This sensing behavior was ascribed due to the possible NH proton dissociation followed by intramolecular change transfer and induced intermolecular hydrogen bonding between **12** and HSO_4^- ion. Moreover, amidst various metal ions, **12** exclusively showed sensing ability with drastic color change towards Cu^{2+} (yellow to green) and Al^{3+} (yellow to dark yellow). Upon successive addition of Cu^{2+} and Al^{3+} into **12** solution separately, the 394 nm absorption band decreasing with concomitant appearance of 420 (red shift ~26 nm) and 430 nm (red shift ~36 nm) absorption bands respectively. The cations (Cu^{2+} and Al^{3+}) complexed to the **12**, perturbed the electronic ground and excited state of **12** as compared to the corresponding electronic state of uncomplexed receptor **12** and possibly the charge transfer from metal to ligand occurred via π-conjugation due to the rich electron density of metal ion in metal complexes. The complexation ratio of **12** with HSO_4^-, Cu^{2+} and Al^{3+} found to be 1:1 by Job's plot and HRMS data and having low LOD values (7.96 μM for HSO_4^-, 0.4 μM for Cu^{2+} and 15.7 μM for Al^{3+}).

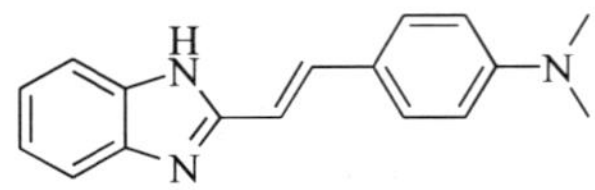

12

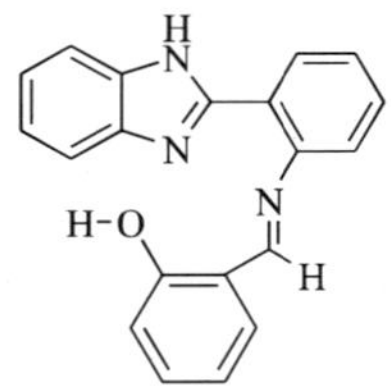

13

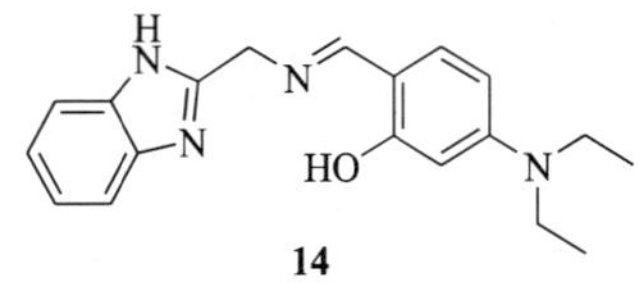

14

Liu et al. prepared a simple Schiff base as an efficient fluorescent chemosensor **13** for Al^{3+} with fluorescence bathochromic shift and Cu^{2+} with fluorescence quenching having detection limits 3.1 x 10^{-7} M in CH_3OH/PBS and 5.4 x 10^{-7} M in CH_3OH/Tris-HCl, respectively, over various metal ions like Mg^{2+}, Ca^{2+}, Mn^{2+}, Co^{2+}, Fe^{2+}, Fe^{3+}, Cr^{3+}, Zn^{2+}, Hg^{2+}, Pb^{2+} and Cd^{2+} ions [36]. The obtained chemosensor **13**-Al^{3+} and **13**-Cu^{2+} could be effectively used for recognition of F^- and S^{2-} respectively as cascade sensor. Furthermore, **13** showed low cytotoxicity in MTT cytotoxicity assay in HeLa cells. It also explored as potential tool for tracking Al^{3+}/Cu^{2+} cascade sensors could respectively sense F^- and S^{2-} in living cells. In the course of recognizing Al^{3+} the observed remarkable fluorescence enhancement (at 485 nm) was attributed to the inhibition of ESIPT phenomenon due to the chelation with Al^{3+} rendering the prevention of –CH=N- isomerization mechanism whereas the CHEQ behavior of **13** in the presence of Cu^{2+} could be attributed to the paramagnetic effect from spin-orbit coupling of Cu^{2+} leading to the formation of the non-fluorescent complex.

Kim et al. designed and developed a dual mode chemosensor **14** based on Schiff base with benzimidazole for the chromogenic detection of $Fe^{2+}/^{3+}$ and fluoregenic detection of Zn^{2+} in 2:1 stiochiometric ratio in aqueous media [37]. Chemosensor **14** showed high selectivity towards $Fe^{2+}/^{3+}$ with a color change from colorless to dark green (due to MLCT) and had very low detection limits (1.21 μM for Fe^{3+} and 1.18 μM for Fe^{2+}). Furthermore, **14** displayed exclusive detection ability for Zn^{2+} by "Off-On" fluorescent response in aqueous solution with detection limit 1.05 μM for Zn^{2+}. All the detection limits of **14** were far below the guidelines of the WHO. Chemosensor **14** was successfully used to real samples for the assessment of Fe^{3+} and Zn^{2+} both quantitatively and qualitatively with great recovery and RSD values. Moreover, **14** could be recycled by treating with an appropriate reagent such as EDTA. The sensing behavior of **14** towards Zn^{2+} was attributed to the rigid structure of Zn^{2+}-2.**14** complex (CHEF) which might inhibit the non-radiative process such as C=N isomerization and the ESIPT phenomenon and causes the fluorescence enhancement. It was further supported by the theoretical calculations.

Horak and coworkers prepared a benzimidazole based Schiff base as a single colorimetric and fluorimetric receptor **15** for different metal analytes [38]. The binding interactions between **15** and metal ions were investigated through absorption and fluorescence spectroscopic techniques and computational modeling. To explore the chemosensing capabilities of **15** (10μM), UV-Vis. absorption and fluorescence titration experiments were conducted with incremental concentrations of Al^{3+} (upto 10 equiv.) in EtOH. The UV-Vis. absorption spectra of receptor **15** showed strong absorption band at 425 nm, when Al^{3+} added, new absorption maxima observed at 455 nm ($\Delta\lambda$ = 30 nm) accompanied by an isosbestic point at 439 nm. This complexation of **15** with Al^{3+}, also observable by a naked eye via color change from light to dark yellow. This effect was ascribed to the di-deprotonation of **15** in the presence of trivalent Al^{3+} ions resulting in the $[\mathbf{15}^{2-}.Al]^{+}$ complex, due to which electronic structure modified and inhibition of ESIPT process occurred in the absence of acidic hydrogen atoms. Whereas out of the metal ions namely K^+, Na^+, Ag^+, Li^+, Zn^{2+}, Co^{2+}, Hg^{2+}, only Al^{3+} caused increase in fluorescence intensity in fluorescence spectra of **15**. Rest of the metal ions could not perturb the fluorescence spectra very much, except the Cu^{2+} and Fe^{3+} metal ions whose chelation induced a noticeable decrease (~ 15 fold) in the fluorescence intensity at double emission wavelength i.e., 468 and 497 nm in EtOH-H_2O (95:5, v/v) solution. The binding constants and limits of detection for receptor **15** towards Al^{3+} were calculated to be 3.5 X 10^4 M^{-1} (Cu^{2+} 1.5 X 10^3 M^{-1}, Fe^{3+} 1.2 X 10^9 M^{-1}), 3.7 X 10^{-7} M (Cu^{2+} 1.3 X 10^{-7} M) and stoichiometric ratio **15**:Al^{3+},1:1 respectively. Computational analysis results using B3LYP-DFT functional and implicit SMD solvation were also in accordance with the experiments performed.

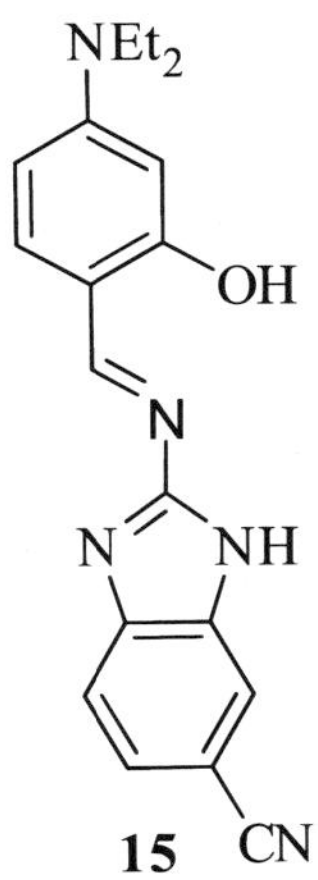

15

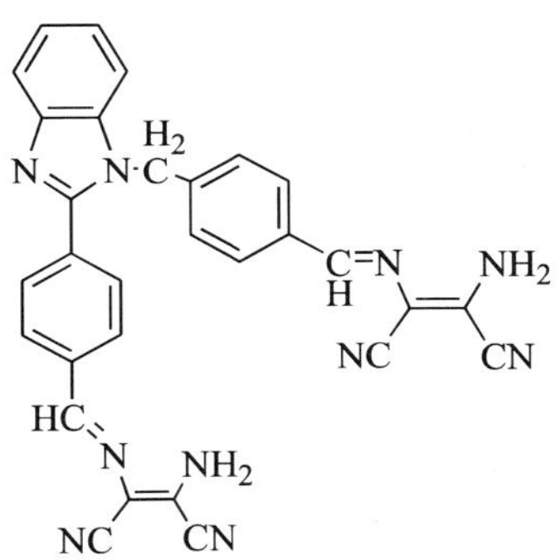

16

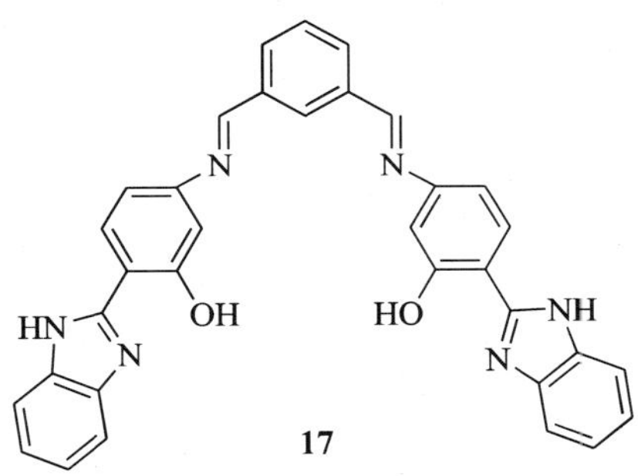

17

Wang et al. developed a fluorescent (OFF-ON) and ratiometric absorption chemosensor **16** for Cu^{2+} ion by exploiting benzimidazole schiff base and bidentate diaminomaleonitrile ligands [39]. The ion recognition properties of chemosensor **16** were explored via both UV-Vis. absorption and fluorescence spectroscopies in aqueous CH_3CN solution. Out of the various metal ions, **16** exhibited high selectivity and sensitivity towards Cu^{2+} ions. In UV-Vis. absorption spectra, **16** showed absorption maxima at 256 and 366 nm. When Cu^{2+} ion was gradually added, ratiometric pattern was observed via regular decrease in absorption peak at 366 nm and simultaneous emergence of a new band 286 nm with an isosbestic point at 327 nm. In fluorescence spectra, **16** showed poor fluorescence emission around 330-400 nm possibly due to the C=N bond isomerisation and PET process from nitrogen atoms of bidentate ligand to benzimidazole. To investigate the fluorescence of **16** towards various metal ions were recorded namely Cr^{3+}, Fe^{3+}, Al^{3+}, Zn^{2+}, Cd^{2+}, Co^{2+}, Mg^{2+}, Ba^{2+}, Mn^{2+}, Hg^{2+}, Pb^{2+}, Ca^{2+}, Sr^{2+}, Fe^{2+}, Ni^{2+}, K^+, Ag^+ and Cu^{2+}. After addition of Cu^{2+} metal ion only, the fluorescence of **16** was drastically enhanced (~ 155 fold) at 405 nm. The fluorescence pictures of the solutions having Cu^{2+} ion displayed bright blue color when illuminated under UV light. Further, the possible sensing mechanism was proposed after Job's plot, Mass spectra and DFT calculation analysis than **16** and Cu^{2+} complex in 1: 2, **16**: Cu^{2+} ratio. As a result the PET process now operated from diaminomaleonitrile group to Cu^{2+} metal ion instead of benzimidazole group, which was responsible for the drastic fluorescence enhancement of **16** in the presence of Cu^{2+} ion. Moreover, in vitro fluorescence imaging of **16** in liver cancer HepG2 cell demonstrated that **16** had wonderful cell permeability and suitable to be used for biological applications.

Jang and his co-workers prepared a benzimidazole based chemosensor **17** capable of showing excited state proton transfer (ESIPT) for Zn^{2+} ions [40]. Chemosensor **17,** in DMSO-CH_3CN (1:9, v/v), exhibited a fluorescence emission at 443 nm which shifted to shorter wavelength (397 nm) upon addition of Zn^{2+} ions, while the individual additions of Cu^{2+}, Fe^{3+}, Co^{2+}, and Ni^{2+} caused quenching of fluorescence intensity at 443 nm. The addition of other metal ions such as Na^+, K^+, Mg^{2+}, Ba^{2+}, Ca^{2+}, Ag^+ and Cd^{2+} caused no changes in the fluorescence profile. The 1:1 stoichiometry of the **17**-Zn^{2+} complex was determined by Jobs plot method. The k_a value of the **17**-Zn^{2+} complex was calculated to be $1.4 (\pm 0.3) \times 10^4 M^{-1}$. The chemosensor **17** was also used to recognize Zn^{2+} in the cytoplasm of *Saccharomyces cerevisiae*.

Tabular comparison of chemosensing properties of chemosensors **13-17** using INHIBIT-ESIPT process as signaling mechanism is given in Table 1.

Sessler and co-workers synthesized benzimidazole embedded N-fused aza indacenes optical based fluorophores **18a** and **18b,** which underwent optical changes with the addition of dissolved CO_2 [41]. The UV-vis. spectra of **18a** and **18b** in CH_3CN displayed an intense band at 277 nm and broad bands at 315 nm, 360 nm, 464 nm, and 468 nm (for **18a)** and strong bands at 289 nm, 313 nm, 459 nm, and 482 nm (for **18b**). The fluorescence spectra of **18a** and **18b** showed strong fluorescence signals at 493 nm and 522 nm, respectively. Upon addition of TBAF to CH_3CN solution of **18b**, the broad absorption band diminished with bathochromic shift and fluorescence was quenched. On the addition of CO_2 gas to CH_3CN solution of **18b.**F^-, both absorption and emission spectra were changed and the fluorescence of **18b** regenerated. The detection limit for CO_2 for 10 μM solution of **18b.**F^- in CH_3CN was 4.1×10^{-7} M.

Table 1. Analyte, K_a and LOD values of benzimidazole based Schiff Bases (13-17) based on INHIBIT-ESIPT Signaling Mechanism

Sr. No.	Chemosensor	Analyte	Solvent	K_a (M^{-1})	LOD (μM)
1.	13	Al^{3+} Cu^{2+}	CH_3OH-PBS (95:5, v/v)	2.61 x 10^5 5.84 x 10^5	3.1 5.4
2.	14	Fe^{2+} Fe^{3+} Zn^{2+}	DMF:bis-tris (1:1, v/v)	3.0 x 10^9 6.0 x 10^9 4.0 x 10^9	1.18 1.21 1.05
3.	15	Al^{3+} Cu^{2+} Fe^{3+}	C_2H_5OH	3.5 x 10^4 1.5 x 10^3 1.2 x 10^9	0.37 0.13
4.	16	Cu^{2+}	CH_3CN:Tris (1:1, v/v)	2.5 x 10^8 M^{-2}	0.49
5.	17	Zn^{2+}	DMSO:CH_3CN (1:9, v/v)	1.4 x 10^4	3.0

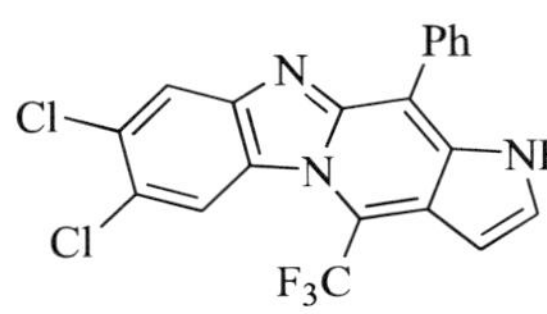

18a

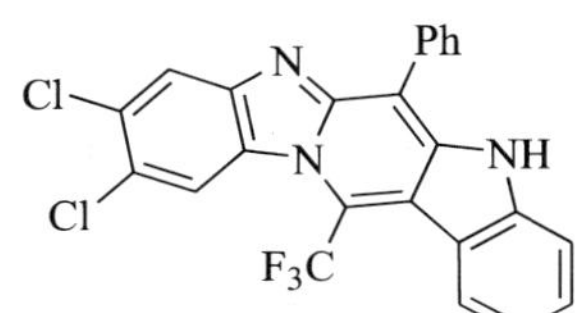

18b

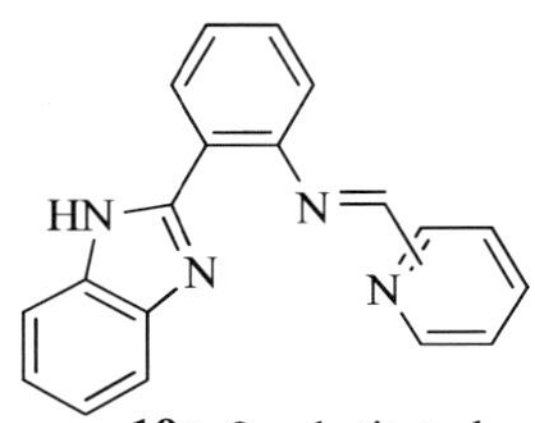

19a. 2-substituted
19b. 3-substituted
19c. 4-substituted

A set of benzimidazole based fluorescent receptors **19a-19c**, having both hydrogen donor and hydrogen acceptor sites, were synthesized by Singh et al. [42]. The fluorescence behavior **19a-19c** was studied in a HEPES buffered CH_3CN-H_2O (8/2, v/v) solvent system upon excitation at 352, 340 and 334 nm, respectively. Chemosensors **19a** and **19b** showed fluorescence emission at 420 nm that was quenched upon addition of Cr^{3+} ion while new emission band was formed at 470 nm. The selective recognition of Cr^{3+} over other metal ions was due to the compatible binding sites and the optimum size of pseudocavity of receptors. Chemosensor **19c** showed no specific binding with any of metal ions tested, including Cr^{3+}. The values of binding constant for **19a**.Cr^{3+} and **19b**.Cr^{3+} complexes with stoichiometry ratio 1:1 were found to be (7.67±0.30) × 10^4 M^{-1} and (7.09±0.35) × 10^4 M^{-1}, respectively. The fluorescence of resultant **19a.**Cr^{3+} complex was quenched at 420 nm, while enhanced at 475 nm with HSO_4^- ion. On the other hand, F^- addition to **19a.**Cr^{3+} complex resulted in 'turn on' fluorescence emission at 420 nm. The values of binding constant of **19a**.Cr^{3+} for HSO_4^- and F^- were found to be (2.89±0.25) × 10^5 M^{-1} and (2.30±0.20) × 10^6 M^{-1}, respectively.

Singh et al. developed a benzimidazole based chemosensor **20** which exhibited specific fluorescence and absorbance changes with addition of Cu^{2+} amongst other metal ions viz. Na^+, Mn^{2+}, Cu^{2+}, Zn^{2+}, Fe^{3+}, Co^{2+}, Pb^{2+}, Ag^{2+}, Hg^{2+} and Cd^{2+} in HEPES buffered CH_3CN/H_2O (8:2, v/v) [43]. The addition of Cu^{2+} ions caused significant quenching at 420 nm in the fluorescence spectra of **20** due to PET process. The value of association constant of **20** with Cu^{2+} (1:1) complex was determined to be 5.34 × 10^3 M^{-1}. Furthermore, on the addition of different anions like F^-, Cl^-, Br^-, I^-, CN^-, ClO_4^-, AcO^-, HSO_4^- and PO_4^{3-} in HEPES buffered CH_3CN-H_2O solution of **20.**Cu^{2+} complex, only PO_4^{3-} restored the fluorescence intensity of **20** at 420 nm by the metal displacement approach mechanism. The value of detection of **20**.Cu^{2+} for PO_4^{3-} was determined to be 2.8 μM.

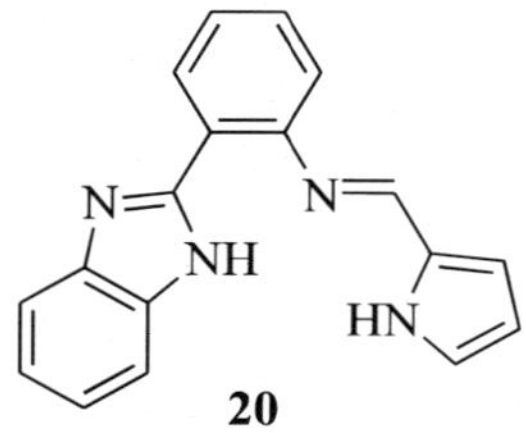

20

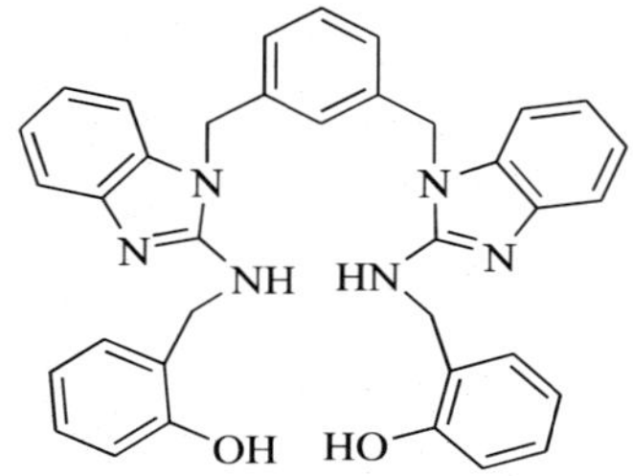

21

Jang et al. synthesized a benzimidazole based ratiometric fluorescent receptor **21** which underwent significant fluorescence changes with Cu^{2+} and Fe^{3+} ions [44]. Amongst different metal ions, alkali metal ions, alkaline earth metals etc., only addition of Cu^{2+} and Fe^{3+} ions perturbed the emission intensity of **21** through quenching at 320 nm and 420 nm and appearance of new band at 490 nm due to charge transfer mechanism. The stoichiometry ratio was confirmed 1:1 by Job's plot method and the binding constant values for **21**.Cu^{2+} and **21**.Fe^{3+} were found to be (3.9± 0.1) × 10^3 $M^{-1}$and (5.1± 0.2) × 10^3 M^{-1}, respectively.

Benzimidazole based chemosensors **22a-22c** bearing imine linkages were prepared by Jang and his co-workers [45]. The fluorescence spectrum of chemosensor **22a** exhibited emission band at 412 nm in CH_3CN-H_2O (95:5, v/v). Upon addition of Fe^{3+} ion, the intensity of emission band at 412 nm decreased along with the appearance of a new red-shifted emission band at 475 nm. No such shifted emission band was observed upon addition of other metal ions such as alkali, alkaline earth metal and transition metal ions. Cu^{2+} ions also caused quenching of fluorescence intensity at 412 nm but did not cause any shift in the emission band. The stoichiometry of the complex formed was found to be 1:1 with k_a value to be (2.90 ± 0.2) × 10^3 M^{-1}. Chemosensor **22b** gave non-selective binding affinity towards different metal ions pointing towards the importance of both imine linkage and benzimidazole platform.

On the other hand, the UV-vis. spectrum of chemosensor **22c** displayed well defined bands at 300, 350 and 400 nm. Upon addition of Mg^{2+} to a solution of chemosensor **22c**, decrease in absorbance at 400 nm and increase at 350 nm was observed. Both Job's plot and mass spectrometry revealed 1:1 complex formation with association constant 3.2 (± 0.1) × 10^2 M^{-1}. The fluorescence spectrum was perturbed only with addition of Cr^{3+} ions i.e., quenching was observed at 415 nm along with enhancement at 475 nm. The Job's plot confirmed the formation of 2:1 complex with association constant 1.6 (± 0.1) × 10^4 M^{-2}.

DFT calculations supported these spectral changes. The detection range for both **22c**.Cr^{3+} and **22c**.Mg^{2+} found to be 3.2-100 μM and 79.4-340 μM, respectively.

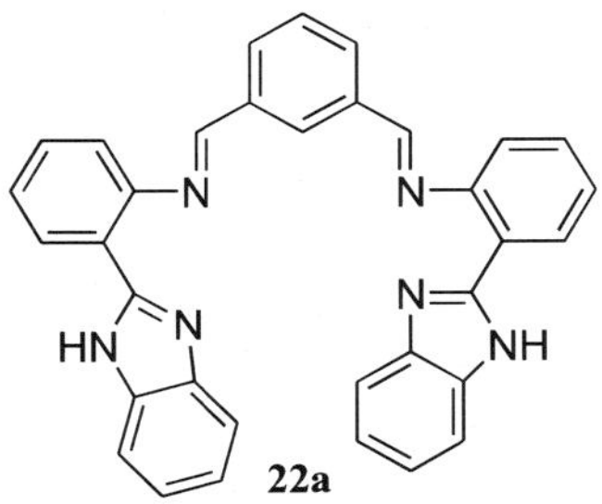

22a

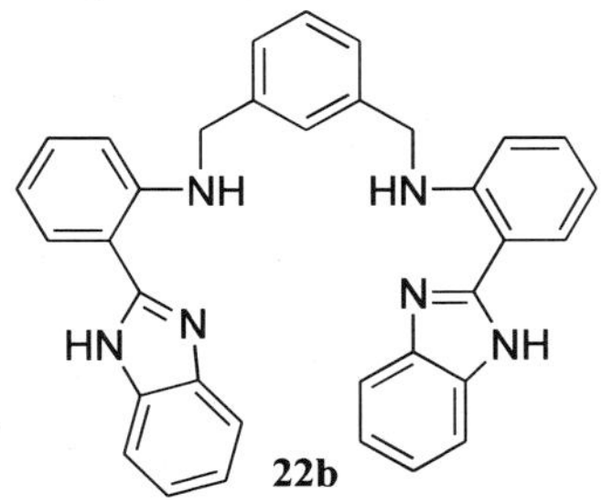

22b

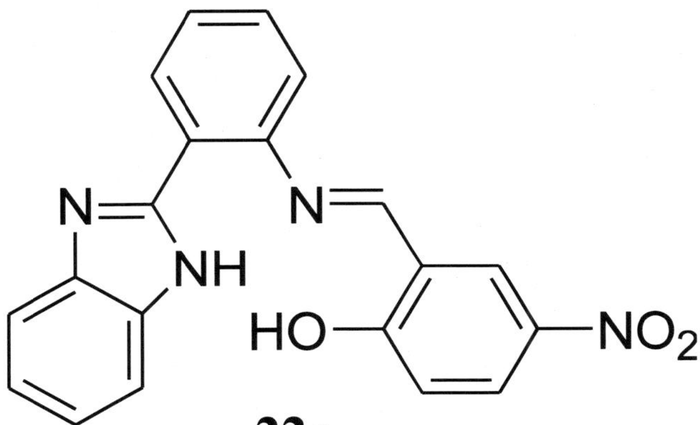

22c

Singh et al. developed a benzimidazole based dipodal receptor **23** which was capable of recognizing Cu^{2+} exclusively amongst other different cations such as Zn^{2+}, Ag^{+}, Ba^{2+}, Ca^{2+}, Cd^{2+}, Co^{2+}, Fe^{3+}, Mg^{2+}, Ni^{2+}, Na^{+}, K^{+}, Hg^{2+} in aqueous media [46]. To appraise the binding behavior of receptor **23** for Cu^{2+} the spectral changes were monitored. Receptor **23** displayed strong emission centered at 451 nm. After addition of Cu^{2+} ions, a remarkable quenching (~90%) was observed via the photoinduced electron transfer (PET) process. Due to the paramagnetic behavior of Cu^{2+} ions, when reacted with receptor **23**, triggered the radiationless decay of its excited state leading to dramatic fluorescence quenching. Thus obtained (FL-OFF **1**-Cu^{2+}) complex exhibited exclusive recognition ability towards $H_2PO_4^-$ through a metal displacement process. The addition of $H_2PO_4^-$ to the **23**-Cu^{2+} complex solution resulted in complete revival of the fluorescence intensity at 451 nm. The binding constants, LOD and stoichiometric ratio for **23** towards Cu^{2+} and $H_2PO_4^-$ were calculated to be 2.30 x 10^4 M^{-1} (Cu^{2+}) and 6.60 x 10^3 M^{-1} ($H_2PO_4^-$), LOD 9.76 μM (Cu^{2+}) and 1.41 μM ($H_2PO_4^-$) and 1:1 for both Cu^{2+}/$H_2PO_4^-$ ions, respectively.

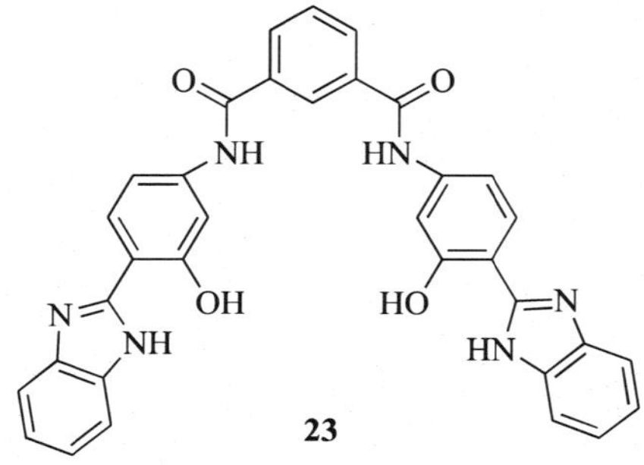

23

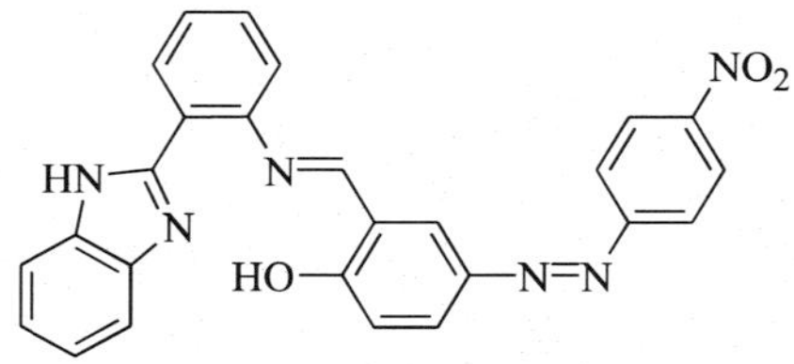

24

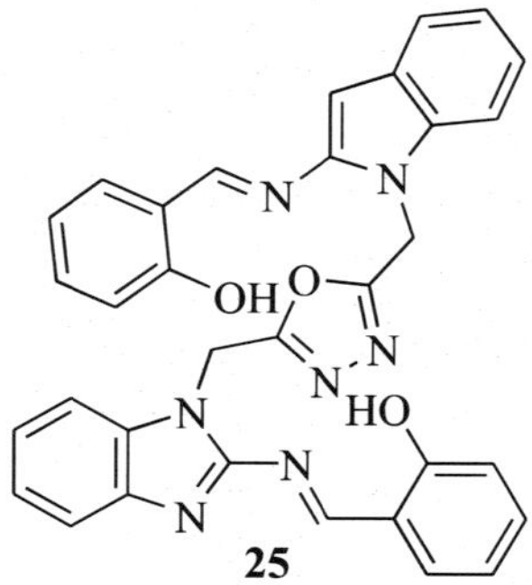

25

Singh and co-workers synthesized an azo dye coupled benzimidazole based receptor **24** which underwent significant change in the UV-vis. spectra upon addition of Cu^{2+} ions [47]. The absorption spectra of **24** in HEPES buffered MeOH-H_2O (1:1, v/v, 20 mM, pH=7.0) exhibited absorption maxima at 365 nm and 490 nm. On the addition of different cations in HEPES buffered solution of **24**, only Cu^{2+} caused a bathochromic shift in absorption maxima from 365 to 415. The value of binding constant and detection limit was found to be $(1.10\pm0.1) \times 10^6$ M^{-1} and 0.56 μM, respectively. Furthermore, the resultant **24**.Cu^{2+} complex was used as selective chromogenic receptor for oxalic acid in semi-aqueous medium over different dicarboxylic acids and anions. On the addition of oxalic acid to the solution of **24**, absorption spectra of **24**.Cu^{2+} gradually shifted from 415 nm and absorbance increased at 375 nm. The value of binding constant and detection limit of **24**.Cu^{2+} complex with oxalic acid was found to be $(6.25\pm0.3) \times 10^7$ M^{-1} and 70.8 μM, respectively.

Jang and team synthesized a benzimidazole based imine linked chemosensor **25** which exhibited significant fluorescence enhancement for Zn^{2+}; while other metals quenched the receptor fluorescence [48]. On the addition of different cations to the solution of receptor **25** in DMSO-CH_3CN (1:9, v/v), only Zn^{2+} coordination led to enhanced fluorescence of receptor **25** at 440 nm. The binding constant for **25**.Zn^{2+} complex was found to be $(7.9\pm0.9) \times 10$ M^{-1} by using the Benesi–Hildebrand method. The supramolecular assembly of $(\mathbf{25}.Zn^{2+})_n$ was used for recognition of anions/dicarboxylic acids/thiols etc. On the addition of various anions/ dicarboxylic acids /amino acids/ thiols to the assembly of $(\mathbf{25}.Zn^{2+})_n$, only L-cysteine exhibited a four-fold enhancement in intensity at 440 nm of $(\mathbf{25}.Zn^{2+})_n$.

Comparison of counter ion displacement data of **23**, **24** and **25** is listed in Table 2.

Table 2. Analyte, K_a and LOD values of Benzimidazole based Chemosensors (23-25) using Counter Ion Displacement Approach

Sr. No.	Chemosensor	Analyte	Solvent	K_a (M^{-1})	LOD (μM)
1.	23	Cu^{2+} $H_2PO_4^-$	DMSO-H_2O (7/3, v/v)	2.30 x 10^4 6.60 x 10^3	9.76 1.41
2.	24	Cu^{2+} Oxalic acid	CH_3OH-H_2O (1:1, v/v)	1.10 x 10^6 6.25 x 10^7	0.56 70.8
3.	25	Zn^{2+}L-Cystein	DMSO-CH_3CN (1/9, v/v)	7.9 x10	-

Mahiya et al. developed a novel fluorescent probe based on bis benzimidazoly diamide **26** which underwent significant changes in fluorescence and UV-vis. spectra with addition of Cu^{2+} ions [49]. The UV-vis. spectra of **26** exhibited two sharp peaks at 276 and 284 nm. Upon the addition of different metal ions, only Cu^{2+} caused small blue shift in MeOH and MeOH-water mixture; while other metal ions showed insignificant changes. The fluorescence spectra of **26** in MeOH showed an emission band at 300 nm. On the addition of various cations to the solution of **26** in MeOH and MeOH-H_2O mixture (9:1), only Cu^{2+} ion caused significant enhancement in fluorescence intensity at 375 nm and quenching of emission intensity at 300 nm.

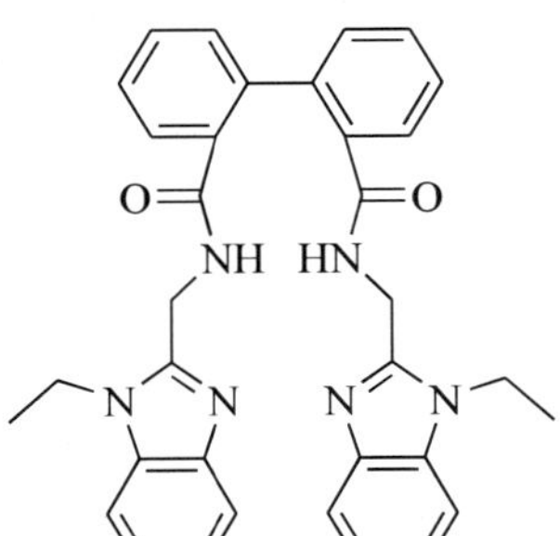

26

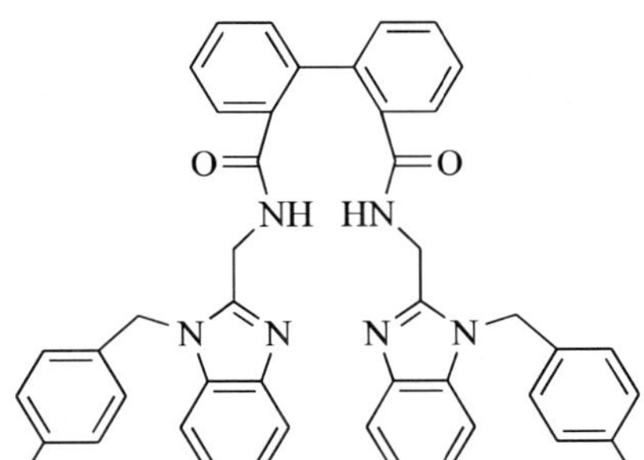

27

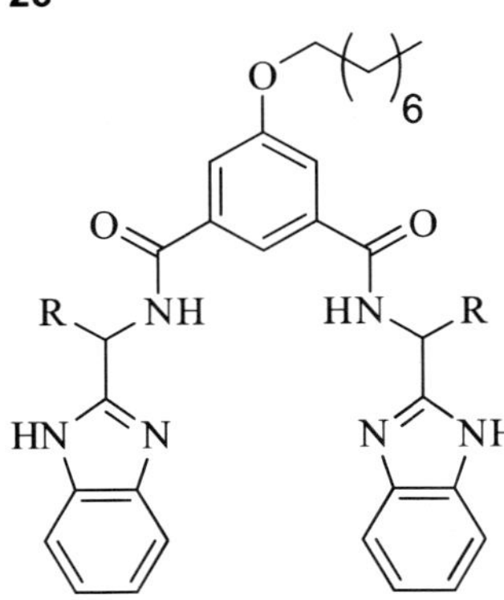

28a. R = $CH(CH_3)_2$
28b. R = H

Mathur and coworkers synthesized bis-benzimidazole diamide 27 similar to 26 and its trinuclear copper (II) complex {$[Cu_3(27)_3Cl_3]_2.6Cl.13CH_3CN.2H_2O$} [50]. The fluorescence spectra of 27 exhibited an emission centred at 300 nm due to benzimidazole moiety in MeOH. Upon the addition of different metal ions to the solution of 27, only addition of Fe^{3+}, Cu^{2+} and Ag^{+} perturbed its fluorescence spectra. With increasing amount of Fe^{3+}, slight red shift of ligand band along with appearance of new band with increasing intensity at 375 nm was observed. Moreover, the emission at 300 nm was quenched due to chelation enhanced quenching (CHEQ) effect. In case of Cu^{2+} and Ag^{+} ions, only quenching at 300 nm was observed.

Ghosh et al. synthesized benzimidazole based receptors 28a and 28b, out of which receptor 28b resulted in non-selective binding with different metal ions due to the absence of isopropyl groups [51]. On the addition of different metal ions to the solution of 28a in CH_3CN containing 0.2% DMSO, only Hg^{2+} ions exhibited significant enhancement in emission at 336 nm. The association constant for 28a.Hg^{2+} (1:1) complex determined to be (3.50±0.90) x 10^3 M^{-1}. The reversibility of complexation of 28a.Hg^{2+} was confirmed by addition of thiol group containing amino acids like cystein, homocystein and glutathione over other amino acids with no thiol group in aqueous DMSO (DMSO:H_2O = 4:1,v/v).

Sheng et al. synthesized rhodamine based chemosensor 29 which underwent fluorogenic, absorbance and colorimetric changes on the addition of Hg^{2+} ions [52]. The chemosensor 29 (10 μM) in CH_3CN showed very weak fluorescence at 557 nm. On the addition of different metal ions, addition of only Hg^{2+} ions caused enhancement in fluorescence intensity at 557 nm with color change from colorless to green.

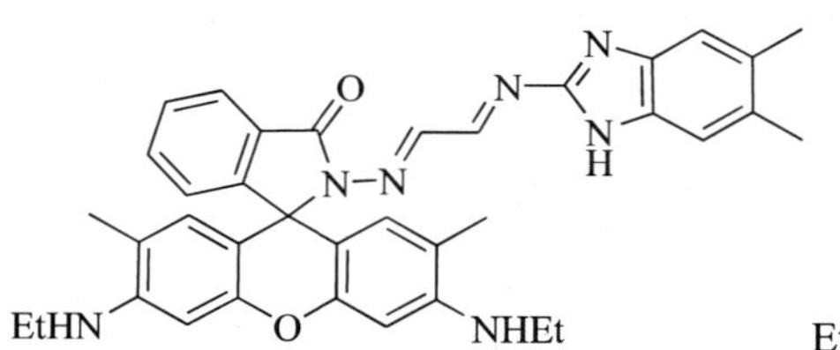

29

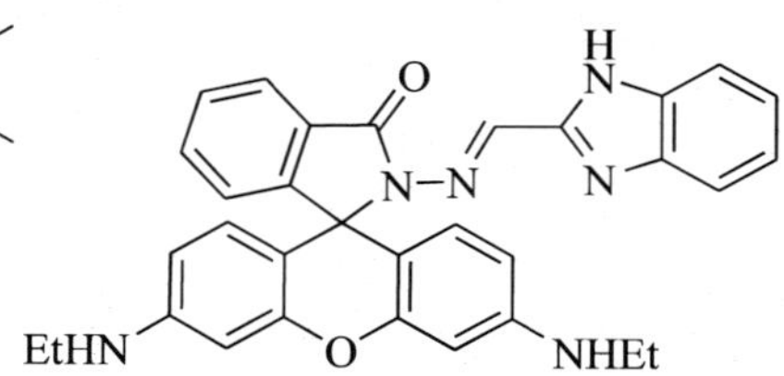

30

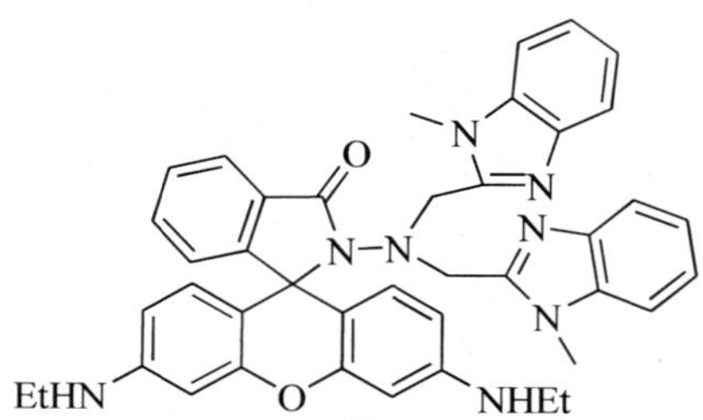

31

Pan and coworkers developed a novel rhodamine benzimidazole conjugate **30** which gave selective and sensitive recognition of Fe^{3+} ion [53]. On the addition of different cations to the solution of **30** in CH_3CN, only Fe^{3+} exhibited significant fluorescence enhancement at 582 nm over other tested cations. The stoichiometry was found to be 1:1

by Job's plot method with binding constant of 1.01×10^{4} M^{-1}. The competition experiment was performed by adding Fe^{3+} to the solution of **30** in the presence of other metal ions, which showed negligible interference except Cu^{2+} with some extent. The detection limit was calculated to be 1.5×10^{-8} M.

Goswami et al. conceptualized the modulation of fluorescence resonance energy transfer (FRET) phenomenon in a multi-fluorophore system **31** consisting of benzimidazole unit as donor and rhodamine unit as acceptor connected through multi-chelating site for Cu^{2+} detection both *in-vivo* and *in-vitro* medium [54]. All the spectroscopic studies were carried out in ACN-HEPES buffer solution (1 mM, pH 7.4, 8:2, v/v) employing **31** (10 μM) and 3 equiv. of competing metal ions like Mn^{2+}, Cr^{3+}, Pb^{2+}, Fe^{2+}, Co^{2+}, Ni^{2+}, Zn^{2+}, Cd^{2+} and Hg^{2+}. In UV-vis. absorption spectroscopy, **31** displayed an absorption band centered at 315 nm and another absorption band around 450-650 nm. When Cu^{2+} metal ions were added, a remarkable gain in absorbance intensity at 556 nm was noticed causing the color changed to pink with the diminished absorbance at 315 nm, ascribed to the Cu^{2+} triggered spirolactam ring opening of the rhodamine unit of **31.** The fluorescence emission behavior of the **31** was also investigated upon excitation at 315 nm. Initially, the emission maximum was found at 490 nm which upon incremental addition of Cu^{2+} ion underwent continuous quenching along with emergence of a new emission peak at 580 nm and intense reddish orange color (I_{580}/I_{490} = 0.28 to 15.1) attributed to the facilitated FRET process. The stoichiometry, association constant (K_a) and detection limit were calculated to be 1:1 (Job's plot), 1.5 x 10^{4} M^{-1} and 3.1 μM respectively. The chemosensor **31** (50 μM) was also explored as a cytoplasmic marker of Cu^{2+} ions in HeLa cells showing good cell permeability and selectivity.

Ghosh et al. synthesized anthracene based chemosensors **32** and **33** which underwent absorption and fluorescence changes on the addition of Cu^{2+} ions [55-56]. The UV-vis. spectra of **32** in CH_3CN containing 0.04% DMSO underwent significant red shift upon addition of Cu^{2+} ions; whereas all other metal ions showed minor changes in the spectra. Fluorescence spectra of **32** resulted in significant enhancement of emission intensity of 418 nm band in the presence of Cu^{2+} ions. On the other hand, the fluorescence spectrum of **33** in $CHCl_3$ displayed excimer band at 520 nm and monomer emission at 420 nm (λ_{ex} 350 nm). Upon addition of different acids like TFA, S-Mandelic acid, methane sulfonic acid, p toluene sulfonic acid, emission was observed at only 419 nm. The change in emission was more significant for sulfonic acids over TFA and S-Mandelic acid. Furthermore on the addition of different metal ions in the solution of **33** in CH_3CN, emissions of both monomer and excimer decreased upon addition of Cu^{2+}, Co^{2+}, and Ni^{2+} ions. From the stern-volmer plots, it was clear that the **33** had greater selectivity for Cu^{2+} ions.

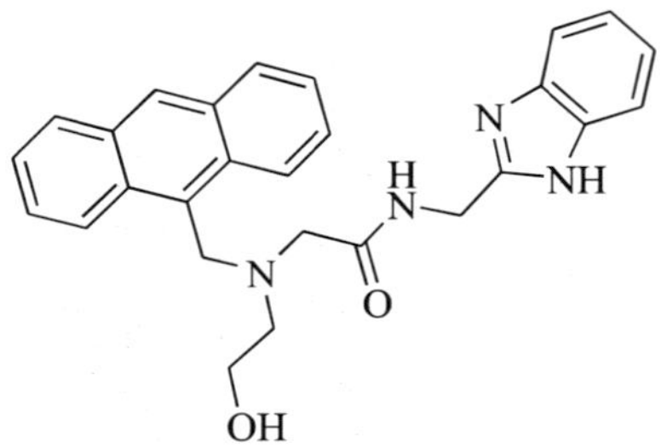

32

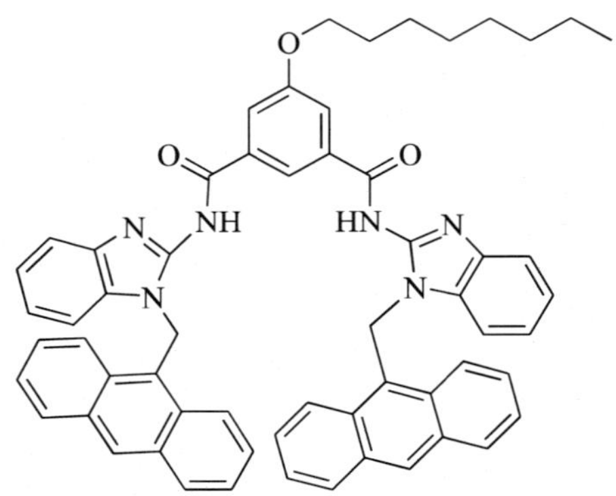

33

Tang et al. developed benzimidazole containing quinazoline based probe **34** which underwent selective and sensitive recognition of Cu^{2+} and CN^- ion [57]. On the addition of different cations to the solution of **34** in Ethanol/ water (1/1, v/v, HEPES 20 mM, pH 7.0), only Cu^{2+} addition exhibited significant fluorescence enhancement over other cations. The stoichiometric for **34** and Cu^{2+} complexation was found to be 2:1 by Job's plot method and the value of association constant was 2.16×10^5 M^{-2}. Furthermore, the **34**.Cu^{2+} complex was also used to recognise CN^- ion. On the addition of various anions, only CN^- ion addition caused strong fluorescence enhancement with slight hypsochromic shift to 419 nm. The stoichiometry of **34**.Cu^{2+} and CN^- ion was found to be 1:2 with binding constant of 3.09×10^9 M^{-2}. Also the detection limit of **34**.Cu^{2+} for CN^- was calculated to be 1.74×10^{-6} M.

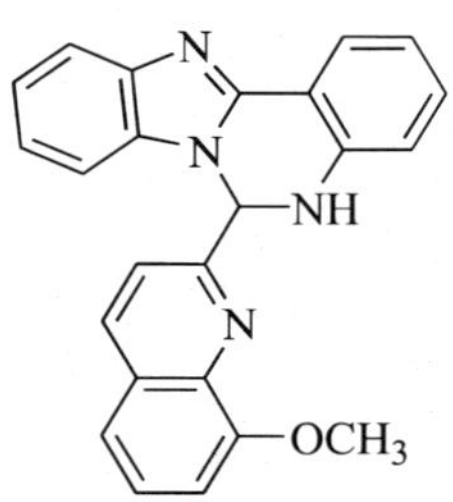

34

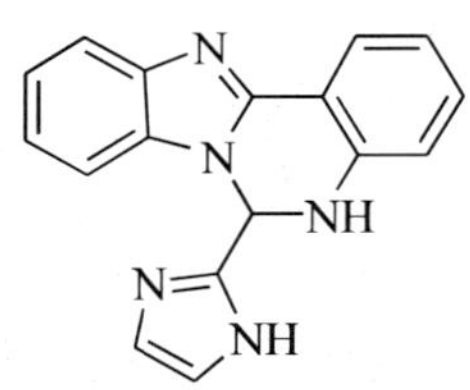

35

Chellappa and co-worker prepared a chromogenic and fluorogenic ratiometric chemosensor **35** for Al^{3+} ion detection in the mixed solvent media DMSO:H_2O (1:9, v/v, pH 7.54, phosphate buffer 100 mM) amongst other ions viz. Li^+, Na^+, K^+, Ca^{2+}, Mg^{2+}, Cu^{2+}, Co^{2+}, Ni^{2+}, Mn^{2+}, Zn^{2+}, Cd^{2+}, Hg^{2+}, Pb^{2+} and Fe^{3+} ions [58]. The UV-vis. spectrum of **35** (10 μM) exhibited an absorption maximum at 354 nm due to imine linkage. Upon addition of Al^{3+} ions to solution of **35,** concomitant decrease in the original absorbance at 354 nm and increase in the new absorbance at 409 nm occurred with a noticeable color change to bluish green due to the ring opening on an account of solvent supported 1, 5-sigmatropic shift. The emission maximum of **35** (ϕ_s = 0.34 in DMSO) was centered at 415 nm (λ_{ex} =

350 nm), which underwent a red shift of about 57 nm to the new emission band at 472 nm in the presence of Al^{3+} metal attributed to the intramolecular charge transfer. The association constant (K_a) and the detection limit of **35** for Al^{3+} ion were estimated to be 1.6 x 10^4 M^{-1} and 5.3 x 10^{-7} M, respectively from the fluorescence titration data. The proposed mechanism was supported by ^{1}H NMR titration (in DMSO-d_6) showing downfield shifting of aromatic protons and upfield shifting of –NH proton to get completely vanished eventually (on account of deprotonation) upon addition of Al^{3+} (0, 0.5, 1.0 equiv.). The mass spectrometric detection and IR data were also in great support of this mechanism. The mode of interaction was also supported by theoretical calculations like DFT and TD-DFT.

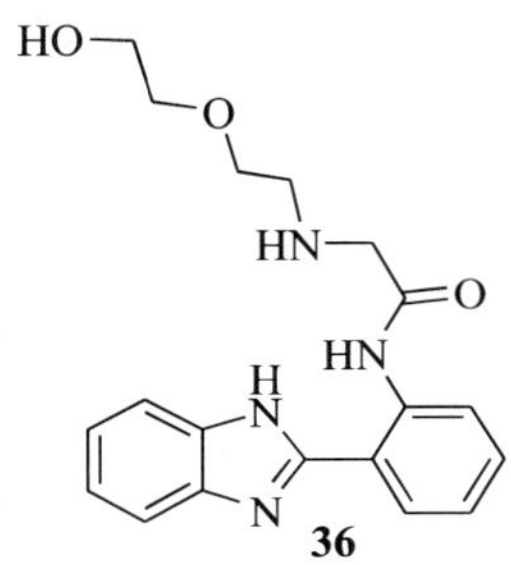

36

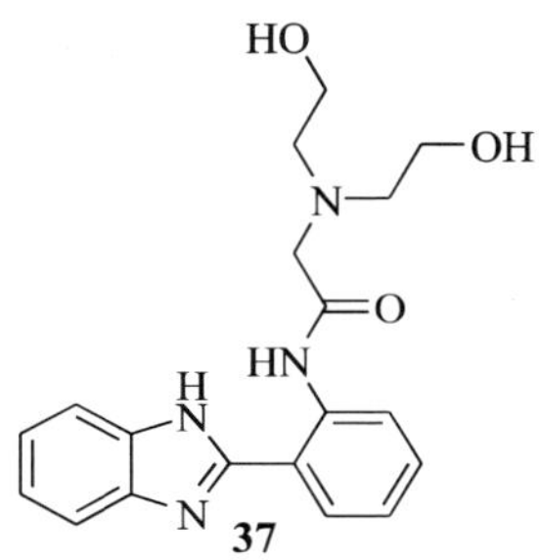

37

Tang et al. developed a novel benzimidazole derivative **36** and **37** which recognized Cu^{2+} or Zn^{2+} ions in 100% water solution [59-60]. On the addition of different cations to the solution of **36**, only Cu^{2+} exhibited maximum fluorescence quenching at 475 nm over other cations. The stoichiometric for **36** and Cu^{2+} found to be 1:1 by Job's plot method and the value of association constant was calculated to be 4.3×10^4 M^{-1}. Furthermore, the **36**.Cu^{2+} complex was used to recognise S^{2-} ion, in which, S^{2-} addition resulted in strong fluorescence emission at 475 nm that was identical with the emission wavelength of free **36**. The detection limit of **36**.Cu^{2+} to S^{2-} ion was found to be 1.35×10^{-6} M. On the contrary, the fluorescence spectra of **37** exhibited weak fluorescence at 367 nm. On the addition of only Zn^{2+} ions, fluorescence enhancement was observed at longer wavelength of 426 nm. The value of association constant and detection limit of **80**.Zn^{2+} (1:1) complex was 1.4×10^5 M^{-1} and 3.31×10^{-7} M, respectively.

A set of five emissive novel benzocrown ether possessing chemosensors **38a-38e** were synthesized and evaluated as fluorogenic chemosensors for transition metal ions by Raposo et al. in CH_3CN-DMSO solution (99:1) [61]. A CHEQ phenomenon (chelation enhanced fluorescence quenching) was observed for the receptors **38a-38e** in the presence of Ni^{2+} and Pd^{2+} ions. On the other hand, all the receptors remained emissive (CHEF effect) upon coordination of Hg^{2+} ions. Interference studies carried out in the presence of Na^+ did not modify the emission of the chemosensors, but the extent of quenching observed after complexation with Ni^{2+} and Pd^{2+} ions was reduced in the presence of Na^+ ion. This effect could be described by a conjugation of the raise of the ionic strength in the medium and the rigidity forced by the coordination of the crown ether to Na^{+} ions. Chemosensors **38a** and **38b** were found to be finest receptors for Pd^{2+} while receptor **38e** was found to be more efficient for Hg^{2+} ion detection.

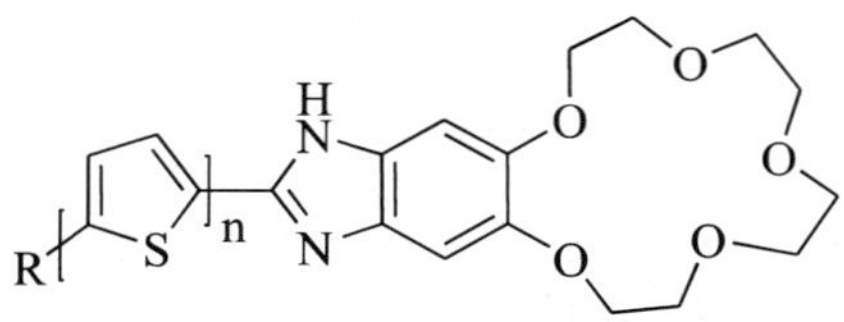

38a. R = H, n =1
38b. R = H, n =2
38c. R = MeO, n =2
38d. R = CN, n =2
38e. R = H, n =3

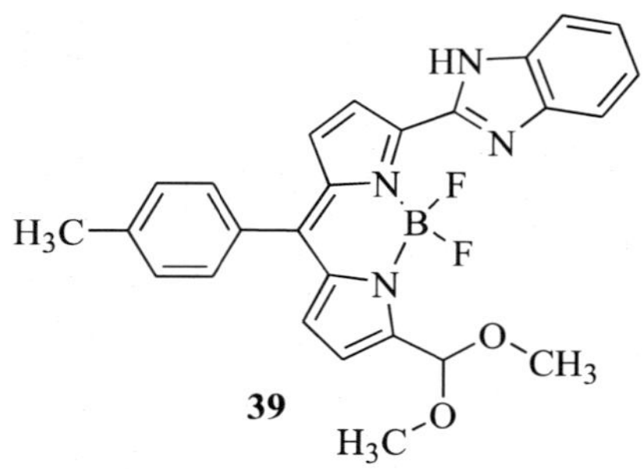

39

Ravikanth and coworkers developed a benzimidazole based BODIPY framework chemosensor **39** which was used to recognize Hg^{2+} ions *in vitro* and *in vivo* [62]. The UV-vis. spectra of **39** exhibited an absorption peak at 570 nm with a distinct shoulder at 545 nm and the emission spectra of **39** exhibited maxima at 603 nm. On the addition of various ions to the solution of **39** in CH_3CN/PBS at pH 7.0, only in the presence of Hg^{2+} ion, the emission spectra exhibited prominent blue shift of transition energies and a simultaneous increase of emission efficiency. The binding constant and limit of detection for Hg^{2+} ion was found to be 6.18×10^{6} M^{-1} and 0.77 μM, respectively. The utility of the chemosensor **39** for detection of Hg^{2+} ions *in vivo* was demonstrated by labelling human breast adenocarcinoma cells with BODIPY **39.**

Wu and his co-workers prepared a ratiometric fluorescence chemosensor **40** bearing two benzimidazole groups [63]. The fluorescence spectra of **40** displayed bands at 443 and 380 nm whose ratios of emission intensities at 443 nm and 378 nm was enhanced by Cr^{3+} ion from 0.17 to 3.21 and from 0.40 to 2.09 by Fe^{3+} ions. No change was observed in fluorescence spectrum with other metal ions.

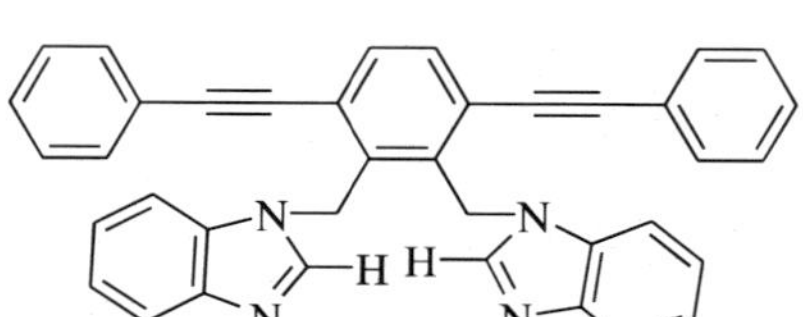

40

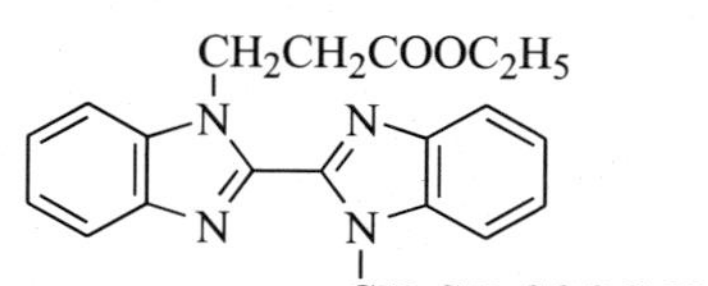

41

Zhang et al. developed 2, 2'-bisbenzimidazole derivatives **41** as Fe^{3+} selective fluorescent probe for in aqueous media (H_2O/DMSO, 2:8, v/v) [64]. The fluorescence spectra of **41** (20 μM) showed a broad band ranging from 348 to 440 nm (λ_{ex} = 328 nm). Upon addition of 20 equiv. of Fe^{3+} ions, fluorescence quenching was observed along with remarkable color change from bright blue to colorless out of other relevant metal ions like Hg^{2+}, Ag^{+}, Ca^{2+}, Cu^{2+}, Co^{2+}, Ni^{2+}, Cd^{2+}, Pb^{2+}, Zn^{2+}, Cr^{3+}, Al^{3+}, Fe^{2+} and Mg^{2+} ions. This fluorescence quenching at 369 nm was ascribed to the chelation of **41** with Fe^{3+} metal ions through its carbonyl oxygen and imidazole nitrogen atoms causing non planarity in the molecule resulting in diminished strong bright blue fluorescence emission. The stoichiometric ratio and detection limit were found to be 1:2 (H:G) and 2.1×10^{-7} M respectively.

Yang and Peng et al. developed the novel polyfluorenes **42a-42d,** which were used for recognition of metal ions and H^+ through absorption and emission spectra in THF solution [65]. The fluorescences of all chemosensors **42a-42d** (5 μM) were efficiently quenched in the presence of Co^{2+}, Ni^{2+}, Fe^{3+} and Ag^+ ions owing to the enhanced electronic communication properties of the conjugated polymers **42a-42d**. The fluorescence of **42a** and **42d** were quenched to 50 (I_o/I; $K_{SV} = 1.7 \times 10^7$) and to 22 (I_o/I; $K_{SV} = 4.05 \times 10^6$) when a Ni^{2+} solution of 3.2×10^{-6} M and 5.0×10^{-6} M added, respectively, due to the diverse conjugated backbone. **42b** exhibited good selectivity to Ni^{2+} (at low quencher concentration of about 5 ppm), was ascribed to the different chelating skill of receptor with ion. Also complete quenching of fluorescence (attributed to the different receptors in side chain) was observed in the case of **42b** ($I_o/I = 262$) upon addition of Al^{3+} solution of 1×10^{-4} M; while fluorescence of **42c** was quenched to $I_o/I = 3.2$ in same concentration of Al^{3+}.

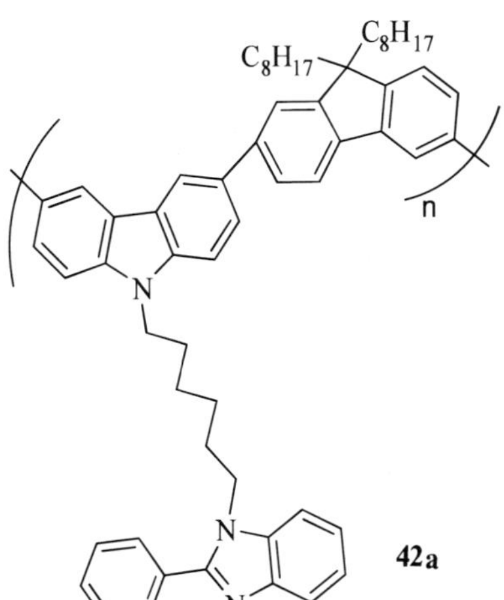

42a

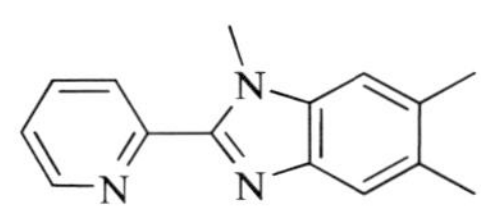

42b

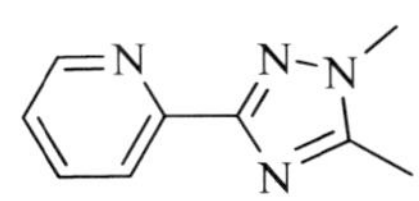

42c

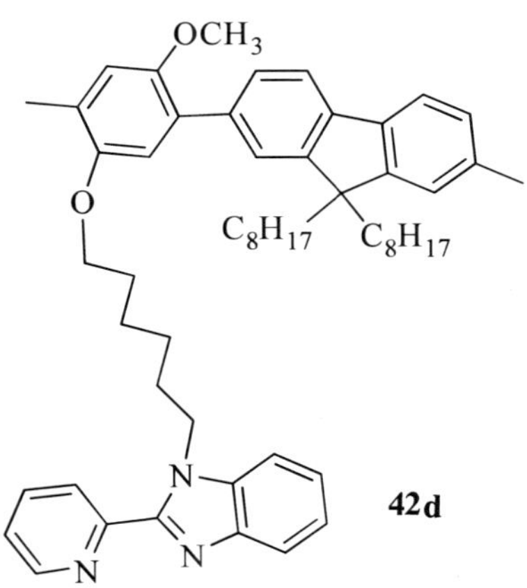

42d

BENZIMIDAZOLE BASED ANION CHEMOSENSORS

Martinez et al. strategized to combine the benzimidazolium and urea scaffoldings linked to the fluorophore (anthracene/acridine) yielding macrocyclic receptors **43a-43d** as ratiometric fluorescent chemosensors for $H_2PO_4^-$ ion in CH_3CN among other relevant anions such as F^-, Cl^-, Br^-, I^-, AcO^-, SCN^-, ClO_4^- ions [66]. Chemosensor **43a** exhibited 68% quenching of fluorescence intensity at 430 nm and 4.3 fold fluorescence enhancement at 501 nm in the presence of 3 equiv. of $H_2PO_4^-$ ion along with change in color from blue to green (λ_{ex} = 365 nm). On the other hand, chemosensor **43b** required 4 equiv. of $H_2PO_4^-$ to display about half of the fluorescence enhancement (2.6 fold) at 501 nm. Furthermore, in the case of chemosensor **43c**, similar drop in fluorescence intensity at 428 nm (monomer peak) and rise at 500 nm (excimer peak) were observed when $H_2PO_4^-$ ion added. Conversely, another anthracene based receptor i.e., **43d** showed diminishing of fluorescence intensity in longer wavelength region with addition of $H_2PO_4^-$ ion. The binding constants for chemosensors **43a**, **43c** and **43d** for $H_2PO_4^-$ ions were determined to be 2.9 x 10^6 M^{-1}, 2.6 x 10^5 M^{-1} and 1.6 x 10^5 M^{-1}, respectively. On the contrary, binding constant for chemosensor **43b** could not be determined owing to the two step process of the titration profile. Thus chemosensors **43a** and **43c**, having benzimidazole and urea coordinating sites, performed better in $H_2PO_4^-$ sensing than **43b** and **43d** ascribed to the urea-assisted synergistic binding effect.

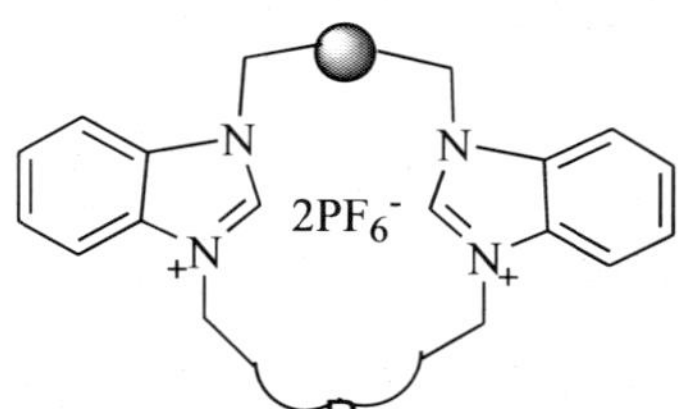

43a. R = $CH_2NHCONHCH_2$; = acridine
43b. R = CH_2CH_2 ; = acridine
43c. R = $CH_2NHCONHCH_2$; = anthracene
43d. R = CH_2CH_2 ; = anthracene

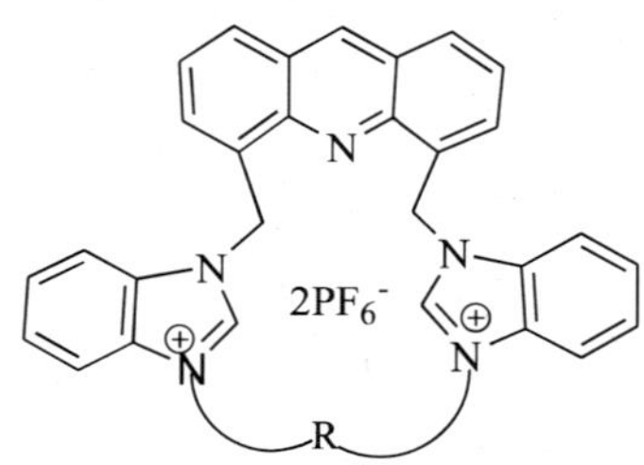

44a. R = $(CH_2)_4$
44b. R = $(CH_2)5$
44c. R = $(CH_2)_6$
44d. R = CH_2-Ph-CH_2

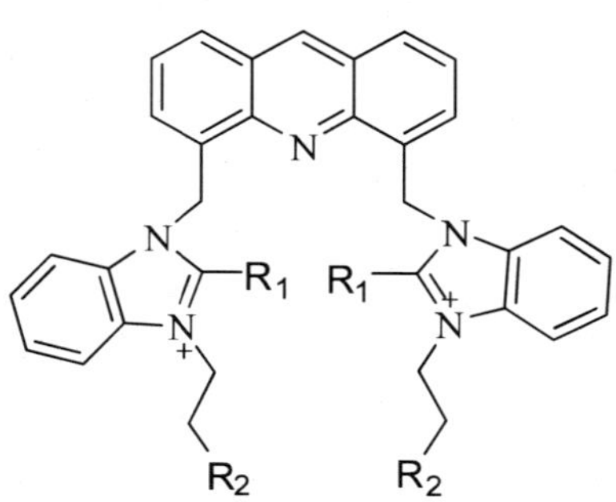

45a. R_1 = H; R_2 = NH-CO-NH_2
45b. R_1 = H; R_2 = CH_2CH_3
45c. R_1 = CH_3; R_2 = NH-CO-NH_2

Similar type of $H_2PO_4^-$ sensing was done by acridine based bis benzimidazolium macrocyclic fluorescent chemosensors **44a-44d** and **45a-45c** [67-68]. On the addition of various anions to the solutions of **44a-44d** in CH_3CN, only addition of $H_2PO_4^-$ resulted in 'turn-off' fluorescence at 430 nm and 'turn-on' fluorescence emission at 556 nm with fluorescence color change from blue to orange red, orange, green and blue green respectively. On the other hand, chemosensor **45a,** having both benzimidazolium and urea group as binding sites, exhibited better anion sensing than **45b** and **45c,** containing only one kind of binding site. **45a** in CH_3CN exhibited significant changes in fluorescence spectra in the presence of $H_2PO_4^-$ and HSO_4^- ions amongst other anions. The addition of $H_2PO_4^-$ gave 'turn-off' fluorescence emission at 430 nm and a 'turn-on' fluorescence emission at 480 nm with change in color from blue to green. While the addition of HSO_4^- resulted in fluorescence quenching because of PET effect.

Ghosh et al. synthesized ortho phenyldiamine based fluorescence cleft **46** which showed significant interactions with $H_2PO_4^-$ and F^- in CH_3CN [69]. On the addition of different anions like AcO^-, propanoate, benzoate, HSO_4^-, $H_2PO_4^-$, F^-, Cl^-, Br^-, and I^- to **46** in CH_3CN, only $H_2PO_4^-$ and F^- caused quenching of emission intensity of receptor **46** at 419 nm upon complexation. The preferable binding of $H_2PO_4^-$ with binding **46** with stoichiometry ratio 2:1 was attributed to hydrogen bonding (NH----O, C-H----O, C^+H---O) and PET mediated quenching. On the other hand, with addition of all these anions in aqueous methanol solution of **46**, only HPO_4^{2-}, PO_4^{3-} and HSO_4^- caused increase in emission intensity of anthracene at high concentrations of guests and showed preference for HSO_4^- with 1:1 stoichiometry.

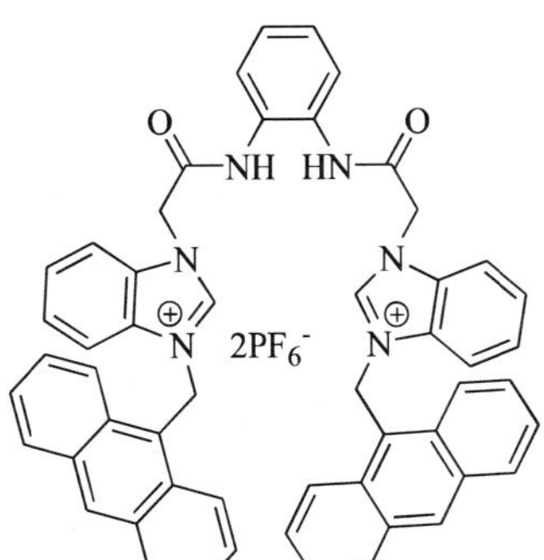

46

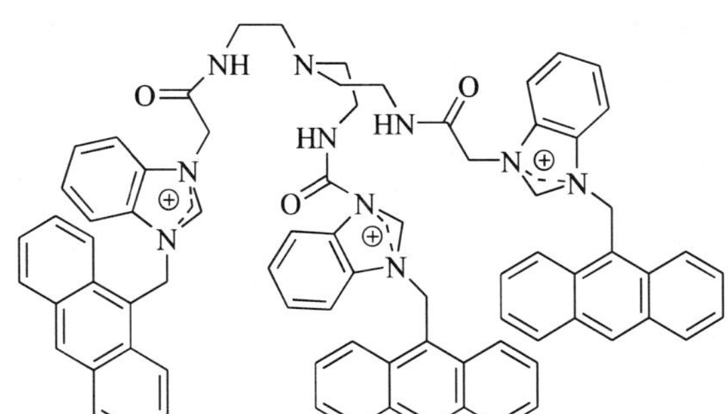

47

Ghosh et al. synthesized a tris-amine-coupled benzimidazolium based fluorescent chemosensor **47** which underwent selectively binding $H_2PO_4^-$ among other anions [70]. The UV-vis. spectra of **47** exhibited an absorption band at 370 nm, whose intensity decreased with red shift upon complexation of $H_2PO_4^-$ ion. The fluorescence spectra of **47** displayed emission bands at 400, 418 and 440 nm upon excitation at 370 nm in CH_3CN. On the addition of $H_2PO_4^-$, large decrease in emission intensity of excimer peak at 500 nm was observed. The stoichiometry ratio of **47** with $H_2PO_4^-$ found to be 3:1.

Baitalik et al. designed and synthesized a series of bis-tridentate Ru(II) and Os(II) complexes **48a-48d** having ditopic ligands along with other tridentate ligands with improved luminescence and longer excited state lifetime [71]. The spectrophotometric data such as UV-vis. absorption, steady-state and time-resolved luminescence and cyclic voltametry of complexes were recorded using 20 μM solutions of **48a-48c** (in DMSO) and **48d** (in CH_3CN). The MLCT band of **48a-48c** was located around 504-508 nm that underwent bathochromic shift to 530-542 nm upon addition of $F^-$and CN^- ions; whereas similar spectral and color change, but to a lesser extent (~6-8 nm red shift) was observed with AcO^-. The presence of Cl^-, Br^-, I^-, NO_3^- and ClO_4^- did not cause any prominent color and spectral changes in the absorption spectra of **48a-48c**. The association constants (M^{-1}) for **48a** towards these anions were determined 2.97 x 10^5 for F^-, 2.68 x 10^5 for AcO^- and 3.69 x 10^5 for CN^-. In case of **48b**, the values were found to be 2.50 x 10^3 M^{-1} for F^-, 2.17 x 10^5 for AcO^- and 3.54 x 10^5 for CN^-. In case of **48c**, the association constant values were 2.27 x 10^5 for F^-, 1.76 x 10^5 for AcO^- and 2.85 x 10^5 for CN^-. The anticipated sensing mechanism behind these observations was hydrogen-bonding or proton transfer between the imidazole –NH protons and the anions. In case of metalloreceptor **48b** having two different ligands with imidazole NH protons exhibited slightly different behaviour. The UV-vis. spectra of **48d** in CH_3CN was altered in the presence of F^-, CN^-, AcO^- in terms of color change and absorption band (MLCT band at 493 nm) shifting towards longer wavelength i.e., from 493 nm to a range of 522-537 nm. Addition of any other anions caused no change in the absorption spectra. The successive addition of these anions into the solution of **48d** led to two step deprotonation of imidazole protons. In the first step change in UV-vis. spectra were observed upon addition of said anion up to 1 equiv. showing 1:1 stoichiometric ratio. On the other hand, the second step took place when excess of these anions were added, showing two different association constants (M^{-1}) K_1 and K_2 i.e., 3.11 x 10^6 and 3.84 x 10^6 for CN^-, 2.84 x 10^6 and 3.80 x 10^6 for F^-, 2.12 x 10^6 and 2.61 x 10^6 for AcO^-. In the fluorescence spectra, emission bands for **48a**, **48b** and **48c** were observed in DMSO at 665 nm, 668 nm and 761 nm, respectively. In the presence of 10 equiv. of F^- and CN^- ions, intensities of emission maxima underwent complete quenching, while AcO^- did not quench completely. However, addition of F^-, AcO^- and CN^- ions to the CH_3CN solution of **48d** led to complete quenching of emission maxima with concomitant red shift from 682 nm to 820 nm. The time-resolved fluorescence responses of these complexes also showed selectively towards F^-, AcO^-, and CN^- ions. These complexes showed longer excited-state lifetimes in a range of 3.8 to 161.1 ns. Upon addition of F^-, AcO^-, and CN^- anions the luminescence lifetime of **48a-48d** underwent decay process. The anions induced deprotonation of imidazole NH was further supported by cyclic voltametry and ^1H NMR titration and X-ray crystallography.

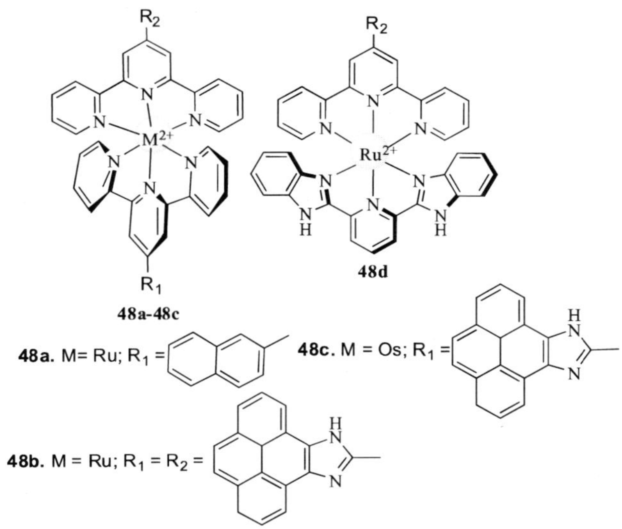

Wang and coworkers developed Ru^{2+} complex **49** which underwent significant changes in UV-vis. and fluorescence spectra on addition of $H_2PO_4^-$, F^- and AcO^- ions [72]. The UV-vis. of **49** in CH_3CN exhibited a broad metal to ligand charge transfer band at 460 nm and two $\pi \rightarrow \pi^*$ bands at 328 nm and 288 nm. On the addition of various anions, **49** served as 'turn off' sensor for F^- and AcO^- ions; while 'turn on' emission sensor for $H_2PO_4^-$ ions. The stoichiometry ratios between **49** and $H_2PO_4^-$/ F^-/AcO^- were found to be 1:1, 1:1 and 1:2, respectively confirmed by Job's plot. Furthermore, on the addition of various cations to the solution of **49** in a CH_3CN-HEPES buffer aqueous solution (1/71,v/v, pH 7.2), only Fe^{2+} ion addition resulted in new strong band at 575 nm with color change from pale yellow to light red purple.

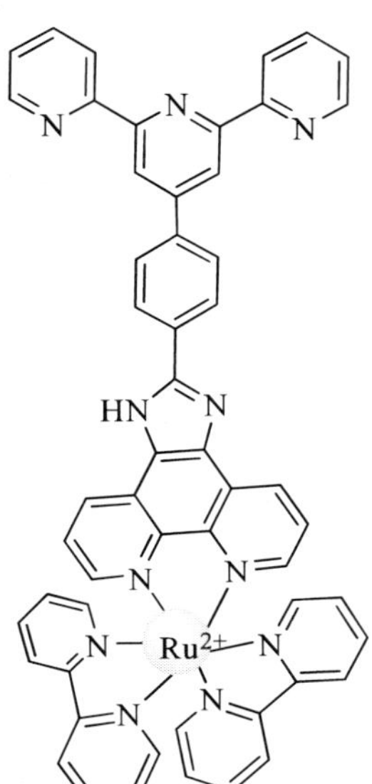

49

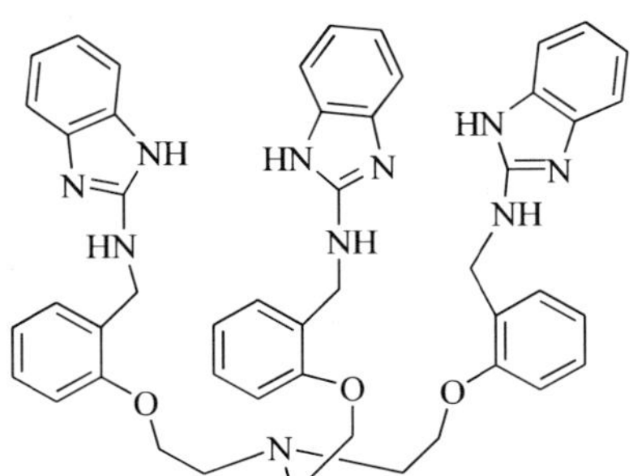

50

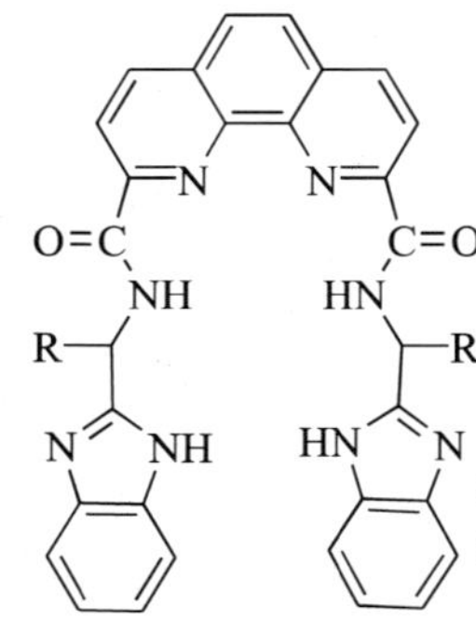

51a. R = H
51b. R = CH_3

Jang et al. synthesized tripodal fluorescent chemosensor **50** made up of three benzimidazole scaffolding for selective detection of I^- ion, over a wide range of other anions such as F^-, Cl^-, Br^-, HSO_4^-, NO_3^-, CH_3COO^- and $H_2PO_4^-$ in CH_3CN-H_2O (9:1, v/v, pH 7.0) [73]. The chemosensor **50** exhibited emission band at 495 nm, which underwent quenching with addition of only I^- ions. This selectivity for I^- was attributed to the flexibility of the chemosensor to accommodate the large sized anion i.e., I^-. The 1:1 stoichiometric complex formed by chemosensor **50** and I^- had association constant of 2.2 x 10^5 M^{-1}.

Different fluorescence responses with different anions i.e., 'turn-*on*' with Cl^-, Br^- and I^- while 'turn-*off*' with F^- and AcO^- ions using benzimidazole chemosensors **51a** and **51b** were observed by Lin and group [74]. In DMSO, fluorescent enhancement was observed by addition of Cl^-, Br^- or I^- to solution of **51a-51b** due to increased rigidity of the host molecule upon anion binding. On the other hand, fluorescence quenching caused by F^- and AcO^- was due to PET mechanism.

Jang and co-workers synthesized chemosensors **52a** and **52b** bearing benzimidazole and amide functionality [75]. The UV-vis. spectrum of chemosensor **52a**, in CH_3CN/DMSO (99:1, v/v), showed two peaks at 277 nm and 339 nm. Upon addition of F^- and AcO^-, a new peak appeared at 429 nm along with naked eye color change from colorless to yellow. This was due to the charge transfer excitation from donor to the acceptor substituents. On the other hand, there was no change in absorption spectra as well as color when Cl^-, Br^-, I^-, NO_3^- and $H_2PO_4^-$ were added to the solution of chemosensor **52a**. The appearance of peak at 429 nm due to the AcO^- complexation was more intense than F^- complexation which indicated stronger complex formation with AcO^-. The complexation of chemosensor **52a** with $H_2PO_4^-$ showed a very light yellow color. However, the chemosensor **52b**, devoid of $-NO_2$ group, led to the absence of color development upon additions of any of mentioned anions.

Zhang et al. developed a long wavelength emission fluorescent and colorimetric chemosensor **53** whose Zn^{2+} complex was further used for recognition of $H_2PO_4^-$ with high selectivity by TICT (twisted intramolecular charge transfer) mechanism [76]. The absorption spectra of **53** underwent red shifting to 416 and 275 nm upon addition of Zn^{2+} ions. The stoichiometric ratio of **53.Zn^{2+}** complex was 2:1 with association constant to be 5.76×10^6 M^{-2}. On the addition of various anions to the solution of **53.Zn^{2+}** in DMSO-H_2O

(8:2, v/v) HEPES buffer, only $H_2PO_4^-$ resulted in the hypsochromic shift of the two absorption maxima i.e., band at 416 nm to 400 nm and 275 nm to 264 nm with dramatic color change from pink to yellow. In fluorescence spectra, **53.Zn^{2+}** complex exhibited a red shifted band at 600 nm due to the interaction between Zn^{2+} ions and imidazole nitrogen and N of pyridine moiety. On the addition of various anions to solution of **53.Zn^{2+}**, only $H_2PO_4^-$ exhibited blue shift in emission from 600 nm to 540 nm with detectable colour change from red to yellow. Furthermore test strips based on **53.Zn^{2+}** complex were fabricated for the practical application of sensor **53.Zn^{2+}** sensing for $H_2PO_4^-$ ions.

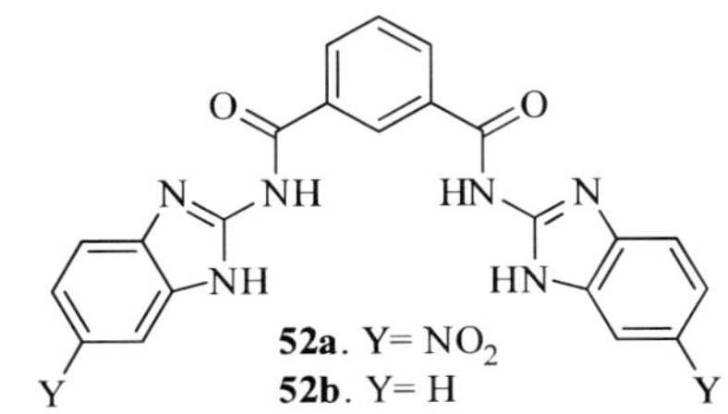

52a. Y= NO_2
52b. Y= H

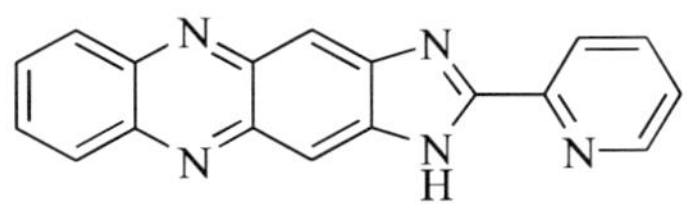

53

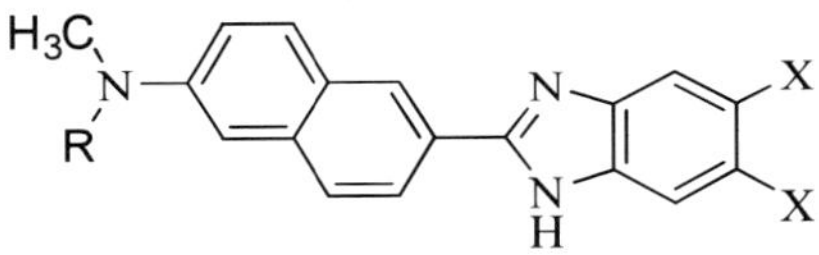

54a. R = H; X = H
54b. R = H; X = F
54c. R = H; X = OMe
54d. R = $CH_2CONHCH_2CH_2N(Me)_2$; X = H

Kim and coworkers synthesized benzimidazole based ratiometric fluorescent probes **54a-54d,** which were used to recognise acidic pH values [77]. At pH 7.2, **54a** exhibited absorption maxima at 337 nm and fluorescence emission maxima at 455 nm with Stoke's shift 118 nm. Similar results were observed for basic medium but **54a-54d** showed a marked blue to green emission color change in response to a change in pH from neutral to acidic medium. **54d** was found to be capable of monitoring the change of intralysosomal pH values along with their transportation in living cell and in live rat brain tissues with the help of two photon microscopy.

Li et al. developed a fluorescent chemosensor **55** based on benzimidazole group whose fluorescence and UV-vis. spectra changed upon addition of CN^- ions [78]. On the addition of different anions to the solution of **55** in H_2O-DMSO (8:2, v/v), only presence of CN^- resulted in appearance of new bands at 328 and 388 nm in absorption spectrum and increased fluorescence intensity at 450 nm in fluorescence spectrum along with dramatic color change from pale blue to mazarine. Furthermore, test kits were also prepared which were used as practical fluorescence sensor for CN^- detection.

Kumar and co-workers developed colorimetric and ratiometric fluorescence chemosensors **55** and **56** which exhibited selective changes with F^- amongst other tested anions [79]. Upon addition of F^- ions to CH_3CN solution of **55**, a new band appeared between 400 and 440 nm with visible color change from colourless to yellow. Also the absorbance of **56** with F^- exhibited a gradual increase in absorbance at 290, 340 and 370 nm. In the fluorescence spectra, on the addition of Cl^-, Br^-, I^-, NO_3^-, F^-, CH_3COO^-, $H_2PO_4^-$

and ClO_4^- anions to **55**, the emission intensity of 586 nm band gradually decreased with appearance of a new blue shifted band at 515 nm, with increasing concentration of F^- ions.

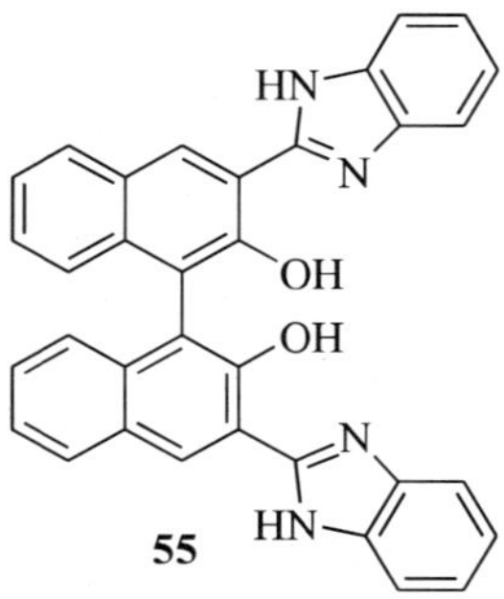

55

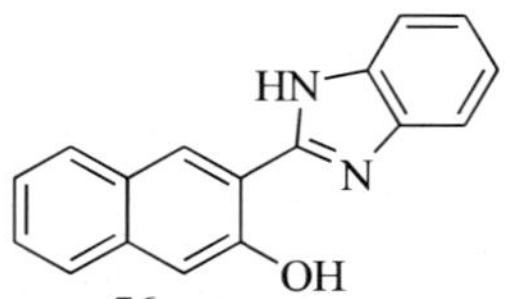

56

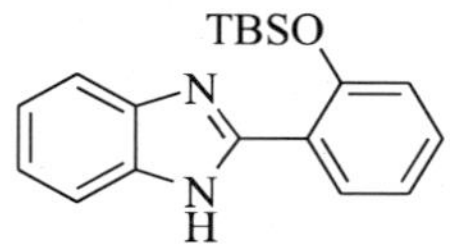

57

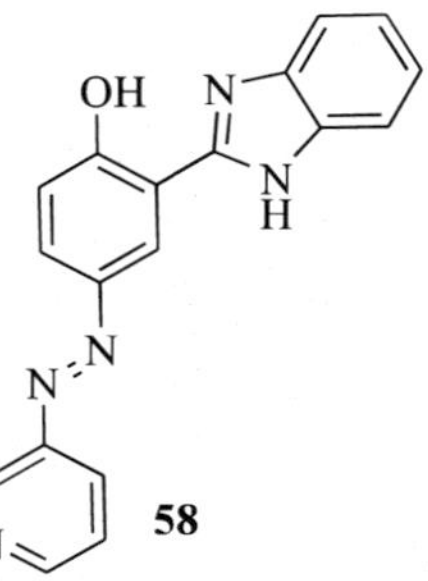

58

Yang et al. synthesized a ratiometric fluorescent probe **57** based on modulation of ESIPT process through protection /deprotection of functional hydroxy group [80]. In the UV-vis. spectra of **57**, with an increasing amount of F^-, there was decrease in absorption intensity of band at 293 nm; while two new bands appeared at 315 and 324 nm. On the addition of various anions to the solution of **57** in aqueous DMF solution, only addition of F^- gradually decreased the intensity of emission band at 360 nm along with growth of a new emission band at 454 nm. The detection limit was determined to be 0.19 μM.

Kaur et al. developed a colorimetric chemosensor consisting of benzimidazole and azo dye moieties to investigate anions such as AcO^-, CN^-, F^-, $H_2PO_4^-$ over a wide range of other related anions like Cl^-, Br^-, I^-, HSO_4^- ions [81]. The UV-vis. absorption spectra of **58** (10 μM) in CH_3CN displayed an absorption band at 315 nm along with a nearby shoulder at 365 nm. When CN^- ions were added even in small amount (0.2 equiv.), emergence of a new absorption maximum at 470 nm was monitored due to intramolecular charge transfer (ICT). After addition of 3 equiv. of CN^-, a distinct enhancement (~30 fold) in absorption intensity of the new band occurred eventually with simultaneous decrease in absorption maximum at 315 nm, which reached to its saturation upto 5 equiv. of CN^-. The bathochromic shift (105 nm) and increased absorbance rendered a conspicuous color change to the solution from colorless to dark yellow. Similar behavior was found for the rest three anions viz. AcO^-, F^-, $H_2PO_4^-$. The stoichiometric ratio was estimated to be 1:2 (H:G) for all the four anions. The strength of binding interaction followed the order $AcO^- > CN^- > F^- > H_2PO_4^-$. This was ascribed to the tendency of oxygen atom of acetate to stabilize the negative charge more compared to the benzimidazole nitrogen. Rest three ions followed their basicity order.

Singh et al. synthesized a benzimidazole based Co^{3+} complex **59** which recognized I^- and HSO_4^- ions by giving selective changes in cyclic voltammetry and UV-vis. spectra [82]. The differential pulse voltammetry (DPV) of complex **59** in MeOH/H_2O (8:2, v/v), tetrabutyl ammonium perchlorate as electrolyte resulted an oxidation peak at -0.235V. On the addition of various ions like F^-, Cl^-, Br^-, I^-, NO_3^-, CN^-, HSO_4^-, and $H_2PO_4^-$ to the solution of **59**, only I^- caused a shift in oxidation potential from -0.235 to -0.512 V and after addition of 40 equivalent of I^-, potential was constant. The addition of HSO_4^- to **59**, did not cause any change in electrochemical signal; whereas selective changes in UV-vis. spectra with visible color change was observed. The UV-vis. spectra of **59** with HSO_4^- resulted in a new band at 400 nm with simultaneous decrease of band at 347 nm. Thus **59** showed high selectivity for HSO_4^- compared to other tested anions and the detection limit and association constant values were found to be 63 μM and $(1.2\pm0.12)\times10^3$ M^{-1}, respectively.

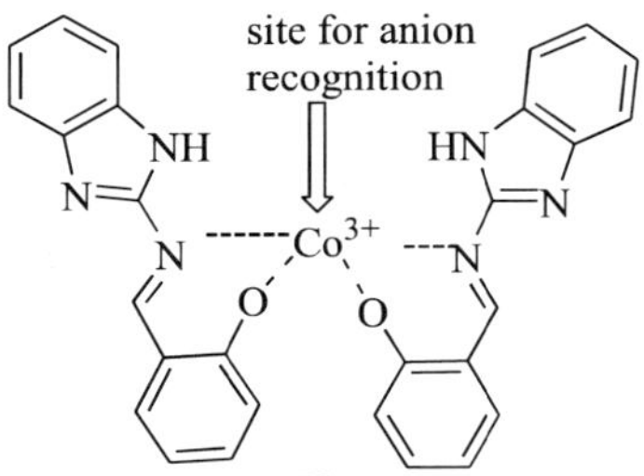

59

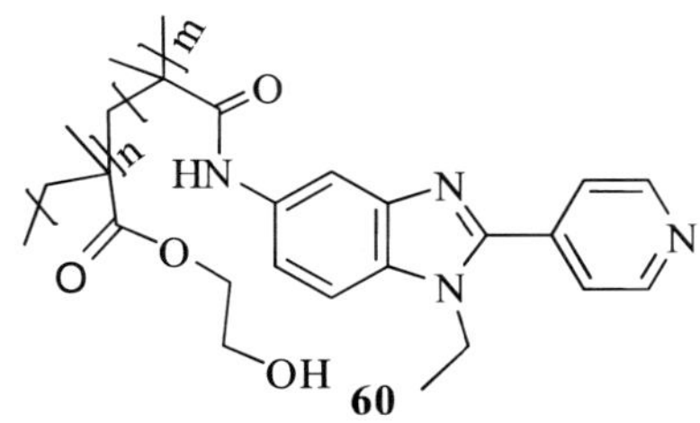

60

Zhu et al. synthesized a ratiometric hydrophilic copolymer **60**, which underwent changes in UV-vis. and fluorescence spectra with the variation of pH within 1.7-4.5 [83]. With the continuous addition of TFA to **60**, absorption band became flat and shifted to longer wavelength with a new absorption band at 337 nm. In fluorescence spectra, addition of aliquots of TFA caused decrease in emission band at 423 nm along with appearance of new emission band at 440 nm.

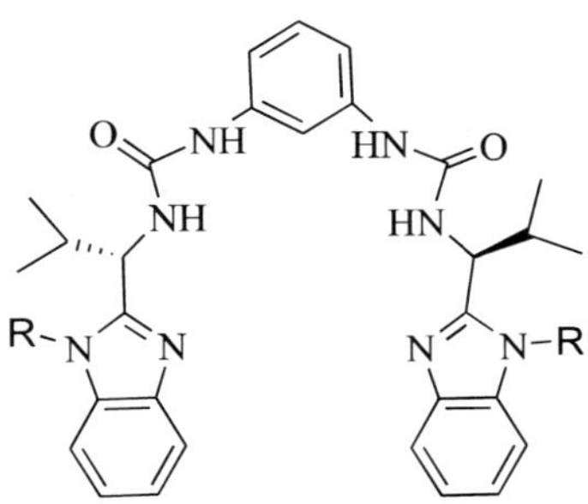

61a. R = $(CH_2)_7CH_3$
61b. R = H

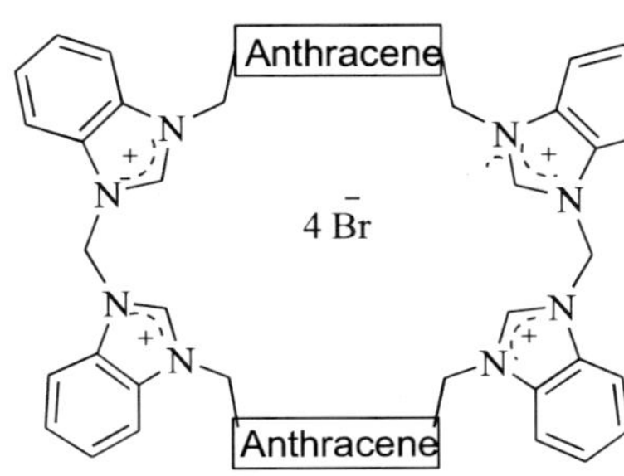

62a. 1,8 substituted anthracene
62b. 9,10 substituted anthracene

Ghosh et al. synthesized L-valine derived benzimidazole based two receptors **61a** and **61b** which underwent enantioselective fluorescence sensing of L-tartrate and D-tartrate, respectively [84]. The UV-vis. spectra of **61a** with all enantiomers of tartaric acids and mandelic acids in DMSO exhibited minor change in absorbance. However, in the fluorescence spectra, upon addition of enantiomers of tartaric acids and mandelic acids to solution of **61a** in DMSO, only L-tartrate quenched the emission intensity at 315 nm more effectively. In comparison, sterically less crowded receptor **61b** gave more preference for

D-tartrate over L-tartrate in DMSO. The values of binding constant for L-tartrate and D-tartrate were found to be $(1.30\pm0.13) \times 10^4$ M^{-1} and $(4.81\pm0.36) \times 10^4$ M^{-1} respectively.

Kim et al. developed two new fluorescent benzimidazolium – anthracene based cyclophane receptors **62a** and **62b**, which were used for recognition of I^- and GTP anions [85]. On the addition of different anions to the aqueous solution of **62a**, only GTP caused fluorescence quenching because of chelation enhanced fluorescence quenching effect. Other anions did not exhibit significant changes in the fluorescence spectra of **62a**. The Job's plot confirmed 1:1 stoichiometric ratio with binding constant 1.67×10^4 M^{-1} for **62a.**GTP and detection limit was found to be 4.22×10^{-5} M. Furthermore, on the addition of various anions and halides to the solution of **62b**, only I^- caused maximum chelation enhanced fluorescence quenching over other ions because of heavy atom effect. The stoichiometric ratio for **62a.**I^- was found to be 1:2 with the values of binding constants (K_1/K_2) and detection limit to be $7.6\times10^2/5\times10^4$ M^{-1} and 2.67×10^{-5} M, respectively. Furthermore the ^{1}H NMR and theoretical calculations were also examined for their binding modes.

Wang and co-workers invoked the use of Eu(III) thenoyltrifluoroacetonate complex as the initial building block to attach it to derivatized phen ligand to retain high luminescence intensity and also to meet the requirement of a high coordination number (C.N) of Eu(III) ion **63b** [86]. The absorption behavior of the receptor **63a** was investigated in DMSO, with F^- and other anions. Upon addition of F^- ion, concomitant decreases of the peak at 333 nm along with appearance of a new band at 344 nm was observed. Rest the halide anions could not alter UV-vis. spectra of **63a**. On the contrary, luminescent Eu(III) complex, **63b**, showed emission peaks in the red region at 583, 590, 615, 650 and 700 nm ($\lambda_{ex} = 350$ nm), which underwent quenching with addition of different anions, giving rise to varied luminescence colors. The changes with F^- and AcO^- were attributed to the hydrogen bonding interactions of the F^-/AcO^- with NH of the receptor. The addition of $H_2PO_4^-$ decreased the emission of Eu^{3+} ions with simultaneous increase in the broad band of phen in the blue region which was attributed to displacement of phen ligand by $H_2PO_4^-$, hence, displaying phen's own blue emission. Upon embedding **63b** into the polymeric hosts, the thick films like PMMA [Poly(methyl methacrylate)] were formed on the glass substrate, which increased the quantum yield of Eu(III) ions and energy transfer efficiency.

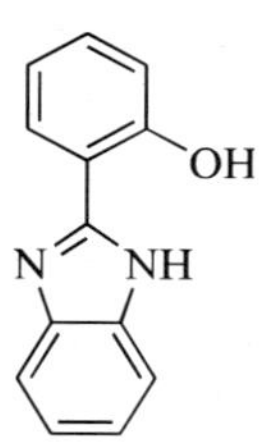

63a

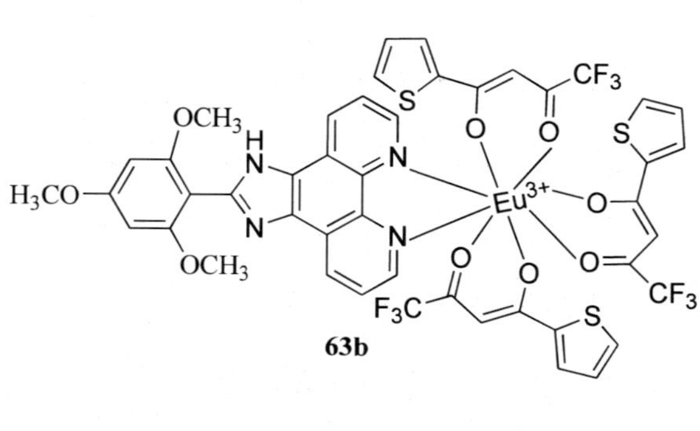

63b

Tian et al. designed and synthesized a naphthalene benzimidazole (NBI) based receptor **64** for F^- ion recognition [87]. Among various anions (F^-, Cl^-, Br^-, I^-, HSO_4^-, AcO^-, $H_2PO_4^-$ and ClO_4^-), **64** could sense F^- ion with a drastic color change from yellow to blue (visible region) along with a 225 nm bathochromic shift (λ_{max} 395 to 620 nm) in DMSO. Furthermore, uncomplexed **64** showed a strong fluorescence emission at 530 nm. When F^- ion added, the fluorescence intensity of **64** at 530 nm significantly quenched. The tremendous selectivity and sensitivity (LOD 3.2 X 10^{-9} M/L) for F^- detection was attributed to the F^- triggered Si-O bond cleavage (desilylation) and intramolecular charge-transfer mechanism. This sensing mechanism was supported by ^{1}H NMR and HRMS-ESI data.

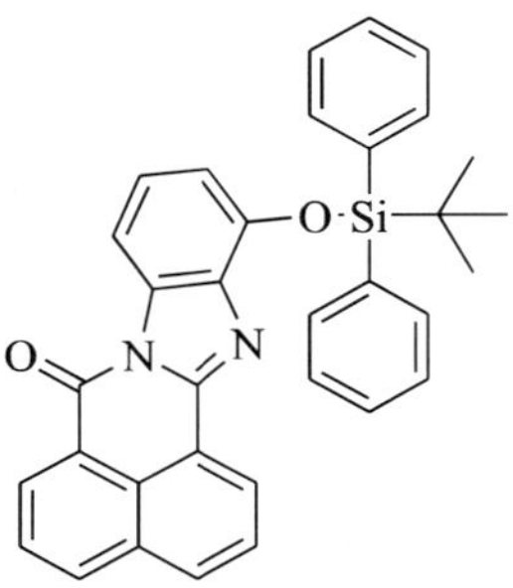

64

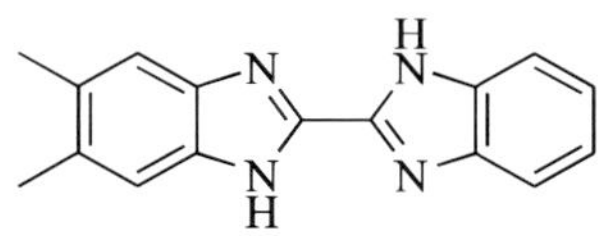

65

Chaudhari and co worker prepared benzimidazole based solid state sensor **65** for naked eye detection of F^-/AcO^- ions selectively accompanied by a dramatic color change of **65** from yellow to deep green and ultimately to brown both in solid state and in solution phase [88]. The sensor **65** displayed four absorption maxima at 292, 327, 414 and 469 nm. After addition of F^-/AcO^- salt solution to **65**, in CH_3CN, exhibited red shifted absorption maxima from 212 nm to 215 nm and 327 nm to 333 nm. Two new bands, noticed at 427 nm and 441 nm, were marked as the charge transfer bands. These red shift and the color change effects were attributed to the hydrogen bonding interaction of F^- with **65** which gradually resulted in complete deprotonation. These results were further supported by photophysics, ^{1}H NMR, CV and DFT studies. The binding constant for **65** with F^- (30.75 M^{-1}) and AcO^- (81.65 M^{-1}), limit of detection for F^- (0.38 mM) and stoichiometric ratio (1:2) were also determined.

REFERENCES

[1] Beer, P. D. (1998) *Acc. Chem. Res.,* 31: 71-80.

[2] Bargossi, C., Fiorini, M. C., Montalti, M., Prodi, L. and Zaccheroni, N. (2000) *Coord. Chem. Rev.,* 208: 17-32.

[3] de Silva, A. P., Gunaratne, H. Q. N., Gunnlaugsson, T., Huxley, A. J. M., McCoy, C. P., Rademacher, J. T. and Rice, T. E. (1997) *Chem. Rev.*, 97: 1515-1566.

[4] Orellana, G. and Moreno-Bondi, M. C. (1005) *Frontiers in Chemical Sensors: Novel Principles and Techniques*, Springer: New York.

[5] (a) de Silva, A. P., Mc Caughan, B. O. F. and Querol, M. (2003) *Dalton Trans.*, 1902-1913; (b) Anslyn, E. V. (2001) Angew Chem. *Int. Ed. Engl.*, 40: 3119-3130; (c) Amendola, V., Fabbrizzi, L., Mangano, C. and Pallavicini, P. (2001) *Acc. Chem. Res.*, 34: 488-493; (d) Czarnik, A. W. (1994) *Acc. Chem. Res.,* 27: 302-308.

[6] de Silva, A. P., Fox, D. B., Huxley, A. J. M. and Moody, T. S. (2000) *Coord. Chem. Rev.,* 205: 41-57.

[7] Zhou, Y., Zhang, J. F. and Yoon, J. (2014) *Chem. Rev.*, 114: 5511-5571.

[8] Martinez-Manez, R. and Sancenon, F. (2003) *Chem. Rev.,* 103: 4419-4476.

[9] Bissell, R., de Silva, A. P., Gunaratne, P. H. Q. N., Lynch, P. L. M., Maguire, G. E. M. and Sandanayake, K. R. A. S. (1992) *Chem. Soc. Rev.*, 21: 187-195.

[10] Wiskur, S. L., Ait-Haddou, H., Lavigne, J. J. and Anslyn, E. V. (2001) *Acc. Chem. Res.*, 34: 963-972.

[11] Martinez-Manez, R. and Sancenon, F. (2006) *Coord. Chem. Rev.*, 250: 3081-3093.

[12] Prodi, L., Bollta, F., Montalti, M. and Zaccheroni, N. (2000) *Coord. Chem. Rev.*, 205: 59-83.

[13] Kaur, B., Kaur, N. and Kumar, S. (2018) *Coord. Chem. Rev.,* 358: 13-69.

[14] Anslyn, E. V. (1999) *Curr. Opin. Chem. Biol.*, 3: 740-746.

[15] Henary, M. M. and Fahrni, C. J. (2002) *J. Phys. Chem. A,* 106: 5210-5220.

[16] (a) Shiraishi, Y., Zhang, X. and Hirai, T. (2007) *Tetrahedron Lett.*, 48: 5455-5459; (b) Thennarasu, S., Mandal, A. B., Chereddy, N. R., Suman, K. and Korrapati, P. S. (2012) *Dyes Pigm.,* 95:, 606-613.

[17] Son, Y. A., Lee, S. and Rao, B. A. (2014) *Sensors Actuat. B-Chem.,* 196: 388-397.

[18] Bera, M., Haldar, S. and Giri, G. C. (2016) *Int. J. Res. Soc. Nat. Sci.,* 1: 1-10.

[19] Matalobos, J. S., Das, D., Banerjee, A., Sahana, A., Guha, S., Lohar, S., Hauli, I. and Mukhopadhyay, S. K. (2012) *Inorg. Chem.*, 51: 5699-5704.

[20] Gunnlaugsson, T., Duke, R. M., Veale, E. B., Pfeffer, F. M. and Kruger, P. E. (2010) *Chem. Soc. Rev.,* 39: 3936-3953.

[21] Choi, K. and Hamilton, A. D. (2001) *J. Am. Chem. Soc.*, 123: 2456-2457.

[22] Gale, P. A., Chen, Z., Drew, M. G. B., Heath, J. A. and Beer, P. D. (1998) *Polyhedron,* 17: 405- 412.

[23] Ma, J. C. and Dougherty, D. A. (1997) *Chem. Rev.*, 97: 1303-1324.

[24] He, Y., Bing, Q., Wei, Y., Zhang, H. and Wang, G. (2019) *luminescence*, 34: 153-161.

[25] Li, C., Fang, M., Zhu, W., Zhang, Q., Xu, Y. and Wu, Z. (2018) *J. Chin. Chem. Soc.*, 65: 597-605.

[26] Liu, Q. X., Hu, Z. L. and Zhao, Z. X. (2018) *New J. Chem.,* 42: 20049-20055.

[27] Jayabharathi, J., Thanikachalam, V., Jayamoorthy, K. and Sathishkumar, R. (2012) *Spectrochim. Acta Mol. Biomol. Spectrosc.*, 97: 384-387.

[28] Mondal, T. K., Sarkar, D. and Pramanik, A. K. (2016) *Spectrochim. Acta Mol. Biomol. Spectrosc.*, 153: 397-401.

[29] Luxami, V., Gupta, A. S. and Paul, K. (2016) *Inorg. Chim. Acta.*, 443: 57-63.

[30] Krishnamoorthy, G., Chipem, F. A. S. and Behera, S. K. (2014) *Sensor Actuat. B-Chem.*, 191: 727-733.

[31] Gupta, T., Kumar, A. and Chhatwal, M. (2012) *Tetrahedron Lett.*, 53: 5691-5694.

[32] Wei, T. B., Zhang, Y. M., You, X. M., Yao, H., Guo, Y., Zhang, P., Shi, B. B., Liu, J. and Lin, Q. (2014) *Supramol. Chem.,* 26: 39-47.

[33] Lin, Q., Wei, T. B., Yao, H., Wang, J., Zhou, Q., Guan, X. W., Fan, Y.-Q. and Zhang, Y.-M. (2018) *Soft Matter, 14*, 8390-8394.

[34] Cui, G. H., Li, J. X., Qin, Z. B. and Li, Y. H. (2018) *Polyhedron*, 151: 530-536.

[35] Das, D., Phapale, D. and Kushwaha, A. (2019) *Spectrochim. Acta Mol. Biomol. Spectrosc.*, 214: 111-118.

[36] Jiang, Y., Tan, Y., Liu, H., Zhang, B., Tan, C., Liu, Feng. and Cao, J. (2016) *Talanta,* 161: 309-319.

[37] Kim, C., Kim, Y. S., Lee, J. J., Lee, S. Y. and Jo, T. G. (2016) *RSC Adv.*, 6: 61505-61515.

[38] Horak, E., Hranjec, M., Vianello, R. and Steinberg, I. M. (2018) *Supramol. Chem.*, 30: 891-900.

[39] Wang, G., Bing, Q., Wang, L. and Li, D. (2018) *Spectrochim. Acta Mol. Biomol. Spectrosc.,* 202: 305-313.

[40] Jang, D. O., Kim, M. J., Kaur, K. and Singh, N. (2012), *Tetrahedron*, 68: 5429-5433.

[41] Sessler, J. L., Kim, D., Yoon, J., Ishida, M., Kim, P. and Choi, J. (2013) *Chem. Commun.*, 49: 6950-6952.

[42] Jang, D. O., Singh, N., Saluja, P. and Kaur, N. (2012) *Tetrahedron,* 68: 8551-8556.

[43] Jang, D. O., Singh, N., Saluja, P. and Kaur, N. (2012) *Tetrahedron Lett.,* 53: 3292-3295.

[44] Jang, D. O., Jung, H. J., Singh, N. and Lee, D. Y. (2009) *Tetrahedron Lett.,* 50: 5555-5558.

[45] Jang, D. O., Jung, H. J. and Singh, N. (2008) *Tetrahedron Lett.,* 49: 2960-2964; (b) Jang, D. O., Singh, N., Saluja, P., Sharma, H. and Kaur, N. (2012) *Tetrahedron,* 68: 2289-2293.

[46] Jang, D. O., Singh, N., Goh, H., Ko, Y. G., Nam, T. K. and Singh, A. (2016) *Tetrahedron Lett.,* 57: 4435-4439.

[47] Jang, D. O., Singh, N., Saluja, P., Kaur, N. and Kang, J. (2013) *Tetrahedron*, 69: 9001-9006.

[48] Jang, D. O., Singh, N., Kim, M. J. and Sharma, H. (2013) *Inorg. Chem. Commun.*, 36: 96-99.

[49] Mathur, P. and Mahiya, K. (2013) *J. Fluoresc.*, 23: 767-776.

[50] Mathur, P. and Mahiya, K. (2013) *Spectrochim. Acta Mol. Biomol. Spectrosc.,* 113: 386-392.

[51] Ghosh, K. and Sarkar, T. (2012) *Supramol. Chem.,* 24: 748-754.

[52] Son, Y. A., Sheng, W. and Yu, H. (2012) *Mol. Cryst. Liq. Cryst.*, 568: 117-124.

[53] Pan, Z., Yu, X., Li, J., Hu, Q., Zeng, Y., Cao, C., Liu, X. and Guo, J. (2011) *J. Fluoresc.,* 21: 2005-2013.

[54] Goswami, S., Maity, S., Maity, A. C., Maity, A. K., Das, A. K. and Saha, P. (2014) *RSC Adv.,* 4: 6300-6305.

[55] Ghosh, K. and Sarkar, T. (2011) *Supramol. Chem.,* 23: 435-440.

[56] Ghosh, K., Sen, T. and Patra, A. (2010) *New J. Chem.,* 34: 1387-1393.

[57] Tang, L., Zhou, P., Zhong, K. and Hou, S. (2013) *Sensor Actuat. B-Chem.,* 182: 439-445.

[58] Chellappa, D., Jeyanthi, D., Iniya, M. and Krishnaveni, K. (2013) *RSC Adv.,* 3, 20984-20989.

[59] Tang, L., Nandhkumar, R., Cai, M., Huang, Z., Zhong, K., Hou, S. and Bian, Y. (2013) *Sensor Actuat. B-Chem.,* 185: 188-194.

[60] Tang, L., Zhong, K., Cai, M., Hou, S. and Bian, Y. (2014) *Bull. Korean. Chem. Soc.,* 35: 489-493.

[61] Raposo, M. M. M., Batista, R. M. F., Oliveira, E., Costa, S. P. G. and Lodeiro, C. (2011) *Talanta,* 85: 2470-2478.

[62] Ravikanth, M., Chowdhury, A., Madhu, S., Sharma, D. K., Basu, S. K. and Jadhav, S. (2013) *Inorg. Chem.,* 52: 11136-11145.

[63] Wu, A., Xue, W., Wang, J. and Wang, M. (2013) *Dyes Pigm.,* 97: 475-480.

[64] Zhang, Y., Wei, T., Liu, J., Yao, H., Lin, Q., Xie, Y., Shi, B., Zhang, P. and You, X. (2013) *Chin. J. Chem.,* 31: 515-519.

[65] Yang, W., Peng, J., Du, B., Liu, R., Zhang, Y., Sun, W., Sun, M. and Cao, Y. (2007) *Polymer,* 48: 1245-1254.

[66] Gao, G., Martinez, A., Zhang, D., Jiang, X., Yang, H., Su, Z. and Gao, E. (2013) *Chem. Commun.,* 49: 6149-6151.

[67] Gao, G., Martinez, A., Zhang, D., Jiang, X., Yang, H., Feng, M. and Dong, Z. (2013) *Org. Biomol. Chem.,* 11: 3375-3381.

[68] Gao, G., Martinez, A., Zhang, D., Jiang, X., Dong, Z. and Yang, H. (2013) *Tetrahedron,* 69: 10457-10462.

[69] Ghosh, K., Saha, I. and Patra, A. (2009) *Tetrahedron Lett.,* 50: 2392-2397.

[70] Ghosh, K. and Saha, I. (2011) *Supromol. Chem.,* 23: 518-526.

[71] Baitalik, S., Maity, D., Bhaumik, C. and Mondal, D. (2013) *Inorg. Chem.,* 52: 13941-13955.

[72] Wang, K. Z., Zheng, Z. B., Duan, Z. M. and Ma, Y. Y. (2013) *Inorg. Chem.,* 52: 2306-2316.

[73] Jang, D. O. and Singh, N. (2007) *Org. Lett.,* 9: 1991-1994.
[74] Lin, H., Shao, J., Qiao, Y. and Lin, H. (2009) *J. Fluoresc.,* 19: 183-188.
[75] Jang, D. O., Moon, K. S., Singh, N. and Lee, G. W. (2007) *Tetrahedron,* 63: 9106-9111.
[76] Zhang, Y., Shi, B., Wei, T., Lin, Q., Yao, H., Zhang, P. and You, X. (2014) *Sensor Actuat. B-Chem.*, 190: 555-561.
[77] Kim, H. M., Kim, H. J. and Heo, C. H. (2013) *J. Am. Chem. Soc.*, 135: 17969-17977.
[78] Hu, J. H., Li, J. B., Chen, J. J. and Qi, J. (2014) *Spectrochim. Acta Mol. Biomol. Spectrosc.*, 133: 773-777.
[79] Kumar, S. and Luxami, V. (2007) *Tetrahedron Lett.,* 48: 3083-3087.
[80] Yang, X. F., Qi, H., Wang, L., Su, Z. and Wang, G. (2009) *Talanta,* 80: 92-97.
[81] Kaur, N., Dhaka, G. and Singh, J. (2015) *Tetrahedron Lett.,* 56: 1162-1165.
[82] Singh, N., Jang, D. O., Sharma, H., Bhardwaj, V. K. and Kaur, N. (2013) *Tetrahedron Lett.,* 54: 5967-5970.
[83] Zhu, W., Shen, L. and Zhao, P. (2011*) Dyes Pigm.,* 89: 236-240.
[84] Ghosh, K. and Sarkar, T. (2013) *Tetrahedron Lett.,* 54: 4568-4573.
[85] Kim, K. S., Ahmed, N., Shirinfar, B., Youn, S. and Yousuf, M. (2013) *Org. Biomol. Chem.,* 11: 6407-6413.
[86] Wang, Q., Zheng, Y. and Tan, C. (2012) *Luminescence,* 27: 302-306.
[87] Tian, Z., Li, D., Zhong, Z. and Zheng, G. (2017) *Spectrochim. Acta Mol. Biomol. Spectrosc.,* 185: 173-178.
[88] Chaudhari, T., Mondal, A. and Mukhopadhyay, C. (2018) *J. Mol. Liq.,* 251: 35-39.

In: Benzimidazole
Editor: Aksel A. Vestergaard
ISBN: 978-1-53616-986-7

Chapter 6

MULTIFUNCTIONAL COMPOUNDS: A NEW JOURNEY OF BENZIMIDAZOLE

Richa Minhas, Ankit Singhal, Gulshan Bansal and Yogita Bansal*

Department of Pharmaceutical Sciences and Drug Research, Punjabi University, Patiala, Punjab, India

ABSTRACT

Benzimidazole has evolved as an important heterocyclic system for the development of diversely active molecules over several decades of research worldwide. Starting from its exploration for developing physiologically active molecules (such as vitamin B12 derivative) to single-target drugs (such as omeprazole, astemizole, enviradene, telmisartan and others), it has emerged as a nucleus of choice for the development of multifunctional drugs for treatment and/or management of multifactorial diseases such as cancer, rheumatoid arthritis, multiple sclerosis, etc. Various reviews explaining the therapeutic journey of benzimidazole derivatives as single-target drugs have appeared in literature during last few years. However, despite the exploration of benzimidazole scaffold for development of multifunctional molecules during last decade by different research groups, no review on this multifaceted ability of benzimidazole has appeared so far. Also, numerous specialized synthetic methods to introduce different chemical descriptors at various positions of benzimidazole are suggested in literature that assist a medicinal chemist to synthesize complex multi-target drugs. Herein, we have searched, reviewed, analyzed and compiled various reports on the development of benzimidazole-based multifunctional molecules to provide medicinal chemists a comprehensive information on the subject till date. The present chapter is focused on important methodologies used for

* Corresponding Author's Email: yogitabansalp@rediffmail.com,yogitabansal101@gmail.com.

the synthesis of variedly substituted benzimidazoles along with their multi-target therapeutic uses.

Keywords: benzimidazole, multifactorial diseases, multi-target, multifunctional

ABBREVIATIONS

EGFR: Epidermal growth factor receptor, VEGFR2 :Vascular endothelial growth factor receptor 2, BZ: Benzimidazole, MFCs: Multifunctional compounds, FDA: Food and Drug Administration, HER2: Human epidermal growth factor receptor 2, *PDGFR: Platelet-derived growth factor receptors,* COX: Cyclooxygenase, LOX: Lipoxygenase, TNFα: Tumor Necrosis Factor α, NSAIDs: Non-steroidal Anti-inflammatory Diseases, MAO: Monoamine Oxidase, 5-HT: 5-hydroxytryptamine or serotonin, D2: Dopamine receptor, CNS: Central Nervous System, NE: Norepinephrine, AChE: Acetylcholinesterase, BChE: Butyrylcholinesterase, PPAR: Peroxisome proliferator-activated receptor, HCMV: Human cytomegalovirus, DCC: Dicyclohexyldicarbodiimide, PPA: Polyphosphoric acid, DMAP: 4-dimethylaminopyridine, SAR: Structural Activity Relationship

INTRODUCTION

Compounds that act at multiple targets often deliver superior therapeutic efficacy against complex multifactorial diseases as compared to compounds with high specificity for a single target. Keeping this in mind, medicinal chemists have been working since last few years for the development of multiple targeting compounds. A multifunctional compound (MFC) is a chemical compound obtained by joining through an appropriate linker or fusing two or more discrete molecules, and interacts independently with different biological targets (East et al., 2013). MFCs contain specific structural components that have the abilities to interact with not only a single target but multiple targets of interest (Figure 1). That's why this approach of using MFCs is helpful in combating complex disorders like rheumatoid arthritis, inflammation, cancer, cardiovascular and neurodegenerative diseases (Chen et al., 2017). Although MFCs have several preferences over single-target ligands (Abdolmaleki et al., 2017 and Kaur et al., 2014), yet a very few multi-target molecules have found their way into clinics till date. Nevertheless, these are being developed at a fast pace and getting regulatory nodes for clinical uses. Recently, about 21% of the drugs approved by FDA are reported to be MFCs (Ramsay et al., 2018), though majority of these are peptidic in nature and suffer from varied disadvantages like cost-effective and bioavailability issues (Table 1).

The major parameter that governs the development of multi-functional drugs is the selection of a suitable scaffold, which has an ability to bind with multiple pathological targets depending on the substituent present on it. Benzimidazole (BZ) is one of the pioneer nitrogen-based benzo-fused heterocyclic molecules, which was first synthesized by Hoebrecker and subsequently explored by Ladenberg and Wundt during the late 1870s (Keri et al., 2016). BZ is weakly basic, slightly less basic than imidazole, therefore is are generally soluble in dilute acids. Earlier studies suggested that physiological properties of BZ derived molecules are due to structural similarity of BZ with nucleotides present in the body (Woolley et al., 1994).

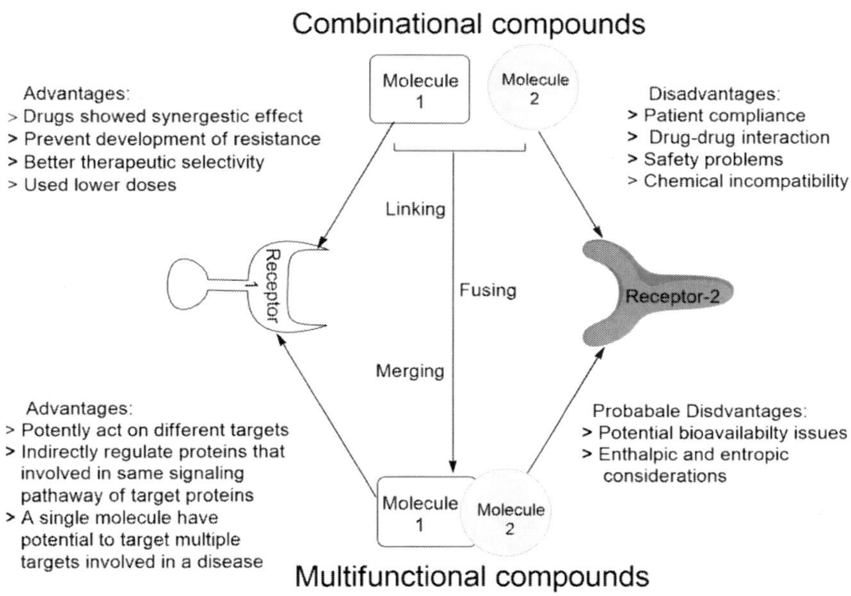

Figure 1. Advantages and disadvantages of different approaches for the treatment of multi-factorial diseases.

BZ is being therapeutically explored since the early 20th century, and is instrumental in generating numerous pharmaceutical molecules with a wide range of biological properties (Valdez et al., 2002). Numerous reports exist in literature that evidence the ability of BZ nucleus to act through a number of pharmacological and molecular mechanisms and exhibit significant pharmacological activities such as anti-ulcer, antihelminthic, anti-coagulant, anti-microbial, anti-inflammatory, anti-protozoal, anti-tumor, anti-HIV, and human cytomegalovirus (HCMV) agents (Keri et al., 2016). Many review articles have compiled the reports on BZ derivatives possessing different biological actions (Akhtar et al., 2017, Bansal et al., 2012, Narasimhan et al., 2012 and Shaharyar et al., 2017). Few reviews have also disclosed the scientific data available on BZ nucleus for particular activities (Gaba et al., 2014, Tahlan et al., 2019 and Keri et al., 2016). Exhaustive exploration of various substituents on BZ nucleus has led to the identification of a number of FDA-approved drugs (Figure 2). Due to this multifaceted ability, BZ nucleus attracts researchers to use this as a scaffold for the designing of compounds to treat complex multi-factorial disorders. The focus of this chapter is on detailed information about BZ derivatives reported as multi-targeting agent or MFCs, and various methodologies used for their synthesis.

Table 1. FDA approved MFCs

S. No.	MFC	Target	Disease Indication
1	Afatinib	EGFR, HER2	Metastatic non-small cell lung cancer
2	Bosutinib	ABL, Src	Ph+ chronicmyelogenous leukemia
3	Cabozantinib	MET, VEGFR2	Metastatic medullary thyroid cancer
4	Dasatinib	ABL, Src	Chronic myelogenous and acute lymphoblastic leukemia
5	Imatinib	ABL, PDGFR-β, c-Kit	Ph+ chronic myelogenous leukemia
6	Lapatinib	EGFR, HER2	HER2-positive metastatic breast cancer
7	Lenvatinib	VEGFR2, c-Kit	RAI-refractory differentiated thyroid cancer
8	Nintedanib	VEGFR2, PDGFR-α and β	Idiopathic pulmonary fibrosis
9	Palbociclib	CDK4, CDK6	ER-positive and HER2-negative breast cancer
10	Pazopanib	VEGFR2, c-Kit, PDGFR-β	Renal cell carcinoma and soft tissue sarcoma
11	Ponatinib	ABL, c-Kit	Chronic myeloid and acute lymphoblastic leukemia
12	Regorafenib	VEGFR2, c-Kit	Metastatic colorectal cancer and advanced gastrointestinal stromal tumor
13	Sorafenib	RET, VEGFR2, PDGFR-β, B-Raf, c-Kit	Renal cell carcinoma and unresectable hepatocellular carcinomas
14	Sunitinb	VEGFR2, c-Kit, PDGFR-β	Renal cell carcinoma and imatinib-resistant gastrointestinal stromal tumor
15	Vandetanib	EGFR, VEGFR2	Medullary thyroid cancer
16	Cariprazine	D2 and D3 receptor agonist	Schizophrenia and bipolar disorders
17	Safinamide	MAO-B and dopamine reuptake inhibitor	Parkinson's Disease
18	Midostaurin	PKC-α, VEGFR2, Kit PDGFR, FLT3 tyrosine kinases	Acute myeloid leukemia
19	Ziprasidone	D2, 5HT2A, and 5HT1D receptors	CNS disorders
20	Duloxetine	Selective serotonin (5-HT) and norepinephrine NE) reuptake inhibitors (SSNRIs)	CNS disorders
21	Ladostigil	Acetylcholinesterase, butyrylcholinesterase, and monoamine oxidases A and B	CNS disorders
22	M-30	Multiple kinases	Pro-Cognitive and Anti-Inflammatory Effects
23	Netoglitazone	PPAR α and PPAR β agonist	Diabetes
24	Omapatrilat	Neprilysin and angiotensin-converting enzyme	Hypertension

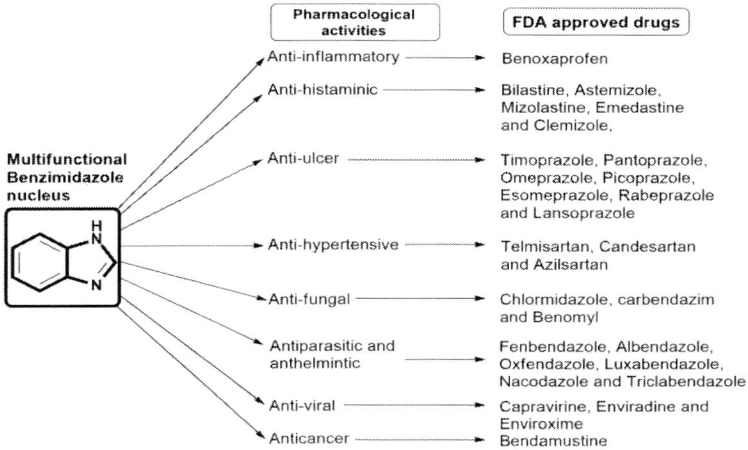

Figure 2. Different benzimidazole derived FDA-approved drugs having different pharmacological activities.

Benzimidazole Derivatives as Multifaceted Molecules

Molecules Targeting Pathological Indications of Inflammatory Disorders

Several BZ derivatives are reported in the literature, which elicit their anti-inflammatory actions by acting on a single inflammatory target or a single step of the complex inflammatory pathway (Gaba et al., 2014 and Jacob P et al., 2018). This approach produces only symptomatic effects but failed to combat inflammatory disease completely. This shortcoming of single-target therapy forced the medicinal chemists to develop BZ derivatives that can act through multiple targets. Very few reports related to the use of BZ derivatives as multifaceted molecules are available. Bansal et al. designed and synthesized a series of BZ and NSAIDs fused compounds and evaluated them for their poly-functional potential such as antioxidant, immunomodulatory and anti-inflammatory activities (*in vitro* and *in vivo*). The molecules were designed by incorporating two pharmacophores responsible for different activities *i.e.,* BZ as an isostere of guanidine moiety from immunomodulatory molecules, and NSAIDs with strong anti-inflammatory and mild antioxidant activities. Two different series were designed i.e., the chimeric molecules formed by the fusion of pharmacophores, and hybrid molecules formed by linking the pharmacophores with an amide linker. The chimeric molecules, also termed as NSAID-conjugated compounds, were prepared by refluxing *o*-phenylenediamine with NSAIDs (ibuprofen, diclofenac, salicylic acid, mefenamic acid, indomethacin, flurbiprofen, naproxen, mesalamine, ketoprofen, felbinac) in the presence of polyphosphoric acid as shown in Figure 3. Almost every chimeric and fused compound showed anti-inflammatory potential as revealed by inhibition in paw edema. Immunomodulatory activity and antioxidant potential were measured by using haemagglutination antibody titre (HAT) model, and *in vitro* and *in vivo* antioxidant models, respectively. Among all conjugates of BZ and NSAIDs, BZ and mesalamine hybrid 1 was maximally potent with 54.6% inhibition in paw volume, antibody titre value 358.4, and antioxidant EC_{50} value 0.03 μM (Bansal et al., 2013).

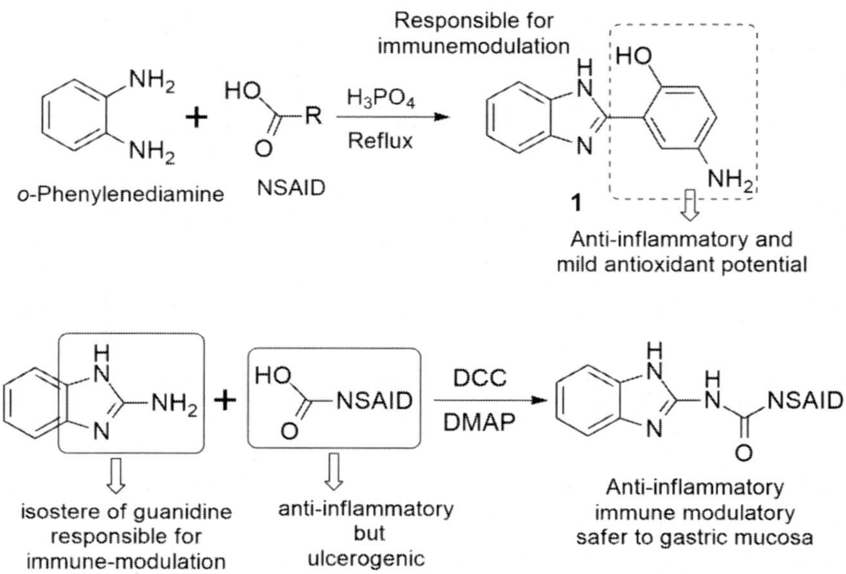

Figure 3. BZ-NSAID conjugates as anti-inflammatory agents.

2-Aminobenzimidazole-NSAID conjugates were synthesized by linking NSAIDs with amino nitrogen at 2-position of BZ through an amide linker. The target compounds were synthesized by the reaction of 2-aminobenzimidazole and NSAIDs in the presence of a coupling agent and a base as given in Figure 3. These compounds were assayed for anti-inflammatory, immunomodulatory and antioxidant activities. Results of the anti-inflammatory activity suggested that all target compounds retained the anti-inflammatory activity of the NSAID. Immunomodulatory assay of these compounds showed that a few synthesized compounds are immunosuppressant whereas others have immune-stimulatory activity. Among all synthesized derivatives, compound 2 (BZ-ibuprofen conjugate) was the most potent immunostimulant (HAT 7.2, and carbon clearance index, K 0.03) and 3 (BZ-mesalamine conjugate) was potent immunosuppressant (HAT 3.4, and K 0.012). These two compounds, 2 and 3, inhibited rat paw edema by 54.6% and 51.6%, respectively and were observed to be safe on the gastric mucosa. In addition, the synthesized compounds possessed drug-like properties in accordance with the Lipinski's parameters. The study reports these compounds as MFCs, which can be used for the treatment of complex immune-inflammatory disorders (Bansal et al., 2014).

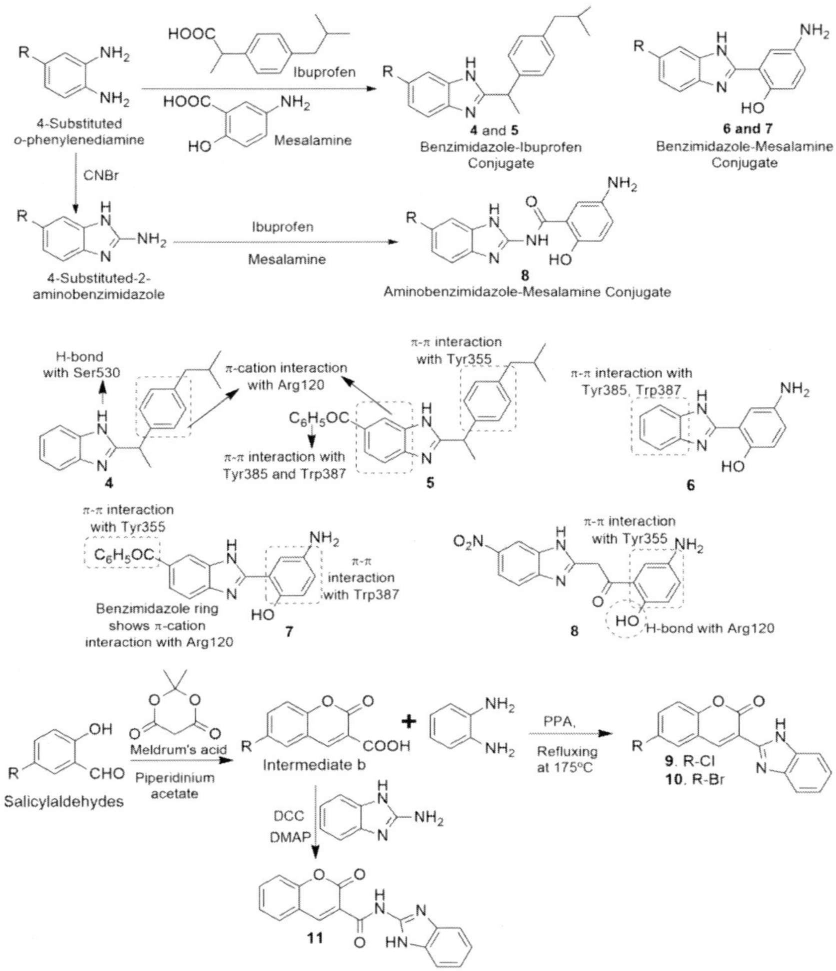

Figure 4. BZ hybrids having anti-inflammatory potential.

The potency of these compounds was further refined by incorporating the substituent at 5-position of BZ nucleus. Compounds 4-7 were synthesized by reacting 4-substituted *o*-phenylenediamine with NSAIDs (ibuprofen and mesalamine) in the presence of polyphosphoric acids as given in Figure 4. Compound 8 was synthesized by coupling intermediate (5-(un)substituted-2-aminobenzimidazole, which was prepared by stirring 4-substituted *o*-phenylenediamine with cyanogen bromide) with NSAIDs (ibuprofen/mesalamine) in the presence of dicyclohexyldicarbodiimide (DCC). SAR studies revealed that 5-substituted derivatives with a benzoyl, methyl or methoxyl group have higher activity, whereas 5-unsubstituted or 5-nitro substituted derivatives show decreased activity. Anti-inflammatory and immunomodulatory activities were different for BZ and 2-aminobenzimidazole derivatives. Benzoyl group increases immunostimulatory activity due to H-bonding and lipophilic interactions between the target and bulky phenyl group, whereas electron-releasing groups such as methyl and methoxyl are unfavorable for the stimulant action on the immune system. Small electron-withdrawing groups such as -NO_2, -CH and halogen at 5-position of 2-aminobenzimidazole/BZ induces immunoinhibitory action. In molecular docking study, all the synthesized compounds were docked in the active sites of COX-1 and COX-2. Based on the docking results, it was suggested that compounds were COX-2 selective and protective to gastric mucosa. The study concluded that compounds 4-7 show noticeable immunostimulatory and anti-inflammatory activities, but compound 8 shows maximum immunosuppressive response. Hence, compound 8 was considered as a potential candidate for the treatment of multifactorial diseases such as ulcerative colitis, rheumatoid arthritis, and other autoimmune disorders (Bansal et al., 2015).

Another similar study was carried out on coumarin-BZ coupled derivatives (either directly or through amide linkage). These were synthesized and estimated for their anti-inflammatory and antioxidant potentials. For the synthesis of target compounds, 6-substituted coumarin-3-carboxylic acids (intermediate b) was prepared by reacting substituted salicylaldehydes with Meldrum's acid in the presence of piperidinium acetate. Finally, the target compounds (9 and 10) were synthesized by refluxing the intermediate b with *o*-phenylenediamine at 175° C in presence of a catalytic amount of polyphosphoric acid. The compound 11 was synthesized by coupling the intermediate b with 2-aminobenzimidazole in the presence of 4-dimethylaminopyridine (DMAP) and DCC (as shown in Figure 4). SAR studies were carried out, and it was found that the presence of an electron-withdrawing group, such as -Cl and -Br, increases anti-inflammatory and antioxidant activities whereas the presence of an electron-releasing group, such as -OCH_3, decreases anti-inflammatory and antioxidant activities. It was also observed that an amide linkage between coumarin and BZ is unfavorable for anti-inflammatory activity, but favorable for antioxidant activity of the compounds. Compounds 9 and 11 exhibited the highest antioxidant activity than other compounds (EC_{50} 13.92 μmol/L and 1.2 μmol/L, respectively). These compounds also proved safe to gastric mucosa. SAR studies revealed that 6-substituted coumarin derivatives showed NSAIDs like activities, and the directly coupled compounds (9 and 10) show better anti-inflammatory response than amide-linked compound (11). During biological evaluations, it was found that compounds 10, 11 and 12 show maximum potency against paw edema with 46.75%, 45.45%, and 42.85% inhibition, respectively (Arora et al., 2014).

Sharma et al. (2017) used structural features of anti-inflammatory and antioxidant activities, and designed a series of 5-methanesulphonamido BZ derivatives based on physiochemical and 3D similarity studies. The series was synthesized by methanesulphonylation of 1-butyl-2-n-propyl-1*H*-benzo[d]imidazol-5-amines with methanesulphonyl chloride in the presence of pyridine and dichloromethane as exhibited in Figure 5. SAR study described that BZ nucleus is essential for anti-inflammatory activity. Variation in N-alkyl chain does not exhibit any significant change in the activity. But the presence of methanesulphonyl group shows marked influence on activity, and dominates over the effect of N-alkyl substituents. Results of biological evaluations show that all methanesulphonamido derivatives, irrespective of nature of alkyl group at N-1 of BZ, (12, 13, 14) show good anti-inflammatory and antioxidant activities (Sharma et al., 2017).

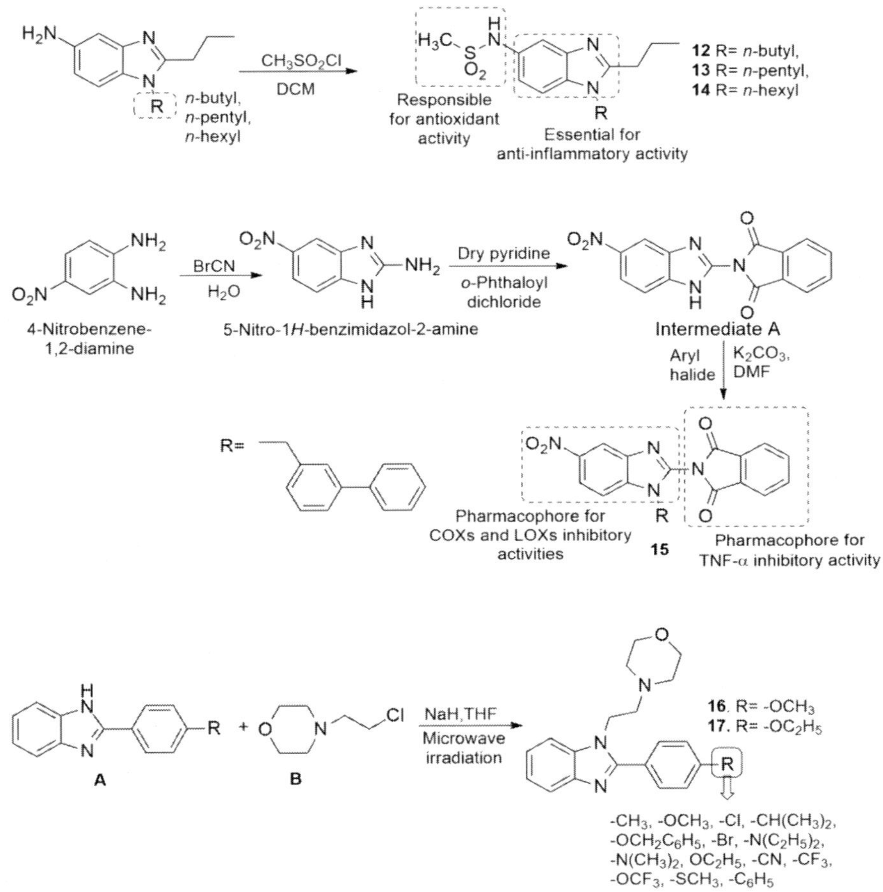

Figure 5. 1,2,5-Substituted benzimidazole derivatives with anti-inflammatory activity.

Kaur et al. (2018) designed and synthesized a series of BZ scaffold-based hybrid derivatives by joining BZ (pharmacophore having COXs and LOXs inhibitory activities) with phthalimide subunit of thalidomide (pharmacophore responsible for TNF-α inhibitory activity). The hybrids were evaluated for their inhibitory activities against COXs and LOXs, and TNF-α. The desired compounds were synthesized in multi-step reactions as given in Figure 5. In the first step, 5-nitro-1*H*-benzimidazol-2-amine was synthesized from 4-nitrobenzene-1,2-diamine through cyclization reaction with cyanogen bromide in the presence of water. Subsequently, 5-nitro-1*H*-benzimidazol-2-amine was reacted with *o*-phthaloyl dichloride in the presence of pyridine to give 2-(5-nitro-1*H*-benzimidazol-2-yl)-1*H*-isoindole-1,3(2*H*)-dione intermediate A. Compound 15 was then synthesized by reacting intermediate A with potassium carbonate or DMF. SAR studies described that the presence of a benzyl group at N-1 of BZ is essential for inhibitory activity against COX and LOX. 5-nitro substituted BZ is selectivity against COX-2. The phthalimide group at C-2 of BZ is reported to be responsible for COX-2 selective and TNF-α inhibitory activities. Molecular docking studies were carried out for all the synthesized compounds, but it was found that compound 15 shows a good binding affinity for COX-2, 5-LOX and 15-LOX. The study of binding interactions with COX-2 suggests that the nitro group in compound 15 interact with Phe518 and Arg513, and biphenyl rings interacted through π-π interactions with Tyr355 and Arg120 amino acid residues. The docking interactions with 5-LOX showed that nitro group in compound15 interacted with His432 of the active site, and the nitrogen atom of BZ formed H-bond with Gln557 and biphenyl ring and BZ nucleus interacted through π-π interactions with His600 and His367 of 5-LOX active sites. The binding interactions with 15-LOX showed π-π interaction between the aryl ring of BZ and Val348, and between the aryl ring of thalidomide and Val116. Nitro group also showed π-π interaction with Tyr348 of the active site. Results of *in vivo* biological evaluation revealed that compound 15 has significant anti-inflammatory activity (81.88% reduction in edema) and good analgesic activity (77.14% inhibition in writhing response). It also showed less severity index against ulcers than reference drug (indomethacin) and lower lipid peroxidation content (1.17 nmoles of MDA/100 mg of tissue), which suggest that it causes less mucosal damage.

Can et al. (2017) synthesized a series of 2-(4-substituted phenyl)-1-[2-(morpholin-4-yl)ethyl]-1*H*-benzimidazole derivatives, and hypothesized these to possess multiple activities due to the presence of BZ and morpholine nuclei. The study evaluated their multi-target ability *i.e.,* inhibitory activities against AChE, MAO-A, MAO-B, COX-1 and COX-2 along with the antimicrobial activity. The target compounds were synthesized by reacting 2-(4-substitutedphenyl)-1*H*-benzimidazole (A) with 2-(morpholin-4-yl)ethylchloride (B) in the presence of sodium hydrides as given in Figure 5. SAR studies were carried out and it was found that methoxyl substituted derivative (16) showed 16 times higher antimicrobial activity than the reference compound chloramphenicol (which is a well-known antimicrobial agent). Trifluoromethoxy, isopropyl, and chlorine substituted derivatives also showed good antimicrobial activity. It was found that methoxyl (16) and ethoxyl (17) substituted derivatives have good COX-1 (IC_{50} 8.096 and 8.369 μM, respectively) and COX-2 (IC_{50} 14.18 and 13.09 μM, respectively) inhibitory activities than the chlorine and isopropyl substituted derivatives. Compound 16 was also found to be the most potent antimicrobial compound (MIC_{50} 0.0156 μg/mL). But these compounds did not show any promising inhibitory activities against other explored targets *i.e.,* AChE, MAO-A and MAO-B.

Molecules Targeting Pathological Indications of Alzheimer's Disease

Sari et al. (2017) designed and synthesized a series of 2-[4-(4-substituted piperazin-1-yl)phenyl]benzimidazole derivatives, and ascertained their multi-targeting potential by evaluating them against BChE and Aβ aggregation inhibitory activities. The synthetic scheme (as shown in Figure 6) started with generation of 4-(4-substituted piperazin-1-yl) benzaldehydes (a) through sonication of 4-fluorobenzaldehyde with appropriate piperazines in the presence of DMSO and K_2CO_3. Subsequently, these benzaldehydes (a) were reacted with sodium bisulfite to yield sodium hydroxy[4-(4-substitutedpiperazin-1-yl)phenyl]methanesulfonate salts (b). Lastly, the target compounds (c) were obtained by treating *o*-phenylenediamines with corresponding salts. SAR studies suggested that phenyl substituted derivatives show no inhibitory activity against BChE and AChE, but alkyl-substituted derivatives 18-22 inhibit BChE better (IC_{50} value 5.18-20.21 μM) than AChE (IC_{50} value 34.85-229.30 μM). Among all alkyl-substituted derivatives, ethyl and benzyl substituted derivatives were found to be the most potent inhibitors of BChE with (IC_{50} values 5.18 and 5.22 μM, respectively). Kinetic studies were also carried out to identify the inhibitory mechanism, and it was found that inhibition is partially competitive type with Ki value 14.5 μM and β value 0.25. In biological evaluations, Thioflavin T (ThT) fluorescence assay was carried out for the estimation of compounds inhibitory action on self-mediated Aβ_{1-40} aggregation, and MTT assay was carried out on the SH-SY5Y cell line to estimate the neuroprotective activity of compounds against H_2O_2 and Aβ_{1-40} induced cell death. Compounds 18-21 were found to be potent neuro-protective agents with cell viability ranging from 81% to 130% at 10 μM. Docking study confirmed that protonated amine of alkyl-substituted derivatives are present close to Glu197 amino acid residue of BChE. π-π interaction was observed between one of the three aromatic residues (Tyr72, Trp286, and Tyr341) anionic sites and phenyl group attached at 2-position of BZ. The study implies that compound 19 can be used as a lead for the development of multi-targeting agents against Alzheimer's disease.

Tan et al. (2017) synthesized another similar series of 2-(4-substituted phenyl)-1*H*-benzimidazole derivatives and evaluated for inhibition of AChE and BChE, Aβ anti-aggregation, and neuroprotective action. The synthetic scheme of target compounds is composed of steps similar to that given in Figure 4. SAR study suggested that compounds with 4-methyl piperidinyl moiety at 2-position of BZ show good activity and selectivity profile. Further, methoxyl-substituted BZ has better selectivity and increased inhibitory activity towards BChE. Compounds 23 showed the most potent BChE inhibitory activity, whereas compound 24 was found to be neuroprotective when evaluated against H_2O_2 induced cytotoxicity. Compounds 24 and 25 showed the most potent Aβ anti-aggregation activity. Based on molecular docking studies, it was found that compound 25 is the most potent compound from the series that shows best binding interactions with the active catalytic site of BChE. Binding interactions of compound 23 shows that the methoxyl group form H-bond with Ser198, the hydroxyl group and BZ ring show CH-π interactions (weak H-bonding) with Phe329 and was observed to be responsible for inhibitory activity.

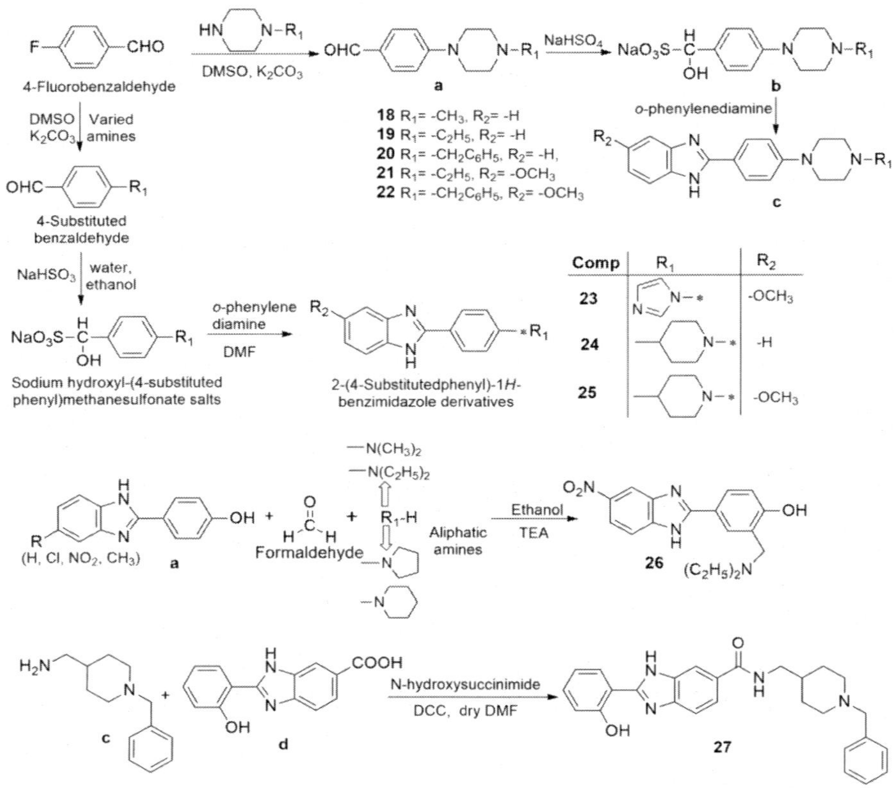

Figure 6. Anti-Alzheimer benzimidazole derivatives.

BZ derivatives have been reported as cholinesterase inhibitors and phenols are reported to possess good antioxidant activity. Taking this into consideration, Alpan et al. (2017) designed and synthesized BZ derivatives having phenolic moiety at 2-position of BZ. The phenol was derivatized at 2-position by various amino substituents (R_1 = dimethylamine, diethylamine, pyrrolidine, and piperidine), and at the same time different electronic groups (R = H, Cl, CH_3, NO_2) were placed at 5-position of BZ. This was achieved by reacting 4-(1*H*-benzimidazole-2-yl)phenol derivatives (a) with various aliphatic amines and formaldehyde in the presence of TEA as catalyst using ethanol as solvent as given in Figure 6. The synthesized compounds were evaluated *via in vitro* antioxidant and AChE/BChE inhibitory activity. Results of biological evaluations showed that among all synthesized compounds, the compound 26 having $-NO_2$ at 5-position of BZ and diethylamine group at 2-position of phenolic residue incur maximum AChE inhibitory activity (0.93 μM) with higher selectivity (13.68) over BChE. Although the majority of compounds were potent AChE inhibitor but none of the target molecules delivered potent BChE inhibition. SAR studies revealed that pyrrolidine and piperidine are less potent functional groups for incurring anti-cholinesterase activity. In lipid peroxidation assay, almost all the compounds showed good to moderate activity in a dose-dependent manner. Molecular docking studies were also carried out to understand the interactions between BZ derivatives and active site of AChE pocket (1EVE) and homology model of BChE using PDB 4TPK. Results of the docking studies supported the outcomes of *in vitro* activity as it proved the compound 26 as the best-fit molecule. In the active pocket of AChE, the $-NO_2$ substituted ring lies in the quaternary ammonium binding domain formed by Trp84 and Tyr130. The phenolic moiety of compound 26 slips into a hydrophobic pocket constituted by Tyr121, Phe290, Phe330, Phe331, and Tyr334. In the case of BChE, BZ and alkyl chain of compound 26 interact with Ser198, Glu325, and His438 amino acid residues. Molecular simulation studies were also carried out for both the enzymes for 42 ns, and results proved that these compounds form stable complexes with enzymes.

Piemontese et al. (2018) hybridized benzylpiperidine or benzylpiperazine moieties with different heterocyclics like BZ or benzofuran to synthesize some anti-Alzheimer molecules. The selection of these nuclei for the hybridization was done in such a way that the target molecules mimic the structure of donepezil (a clinically used AChE inhibitor). The synthetic scheme employed for the synthesis of target molecules is given in Figure 6. It involved the coupling of two intermediates, c and d, in the presence of DCC and N-hydroxysuccinimide. All the target molecules showed mild to moderate inhibitory activities against AChE and Aβ aggregation, with additional antioxidant capacity. Compound 27 is reported as the best Aβ aggregation inhibitor (36.3% at 80 μM concentration) with moderate AChE inhibition (IC_{50} 4.2 μM).

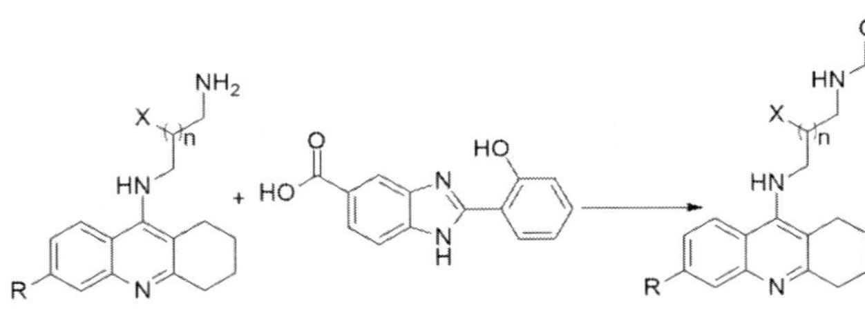

Comp	R	n	X
28	-H	1	-H
29	-Cl	0	-H
30	-H	1	-OH
31	-H	0	-H
32	-Cl	2	-OH

Figure 7. Benzimidazole-acridine hybrids as Anti-Alzheimer agents.

Another series of 2-hydroxyphenyl-benzimidazole based hybrids was synthesized by selecting tacrine as AChE pharmacophore (Figure 7). The hybrids were evaluated for AChE inhibitory, Aβ aggregatory, radical scavenging and neuroprotective activities. It has been reported that compound 29 shows the best inhibitory activity against AChE with IC_{50} value of 6.3 nM, whereas compound 28 shows the best inhibitory activity against self- and Cu-induced Aβ aggregation at 74.6% and 77.3%, respectively. Molecular docking studies suggested that compound 29 dually binds with CAS and PAS through π-π interactions with Trp279, Tyr70, and Tyr121, and through H-bond interaction with Tyr121 for inhibition of AChE. Compound 30 showed H-bond with Asp72 residue. Compound 30 also showed the best antioxidant activity with an IC_{50} value of 564 μM (Hirematthad et al., 2018).

Chaves et al. (2018) further evaluated Cu- and Zn-chelating capacities, and the role of Cu in A*β* aggregation inhibitory potency of tacrine/donepezil and hydroxyphenylbenzimidazole derivatives. During the analysis of Cu/Zn chelating capacity of series of compounds, it was found that all the compounds show good chelating capacity for Cu but moderate capacity for Zn. It was also found from the study that tacrine-based derivatives, 31 and 32 (Figure 7), show good self-induced A*β* aggregation inhibition than donepezil-based hybrids.

Dolles et al. (2016) designed a series of compounds by joining common pharmacophores, such as piperidinepropoxyphenyl group for blocking CB2R ligand and varied heterocyclic for cholinesterase inhibition. Molecules were evaluated for their inhibitory activity against AChE and BChE through CB2R ligand. The target compound 33 was synthesized (Figure 8) by cyclization of thiourea derivatives (a) by using DCC and EDCI. SAR studies described that, for dual action, four major structural changes are required in BZ-based derivatives *i.e.,* diethylamide moiety replaced by hydrogen for the conformational requirement of CB2R ligand and BChE pharmacophore, a basic moiety to be introduced into the lipophilic alkyl chain linked with N-atom of imidazole, modify imidazole attached moieties, and changes in the heterocyclic ring. Replacement of the diethylamide group with another group resulted in the loss of antagonistic activity for CB2R ligand. BZ may be replaced with imidazopyridine or 3-aminoindazole without any loss of activity. Compound 33 showed the best selectivity and moderate affinity for CB2R ligand. Molecular dynamics and molecular docking studies were also carried out for 33, wherein it shows the most stable binging conformation against hCB2R. The oxygen atom of amide formed a stable H-bond with Thr3.33, and N-atom interacts with Ser287 and Pro285. The study also confirmed that compound 33 binds in the TMIII, TMIV and TMVI pockets of cannabinoid receptors.

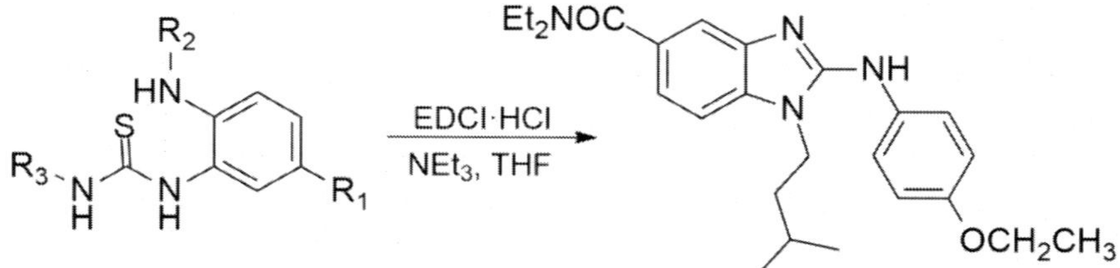

a

33

Figure 8. Anti-Alzheimer benzimidazole derivatives.

Molecules Targeting Pathological Indications of Cancer

Virtual screening of the PubChem database for SVM models of EGFR, VEGFR2 and PDGFR led to the identification of 2-aryl benzimidazole, which is a common scaffold for all three inhibitors. Thus, Li et al. (2011) synthesized a series of multi-target 2-aryl benzimidazole derivatives for the treatment of cancer. Molecules were evaluated for their biological activity against HepG2 tumor cell line and specific kinases EGFR, VEGFR-2, and PDGFR. For the synthesis of target compounds, initially, 4-nitro-*o*-phenyl-enediamine was condensed with *p*-tolualdehyde to give an intermediate (a), which was further refluxed with Tin (II) chloride anhydrous and ethanol to obtain 2-*p*-tolyl-1*H*-benzo[d]imidazol-5-amine (b). It was then treated with chloroacetyl chloride to yield an intermediate (c), which was finally refluxed with KI in THF to yield the target compounds (Figure 9). SAR studies suggested that the deactivated phenyl ring shows higher cytotoxicity, and length of the chain between carbonyl and the end group (like piperidine, pyrrole) affects the cytotoxicity of the compound (shorter chain length compound shows high inhibitory activity). Some of the functional groups, such as piperidine and pyrrole, also increase the inhibitory activity of compounds. Compounds 34 and 35 show the best anti-proliferative activity in HepG2 cells (IC_{50} 2μM and 1.8 μM, respectively). These compounds also showed good cytotoxicity in kinase inhibitor assay. Compound 35 showed better cell apoptosis in the apoptosis test (Annexin-V/PI binding assay). Molecular docking studies revealed that compound 34 shows good binding interaction with EGFR active site (H-bond with Met793), but compound 35 shows weaker binding interactions with EGFR active site. Compound 34 exhibits good binding interaction with VEGFR-2 active site (two H-bonds with Cys917 and Asp1044). The study suggested that compounds 34 and 35 are the most potent compounds of the series, and can be used as multi-targeting agents for cancer.

Chu et al. (2015) investigated the mechanism of EGFR and HER2 inhibitory activities of compounds 34 and 35. The authors suggested that these compounds act *via* decreasing the EGFR and HER2 tyrosine phosphorylation, and preventing the downstream activation of PI3K/Akt and MEK/Erk pathways *in vitro* and *in vivo*. These compounds also inhibited the phosphorylation of FOXO, and promoted the FOXO translocation from cytoplasm into nucleus, resulting in the arrest of G1-phase of cell cycle and apoptosis.

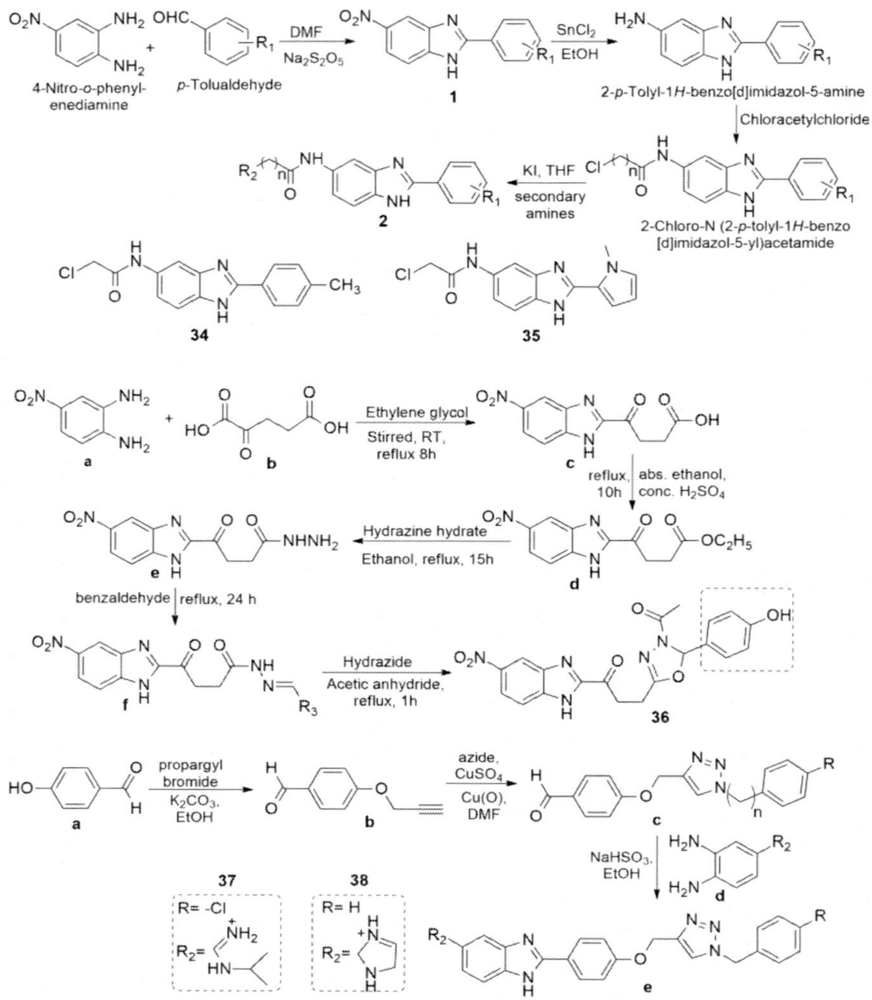

Figure 9. Anti-cancer benzimidazole derivatives.

In the year 2018, Ibrahim et al. synthesized three series of 2-substituted-5-nitrobenzimidazole derivatives by adopting the analog-based approach. Different heterocyclic fragments, such as substituted oxadiazole, piperazine, and triazolothiadiazole, were placed at the 2-position of 5-nitrobenzimidazole, and were evaluated for their inhibitory activity against VEGFR-2 and c-Met kinase. For the synthesis, firstly 4-(5-nitro-1*H*-benzo[d]imidazol-2-yl)-4-oxobutanoic acid (c) was synthesized by refluxing 4-nitro-*o*-phenylenediamine (a) with α-keto-glutaric acid (b) in ethylene glycol. Then, ethyl 4-(5-nitro-1*H*-benzo[*d*]imidazol-2-yl)-4-oxobutanoate (d) was synthesized by refluxing (c) with absolute ethanol in the presence of sulphuric acid. After this, 4-(5-nitro-1*H*-benzo[*d*]imidazol-2-yl)-4-oxobutanehydrazide (e) was synthesized by refluxing the intermediate (d) with a solution of hydrazine hydrate and ethanol. Final compounds were synthesized in two steps. The first step involved the synthesis of intermediate (f), and the second step involves the synthesis of the final compound by refluxing (f) with acetic anhydride and hydrazide (Figure 9). SAR studies described that 4-ethyl or -phenyl substituted piperazine derivatives, and substituted (ethylbenzene) or unsubstituted oxadiazole derivatives show good inhibitory activity against c-Met but poor inhibitory activity against VEGFR-2 and; methyl and propyl substituted triazolo-thiadiazole derivatives show a wide range of inhibitory activities against c-Met, VEGFR-2, and various cancerous cell lines. Phenoxy substituted triazolo-thiadiazole showed the best inhibitory activity against both c-Met and VEGFR-2 enzymes. Molecular docking was carried out for the most potent compound 36 against VEGFR-2 and c-Met active binding sites. It was found that compound 36 shows as good interactions as native ligand shows. Compound 36 interacted through two H-bonds (one between C=O and Cys919, and the second between oxygen of OCH_2Ph and Asp1046), and through hydrophobic interactions (BZ with Phe918 and Leu840, and thiadiazole with Lys868) in the VEGFR-2 active site. Docking studies for c-Met active site reveals that compound 36 form two H-bonds with Ile10 and Arg1086, and hydrophobic interactions with Tyr1159, Tyr1230, Gly1085 and Met1210, and electrostatic interactions with Asp1164 and Asp1231 (Ibrahim et al., 2018).

BZ and 1,2,3-triazole nuclei are potential inhibitors of different types of protein kinases. Therefore, Bistrovic et al. (2018) synthesized a series of 1,2,3-triazolyl linked 2-aryl benzimidazole derivatives, and evaluated their inhibitory activity against different protein kinases to treat lung cancer cell line (A549). The final compounds were synthesized by microwave and ultrasound irradiation in click reaction. Firstly, 4-hydroxy benzaldehyde was treated with propargyl bromide to prepare 4-*o*-propargylated benzaldehyde that acts as a dipolarophile. It was then reacted with unsubstituted, *p*-halogen or *p*-trifluoromethyl-substituted phenyl azides to obtain intermediate c. This intermediate c was further condensed with *o*-phenylenediamine using $NaHSO_3$ to synthesize the target 5-amidino-substituted benzimidazoles derivatives (Figure 9). SAR study revealed that the presence of different types of cationic moieties on BZ greatly impact the antiproliferative activity. Imidazoline substituted (at R_2) and *p*-chlorophenyl substituted (at R_4) derivatives showed maximum potency. The presence of a linker between BZ and 1,2,3-triazole showed good inhibitory activity against A549 cells. Docking studies were carried out, which revealed that compound 38 show good binding with p38MAPK than compound 37 shows. Docking interactions of compounds 37 and 38 with p38MAPK show H-bonds between triazole moiety and Met109, and between amidino-benzimidazole group and Glu71, His148 and Asp168. Biological evaluations revealed that compounds 37 and 38 show the best inhibitory activity against A549 cancer cell line (IC_{50} 0.05 and 0.07 μM, respectively).

Molecular reciprocity between the Ras/MEK/ERK and PI3K/Akt/mTOR pathways provides cancer cells with an ability to evade the treatment when targeting only one pathway with monotherapy. Working on this lead, Van Dort et al. (2015) reported the designing, synthesis and preliminary biological evaluation of prototypical bifunctional compounds capable of co-targeting the complementary PI3K/Akt/mTOR and

Ras/MEK/ERK. Design strategy involved the development of a dual MEK/PI3K inhibitor through covalent coupling of an optimized allosteric MEK inhibitor with an optimized PI3K inhibitor. The appropriate inhibitors was chosen on the basis of optimal pharmaceutical properties, positive pre-clinical studies, structural characterization by X-ray crystallography and the presence of synthetically tractable handles that would permit facile linking of the two motifs. After critically analysis of structures of compounds responsible for inhibition of each target, ZSTK474 (ATP-competitive pan-PI3K inhibitor) and RO5126766 (an allosteric MEK inhibitor) was chosen as parent skeletons. These two skeletons (A and B) were then necessarily modified and further coupled together in the presence of a base and acetonitrile as a solvent to get potent molecule 39 (Figure 10). Compound 39 showed the maximum *in vitro* PI3K and MEK1 inhibitory activities (IC_{50} 172 nM and 473 nM, respectively).

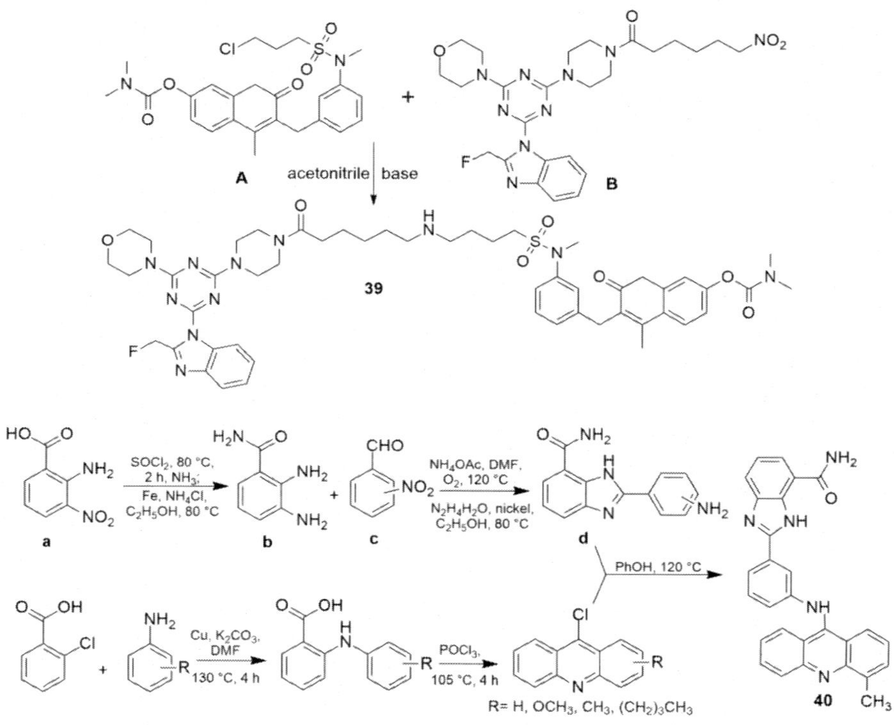

Figure 10. Anti-cancer benzimidazole derivatives.

In the year 2017, Yuan et al. proposed that as PARP-1 could reverse the Topoisomerase (Topo) inhibitors-induced DNA damage, the dual inhibition of Topo and PARP-1 may provide a better alternative for cancer therapy. The acridines derivatives acting as Topo inhibitor were modulated by adding a BZ carboxamide moiety, which is reported to be responsible for PARP inhibition. A series of analogs 40 was designed by changing the substituent patterns on acridine. The synthesis of target molecules was achieved by following the synthetic scheme as shown in Figure 10. First, converting 2-amino-3-

nitrobenzoic acid (a) to 2,3-diaminobenzamide (b) *via* amidation and reduction reactions. Intermediate (b) was then reacted with (c) to give BZ carboxamides, which were reduced to produce intermediate (d). On the other side, acridine derivatives were obtained by carrying the Ullmann reaction and Friedel-Crafts acylation in sequence. Finally, the desired compounds were produced by nucleophilic substitution reactions between the two different scaffolds. Biological evaluations revealed that compound 40 is a potent inhibitor against both PARP and Topo. Besides this, it also exhibited potent antiproliferative activity against MCF-7 cell line. Marked apoptosis was also observed with compound 40 *via* modulating both extrinsic as well as intrinsic pathways. Cell cycle analysis and western blot assay indicated that it causes stable G0/G1 phase arrest *via* different mechanisms. *In vivo* study was carried out to check the growth of MCF-7 xenograft tumors, which also warranted compound 40 as a promising lead compound for cancer therapy (Yuan et al., 2017).

Miscellaneous

Telmisartan is known to lower the blood pressure by antagonizing angiotensin (AT_1) receptors. But recently, it was noted that it also acts as a partial agonist of PPARγ receptor and exerts pleiotropic effects. Inspired from this, Mizuno et al. (2017) designed and synthesized BZ-based compounds having the potential to exert dual activity (*i.e.,* PPARγ agonists/angiotensin II antagonist). Target compound 41 was synthesized by coupling *bis*-benzimidazole derivatives (a) with a fragment (b) in the presence of Cs_2CO_3 and DMF (Figure 11). The biological evaluation provides compound 41 to be a dual inhibitor with moderate PPARγ activity (29%) and affinity for AT_1 receptor ($K_i = 2.5$ μM). Molecular docking studies were carried out to find out the best binding interactions between compound 41 and AT_1 and PPARγ receptors. It was found that the compound 41 binds with AT_1 receptor through hydrogen bonding of its BZ with His256, Asn295 and Asn111, and through π-π interactions of its phenyl side chain with Phe261 amino residues. It also interacts through good binding interactions with Cys285, Leu330, Phe363, and Tyr473 amino residues of the PPARγ receptors.

Spasov et al. (2017) synthesized a series of 2-biphenyl substituted 9*H*-imidazo[1,2-*a*]benzimidazole based derivatives, and evaluated their activity against various targets involved in the treatment of diabetes mellitus. The target biphenyl derivatives were synthesized through a two-step synthetic process. In first step, 1*R*-3-(4-biphenyl)methyl-2-iminobenzimidazoline hydrobromides (b) was obtained by quaternization of *1R*-2-aminobenzimidazoles (a) with 4-(bromoacetyl)biphenyl. In the second step, base-catalyzed cyclization of (b) occurred to synthesize final compound (c) (Figure 11). Biological evaluations indicated that compound 42 showed the best antioxidant activity for ascorbate-dependent LPO inhibition, free radical formation inhibition, and ABTS oxidation inhibition with IC_{50} of 0.19, 0.27, and 22.4 μmol, respectively. Compounds 43 and 45 are the best inhibitor of Cu^{2+}-dependent oxidation of ascorbic acid (IC_{50} 4.6 and 9.9 μmol, respectively), and thus responsible for the anti-glycation activity. Compound 44 exhibits the best anti-platelet activity (18.18% activity). From these studies, it was found that varied 2-biphenyl substituted 9*H*-imidazo[1,2-*a*]benzimidazole derivatives act as a multi-targeting compounds for the cure of multi-factorial diabetic complications.

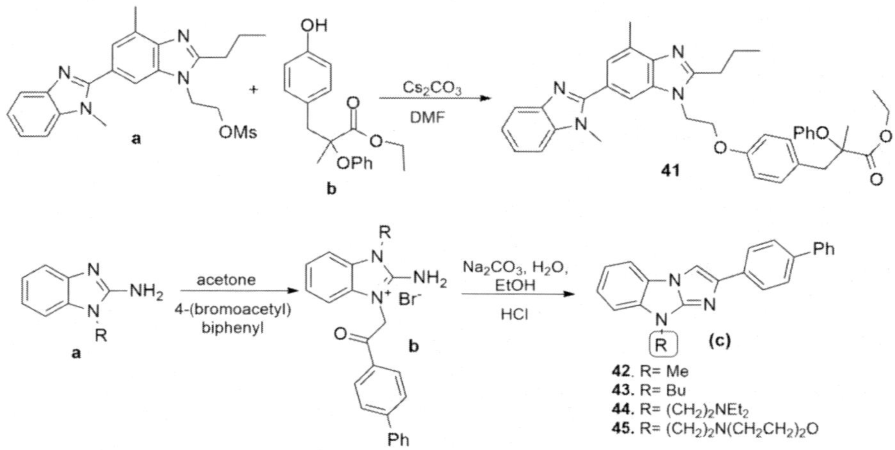

Figure 11. Benzimidazole based MFCs against diabetes.

CONCLUSION

MFCs have emerged as an effective tool to counter complex diseases possessing multiple pathological indications. The key chemical scaffold in designing of these MFCs is the heteronucleus, which forms the core of the molecule. Wisdom of the medicinal chemist lies in the identification of such scaffolds, which can act against individual targets. BZ is a nitrogenous bicyclic heterocyclic nucleus found in numerous therapeutically effective agents, which are either already in use in clinics or are paving their way in to clinics. The existence of a large number of literature reports on the therapeutic efficacy of BZ-based molecules in different disorders proved it as a privileged scaffold for the designing of potential MFCs. Despite this proven acceptability of BZ in the development of therapeutic agents, only limited numbers of reports are available on the exploration of BZ-based ligands as MFCs for the treatment of complex pathological disorders. This chapter highlights the current status of the research reports on BZ-based MFCs and also tries to provide an insight into the synthetic schemes and SAR of these compounds. This information is expected to act as an imperative aid to the drug designer and medicinal chemist for the development of clinically viable BZ-based MFCs.

REFERENCES

Abdolmaleki, A., B Ghasemi, J. and Ghasemi, F. (2017). Computer aided drug design for multi-target drug design: SAR/QSAR, molecular docking and pharmacophore methods. *Current Drug Targets*, 18(5): 556-575.

Akhtar, W., Khan, M. F., Verma, G., Shaquiquzzaman, M., Rizvi, M. A., Mehdi, S. H. and Alam, M. M. (2017). Therapeutic evolution of benzimidazole derivatives in the last quinquennial period. *European Journal of Medicinal Chemistry,* 126: 705-753.

Alpan, A. S., Sarıkaya, G., Çoban, G., Parlar, S., Armagan, G. and Alptüzün, V. (2017). Mannich Benzimidazole Derivatives as Antioxidant and Anticholinesterase Inhibitors: Synthesis, Biological Evaluations, and Molecular Docking Study. *Archiv Der Pharmazie*, 350(7): (1-15) e1600351.

Arora, R. K., Kaur, N., Bansal, Y. and Bansal, G. (2014). Novel coumarin–benzimidazole derivatives as antioxidants and safer anti-inflammatory agents. *Acta Pharmaceutica Sinica B*, 4(5): 368-375.

Bansal, Y. and Silakari, O. (2012). The therapeutic journey of benzimidazoles: a review. *Bioorganic and Medicinal Chemistry*, 20(21): 6208-6236.

Bansal, Y. and Silakari, O. (2014). Synthesis and pharmacological evaluation of polyfunctional benzimidazole-NSAID chimeric molecules combining anti-inflammatory, immunomodulatory and antioxidant activities. *Archives of Pharmacal Research*, 37(11): 1426-1436.

Bansal, Y. and Silakari, O. (2015). 2-Aminobenzimidazole conjugates of NSAIDS: novel compounds with immunomodulatory, anti-inflammatory and antioxidant actions. *Medicinal Chemistry Research*, 24(3): 1170-1179.

Bansal, Y., Kaur, M. and Silakari, O. (2015). Benzimidazole–ibuprofen/mesalamine conjugates: potential candidates for multifactorial diseases. *European Journal of Medicinal Chemistry*, 89: 671-682.

Bistrović, A., Krstulović, L., Harej, A., Grbčić, P., Sedić, M., Koštrun, S. and Raić-Malić, S. (2018). Design, synthesis and biological evaluation of novel benzimidazole amidines as potent multi-target inhibitors for the treatment of non-small cell lung cancer. *European Journal of Medicinal Chemistry*, 143: 1616-1634.

Can, N., Çevik, U. A., Sağlık, B. N., Özkay, Y., Atlı, Ö., Baysal, M. and Can, Ö. D. (2017). Pharmacological and Toxicological Screening of Novel Benzimidazole-Morpholine Derivatives as Dual-Acting Inhibitors. *Molecules*, 22(8): (1-22)1374.

Chaves, S., Hiremathad, A., Tomás, D., Keri, R. S., Piemontese, L. and Santos, M. A. (2018). Exploring the chelating capacity of 2-hydroxyphenyl-benzimidazole based hybrids with multi-target ability as anti-Alzheimer's agents. *New Journal of Chemistry*, 42(20): 16503-16515.

Chen, Z. F., Orvig, C. and Liang, H. (2017). Multi-target metal-based anticancer agents. *Current Topics in Medicinal Chemistry*, 17(28): 3131-3145.

Chu, B., Liu, F., Li, L., Ding, C., Chen, K., Sun, Q. and Jiang, Y. (2015). A benzimidazole derivative exhibiting antitumor activity blocks EGFR and HER2 activity and upregulates DR5 in breast cancer cells. *Cell Death and Disease*, 6(3): (1-10)e1686.

Dolles, D., Nimczick, M., Scheiner, M., Ramler, J., Stadtmüller, P., Sawatzky, E. and Decker, M. (2016). Aminobenzimidazoles and Structural Isomers as Templates for Dual-Acting Butyrylcholinesterase Inhibitors and hCB2R Ligands To Combat Neurodegenerative Disorders. *Chem. Med. Chem.*, 11(12): 1270-1283.

East, S. P. and Silver, L. L. (2013). Multitarget ligands in antibacterial research: progress and opportunities. *Expert Opinion on Drug Discovery*, 8(2): 143-156.

Gaba, M., Singh, S. and Mohan, C. (2014). Benzimidazole: an emerging scaffold for analgesic and anti-inflammatory agents. *European Journal of Medicinal Chemistry*, 76: 494-505.

Hiremathad, A., Keri, R. S., Esteves, A. R., Cardoso, S. M., Chaves, S. and Santos, M. A. (2018). Novel tacrine-hydroxyphenylbenzimidazole hybrids as potential multitarget drug candidates for Alzheimer's disease. *European Journal of Medicinal Chemistry*, 148: 255-267.

Ibrahim, H. A., Awadallah, F. M., Refaat, H. M. and Amin, K. M. (2018). Design, synthesis and molecular modeling study for some new 2-substituted benzimidazoles as dual inhibitors for VEGFR-2 and c-Met. *Future Medicinal Chemistry*, 10(5): 493-509.

Kaur, G. and Silakari, O. (2018). Benzimidazole scaffold based hybrid molecules for various inflammatory targets: Synthesis and evaluation. *Bioorganic Chemistry*, 80: 24-35.

Keri, R. S., Rajappa, C. K., Patil, S. A. and Nagaraja, B. M. (2016). Benzimidazole-core as an antimycobacterial agent. *Pharmacological Reports,* 68(6): 1254-1265.

Li, Y., Tan, C., Gao, C., Zhang, C., Luan, X., Chen, X. and Jiang, Y. (2011). Discovery of benzimidazole derivatives as novel multi-target EGFR, VEGFR-2 and PDGFR kinase inhibitors. *Bioorganic and Medicinal Chemistry*, 19(15): 4529-4535.

Manju, S. L., Ethiraj, K. R. and Elias, G. (2018). Safer anti-inflammatory therapy through dual COX-2/5-LOX inhibitors: A structure-based approach. *European Journal of Pharmaceutical Sciences*, 121: 356-381.

Mizuno, C. S., Chittiboyina, A. G., Shah, F. H., Patny, A., Kurtz, T. W., Pershadsingh, H. A. and Avery, M. A. (2010). Design, synthesis, and docking studies of novel benzimidazoles for the treatment of metabolic syndrome. *Journal of Medicinal Chemistry*, 53(3): 1076-1085.

Narasimhan, B., Sharma, D. and Kumar, P. (2012). Benzimidazole: a medicinally important heterocyclic moiety. *Medicinal Chemistry Research*, 21(3): 269-283.

Ozadali-Sari, K., Küçükkılınç, T. T., Ayazgok, B., Balkan, A. and Unsal-Tan, O. (2017). Novel multi-targeted agents for Alzheimer’s disease: Synthesis, biological evaluation, and molecular modeling of novel 2-[4-(4-substitutedpiperazin-1-yl) phenyl] benzimidazoles. *Bioorganic Chemistry,* 72: 208-214.

Piemontese, L., Tomás, D., Hiremathad, A., Capriati, V., Candeias, E., Cardoso, S. M. and Santos, M. A. (2018). Donepezil structure-based hybrids as potential multifunctional anti-Alzheimer's drug candidates. *Journal of Enzyme Inhibition and Medicinal Chemistry*, 33(1): 1212-1224.

Ramsay, R. R., Popovic-Nikolic, M. R., Nikolic, K., Uliassi, E. and Bolognesi, M. L. (2018). A perspective on multi-target drug discovery and design for complex diseases. *Clinical and Translational Medicine*, 7(1): 1-14.

Shaharyar, M. and Mazumder, A. (2017). Benzimidazoles: A biologically active compounds. *Arabian Journal of Chemistry*, 10: S157-S173.

Sharma, R., Bali, A. and Chaudhari, B. B. (2017). Synthesis of methanesulphonamido-benzimidazole derivatives as gastro-sparing antiinflammatory agents with antioxidant effect. *Bioorganic and Medicinal Chemistry Letters*, 27(13): 3007-3013.

Spasov, A. A., Zhukovskaya, O. N., Brigadirova, A. A., Abbas, H. S. A., Anisimova, V. A., Sysoeva, V. A. and Morkovnik, A. S. (2017). Synthesis and pharmacological activity of 2-(biphenyl-4-yl) imidazo[1,2-a]benzimidazoles. *Russian Chemical Bulletin*, 66(10): 1905-1912.

Tahlan, S., Kumar, S., Kakkar, S. and Narasimhan, B. (2019). Benzimidazole scaffolds as promising antiproliferative agents: a review. *BMC Chemistry*, 13(1): 66.

Unsal-Tan, O., Ozadali-Sari, K., Ayazgok, B., Küçükkılınç, T. T. and Balkan, A. (2017). Novel 2-Arylbenzimidazole derivatives as multi-targeting agents to treat Alzheimer's disease. *Medicinal Chemistry Research,* 26(7): 1506-1515.

Valdez, J., Cedillo, R., Hernández-Campos, A., Yépez, L., Hernández-Luis, F., Navarrete-Vázquez, G. and Castillo, R. (2002). Synthesis and antiparasitic activity of 1H-benzimidazole derivatives. *Bioorganic and Medicinal Chemistry Letters*, 12(16): 2221-2224.

Van Dort, M. E., Galbán, S., Wang, H., Sebolt-Leopold, J., Whitehead, C., Hong, H. and Ross, B. D. (2015). Dual inhibition of allosteric mitogen-activated protein kinase (MEK) and phosphatidylinositol 3-kinase (PI3K) oncogenic targets with a bifunctional inhibitor. *Bioorganic and Medicinal Chemistry*, 23(7): 1386-1394.

Woolley, D. W. (1944). Some biological effects produced by benzimidazole and their reversal by purines. *Journal of Biological Chemistry,* 152: 225-232.

Yuan, Z., Chen, S., Chen, C., Chen, J., Chen, C., Dai, Q. and Jiang, Y. (2017). Design, synthesis and biological evaluation of 4-amidobenzimidazole acridine derivatives as dual PARP and Topo inhibitors for cancer therapy. *European Journal of Medicinal Chemistry*, 138: 1135-1146.

BIOGRAPHICAL SKETCH

Dr. Yogita Bansal

Affiliation: Punjabi University, Patiala (India)

Education: M. Pharm, PhD

Business Address: Department of Pharmaceutical Sciences and Drug Research, Punjabi University, Patiala - 147002 (India)

Research and Professional Experience: 17 years

Professional Appointments:

Aug. 2002	Lecturer at Akal College of Pharmacy and Technical Education, Mastuana Sahib, Sangrur (India)
Sept. 2005	Lecturer at Punjabi University, Patiala (India)
Aug. 2008	Lecturer (Senior Scale) at Punjabi University, Patiala (India)
Aug. 2013	Assistant Professor at Punjabi University, Patiala (India)
Aug. 2016	Associate Professor at Punjabi University, Patiala (India)

Honors: Best NSS Program Officer in 2015

Publications from the Last 3 Years:

Book

Rajeev Sharma, Yogita Bansal. Drug Abuse: Problem, Management and Prevention. RD Publication, India, 2017, ISBN: 978-93-84594-56-5.

Book Chapter

1. Minhas, R., Sunaina., Bansal, G. and Bansal, Y. (2017) Human Topoisomerase I and Its Inhibitors: Recent Developments. In: *Topoisomerase Inhibitors: Classification, Mechanisms of Action and Adverse Effects.* R. Kumar and S. Singh (Eds), Nova Science Publishers, Inc., New York, pp 1-48. (ISBN: 978-1-53611-915-2).
2. Bansal, G., Kaur, J. and Bansal, Y. (2015) Shelf life assessment and clinical trials of herbal drugs: Pre-requisites, challenges and perspectives. In: *Potentials and*

Bottlenecks in Clinical Trials of Herbal Drugs. Eds P. Bansal and S. Eswara Reddy, Gulab Publishers, India, pp 97-119 (ISBN: 978-81-920643-9-0).

Publications

1. Minhas, R., Bansal, Y. and Bansal, G. (2019). Inducible nitric oxide synthase inhibitors: A comprehensive update. *Medicinal Research Reviews*. (In Press).
2. Minhas, R., Bansal, G. and Bansal, Y. (2019). Novel Coupled Molecules from Active Structural Motifs of Synthetic and Natural Origin as Immunosuppressants. *Medicinal Chemistry*. doi: 10.2174/1573406415666190409111459.
3. Bansal, Y., Kaur, M. and Bansal, G. (2019). Antimicrobial Potential of Benzimidazole Derived Molecules. *Mini Reviews in Medicinal Chemistry*, 19(8), 624-646.
4. Sethi, P., Bansal, Y. and Bansal, G. (2018). Synthesis and PASS-assisted evaluation of coumarin–benzimidazole derivatives as potential anti-inflammatory and anthelmintic agents. *Medicinal Chemistry Research*, 27(1), 61-71.
5. Kaur, S. and Bansal, Y. (2017). In silico assisted design, synthesis and evaluations of pluripotent compounds for combating Alzheimer's disease. *Alzheimer's & Dementia: The Journal of the Alzheimer's Association*, 13(7), P268-P269.
6. Minhas, R., Sandhu, S., Bansal, Y. and Bansal, G. (2017). Benzoxazole-Coumarin Derivatives: Potential Candidates for Development of Safer Anti-inflammatory Drugs. *Der Chemica Sinica*, 8(1), 146-157.
7. Jain, A., Kaur, J., Bansal, Y., Saini, B. and Bansal, G. (2017). WHO guided real time stability testing on Shankhpushpi syrup. *Journal of Pharmaceutical Technology, Research and Management Volume*, 2017, 5(1), 1-19.
8. Kaur, J., Bansal, Y. and Bansal, G. (2016). Understanding unconventional routes to impurities from drugs in hydrolytic conditions. *Indian Journal of Pharmaceutical Education and Research*, 50(3s), S161-S190.
9. Kaur, I., Suthar, N., Kaur, J., Bansal, Y. and Bansal, G. (2016). Accelerated stability studies on dried extracts of Centella asiatica through chemical, HPLC, HPTLC, and biological activity analyses. *Journal of Evidence-Based Complementary & Alternative Medicine*, *21*(4), NP127-NP137.
10. Kaushik, D., Kaur, J., Kaur, V. P., Saini, B., Bansal, Y. and Bansal, G. (2016). Forced degradation, LC–UV, MSn and LC–MS–TOF studies on azilsartan: Identification of a known and three new degradation impurities. *Journal of Pharmaceutical and Biomedical Analysis*, *120*, 202-211.

In: Benzimidazole
Editor: Aksel A. Vestergaard
ISBN: 978-1-53616-986-7

Chapter 7

BENZIMIDAZOLE COMPOUNDS WITH ANTI-TUMOR AND ANTIBACTERIAL ACTIVITIES

Christina-Marie Zalaru and Maria Marinescu*
Department of Organic Chemistry, Biochemistry and Catalysis, University of Bucharest, Bucharest, Romania

ABSTRACT

Benzimidazole moiety is one of the prominent heterocycle in medicinal chemistry, with outstanding results in the last decades of research. The pharmaceutical market comprises a series of benzimidazole - based medicines such as bendamustine, selumetinib - antitumorals; pantoprazole, lansoprazole - proton pump inhybitors; bezitramide - analgesic; mebendazole, albendazole - antihelmintics; ridinazole - antibacterial; enviradine - antiviral and candesartan as antihypertensive.

Cancer continues to be one of the worst diseases of the century in which we live, for which research is most advanced for the discovery of new structures with anti-tumor action. Also, antimicrobial resistance is one of the major problems faced by today's medicine that requires the synthesis of new drugs with a wider range of biological activities with a known mechanism of action and less harmful side effects.

This chapter reviews the latest research on anti-tumor and antimicrobial benzimidazole compounds, as well as structure-activity correlations, drug design, clinical and preclinical studies. The review also covers synthetic strategies for the obtaining of the target structures.

Literature shows that there is a need for modification of existing benzimidazole agents that will bring about a big change for improvement of activity or change in activity, due to the problems like resistance, toxicity, or poor bioavailability.

Keywords: benzimidazole ring, antimicrobial, anti-tumor, cytotoxicity, structure-activity relationship

* Corresponding Author's Email: maria.marinescu@chimie.unibuc.ro.

INTRODUCTION

Benzimidazoles has received considerable attention in the past few decades, beying a very important class of heterocycle compounds. Benzimidazole compounds [1], due to the pharmacophore nucleus, possess a wide range of biological activities, such as anticancer [2], antiviral [3], antibacterial [4], antifungal [5], anti-inflammatory [6], antioxidant [7] properties.

Cancer is the second major cause of death worldwide, estimating that by 2030 it will reach approximately 22 million cases [8].

Heterocyclic compounds possessing heteroatoms, N, S and O may form hydrogen bonds with DNA, the interaction force being correlated with anticancer activity. The presence of one or more nitrogen atoms leads to a polarized character and optimal interaction. In recent years, there is considerable interest for synthesis of heterocyclic compounds such as benzimidazoles, benzothiazoles, oxadiazoles, imidazoles, pyrazoles, triazoles, which exhibit therapeutic activity. These N-heterocyclic compounds in various cancer types induce anticancer effects by inhibiting cell growth and inducing cell differentiation and apoptosis [8].

However, it is necessary to develop new methods of synthesis of nitrogen-containing heterocyclic compounds which today become an important objective in modern organic synthesis.

I. Benzimidazoles as Anti-Tumorals

We propose to review methods of synthesis of benzimidazole compounds, their effectiveness in treating various types of cancer [9].

Hoebrecker in 1872 synthesizes for the first time (2,5 or 2,6-dimethylbenzimidazole) by reducing 4-methyl-2-nitroacetanilide (Scheme 1).

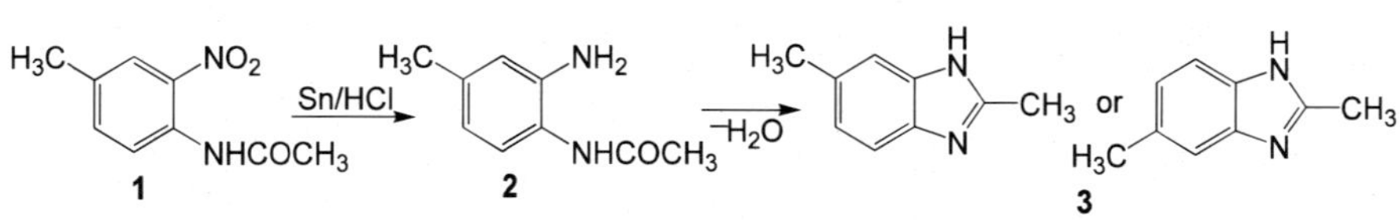

Scheme 1. Synthesis of 2,5 or 2,6-dimethylbenzimidazole by Hoebrecker.

After a few years Ladenburg obtained the same compound by refluxing 3,4-diaminotoluene with acetic acid (Scheme 2) [9]

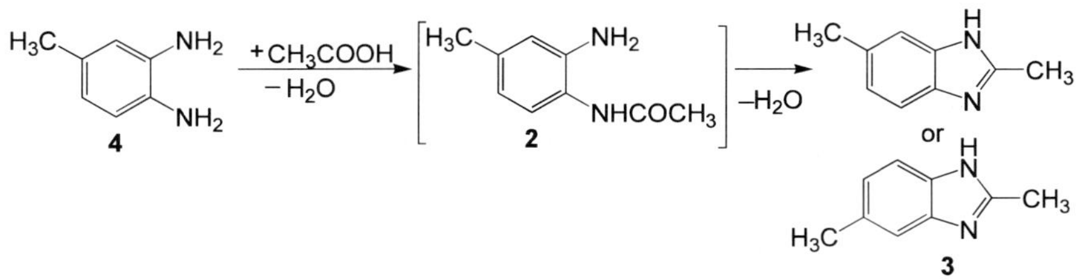

Scheme 2. Synthesis of 2,5 or 2,6-dimethylbenzimidazole by Ladenburg.

Later, in 1928, Phillips describes a method of obtaining the benzimidazole nucleus by condensation o-phenylenediamine with various carboxylic acids in the presence of boiling dilute hydrochloric acid (Scheme 3) [10].

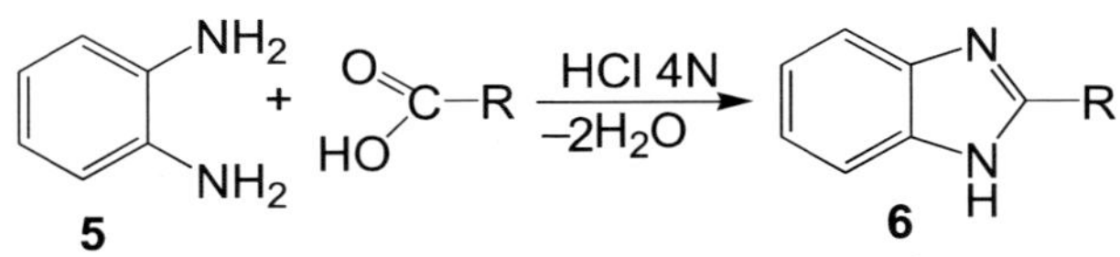

Scheme 3. Phillips synthesis of benzimidazoles.

Numerous synthetic strategies for the synthesis of benzimidazole **6** have been described in Scheme 4 [9]. In recent years a wide range of benzimidazole compounds have been synthesized following synthesis strategies. (Scheme 4) [9].

Nguyen et al. synthesized benzimidazoles **6** using elemental sulfur as an oxidizing agent on alkylamines (**A**) [11].

Hanan et al. obtained benzimidazoles a derivatives starting from *o*-nitroaniline by reducing the nitro group with Fe and ammoniated chloride followed by condensation and cyclization with formic acid in a single step (**B**) [12].

Yang et al. have developed a new method for synthesis of 2-substituted benzimidazoles in one step by reducing *o*-nitroanilines (**C**) in the presence of aldehydes and $Na_2S_2O_4$ [13].

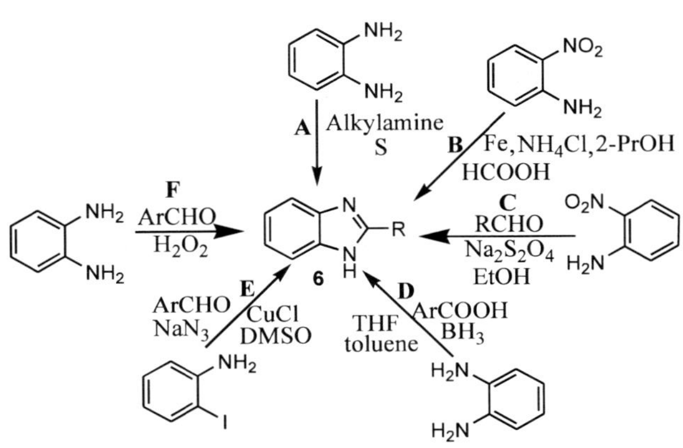

Scheme 4. Synthesis of 2-substituted benzimidazoles.

Cui et al. in 2012 synthesizes benzimidazole derivatives using *o*-phenylenediamines and aromatic carboxylic acids in the presence of toluene and BH_3 in THF (**D**) [14].

Another way to obtain benzimidazole derivatives is due to the studies of Kim et al., which in 2011 in one pot reaction use 2-haloaniline, aldehydes, and NaN_3 and CuCl in catalytic amount, DMSO (**E**) [15].

Bahrami et *al.* in 2007, obtain benzimidazole derivatives by one pot reaction using *o*-phenylenediamines with aryl aldehydes, H_2O_2 and HCl in acetonitrile (**F**) [16].

Fluoro at the R_3 position and a nitro at the R_2 position exhibit growth inhibitory activity. Replacement with only flouro group and nitro group with amine group decreases the activity. Nitro and amine groups at the R_3 position and methoxy group at R_3 position also exhibit low activity. Introduction of tertiary butyl group and fluorine groups improves the cytoxicity but results in loss of activity when fluorine is replaced with chloro and bromo groups (Figure 1) [9].

The biological activity of the benzimidazole nucleus involves various mechanisms, such as oxidative stress, receptor mediatory mechanism, enzymatic action. The presence of the various substituents attached to the benzimidazole nucleus induce some biological activity.

Gowda et *al.* in 2009, prepared new derivatives of 1-(4-methoxyphenethyl)-1H-benzimidazole-5-carboxylic acid derivatives, which are tested for the evaluation of antileukemic activity [17].

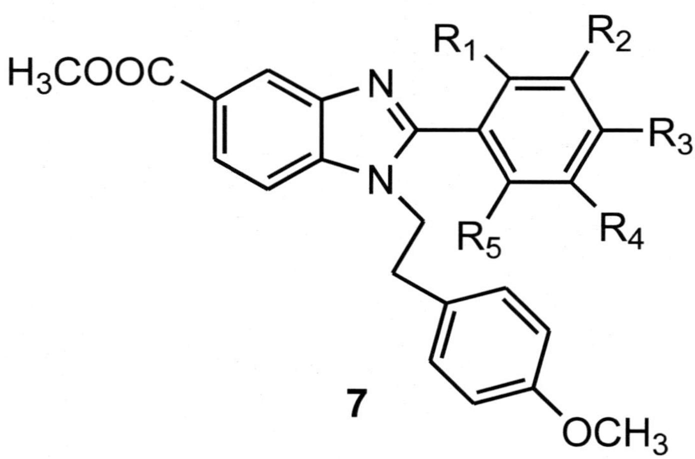

7

Figure 1. SAR of 1-(4-methoxyphenethyl)-1H-benzimidazole-5-carboxylic acid derivatives as antileukemic agents.

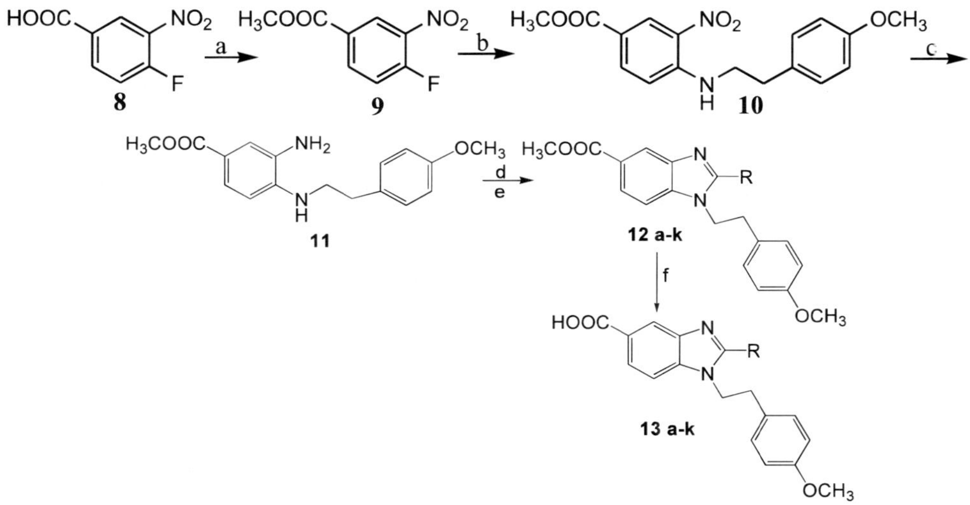

Scheme 5. Synthesis of benzimidazole derivatives **12a–k** and **13a–k**. Reagents and conditions: (a) Thionyl chloride, Methanol, MDC, reflux condition; (b) 2-(4-methoxyphenyl)ethanamine, DIPEA, DMF, rt, 3 h; (c) Fe powder, NH4Cl, IPA, 70 C, 4h; (d) R–COCl, TEA, MDC, rt, 3h; (e) CH_3COOH, 90 C, 8h; (f) LiOH, THF/water (2:1), reflux condition.

Synthesis of 1-(4-methoxyphenethyl)-1H-benzimidazole-5-carboxylic acid derivatives by Gowda et *al.* was described in Scheme 5. To 4-fluoro-3-nitro-benzoic acid **8** dissolved in dichloromethane is added methanol and thionyl chloride to give compound **9**. Compound **10** was obtained following nucleophilic aromatic substitution of fluorine atom with 2-(4-methoxyphenyl)ethanamine. Amine **11** was obtained by reduction of the aryl nitro group with NH_4Cl solution and 20% iron (II) powder. The selective acylation of the resulting compound, followed by cyclization, gave 12a-k precursors and the hydrolysis of **12a–k** gave the final compounds **13a–k** [17].

The compound (methyl 2-(4-fluoro-3-nitro-phenyl)-1*H*-benzimidazole-5-carboxylate) induced maximum cell death in leukemic cells with an IC_{50} value of 3 μM [17]. SAR study is also depicted in Figure 2. Where R is:

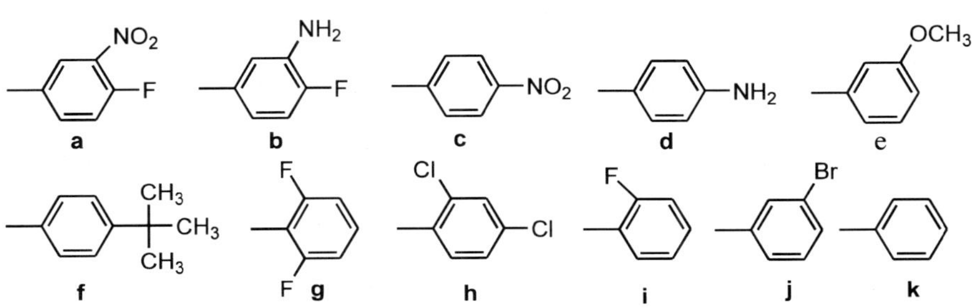

Figure 2. SAR of 1-(4-methoxyphenethyl)-1*H*-benzimidazole-5-carboxylic acid derivatives as antileukemic agents. The structure of compounds **13a-k.**

The presence of the fluorine atom in the poison R_3 and the nitro group at the R_2 position show an inhibitory growth activity. Replacement of the fluorine atom and the nitro group with the amino group leads to decreased activity. The presence of nitro, methoxy in the R_3 position also has low activity.

The introduction of the *tert*-butyl group and the fluorine atom improve the cytotoxicity, but replacing the fluorine atom with chlorine or bromine atoms leads to loss of activity (Figure 2).

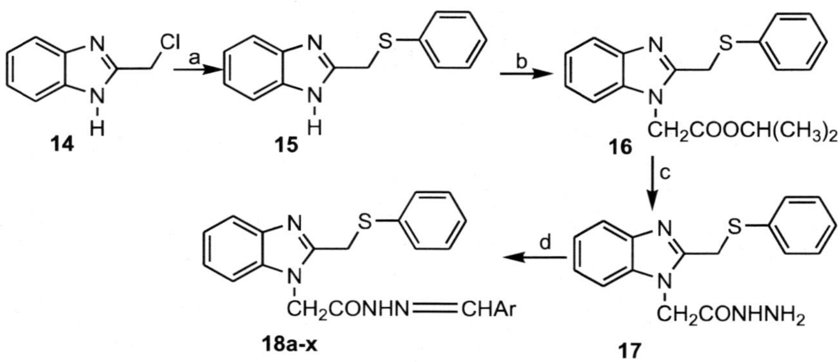

Scheme 6. Synthetic pathway of benzimidazole derivatives (**18a–x**). Reagents and conditions: a K_2CO_3 and benzenethiol, stirred for 0.5 h at rt; then compound 1 in DMF was added, and stirred for 2 h at rt; b K_2CO_3, tert-butyl 2-bromopropanoate, stirred for 6 h at rt; c EtOH, NH_2NH_2 H_2O; d Ar-CHO, reflux for 2 h in ethanol.

A novel class of 2-[2-(phenylthiomethyl)-1H-benzo[*d*]imidazol-1-yl) acetohydrazide derivatives was described by Liu et al., tested for anti-tumor activity against A549, HCT116, HepG2, 9 and A375 determined by the MTT method [18].

The target compounds 18a-x, were synthesized by the general method of preparation shown in Scheme 6.

Where Ar is (Figure 3):

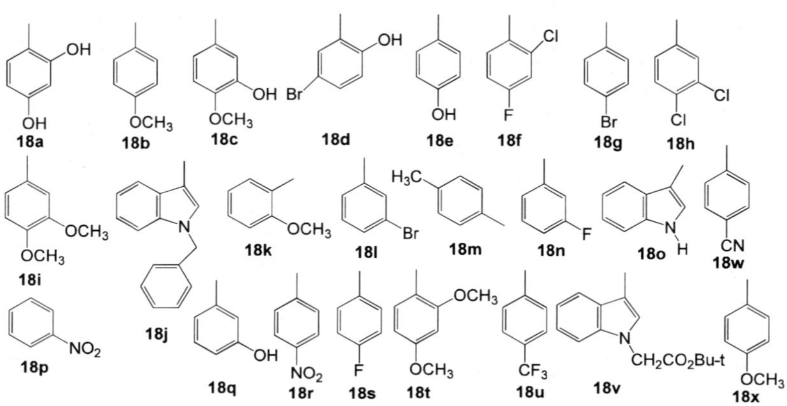

Figure 3. The structure of compounds **18a-x**.

The compound **14** was obtained following the condensation and cyclization reaction of o-phenylenediamine and 2-chloroacetic acid, and the Compound **15** was obtained following the reaction of 2-(chloromethyl)-1*H*-benzo[*d*]-imidazole **14**, and benzenethiolate potassium. Compound **16** was obtained by treating 2-bromoacetyl tert-butyl in dry acetone in the presence of anhydrous K_2CO_3 with compound **15**, which was then treated with 85% hydrazine hydrate to afford compound **17**. Compounds **18a–x** were obtained by the condensation of compound **17** with corresponding phenylcarbaldehyde in the presence of glacial acetic acid.

Liu et al. introduce a hydrazone group at the 1 position of 2-(phenylthiomethyl)-1H-benzo[*d*]-imidazole to assess the biological activity against 5 cancer cells using the MTT method. Of the evaluated series, only 2 compounds **18a** and **18d** show excellent anti-cancer inhibitory activity compared to 5-FU and SU11248 [54]. The SAR study was necessary to highlight the correlation between certain functional groups and their biological activity (Table 1) [18].

Table 1. 2-(Phenylthiomethyl)-1H-benzo[*d*]-imidazole with anti-cancer activity

Compound	Ar	IC$_{50}$ (μM)				
		A549	HCT116	HepG2	PC-9	A375
18a	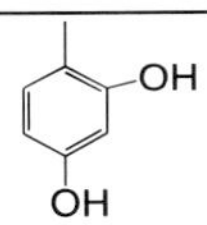	12.36	4.86	>20c	14.63	5.75
18d	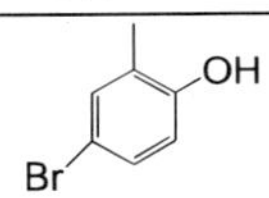	10.75	8.33	17.03	12.20	4.09

Legend: A549, HCT116, HepG2, PC-9 and A375 cancer cell lines.

The method for the synthesis of benzimidazole derivatives has been adopted and modified from the literature previously published by Yoon et *al*. (Scheme 7).

To obtain the target compounds, the starting compound was 4-fluoro-3-nitro benzoic acid which had been etherified with ethanol in the presence of sulfuric acid by refluxing for 8 hours to obtain the corresponding ethyl ester in 75% yield.

Yoon et al. took into account the possibility of basic functional groups to change in terms of lipophilicity capacity. This first transformation reaction of the carboxylic group into the ester group was intended to avoid low solubility in lipid membranes and chemical instability of the drug.

The ethyl benzoate **20** was then treated with ethanolamine and DIPEA in anhydrous dichloromethane at room temperature resulting ethyl 4-(2-hydroxyethylamino)-3-nitrobenzoate **21**, which was then reduced to amine **22** using ammonium formate and Pd/C 10% for 3 hours. Ethyl 4-(2-hydroxyethylamino)-3-aminobenzoate **22** was then refluxed with various substituted bisulfite adducts **24a-l** of aromatic aldehydes **23a–l** in DMF overnight to afford benzimidazole derivatives **25a–l** in good yields (75–95%) [19].

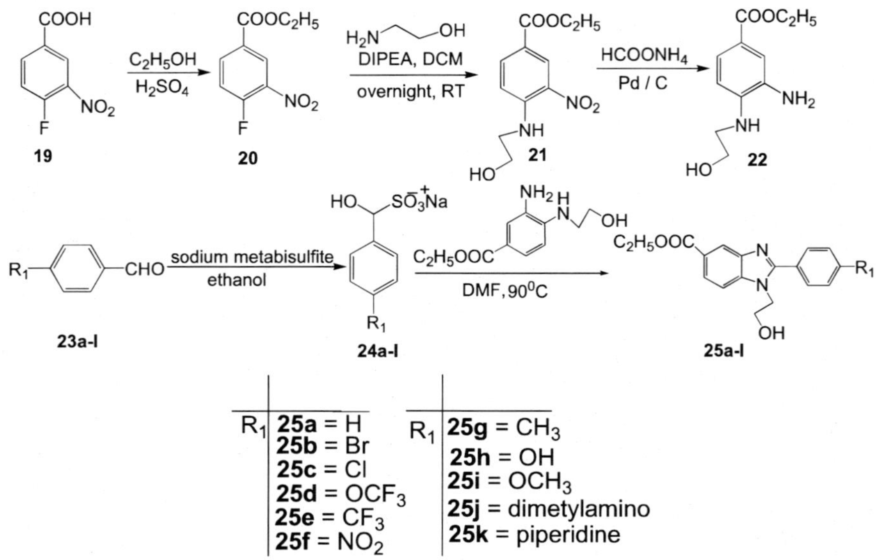

Scheme 7. Reagents and conditions: (a) RNH_2, K_2CO_3, DMSO, rt; (b) H_2 (1 atom), Pd/C, EtOH; (c) 4-benzyloxybenzoyl chloride, cat. DMAP, pyridine–CH_2Cl_2, rt; (d) AcOH, reflux; (e) 4N aq NaOH, EtOH–THF, reflux; (f) H_2 (1 atom), Pd/C, AcOH; (g) $ArCH_2Br$, K_2CO_3, DMF, 85^{0}C.

Table 2. SAR of benzimidazole derivatives as sirtuin inhibitors with antitumor activity

Compound	R_1	IC_{50} SIRT1 inhibition (μM)	IC_{50} SIRT2 inhibition (μM)
25j	Dimethylamino	58.43	45.12
Cambinol		47.90	52.89

Yoon et *al.* tested the antitumor activity of benzimidazole derivatives synthesized as sirtuin inhibitors. Compound **25j** exhibited a good inhibitory activity on both SIRT1 (IC_{50} = 58.43 μM) and SIRT 2 (IC_{50} = 45.12 μM), and cell cytotoxicity assays showed what is a good antitumor compound against two cancer cell lines MCF-7 and MDA-MB-468, derived from breast cancer [19] (Table 2). SAR studies have shown the contribution of each functional group in correlation with antitumor activity.

Kilcigil et *al.* have synthesized a series of benzimidazole compounds containing thiosemicarbazide, triazole, thiadiazole rings. The synthesized compounds were tested *in vitro* as potential inhibitors of EGFR kinase compared to erlotinib, most compounds having significant activities. The compound **31** was most active (Scheme 8) [20].

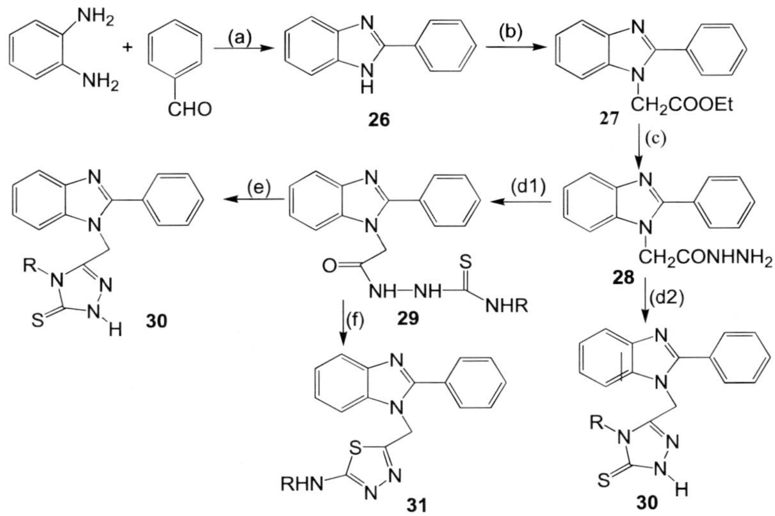

Scheme 8. Synthesis pathway of benzimidazole derivatives. Reagents and conditions (a) $Na_2S_2O_5$ (b) $ClCH_2COOEt$/KOH-DMSO (c) $NH_2NH_2.H_2O$/EtOH.(d1) methyl/ethyl isothiocyanate-abs EtOH 3h. (d2) 2-piperidino/morpholinoethylisothiocyanate-abs EtOH 3h (e) H_2SO_4 (f) NaOH.

Azam et al. synthesizes benzimidazole derivatives using benzene-1,2-diamine and aryl-aldehydes at room temperature. All of the synthesized compounds were tested for their antitumor activities against various human breast adenocarcinoma cell lines (MCF-7), human leukemia cell line (THP-1), human prostate cancer cell lines) and human alveolar basal epithelial cell lines adenocarcinoma (A-549). All compounds exhibiting significant anticancer activity [21].

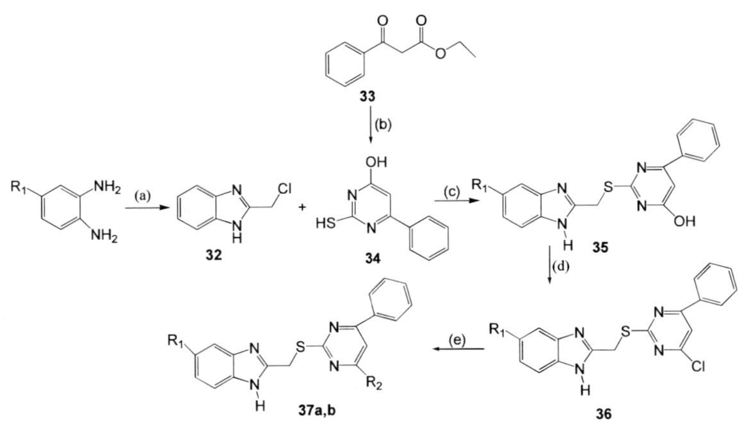

Scheme 9. Reagents and conditions: (a) chloroacetyl chloride, 5 N HCl, reflux, 4 h (85% yield); (b) thiourea, KOH, EtOH, reflux, 10 h (82% yield); (c) KOH, H_2O–acetone (2:1), 55-60^{0}C, 30 min (87–92% yield); (d) $POCl_3$, 90^{0}C, 20 min (89–91% yield); (e) EtOH, reflux, 10 h (65–92% yield).

Shao and colleagues synthesized a series of pyrimidine-benzimidazole hybrid compounds, which they synthesized against four human tumor cell lines MCF-7, MGC-

803, EC-9706 and SMMC-7721. In this series compounds **37a** (R_1 = H; R_2 = 4-CH_3O-C_6H_5-NH–) and **37b** (R_1 = Cl; R_2 = 4-CH_3O-C_6H_5-NH–) were more cytotoxic than 5-fluorouracil against all the human cancer cell lines tested, with IC_{50} values ranging from 2.03 to 10.55 μM and 1.06-12.89 μM, respectively (Scheme 9) [22].

Zhu et *al.* have synthesized a series of 5-fluoro-benzimidazole-4-carboxamides. All compounds were tested to show their inhibitory activity on PARP-1 (Scheme 10 and 11) [23].

The compounds exhibit *in vitro* an intrinsic inhibitory activity of PARP-1, of which compound **47f** is most promising as an anticancer agent with an extracellular activity of inhibiting cells against HCT116 cells with an IC_{50} of 7.4 μM [23]

Synthetic method used by Kamal and his group to give 2-(1,3-diphenyl-1*H*-pyrazol-4-yl)-1*H*-benzo[d]imidazoles (**54-109**) is shown in Scheme 12. The aldehyde key intermediates (**52a-d**) were prepared in two steps. Condensation of the acetophenones (**49a-d**) with phenylhydrazine (**50**) in ethanol led to the corresponding phenyl acetophenone hydrazones (**51a-d**). The phenylacetophenone hydrazones were then cyclized by the Vilsmeir-Haack reaction. The desired target compounds (**54-109**) were obtained in good yields by refluxing the aldehyde intermediates (**52a-d**) with different substituted *o*-phenylenediamines (**53a-d**), the reactions taking place in the presence of ethanol and sodium metabisulfate (Scheme 12) [24].

Kamal et *al.* have been designed, synthesized a series of 40 hybrid pyrazolo-benzimidazole compounds to tested for their potential antiproliferative activity against three human-lung (A549), mammary (MCF-7) HeLa tumor cell lines.

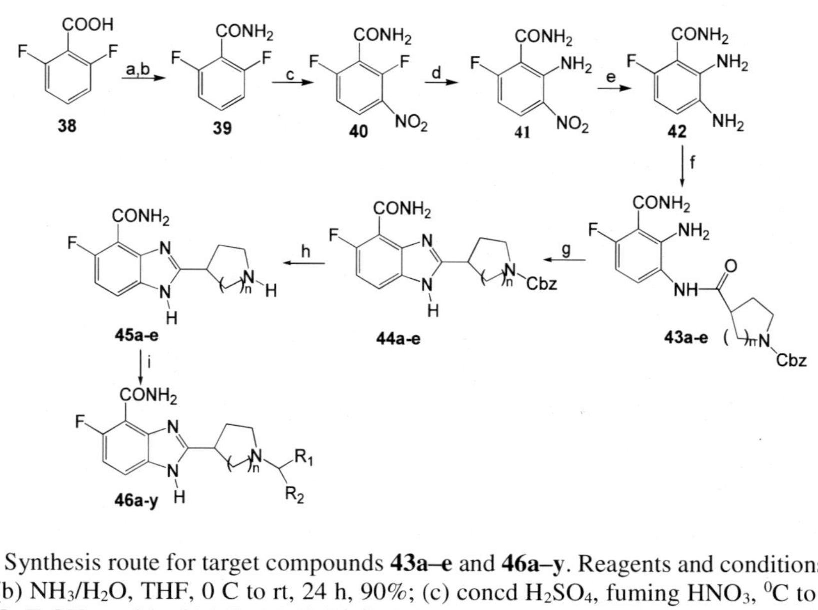

Scheme 10. Synthesis route for target compounds **43a–e** and **46a–y**. Reagents and conditions: (a) $SOCl_2$, reflux, 2h; (b) NH_3/H_2O, THF, 0 C to rt, 24 h, 90%; (c) concd H_2SO_4, fuming HNO_3, ^{0}C to rt, 2h, 92%; (d) NH_3/H_2O, EtOH, rt, 4 h, 64.1%; (e) 10% Pd–C, H_2, MeOH, rt, 12h, 90%; (f) PyBOP, DIEA, DMF, rt, 12h; (g) AcOH, 120^{0}C, 4h, 50–72%; (h) 10% Pd–C, $HCOONH_4$, MeOH, rt, 2h, 52–80%; (i) $R_1R_2C(O)$, $NaBH_3CN$, MeOH, rt, 12h, 60–94%.

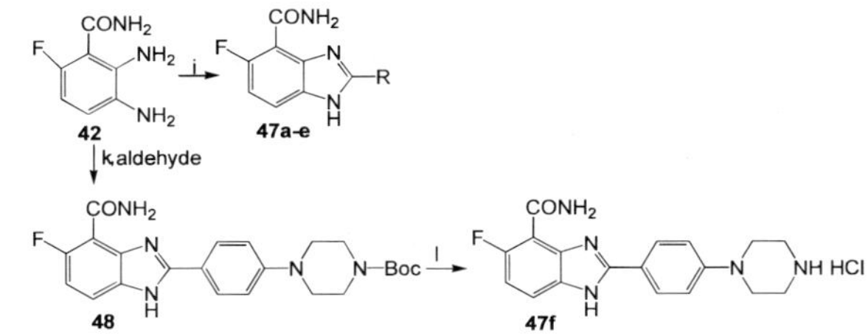

Scheme 11. Synthesis route for target compounds **47a–f.** Reagents and conditions: (j) $NaHSO_3$, CH_3OH, reflux, 12h, 40–67%; (k) $NaHSO_3$, CH_3OH, reflux, 12h, 64%; (l) THF, CH_3OH/HCl, rt, 2h, 63%.

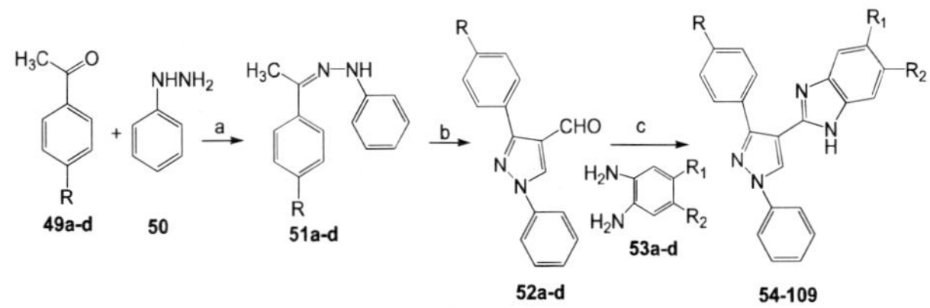

Scheme 12. General synthesis of 2-(1,3-diphenyl-1*H*-pyrazol-4-yl)-1H-benzo[d]imidazole derivatives (**54-109**). Reagents and conditions: (a) ethanol, 50-60^{0}C, 3 h; (b) DMF, $POCl_3$, 50-60^{0}C, 5h; (c) ethanol, $Na_2S_2O_5$, 50-60 C, 5h.

A series of N-(2-phenyl-1*H*-benzo[d]imidazol-5-yl) quinazolin-4-amine derivatives as dual c-Met and VEGFR-2 inhibitors have been synthesized by Xu et *al.* (Scheme 13). Among these quinazoline-benzimidazole compounds, compound **116j** showed the strongest c-Met and VEGFR-2 inhibitory activity with an IC_{50} of 0.05 μM and 0.02 μM, respectively. It also demonstrated the highest anticancer activity against the tested cancer cell lines Hep-G2, having IC_{50} of 1.5 μM vs. MCF-7 and 8.7 μM [25, 26].

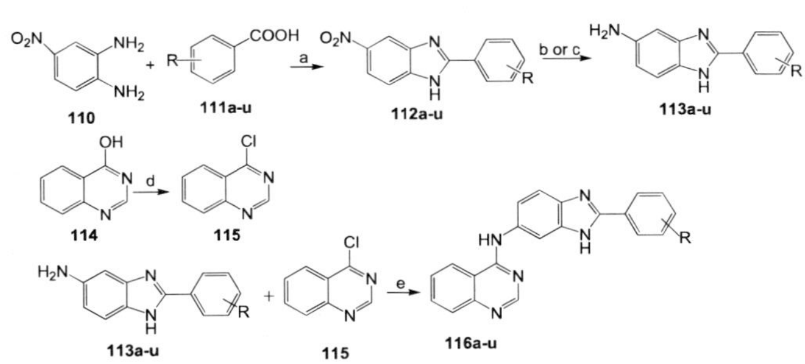

Scheme 13. General procedure for the synthesis of N-(2-phenyl-1*H*-benzo[d]imidazol-6-yl)quinazolin-4-amine derivatives. Reagents and conditions: (a) PPA, 120–150^{0}C, 5h; (b) Pd/C, H_2, CH_3OH, rt, 5h; (c) Fe, AcOH, EtOH, reflux, 2h; (d) $POCl_3$, 100^{0}C, 4h; (e) isopropanol, reflux, 5h Where for **116a-u** R is:

a	R = H	h	R = 2-F	o	R = 2-CF_3
b	R = 2-CH_3	i	R = 3-F	p	R = 3-CF_3
c	R = 3-Ch_3	j	R = 4-F	q	R = 4-CF_3
d	R = 4-CH_3	k	R = 2-Cl	r	R = 2,6-diCl
e	R = 2-OCH_3	l	R = 3-Cl	s	R = 3,4-diCl
f	R = 2-OH	m	R = 4-Cl	t	R = 3,5-diCl
g	R = 4-CH_2CH_3	n	R = 4-Br	u	R = 2-Br-5-F

The route of synthesis of N-(2-phenyl-1*H*-benzo[d]imidazol-5-yl)quinazolin-4-amine derivatives is described in Scheme 13. Preparation of these compounds starts from a series of commercially available benzoic acids (**111a-u**), which is reacted with 4-nitro-*o*-phenylenediamine (**110**) in polyphosphoric acid (PPA) at 120-150°C for 5 hours, resulting 5α-nitrobenzimidazole **112a-u** intermediates. These compounds were reduced with Pd/C/H_2 at normal pressure at room temperature for 5 hours or with Fe/AcOH/EtOH at reflux for 2 hours resulting 5α-amino-benzimidazole **113a-u** intermediates. 4-Chloroquinazoline (**115**) was in turn prepared from quinazolin-4-ol (**114**) in $POCl_3$ at 100°C for 4 hours in refluxing isopropanol. The 5-amino-benzimidazole intermediates **113a-u** were treated with 4-chloroquinazoline to give quinazolines **116a-u** at reflux for 5 hours.

2-Aminobenzimidazole was obtained by amination in position 2 of benzimidazole with sodium amide in the liquid ammonia of *o*-phenylendiamine and then submitted to the reaction with a slight excess of the alkyl halides to give the desired monoalkylated products (Scheme 14) [27].

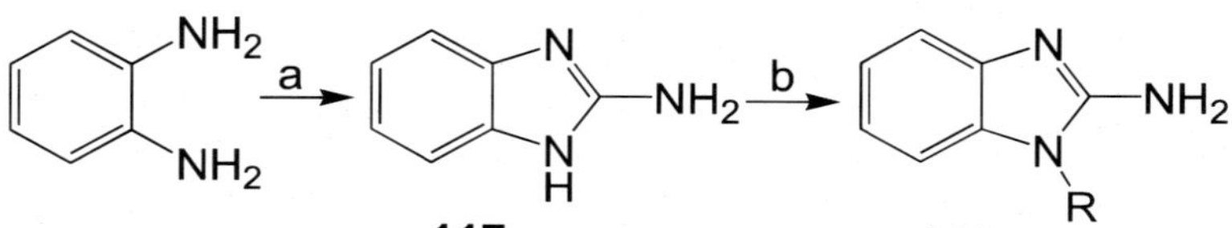

117 **118**

R
H
CH_3
C_2H_5
C_3H_7
C_4H_9
$H_2C—C_6H_5$
$H_2C—C_6H_4F$
$CH_2CH{=}CH_2$

X
Br
I

Scheme 14. Synthesis route for target compounds **117** and **118.** Reagents and conditions: CNBr, RX/KOH, acetone.

II. Benzimidazoles as Antimicrobials

The increase in antimicrobial resistance to existing drugs necessitated the search for new molecules for the treatment of bacterial infections [28]. Currently, a number of benzimidazole containing drugs are available in market namely: galeterone, bendamustine, selumetinib and pracinostat, as antitumor agents; bezitramide as analgesic; mebendazole, albendazole, thiabendazole, flubendazole as antihelmintics; ridinazole as antibacterial, astemizole and bilastine as antihistamines; pantoprazole, lansoprazole, esomeprazole and ilarazole as proton pump inhibitors; enviradine, samatasvir and maribavir as antivirals, candesartan and mibefradil as antihypertensive [28-33]. In the following, we will present recent syntheses of benzimidazoles with antimicrobial properties.

Zhou et al. synthesized 2-arylbenzimidazoles **1a–1l** from various substituted benzaldehyde and *o*-nitroaniline (or 4-methoxy-2-nitroaniline) in the presence of $Na_2S_2O_4$, according to Yang's method (Scheme 15) [34].

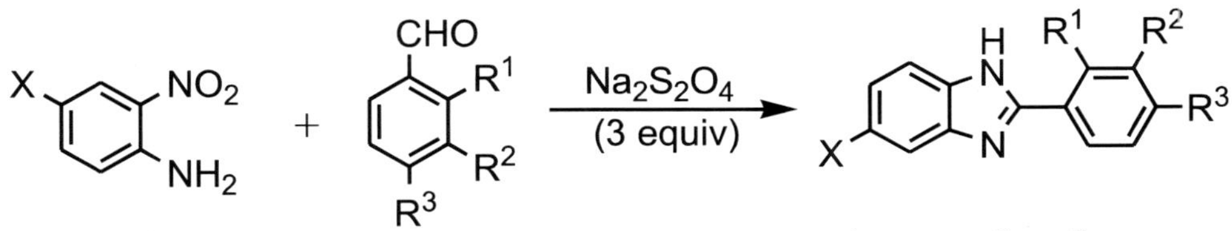

1a: X = H; $R^1 = R^2 = R^3$ = H
1b: X = H; R^1 = OH, $R^2 = R^3$ = H
1c: X = H; $R^1 = R^3$ = H, R^2 = OH
1d: X = H; $R^1 = R^2$ = H, R^3 = OH
1e: X = H; $R^1 = R^2$ = H, R^3 = OMe
1f: X = OMe; $R^1 = R^2 = R^3$ = H
1g: X = OMe; R^1 = OH, $R^2 = R^3$ = H
1h: X = OMe; $R^1 = R^3$ = H, R^2 = OH
1i: X = OMe; $R^1 = R^2$ = H, R^3 = OH
1j: X = OMe; $R^1 = R^2$ = H, R^3 = OMe
1k: X = OMe; $R^1 = R^3$ = OH, R^2 = H
1l: X = OMe; R^1 = H, R^2 = OMe, R^3 = OH

Scheme 15. Synthesis of 2-arylbenzimidazoles **1a–1l** using Yang's method [34].

Table 3. Antibacterial activities of selected 2-arylbenzimidazoles 1a-1l

Compounds	MIC[a] (MBC[b]) (μM)				
	E. coli	*P. aeruginosa*	*S. aureus*	*S. albus*	*B. cereus*
1a	6.25 (100)	>200	100 (>200)	>200	>200
1c	12.5 (100)	>200	0.39 (0.78)	>200	50.0 (200)
1d	25 (100)	>200	12.5 (50.0)	>200	>200
1e	12.5 (50)	>200	1.56 (156)	100 (100)	50.0 (100)
1h	12.5 (100)	>200	100 (200)	>200	25.0 (200)
1i	12.5 (100)	>200	100 (200)	>200	50.0 (200)
1j	6.25(100)	>200	25.0 (200)	100 (>200)	50.0 (200)
1k	6.25 (100)	>200	50.0 (100)	100 (200)	50.0 (100)
Ciprofloxacin	0.39 (0.39)	0.78	0.39 (0.39)	0.39 (0.39)	0.39 (0.39)

2-Arylbenzimidazoles **1a-1l** were screened for their *in vitro* antimicrobial activity against Gram-negative (*Escherichia coli, Pseudomonas aeruginosa*) and Gram-positive (*Staphylococcus aureus, Bacillus cereus, Staphylococcus albus*) bacteria by the conventional agar dilution method. Minimum inhibitory concentration (MIC) and minimum bactericidal concentration (MBC) values were evaluated using ciprofloxacin as a standard reference drug (Table 3). MIC values of 2-arylbenzimidazoles against *E. coli* indicated that some compounds exhibited moderate to good activity.

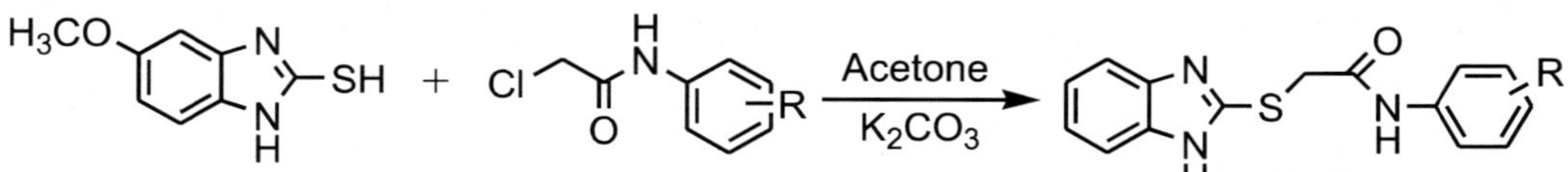

2a: R = *m*-NO_2 **2d**: R = 3-F-4-Cl **2g**: R = *o*-OCH_3 **2j**: R = *o*-NO_2
2b: R = *p*-NO_2 **2e**: R = *H* **2h**: R = *m*-CH_3 **2k**: R = *o*-CH_3
2c: R = *m*-Cl **2f**: R = *p*-F **2i**: R = *p*-OCH_3 **2l**: R = *p*-CH_3
2m: R = *m*-CF_3

Scheme 16. Synthesis of 2-arylbenzimidazoles **1a–1l** using Yang's method [35].

Joshi and Parikh synthesized benzimidazole analogs **2a-2m** by reaction between 2-mercapto benzimidazole and N-arylacetamides (Scheme 16) [35]. All synthesized compounds were screened against two Gram-positive bacteria (*S. aureus, E fecalis*), two Gram-positive bacteria (*E. coli, P. aruginosa*) and two fungi (*C. albicans, A. niger*) using ciprofloxacin and fluconazole as standard drugs. Table 4 shows the antimicrobial activities of the best benzimidazoles: **2a, 2d, 2g, 2i** and **2j.**

Table 4. Antimicrobial activities of selected benzimidazole analogs 2a-2j

Compound	MIC (μM)					
	Gram positive bacteria		Gram negative bacteria		Fungi	
	S. aureus	*E. fecalis*	*E. coli*	*P. aerugi-nosa*	*C. albi-cans*	*A. ni-ger*
2a	250	50	125	62.5	250	250
2d	125	125	62.5	62.5	125	250
2g	31.25	125	62.5	125	250	125
2i	62.5	125	125	62.5	250	250
2j	62.5	125	62.5	125	250	250
Floconazole	-	-	-	-	125	62.5
Ciprofloxacin	62.5	125	125	125	-	-

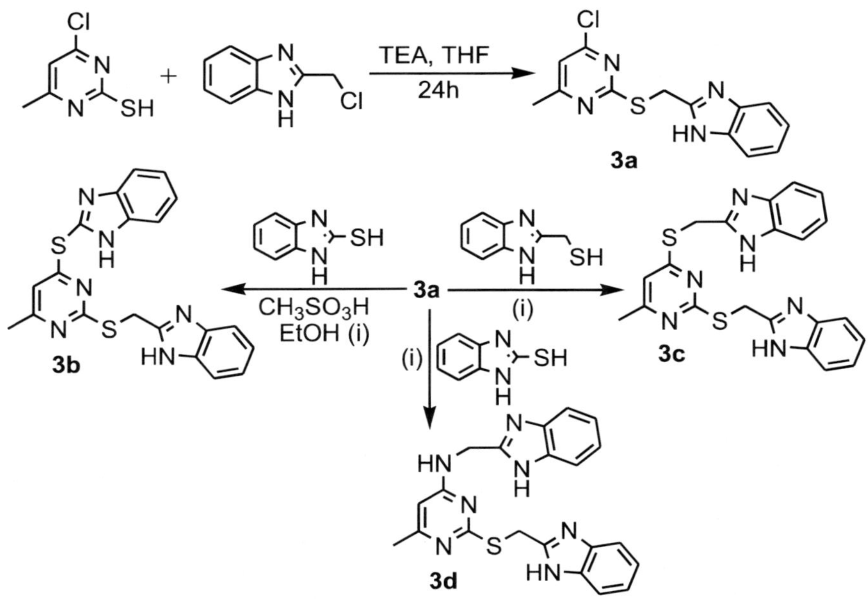

Scheme 17. Synthesis of pyrimidinyl-benzimidazoles **3a-3d.**

Seenaiah et al. synthesized a variety of pyrimidinyl-benzimidazoles **3a-3d** using the reaction steps indicated in Scheme 17 [36]. Compound **3c** possessed an excellent antimicrobial activity, against *Staphylococcus aureus* (29 mm, MIC 12.5 μg/mL) and *Penicillium chrysogenum* (38 mm, MIC 12.5 μg/mL).

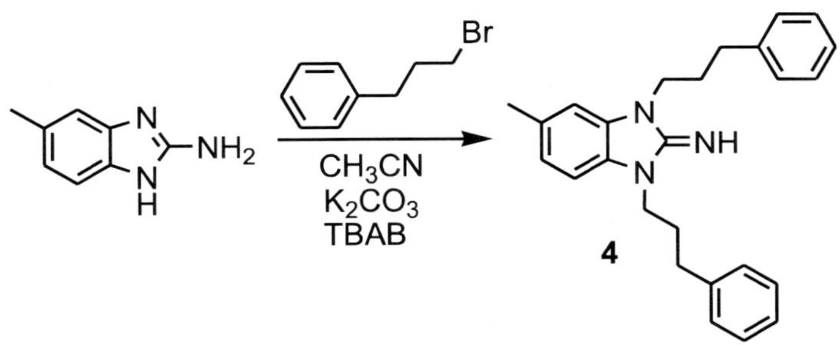

Scheme 18. Synthesis of the benzimidazol-2-imine **4.**

Mavrova et *al*. synthesized diphenylpropyl-5-methyl-1,3-dihydro-2*H*-benzimidazol-2-imine **4** by reaction of 5-methyl-2-aminobenzimidazole with (3-bromopropyl)benzene, anhydrous potassium carbonate and tetrabutylammonium bromide (TBAB) (Scheme 18) [37]. Compound **4** exhibited significant activity against *B. subtilis*, *S. aureus*, *S. abony* and *E. coli* (with MIC values of 0.125, 0.016, 0.50 and 0.50 mg/mL, respectively).

The authors correlated that the best antibacterial activity of the compound 4 with the presence of phenylpropyl groups at positions 1 and 3 of 2-imino-benzimidazole.

Yadav et al. synthesized a series of novel benzimidazole compounds bearing alkyl chain as linker at N-1 position (Scheme 19) and evaluated to their antimicrobial activities against gram positive bacteria (*Staphylococcus aureus* and *Bacillus subtilis*) and some gram-negative bacterial strains (*Escherichia coli, Salmonella typhi*, *Klebsiella pneumoniae* and *Pseudomonas aeruginosa*) [38].

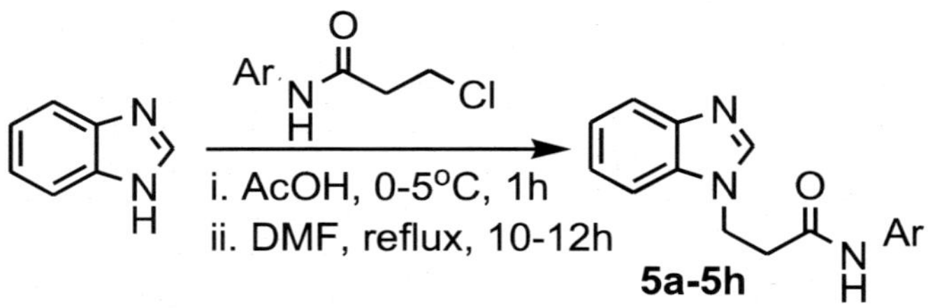

5a: Ar = 2-SHC_6H_4 **5e**: Ar = 4-Br-C_6H_4
5b: Ar = 2-Cl-4-CH_3-C_6H_3 **5f**: Ar = 3,4-CH_3-C_6H_3
5c: Ar = 3-Cl-2-CH_3-C_6H_3 **5g**: Ar = 4-OC_2H_5-C_6H_4
5d: Ar = 3,4-Cl-C_6H_3 **5h**: Ar = 4-Cl-3-CH_3-C_6H_3

Scheme 19. Synthesis of 1-substituted benzimidazoles **5a-5h.**

Compounds **5a-5e** surfaced out as the most effective antibacterials but less effective than the standard ciprofloxacin. Also, compounds **5f-5h** also showed significant activity against almost all the bacterial strains (Table 5). All compounds were found to exhibit potency comparable or more than that of standard drug fluconazole in case of fungal strains like *Candida albicans* and *Aspergillus niger.*

Table 5. Antimicrobial activities of 1-substituted benzimidazoles 5a-5h

Compound	MIC (μM)					
	Gram positive bacteria		Gram negative bacteria			
	S. aureus	*B. subtilis*	*E. coli*	*P. aeruginosa*	*S. typhi*	*K. pneumoniae*
5a	3.125	6.25	6.25	6.25	6.25	6.25
5b	3.125	3.125	6.25	6.25	6.25	6.25
5c	3.125	3.125	6.25	6.25	6.25	6.25
5d	3.125	3.125	6.25	6.25	6.25	6.25
5e	3.125	6.25	6.25	6.25	6.25	6.25
5f	6.25	6.25	6.25	6.25	6.25	6.25
5g	6.25	6.25	6.25	6.25	6.25	6.25
5h	6.25	6.25	6.25	6.25	12.5	12.5
Ciprofloxacin	≤1	≤1	≤1	≤1	≤1	≤1

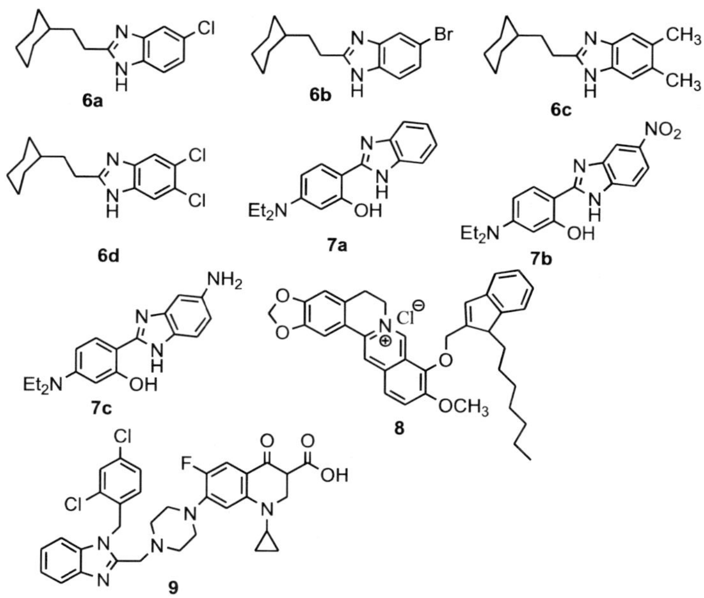

Figure 4. Benzimidazoles **6a-6d, 7a-7c, 8, 9** with antimicrobial activity.

New 1*H*-benzo[d]imidazole derivatives (**6a-6d**) were synthesized and evaluated for tuberculostatic activity [39] (Figure 4). All compounds bearing halogen atoms or methyl groups at the benzimidazole system and cyclohexylethyl substituent at the C-2 position showed an excellent tuberculostatic activity against *Mycobacterium tuberculosis* $H_{37}Rv$, Spec. 192 and Spec. 210 strains and *Mycobacterium bovis* strain, with MIC values ranging from 0.75 to 1.5 µg/mL.

Padalkar et al. reported synthesis of benzimidazole **7a-7c** (Figure 4) with excellent *in vitro* antibacterial activity against *E. coli* and *S. aureus* strains as well as for antifungal activity against *C. albicans* and *A. niger* strains [40].

Jeyakkumar et al. sythesized berberine-benzimidazole hybrids as new type of potentially DNA-targeting antimicrobial agents [41]. Compound **8** (Figure 4) exhibited good anti-Methicillin-resistant *Staphylococcus aureus* N315 (MRSA), anti-*Escherichia coli*, and anti-*Salmonella typhi* activity with low MIC values of 2-8 µg/mL, which were comparable or even superior to reference drugs. Compound **8** was able to efficiently permeabilize the membranes of both Gram-positive (MRSA) and Gram-negative (*E. coli* DH52) bacteria. Also, it was also found that compound **8** could cleavage DNA by agarose gel electrophoresis experiments.

Zhang et al. reported synthesis of a series of novel benzimidazole quinolones which exhibited good or even stronger antimicrobial activities in comparison with reference drugs [42]. Compound **9** proved to be membrane active and did not trigger the development of resistance in bacteria. This compound also show low toxicity against mammalian cells. It inhibited the formation of biofilms and also disrupted the established *Staphylococcus aureus* and *Escherichia coli* biofilms. Compound **9** was able to inhibit the relaxation activity of *E. coli* topoisomerase IV at 10 μM concentration. Experimental investigation and molecular modeling of compound **9** with DNA suggested that this compound could effectively bind with DNA to form a steady 9-DNA complex which might further block DNA replication to exert the powerful bioactivities.

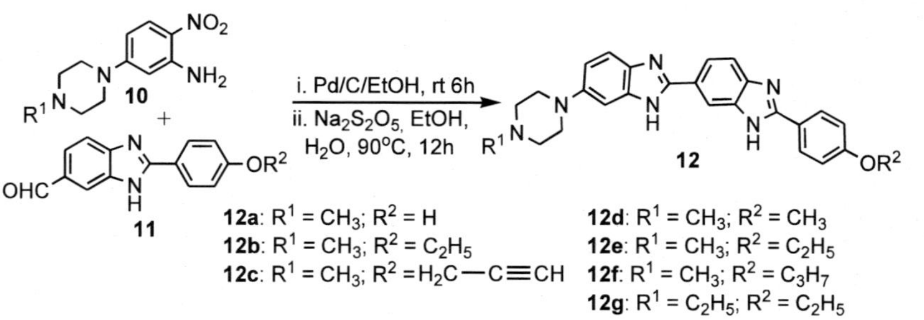

Scheme 20. Synthesis of bis-benzimidazoles **12a-12g.**

Chandrika et al. synthesized a series of bis-benzimidazoles **12a-12g** with remarkable antifungal properties by the reaction of *o*-nitroanilines **10** and 6-benzimidazole-aldehydes **11** (Scheme 20) [43].

Table 6. Antifungal activities of the compounds 12a-12g

Compd	Yeast srains				Filamentous fungi		
	A	B	C	D	E	F	G
12a	7.8	0.975	7.8	0.975	31.2	1.95	7.8
12b	7.8	7.8	7.8	3.9	31.2	3.9	7.8
12c	7.8	7.8	7.8	1.95	>31.2	15.6	15.6
12d	15.6	3.9	31.2	3.9	31.2	15.6	1.6
12e	>31.2	7.8	31.2	7.8	>31.2	>31.2	>31.2
12f	>31.2	>31.2	7.8	31.2	>31.2	>31.2	>31.2
12g	15.6	31.2	15.6	3.9	>31.2	15.6	15.6
FLC	62.5	>31.2	>31.2	1.95	62.5	62.5	62.5
ITC	0.5	7.8	0.48	0.12	0.48	0.195	0.975
POS	0.5	0.12	0.06	<0.03	0.24	0.195	0.48
VOR	0.24	0.06	0.12	<0.03	0.24	0.03	0.12
AmB	3.9	1.95	3.9	1.95	15.6	15.6	3.9

The antifungal activities were screened against a variety of *Candida albicans* and *Aspergillus*: *C. albicans* ATCC 10231 (A), *C. glabrata.* ATCC 2001 (B), *C. krusei* ATCC 6258 (C), *C. parapsilosis* ATCC 22019 (D), *A. flavus* ATCC MYA-3631 (E), *A. nidulans* ATCC 38163 (F), and *A. terreus* ATCC MYA-3633 (G) (Table 4). Compounds **12a**, **12b** and **12c** displayed broad-spectrum antifungal activities against both yeasts and filamentous fungi. In most cases, analogues **12a**, **12b** and **12c** displayed enhanced or comparable antifungal activity against fungal strains when compared to the commercial antifungal drugs FLC (fluconazole), ITC (itraconazole), POS Posaconazole), VOR (voriconazole) and AmB (amphotericin), as an be observed in Table 6.

Chikkula et al. synthesized a series of isoxazole/pyrimidine/pyrazole substituted novel benzimidazole analogs **13a-13g** through multi step synthesis based on hybrid approach, as indicated in Scheme 21 [44]. Four compounds **13a-13d** exhibited very good antimicrobial activity; whereas compounds **13e-13g** displayed moderate antimicrobial activity, as can be seen in Table 7. Compound **13a** displayed equal activity like standard Ciprofloxacin against *S. aureus* (MIC: 7.81 μg/ml), *S. epidermidis* (MIC: 1.95 μg/ml) and *E. coli* (MIC: 7.81 μg/ml). Similarly against *M. luteus* (MIC: 3.9 μg/ml) and *E. coli* (MIC: 7.81 μg/ml), compound **13c**, showed equipotent activity with Ciprofloxacin.

Table 7. Antimicrobial activities of the compounds 13a-13g

Compound	Minimum Inhibitory Concentration (μM)								
	Antibacterial activity							Antifungal activity	
	B.c	*S.a*	*M.l*	*S.e*	*K.p*	*P.a*	*E.c*	*A.f*	*A.n*
13a	15.62	7.81	7.81	1.95	3.9	7.81	7.81	15.62	15.62
13b	15.62	15.62	7.81	3.9	7.81	7.81	15.62	31.25	15.62
13c	15.62	15.62	3.9	3.9	3.9	7.81	7.81	15.62	15.62
13d	31.25	15.62	7.81	7.81	7.81	7.81	15.62	31.25	31.25
13e	62.5	31.25	15.62	15.62	7.81	15.62	31.25	62.5	31.25
13f	62.5	62.5	31.25	31.25	15.62	31.25	62.5	62.5	62.5
13g	62.5	31.25	31.25	15.62	15.62	15.62	62.5	62.5	62.5
Std	7.81	7.81	3.9	1.95	1.95	3.9	7.81	3.9	7.81

B.c: B. cereus; S.a: S. aureus; M.l: M. luteus; S.e: S. pidermitis; K.p: K. pneumoniae; P.a: P. aeruginosa; E.c: E. coli; A.f: A. fumigatus; A.n: A. niger; Std: Standard: Ciproflo-xacin used as a reference standard for bacteria and Ketoconazole used as a reference standard for fungi.

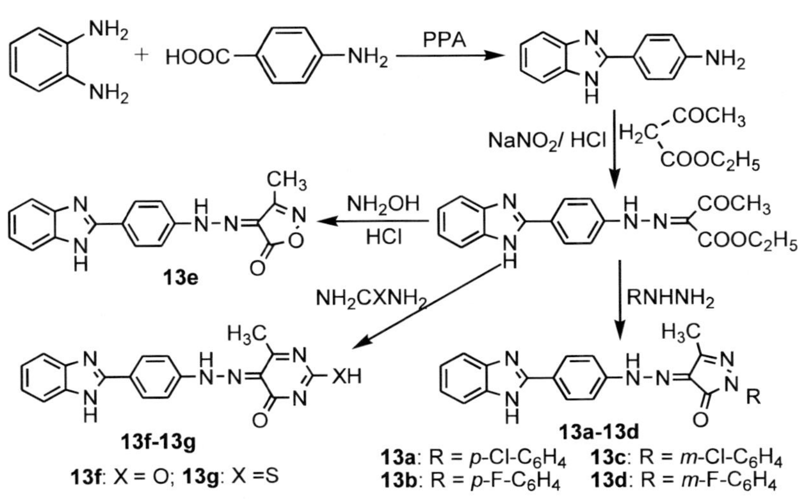

Scheme 21. Synthesis of benzimidazoles **13a-13g**.

El-Gohary and Shaaban reported synthesis of some benzimidazoles by reaction between (5-nitro-1H-benzo[d]imidazol-2-yl)methanethiol and 2-(chloromethyl) -5-nitro-1H-benzo[d]imidazole (Scheme 22) [45]. Compounds **14a** and **14b** show eminent antimicrobial efficacy toward *B. cereus*, *S. aureus*, *C. albicans* and *A. fumigatus*; in addition, **14a** displayed interesting anti-QS activity.

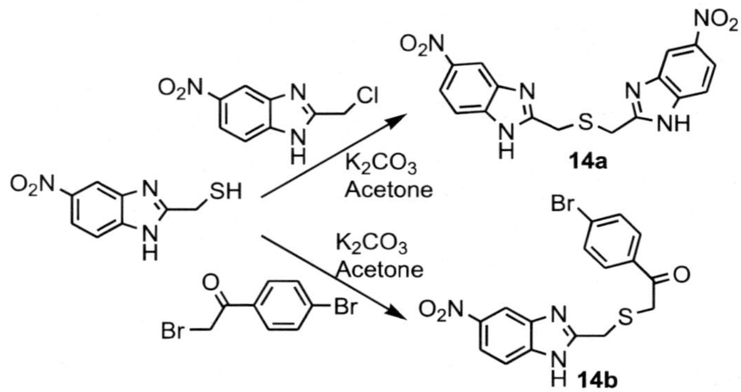

Scheme 22. Synthesis of benzimidazoles 14a-14b.

Zhang et al. reported a novel series of benzimidazole-incorporated sulfonamide analogues [46]. Compound **15a** gave potent activities against Gram-positive bacteria and fungi, and 2,4-dichlorobenzyl derivative **15b** showed good activities against Gram-negative bacteria (Figure 5, Table 8). Both of these two active molecules **15a** and **15b** could effectively intercalate into calf thymus DNA to form compound DNA complex respectively, which might block DNA replication to exert their powerful antimicrobial activity.

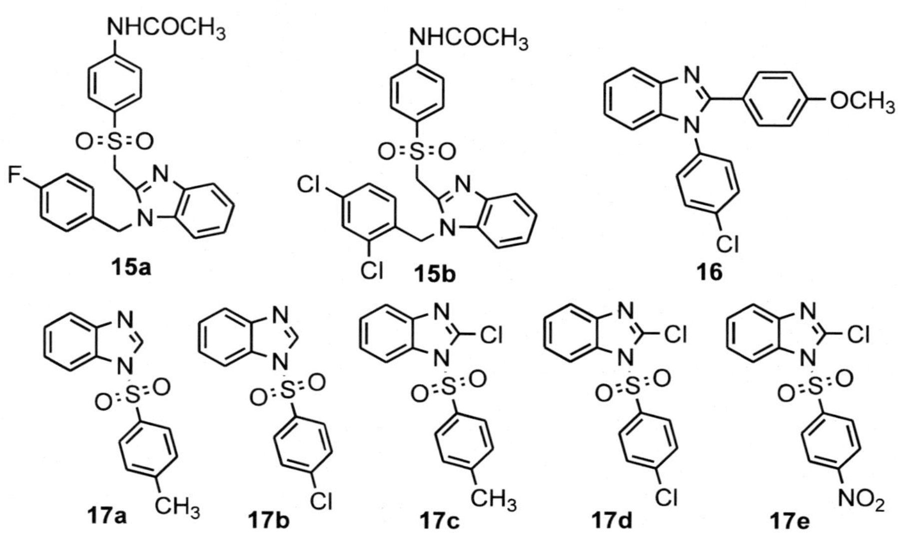

Figure 5. Structures of antimicrobial benzimidazoles **15a-15b, 17a-17e.**

Table 8. Antimicrobial activities of the compounds 15a-15b

Compound	Minimum Inhibitory Concentration (μM)							
	Gram positive bacteria				Gram negative bacteria			
	S. aureus	MRSA	*B. subtilis*	*M. luteus*	*B. proteus*	*E. coli*	*P. aeruginosa*	*B. typhy*
15a	4	16	16	64	32	32	256	16
15b	64	128	16	128	3	8	16	4
Chloromycin	8	16	32	8	32	16	16	32
Norfloxacin	8	1	2	1	4	1	1	1

Wavhale et al. designed and synthesized new promising antitubercular agents as BM212 analogs with lower toxicity and better pharmacokinetic profile [47]. The benzimidazole compound **16** is as potent as the parent molecule BM212 (MIC 2.3 μg/ml vs 0.7–1.5 μg/ml), but is devoid of the toxicity against HepG2 cell lines (IC_{50} 203.10 μM *vs* 7.8 μM).

Naaz et *al.* synthesized and screened for antibacterial activity a new series of heterocyclic molecules bearing sulfonamide linkage **17a-17e** [48]. During antibacterial screening molecules were found to be highly active (MIC value 50–3.1 μg/mL) against different human pathogens, namely *B. cerus*, *S. aureus*, *E. coli*, *P. aeruginosa* and most effective against *E. coli*.

A series of aminopyrimidinyl benzimidazoles were synthesized by Liu et al. The *in vitro* antimicrobial evaluation revealed that some compounds exhibited good bioactivities against the tested strains, comparable or even superior to reference drugs [49]. Particularly, compound 18 effectively inhibit the growth of all the tested bacterial and fungal strains, *A. flavus*, *E. coli* DH52 and MRSA with MIC values of 1, 1 and 8 μg/mL, respectively (Figure 6).

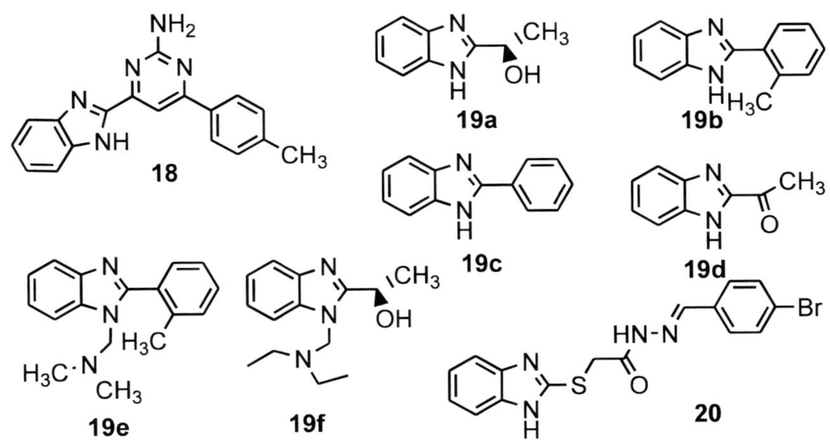

Figure 6. Structures of antimicrobial benzimidazoles **18, 19a-19f, 20.**

Table 9. Anti-biofilm activities of the compounds 19a-19e

Compd	Minimum Inhibitory Concentration (μM)							
	Gram positive cocci				Gram negative bacilli			
	A	B	C	D	E	F	G	H
19a	>1	0.002	0.002	>1	0.01562	>1	0.03125	>1
19b	>1	>1	0.125	>1	0.3125	0.5	>1	>1
19c	>1	>1	>1	>1	0.3125	0.25	>1	>1
19d	>1	0.3125	0.008	>1	>1	>1	>1	>1
19e	>1	>1	0.25	>1	>1	>1	>1	>1
19f	>1	>1	0.008	0.25	0.0625	>1	0.03125	>1

A: S. aureus; B: S. saprophyticus; C: E. faecium; D: E. faecalis; E: P. aeruginosa; F: A. baumanii; G: K. pneumoniae; H: *E.coli.*

Benzimidazoles **19a-19e** with good anti-biofilm activity were studied by Marinescu et al. (Table 9) [50]. The compound **19a** proved to be the most effective anti-biofilm agent, inhibiting the biofilm development of strains belonging to all three bacterial groups, Gram positive cocci, *Pseudomonadaceae* and *Enterobacteriaceae,* the lowest MBEC value (0.008 mg/ml) being recorded against *E. faecium.*

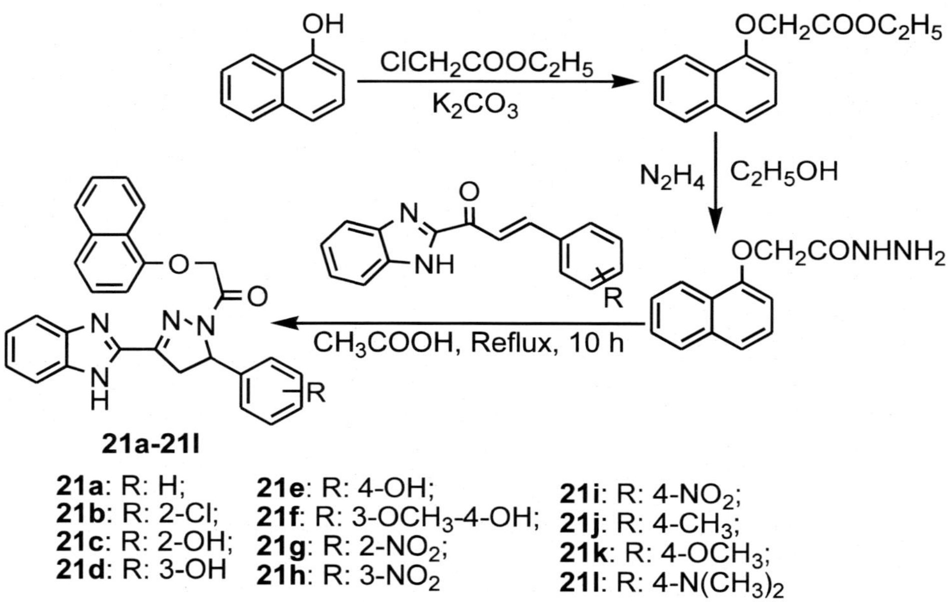

Scheme 23. Synthesis of benzimidazoles 21a-21l.

Yadav et *al.* synthesizes a series of benzimidazole derivatives and evaluated for its in vitro antimicrobial and antitubercular activities [51]. Compound **20** (MIC = 0.0321 μM) showed highest inhibitory action against a Gram negative bacterium, *E. coli* and also possessed the most effective antifungal activity against *C. albicans*, having MIC value of 0.018 μM against *A. niger*.

Table 10. Antibacterial and antifungal activity of the compounds 21a-21l

Com-pound	Minimum Inhibitory Concentration (μg/mL)						
	Bactericidal activity				Fungicidal activity		
	E.c.	*P.a.*	*S.a.*	*S.p.*	*C.a.*	*A.n.*	*A.c.*
21a	250	250	200	250	100	100	100
21b	50	50	100	12.5	500	500	500
21c	100	100	100	100	100	100	100
21d	500	500	500	100	500	100	>1000
21e	500	250	100	100	100	500	100
21f	250	250	250	500	100	500	500
21g	500	250	500	500	500	100	>1000
21h	500	500	250	500	500	1000	>1000
21i	25	12.5	50	25	250	100	250
21j	100	500	500	500	500	50	>100
21k	50	50	100	50	100	50	25
21l	250	250	500	500	100	500	500
A	100	100	250	100	-	-	-
G	-	-	-	-	500	100	100

E.c.: Escherichia coli (MTCC-442), P.a.: Pseudomonas aeruginosa (MTCC-441), S.a.: Staphylococcus aureus (MTCC-96), S.p.: Streptococcus pyogenes (MTCC-443), C.a.: Candida albicans (MTCC-227), A.n.: Asperginus niger (MTCC-282), A.c.: Asperginus clavatus (MTCC-1323), A: Ampicilin, G: Griseofulvin.

Desai et al. reported a series of heterocyclic compounds bearing benzimidazole and pyrazoline motifs **21a–21l** (Scheme 23) with antimicrobial activity against gram positive (*S. aureus* and *S. pyogenes*), gram negative bacteria (*E. coli* and *P. aeruginosa*), and strains of fungi (*C. albicans*, *A. niger*, and *A. clavatus*) (Table 10) [52]. Compound **21i** containing nitro group at *para* position and compound **21b** with chloro group in *ortho* position showed the highest inhibition against bacterial strains. Compounds **21j** and **21k** substituted with methyl and methoxy group at *para* position was found to be the most promising antifungal agents.

Baartzes et al. synthesizes a series of aminoquinoline-benzimidazole hybrid compounds and their bioisosteric ferrocenyl analogues [53]. The compounds display good in vitro activity against the chloroquine-sensitive NF54 and multi-drug resistant K1 strains of *Plasmodium falciparum*. Phenyl-derived **22** and ferrocenyl-derived **23** were assessed for *in vivo* efficacy in *Plasmodium berghei*-infected mice (Figure 7). The most active derivatives **22** and **23** showed inhibition of β-haematin formation in a NP-40 detergent-mediated assay, indicating a possible contributing mechanism of antiplasmodial action.

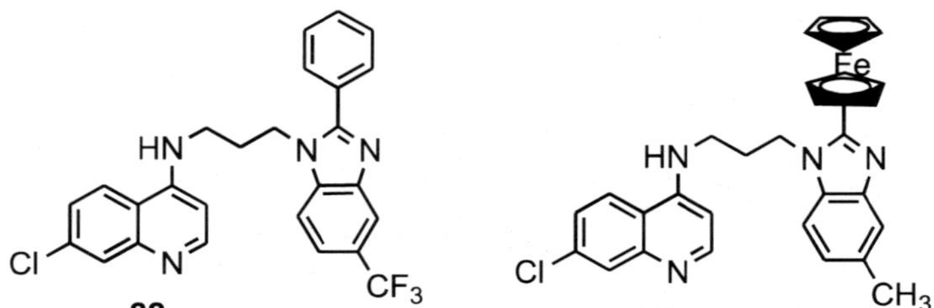

22 23

Figure 7. Structures of antimicrobial benzimidazoles **22** and **23.**

Conclusion

Benzimidazole can be found in numerous therapeutically used agents, like anti-tumorals and antimicrobials. There is a great interest in the synthesis of benzimidazole compounds to be used as therapeutic agents for different diseases [54]. Benzimidazole-based medicinal chemistry as well as designing of drugs will continue to be an overwhelmingly attractive topic for quite a long time [55]. In future research, novel as well as potent chemical entities can be explored by combining benzimidazole scaffolds with enormous potentiality for the treatment of diverse diseases. There is a need for modification of existing benzimidazole agents that will bring about a big change for improvement of activity or change in activity, due to the problems like resistance, toxicity, or poor bioavailability. Studying the structures of the new compounds, it can be concluded that combination of the modifications at positions 1, 2 and 5 of the molecule has provided the most active drugs. Therefore, the 4-, 6-, and 7-positions of benzimidazole need to be explored further for novel entities with exciting biological activities [56]. We hope that this chapter will open new perspectives for researchers interested in the chemistry of benzimidazole-based drugs, so promising for medicinal chemistry.

References

[1] Sharma, M.C. (2016). QSAR studies of novel1-(4-methoxyphenethyl)-1H-benzimidazole-5-carboxylic acid derivatives and their precursors as antileukaemic agents. *Journal of Taibah University for Science* 10: 122–130.

[2] Mann, J. Baron, A. Opoku-Boahen, Y., Johansson, E., Parkinson, G. Kelland, L.R., Neidle, S. (2001). A new class of symmet-ric bisbenzimidazole-based DNA minor groove-binding agents showing antitumor activity. *Journal of Medicinal Chemistry* 44 (2): 138–144.

[3] Ishida, T., Suzuki, T., Hirashima, S., Mizutani, K., Yoshida A., Ando, Ikeda, S., Adachi, T., Hashimoto, H. (2006). Benzimidazoleinhibitors of hepatitis C virus

NS5B polymerase: identification of 2-[(4-diarylmethoxy) phenyl]-benzimidazole. *Bioorganic & Medicinal Chemistry Letters* 16 (7): 1859–1863.

[4] Tuncbilek, M., Göker, H., Ertan, R., Eryigit, R., Kendi, E., Altan-lar, N. (1997). Synthesis and antimicrobial activity of some new anilinobenzimidazoles. *Archiv der Pharmazie (Weinheim)* 330 (12): 372–376.

[5] Venkatesan, P. (1998). Albendazole. *Journal of Antimicrobial Chemotherapy* 41: 45–147.

[6] Labanauskas, L.K., Brukstus, A.B., Gaidelis, P.G., Buchinskaite, V.A., Udrenaite, E.B., Dauksas, V.K. (2004). Synthesis and anti -inflammatory activity of some new 1-acyl derivatives of 2-methylthio-5,6-diethoxybenzimidazole. *Pharmaceutical Chemistry Journal* 34: 353–355.

[7] Ayhan-Kilcigil, G., Kus¸ Coban, C.T., Can-Eke, B, Ozbey, S., Iscan, M. (2005). Synthesis, antioxidant and radical scavenging activities of novel benzimidazoles. *Journal of Enzyme Inhibition and Medicinal Chemistry* 20: 503–514.

[8] Akhtar, J., Khan, A.A., Ali, Z., Haider, R., Yar, M.S. (2017). Structure-activity relationship (SAR) study and design strategies of nitrogen-containing heterocyclic moieties for their anticancer activities. *Europeam Journal of Medicinal Chemistry* 125: 143-189.

[9] Yadav, G., Ganguly, S. (2015). Structure activity relationship (SAR) study of benzimidazole scaffold for different biological activities: A mini-review. *Europeam Journal of Medicinal Chemistry* 97: 419-443.

[10] Phillips, M.A. (1928). The formation of 2-substituted benzimidazoles. *Journal of Chemical Society* 1928: 2393-2399.

[11] Nguyen, T.B., Ermolenko, L., Dean, W.A., Al-Mourabit, A. (2012). Benzazoles from aliphatic amines and *o*-Amino/Mercaptan/Hydroxyanilines: elemental sulfur as a highly efficient and traceless oxidizing agent. *Organic Letters* 4: 5948-5951.

[12] Hanan, E.J., Chan, B.K., Estrada, A.A., Shore, D.G. Lyssikatos, J.P. (2010). Mild and generalone-pot reduction and cyclization of aromatic and heteroaromatic 2-nitroamines to bicyclic 2H-imidazoles. *Synlett* 18: 2759-2764.

[13] Yang, D., Fokas, D., Li, J., Yu, L., Baldino, C.M.(2005). A versatile method for the synthesis of benzimidazoles from *o*-nitroanilines and aldehydes in one step via a reductive cyclization. *Synthesis* 1: 47-56.

[14] Cui, W., Kargbo, R.B., Sajjadi-Hashemi, Z., Ahmed, F., Gauuan, J.F. (2012). Efficient one-pot synthesis of 2-substituted benzimidazoles from triacyloxyborane intermediates. *Synlett* 23: 247-250.

[15]. Kim, M.R. Kumar, N., Park, Y., Lee, H.S. (2011). Copper-catalyzed one-pot three-component synthesis of benzimidazoles by condensation and C-N bond formation. *Journal of Oganic Chemistry* 76: 9577-9583.

[16] Bahrami, K., Khodaei, M.M., Kavianinia, I. (2007). A simple and efficient one-pot synthesis of 2-substituted benzimidazoles. *Synthesis* 2007: 417-427.

[17] Gowda, N.R., Kavitha, C.V., Chiruvella, K.K., Joy, O., Rangappa, K.S., Raghavan, S.C. (2009). Synthesis and biological evaluation of novel 1-(4-methoxyphenethyl)-1Hbenzimidazole-5-carboxylic acid derivatives and their precursors as antileukemic agents. *Bioorganic & Medicinal Chemistry Letters* 19: 4594-4600.

[18] Liu, T., Sun, C., Xing, X., Jing, L., Tan, R., Luo, Y., Zhao, Y. (2012). Synthesis and evaluation of 2-[2-(phenylthiomethyl)-1H-benzo [d]imidazol-1-yl) acetohydrazide derivatives as antitumor agents. *Bioorganic & Medicinal Chemistry Letters* 22: 3122-3125.

[19]. Yoon, Y.K., Ali, M.A., Wei, A.C., Shirazi, A.N., Parang, K., Choon, T.S. (2014). Benzimidazoles as new scaffold of sirtuin inhibitors: green synthesis, in vitro studies, molecular docking analysis and evaluation of their anti-cancer properties. *Europen Journal of Medicinal Chemistry* 83: 448-454.

[20] Celik, I., Kılcıgil, G.A, Guven, B., Kara, Z., -Alp, A.S.G, Karayel, G.A., Onay-Besikci, A. (2019). Design, synthesis and docking studies of benzimidazole derivatives as potential EGFR inhibitors. *Europen Journal of Medicinal Chemistry* 173: 240-249.

[21] Azam, M., Khan, A.A., Al-Resayes, S.I., Islam, M.S., Saxena, A.K., Dwivedi, S., Musarrat, J., Trzesowska-Kruszynska, A., Kruszynski, R. (2015). Synthesis and characterization of 2-substituted benzimidazoles and their evaluation as anticancer agent. *Spectrochimica Acta A.* 142: 286-291.

[22] Shao, K.P., Zhang, X.Y., Chen, P.J., Xue, D.Q., He, P., Ma, L.Y., Zheng, J.X. Zhang, Q.R., Liu, H.M. (2014). Synthesis and biological evaluation of novel pyrimidine-benzimidazol hybrids as potential anticancer agent. *Bioorganic & Medicinal Chemistry Letters*, 24: 3877-3881.

[23] Wang, J., Wang, X., Li, H., Ji, D., Li, Y., Xu, Y., Zhu, Q. (2016). Design, synthesis and biological evaluation of novel 5-fluoro-1H-benzimidazole-4-carboxamide derivatives as potent PARP-1 inhibitors. *Bioorganic & Medicinal Chemistry Letters* 26: 4127-4132.

[24] Reddy, T.S., Kulhari, H., Reddy, V.G., Bansal, V. Kamal, A., Shukla, R. (2015). Design, synthesis and biological evaluation of 1,3-diphenyl-1H-pyrazole derivatives containing benzimidazole skeleton as potential anti-cancer and apoptosis inducing agents. *European Journal of Medicinal Chemistry* 101: 790-805.

[25] Shi, L., Wua, T.T., Wang, Z., Xue, J.Y., Xu, Y.G. (2014). Discovery of quinazolin-4-amines bearing benzimidazole fragments as dual inhibitors of c-Met and VEGFR-2. *Bioorganic & Medicinal Chemistry Letters* 22: 4735-4744.

[26] Akhtar, W., Khan, M.F., Verma, G., Shaquiquzzaman, M., Rizvi, M.A., Mehdi, S.H., Akhter, M., Alam, M.M. (2017). Therapeutic evolution of benzimidazol derivatives in the last quinquennial period. *European Journal of Medicinl Chemistry* 126: 705-753.

[27] Guida, X., Jianhua, H., Li X. (2006). Synthesis and QSAR studies of novel 1-substituted-2-aminobenzimidazoles derivatives. *European Journal of Medicinal Chemistry* 41: 1080–1083.

[28] Tahlan, S., Kumar, S., Narasimhan, B. (2019). Antimicrobial potential of *1H*–benzo[d]imidazole scaffold: A review. *BMC Chemistry*, 13 (18): 1-27.

[29] Bansal, Y., Silakar, O. (2012). The therapeutic journey of benzimidazoles: A review. *European Journal of Medicinal Chemistry*, 20 (21): 6208–6236.

[30] Njar, V.C, Brodie, A.M. (2015). Discovery and Development of Galeterone (TOK-001 or VN/124-1) for the Treatment of All Stages of Prostate Cancer. *Journal of Medicinal Chemistry*, 58 (5): 2077-2087.

[31] Moniruzzaman, R.S., Mahmud T. (2019). Quantum chemical and pharmacokinetic studies of some proton pump inhibitor drugs. *American Journal of Biomedical Sciences &Research*. 2 (1): 3-8.

[32] Scholten, W.K., Christensen, A.E.; Olesen, A.E., Drewes, A.M. (2019). Quantifying the adequacy of opioid analgesic consumption globally: an updated method and early findings. *American Journal of Public Health (AJPH)*, 109 (1): 52-57.

[33] Tahlan, S., Kumar, S., Ramasamy, K., Lim, S.M., Shah, S.A.A., Mani, V., Pathania, R., Narasimhan, B. (2019). Design, synthesis and biological profile of heterocyclic benzimidazole analogues as prospective antimicrobial and antiproliferative agents. *BMC Chemistry*, 13 (50): 1-15.

[34] Zhou, B., Li, B., Yi, W., Bu, X., Ma, L. (2013). Synthesis, antioxidant, and antimicrobial evaluation of some 2-arylbenzimidazole derivatives. *Bioorganic & Medicinal Chemistry Letters* 23 (13): 3759–3763.

[35] Joshi, D., Parikh, K. (2014). Synthesis and evaluation of novel benzimidazole derivatives as antimicrobial agents. *Medicinal Chemistry Research*, 23 (3): 1290-1299.

[36] Seenaiah, D., Ramachandra Reddy P., Mallikarjuna Reddy G., Padmaja A., Padmavathi V., Sivakrishna N. (2014). Synthesis, antimicrobial and cytotoxic activities of pyrimidinyl benzoxazole, benzothiazole and benzimidazole. *European Journal of Medicinal Chemistry*, 77: 1-7.

[37] Mavrova, A.T., Yancheva, D., Anastassova, N., Anichina, K., Zvezdanovic, J., Djordjevic, A., Markovic, D., Smelcerovic, A. (2015). Synthesis, electronic properties, antioxidant and antibacterial activity of some new benzimidazoles. *Bioorganic & Medicinal Chemistry Letters* 23 (19): 6317–6326.

[38] Yadav, G., Ganguly, S., Murugesan, S., Dev, A. (2015). Synthesis, anti-HIV, antimicrobial evaluation and structure activity relationship studies of some novel benzimidazole derivatives. *Anti-Infective Agents*, 13 (11): 65-77.

[39] Gobis, K., Foks, H., Serocki, M., Augustynowicz-Kopec, E., Napiorkowska, A. (2015). Synthesis and evaluation of in vitro antimycobacterial activity of novel 1H-

benzo[d]imidazole derivatives and analogues. *European Journal of Medicinal Chemistry*, 89: 13-20.

[40] Padalkar, V.S., Borse, B.N., Gupta, V.D., Phatangare, K.R., Patil, V.S., Umape, P.G., Sekar, N. (2016). Synthesis and antimicrobial activity of novel 2-substituted benzimidazole, benzoxazole and benzothiazole derivatives. *Arabian Journal of Chemistry*, 9: 1125-1130.

[41] Jeyakkumar, P., Zhang, L., Avula S.R., Zhou, C.-H. (2016). Design, synthesis and biological evaluation of berberine-benzimidazole hybrids as new type of potentially DNA-targeting antimicrobial agents. *European Journal of Medicinal Chemistry*, 122: 205-215.

[42] Zhang, L., Addla, D., Ponman, J., Wang, A., Xie, D., Wang, Y.-N., Zhang, S.-L., Geng, R.-X., Cai, G.-X., Li, S., Zhou, C.-H. (2016). Discovery of membrane active benzimidazole quinolones-based topoisomerase inhibitors as potential DNA-binding antimicrobial agents. *European Journal of Medicinal Chemistry*, 111: 160-182.

[43] Chandrika, N.T., Shrestha, S.K., Ngo, H.X., Garneau-Tsodikova, S. (2016). Synthesis and investigation of novel benzimidazole derivatives as antifungal agents. *Bioorganic & Medicinal Chemistry Letters* 24 (20): 3680–3686.

[44] Chikkula, K.V., Sundararajan, R. (2017). Analgesic, anti-inflammatory, and antimicrobial activities of novel isoxazole/pyrimidine/pyrazole substituted benzimidazole analogs. *Medicinal Chemistry Research*, 26 (11): 3026-3037.

[45] El-Gohary, N.S., Shaaban, M.I. (2017). Synthesis, antimicrobial, antiquorum-sensing and antitumor activities of new benzimidazole analogs. *European Journal of Medicinal Chemistry*, 117: 439-449.

[46] Zhang, H.Z., He, S.C., Peng, Y.J., Zhang, H.J., Gopala, L., Tangadanchu, V.K.R., Gan, L.L., Zhou, C.H. (2017). Design, synthesis and antimicrobial evaluation of novel benzimidazole-incorporated sulfonamide analogues. *European Journal of Medicinal Chemistry*, 136: 165-183.

[47] Wavhale, R.D., Martis, E.A.F., Ambre, P.K., Wanc, B., Franzblau, S.G., Iyer, K.R., Raikuvar, K., Macegoniuk, K., Berlicki, L., Nandan, S.R., Coutinho, E.C. (2017). Discovery of new leads against Mycobacterium tuberculosis using scaffold hopping and shape based similarity. *Bioorganic & Medicinal Chemistry Letters* 25 (17): 4835–4844.

[48] Naaz, F., Srivastava, R., Singh, A., Singh, N., Verma, R., Singh, V.K., Singh, R.K. (2018). Molecular modeling, synthesis, antibacterial and cytotoxicity evaluation of sulfonamide derivatives of benzimidazole, indazole, benzothiazole and thiazole. *Bioorganic & Medicinal Chemistry Letters* 26 (12): 3414–3428.

[49] Liu, H.B., Gao, W.W., Tangadanchu, V.K.R., Zhou, C.H., Geng, R.X. (2018). Novel aminopyrimidinyl benzimidazoles as potentially antimicrobial agents: Design, synthesis and biological evaluation. *European Journal of Medicinal Chemistry*, 143: 66-84.

[50] Marinescu, M, Tudorache, D.G., Marton, G., Zalaru, C.M., Popa, M., Chifiriuc, M.C., Stavarache, C., Constantinescu, C. (2017). Density functional theory molecular modeling, chemical synthesis, and antimicrobial behaviour of selected benzimidazole derivatives. *Journal of Molecular Structure*, 1130: 463-471.

[51] Yadav, S., Narasimhan, B., Lim, S.M., Ramasamy, K., Vasudevan, M., Shah, S.A.A., Mathur, A. (2018). Synthesis and evaluation of antimicrobial, antitubercular and anticancer activities of benzimidazole derivatives. *Egyptian Journal of Basic and Applied Sciences*, 5: 100-109.

[52] Desai, N.C., Pandya, D., Vaja, D. (2018). Synthesis and antimicrobial activity of some heterocyclic compounds bearing benzimidazole and pyrazoline motifs. *Medicinal Chemistry Research*, 27 (1): 52-60.

[53] Baartzes, N., Stringer, T., Seldon, R., Warner, D.F., Taylor, D., Wittlin, S., Chibale, K., Smith, G.S. (2019). Bioisosteric ferrocenyl aminoquinoline-benzimidazole hybrids: Antimicrobial evaluation and mechanistic insights. *European Journal of Medicinal Chemistry*, 180: 121-133.

[54] Akhtar, W., Khan, M.F., Verma, G., Shaquiquzzaman, M., Rizvi, M.A., Mehdi, S.H., Akhter, M., Alam, M.M. (2017). Therapeutic evolution of benzimidazole derivatives in the last quinquennial period. *European Journal of Medicinal Chemistry*, 126: 705-753.

[55] Gaba, M., Mohan, C. (2016). Development of drugs based on imidazole and benzimidazole bioactive heterocycles: recent advances and future directions. *Medicinal Chemistry Research*, 25 (2): 173-210.

[56] Anand, K., Wakode, S. (2017). Development of drugs based on Benzimidazole Heterocycle: Recent advancement and insights. *International Journal of Chemical Studies*, 5 (2): 350-362.

In: Benzimidazole
Editor: Aksel A. Vestergaard

ISBN: 978-1-53616-986-7

Chapter 8

SYNTHESIS OF BENZIMIDAZOLE BASED POLYMERIC FILM FOR APPLICATION IN FUEL CELL

Asit Baran Samui*
Institute of Chemical Technology, Nathalal Parekh Marg,
Matunga, Mumbai, India

ABSTRACT

High-temperature polymer electrolyte membrane fuel cell has been demonstrated as a sustainable green technology for power generation. In fact, some of the problems associated with other membranes, such as CO tolerance, improved kinetic and enhanced water management etc., have been solved by above type of membrane. Polybenzimidazole is considered one of the best membranes, which is extensively studied due to excellent properties most suitable for high temperature applications.

In this chapter benzimidazole based polymers with various architecture and copolymers have been discussed with variation of properties, such as mechanical properties, thermal stability, thermo-oxidative stability and enhanced performance. Most of the studies are based on phosphoric acid dopant. Few variations have also been incorporated. This is followed by discussion on blends and composites, which are expected to augment the properties required for fuel cell performance at high temperature. Nanocomposites are also incorporated in the chapter. Various nanocarbons, nanometal salts etc. have been discussed in the section. The design of nanoporous membranes have been discussed as they provide higher space for acid doping to enhance the performance. Few special cases have been discussed to bring out the efforts in a particular direction. Finally, various characterization techniques have been discussed in brief.

* Corresponding Author's Email: absamui@gmail.com.

INTRODUCTION

The constantly increasing energy demands and the depletion of fossil fuels as sources of energy have prompted research on alternative and sustainable alternative energy sources. Reduction of the greenhouse gas emission (from the use of hydrocarbon based energy sources) is a priority now to get a cleaner atmosphere. Hydrogen energy-based system has been observed to be a suitable candidate to provide clean energy because of the absence of pollutant like SOx, NOx and CO_2, which are present in hydrocarbon-based energy sources. In the polymer electrolyte membrane fuel cell (PEMFC) system, energy is generated by the electrochemical reaction between hydrogen and oxygen fed at anode and cathode, respectively. Hydrogen produces protons and electrons at anode. A membrane placed between anode and cathode allows the diffusion of proton from anode to cathode while electrons move through external electric circuit to cathode. Protons, electrons and oxygen react at the cathode and water is produced as a byproduct. Water is removed from the system by a flow of excess oxygen [1].

PEMFC has been successfully developed by using the perfluorinated polymer membranes having a perfluorinated backbone similar to polytetrafluoroethylene with pendant perfluorinated chains linked to the main chain by ether bonds, and a sulfonated group at the other end of these pending chains [2, 3]. The most widely used membrane is the perfluorosulfonated membranes that belong to polymers of the nafion family, developed by DuPont [4].

Nafion or other perfluorosulfonated membrane based PEMFC suffer from limitations, such as high cost and maintenance of fully hydrated membrane. Humidity management along with the prevention of flooding of the electrodes is a tricky job. The working temperature limit is also affected. It needs to be kept below 100°C to keep the dehydration above this temperature under check so that the conductivity does not decay [5, 6]. The problem is tackled by recent developments of self-humidifying composite polymer electrolyte membranes by using highly dispersed nanometre-size platinum (Pt) and/or metal oxides that afford promising results in this respect [7]. Operating at higher temperatures directly augment the PEMFC performance for several reasons, such as (i) Below150°C, the Pt catalyst used is poisoned by CO present in the H_2 fuel; thus the use of expensive highly pure hydrogen is must in conventional PEMFCs. Above 160°C, the issue of poisoning of Pt is not so important [8]. (ii) Higher temperatures lead to faster electrode reactions. (iii) Temperatures near but below 100°C complicate water management. (iv) Higher temperatures make it easier to take advantage of residual heat to be utilized for other purposes. Along with perfluorinated membranes and derivatives, block copolymers [9] as well as nanocomposite materials [10] have been developed for the purpose.

Many non-fluorinated membranes have been studied, which are based mostly on thermally stable sulfonated polymers, like polysulfone (PSU), poly(ether sulfone) (PES), poly(ether ketone) (PEK), poly(ether ether ketone) (PEEK), poly(phenylene sulfide) (PPS), poly(phosphazene) or polyimide (PI) among many others [11-15]. However, all these sulfonated membranes need to be fully hydrated to maintain high proton conductivity. Furthermore, the maximum working temperatures are still low. Most of these acid–polymer combinations exhibit proton conductivity up to a maximum of 10^{-3} Scm^{-1} at room temperature [16].

By increasing the acid content, the plasticization effect of the excessive acid results in the formation of a soft paste that cannot be processed into membranes. Polybenzimidazole (PBI) was first proposed for preparing acid-doped membranes to overcome the earlier drawbacks [17]. Acid-doped polybenzimidazole (PBI) membranes have promising properties that allow them to be used in PEMFC at temperatures as high as 200°C without humidification [18]. By suitably doping with strong acids, such as H_3PO_4, the proton conductivity of PBI-based membranes can be increased to very high value. In fact, PBI is a basic polymer, which dissociates phosphoric acid releasing protons, as presented in the following reaction:

$$H_3PO_4 + PBI \rightarrow H_2PO_4^- + PBI\text{-}H^+ \quad (1)$$

H^+ ions can migrate through the polymer backbone assisted by phosphate anions.

PBI based fuel cells are technologically most suited for stationary power applications, for example, the systems based on natural gas reforming and combined heat and power (CHP) generation [19]. Automobile applications are more challenging. During 2007 Los Angeles Motor Show, Volkswagen has presented a hybrid concept car with a PBI stack as a charger for batteries to extend the driving range. As small power units, the PBI-based cells can also be integrated with a simplified methanol reformer or metal hydride tank.

Benzimidazole based polymers can form proton conductive films with good thermal and chemical stabilities together with high mechanical strength. Polybenzimidazole (PBI) has a heterocyclic benzimidazole ring with an amphoteric character. (Figure 1) Poly(2,5-benzimidazole) (ABPBI), another variation of ABPBI, can be prepared from 3,4-diaminobenzoic acid (DABA).

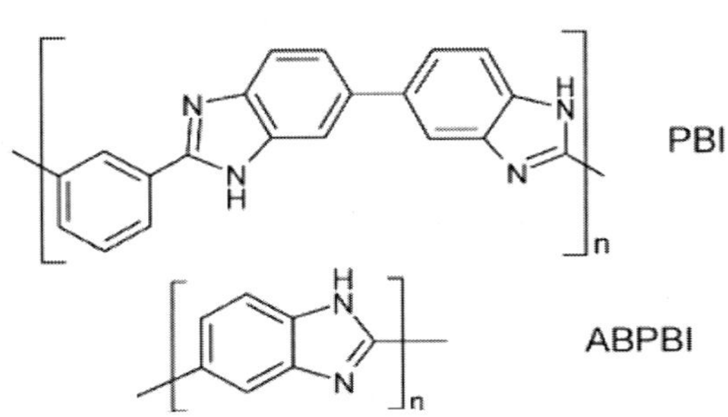

Figure 1. Structure of poly[2,2'-(m-phenylene)-5,5'-bibenzimidazole] (polybenzimidazole, PBI) and poly(2,5-benzimidazole) (ABPBI).

As its synthesis and stoichiometry can be balanced very easily, a high molecular weight ABPBI can easily be prepared that increases the stability of the film. The first fully aromatic polybenzimidazoles are synthesized in 1961 [20]. Tetraamines and diacids are condensed to synthesize these polymers having large number of combinations [21].

Compounds with benzimidazole moieties are generally characterized by high reactivity. Therefore, the physico-chemical properties of the final membrane, including thermo-oxidative stability, mechanical performances, methanol permeability, polymer solubility, and film processing, can be modulated by varying the chemical nature of the starting monomers. Thus, the polymers containing functional groups, such as ether, ketone, sulphone, and fluoroalkyl units, have been developed by various research groups [22, 23]. The crosslinking of PBI can also be performed to enhance the mechanical properties, which is a critical issue for practical application. The crosslinking studies include covalent, ionic, and mixed ionic-covalent crosslinking [24]. Poly(imide benzimidazole)s form another type of polymer film, which is chemically stable in polyphosphoric acid up to 200°C with proton conductivity of 0.26 S cm^{-1} at 170°C and 0% RH [25]. A PBI based fuel cell may be shown as: (Figure 2) [26].

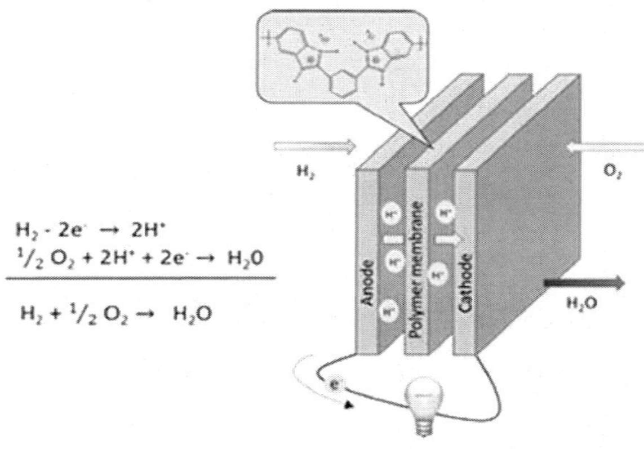

Figure 2. Polymer exchange membrane fuel cell working with H_2 and O_2.

PBI acts as ionic exchange membrane. If doped with phosphoric acid (H_3PO_4), sulfuric acid (H_2SO_4), and nitric acid (HNO_3) it functions as cation exchange membrane. The benzimidazole units in PBI bears the pKa = 5.5 (weak acid character). The acid molecules penetrate the membranes during doping due to the acid-base interaction between them that result in gradually swelling of PBI membrane [26]. The doped membrane can retain the solid electrolyte characteristics in a fuel cell in the temperature range of 100-200°C. Similarly, the benzimidazole based polymers can function as anion exchange membrane in presence of alkali, e.g., KOH. The amphoteric nature is presented in Figure 3 [26].

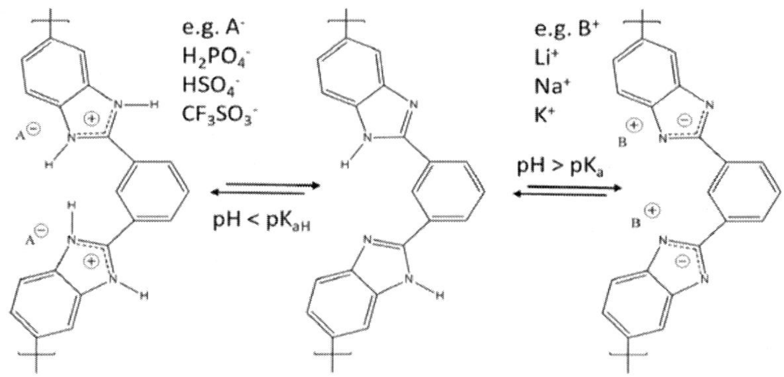

Figure 3. Scheme showing the amphoteric nature of PBI in acidic (left) and alkaline (right) environments.

The synthesis strategies of benzimidazole based polymers together with their performance as membrane for fuel cell will be discussed in the chapter.

Membranes with Dopants Other Than Phosphoric Acid

Alkali as Dopant

Alkaline doped PBI membrane, having high hydroxide ion conductivity and excellent chemical stability in alkaline media, can be used as an anion exchange membrane (AEM) in alkaline direct alcohol fuel cells (ADAFCs) with marked performance [27]. One major challenge in an alkaline doped PBI membrane is the long-term operation, as it can be affected by the progressive release of the doped alkali. Improving the retention of the doped alkali in the PBI skeleton is vital for the long term use of alkaline doped PBI membrane.

Ionic Liquid as Dopant

Although the use of phosphoric acid leads to high proton conductivity, the disadvantages are associated with its chemical stability (pyrolysis of phosphoric acid at 150°C) and the environmental issues [28]. There is possibility of use of ionic liquids (ILs) to replace phosphoric acid. Ionic liquids (ILs) are purely ionic materials having a melting temperature often around room temperature and high temperature stability, non-volatility and non-flammability, and reasonably high ionic conductivity. As both the PBI support and the IL are tailor-made, optimization of the properties of both can be done for the desired application. The thermal and mechanical strength provided by the porous PBI support and the proton conductivity by the IL combines the best properties of both materials.

PBI/ABPBI BASED MEMBRANE

Poly[2,2'-(m-phenylene)-5,5'-bibenzimidazole] (mPBI) is commonly used as the proton exchange membrane. This is prepared by refluxing a mixture of 3,3'-diaminobenzidine and isophthalic acid along with polyphosphoric acid (PPA) under nitrogen atmosphere for long duration. The mPBI films doped with phosphoric acid can be used in fuel cells operating at 100–200°C. Due to the higher acid doping levels (14-26 PA per repeat unit of mPBI) prepared by the PPA process, these membranes exhibit higher conductivities [29].

It is commonly observed that with increase of acid doping the mechanical strength of the membrane decreases. The para-structured PBI (pPBI) exhibits much superior mechanical properties than mPBI and is likely to maintain higher mechanical properties even after acid doping. The pPBI films are prepared by casting the hot polymerization solutions directly on to a stainless steel plate or a glass substrate in air and smoothening the film by using a film applicator at 100°C and then quenching it to −20°C immediately [30]. The X-ray pattern of the pPBI film from direct casting exhibits a broad amorphous halo at 23–28° indicating amorphous pPBI film in spite of the semi-crystalline character of pPBI powder. It exhibits large water uptake characteristics. It maintains reasonably good mechanical stability even at acid doping levels higher than 5 mol H_3PO_4. Tensile strengths of around 60MPa are observed for pPBI membranes from solution casting and direct-casting methods having acid doping levels ranging from 6 to 9 mol. The conductivity values of pPBI membranes are around 0.15 S cm^{-1} at 180°C and 1% relative humidity and 0.18 S cm^{-1} at180°C and 5% relative humidity with doping level of 8.78 mol. The values are much higher than those for mPBI membranes prepared by the conventional method.

Easily Soluble Polybenzimidazoles

The large scale use of PBI faces serious processability problem due to the difficulty in their solubility in low boiling solvents and the uptake of PA. The attachment of soluble functionality in dicarboxylic compound is the most acceptable solution. The PBI backbone structure can also be varied by altering the tetraamine structure. Incorporation of hetero atoms in the polymer backbone and increasing the chain to chain distance in the molecular packing by pushing away the chains are some of the possible strategies, which will enhance the PA loading. A new type of pyridine bridged tetraamine (Py-TAB) can be the answer to three requirements: (I) easily synthesizable cost effective tetraamine monomer and (II) soluble and processable PBI irrespective of dicarboxylic acid (DCA) structure and (III) modified PBI can hold a large quantity of PA (Scheme 1) [31]. A broad XRD peak in all the samples indicates their amorphous nature. These polymers (Py-PBIs) are easily soluble in polar aprotic solvents such as dimethyl formamide (DMF), dimethyl acetamide (DMAc), N-methyl-2-pyrrolidinone (NMP), and dimethyl sulfoxide (DMSO) at room temperature. They are also readily soluble in low boiling acidic solvent like formic acid. The thermal stability results (onset of decomposition >550°C) indicate remarkable thermal stability of the polymers, which is due to the influence of the DCA monomers structure. The Tg of Py-PBI polymers are observed around 400°C. The presence of bulky pyridine groups in the main chain of Py-PBI polymers enhances the oxidative stability. Also, the fluorinated aromatic ring containing copolymers exhibit maximum stability. Under identical doping conditions, the Py-PBI polymers load higher PA than the conventional PBIs [31] (Figure 4). For example, the PA loading of all the Py-PBI polymers studied are >9 mol per repeat unit when dipped in 60 wt % PA bath for three days, whereas the PA loading of PBI is much lower. The proton conductivity of acid-doped fluorine and carbonyl containing Py-PBI membranes at 160°C is nearly 0.01 S cm^{-1}. Furthermore, the heteroatom containing polymers are more conducting compared to iso- and tere-phthalic acid containing polymers.

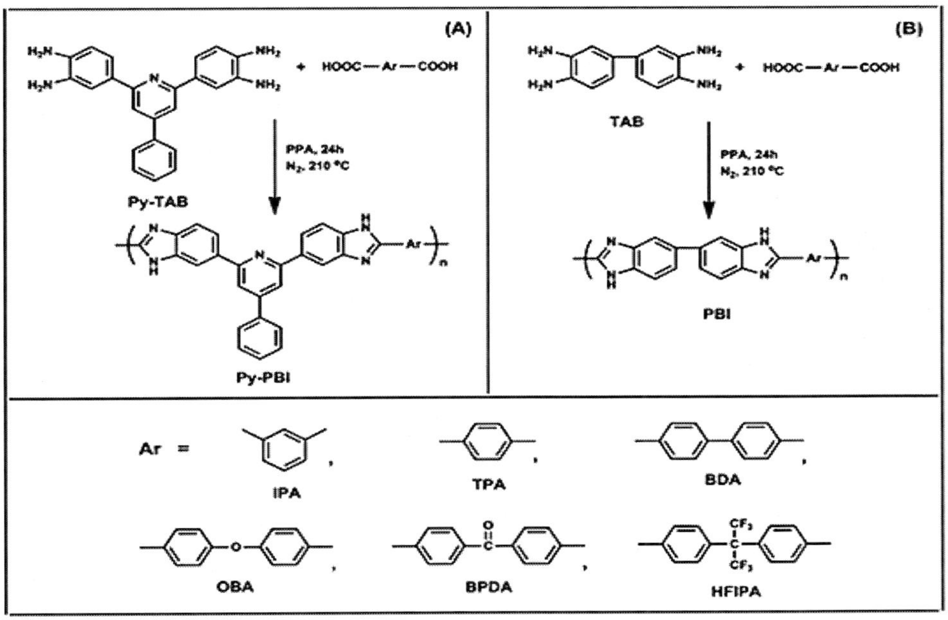

Scheme 1. Synthesis of (A) Pyridine Bridge Polybenzimidazoles (Py-PBI) Polymers from the 2,6-Bis(3′,4′- diaminophenyl)-4-phenylpyridine (Py-TAB) Monomer and (B) Conventional PBI from 3,3′,4,4′-Tetraaminobiphenyl (TAB) Monomer with Varieties of Dicarboxylic Acid Monomers.

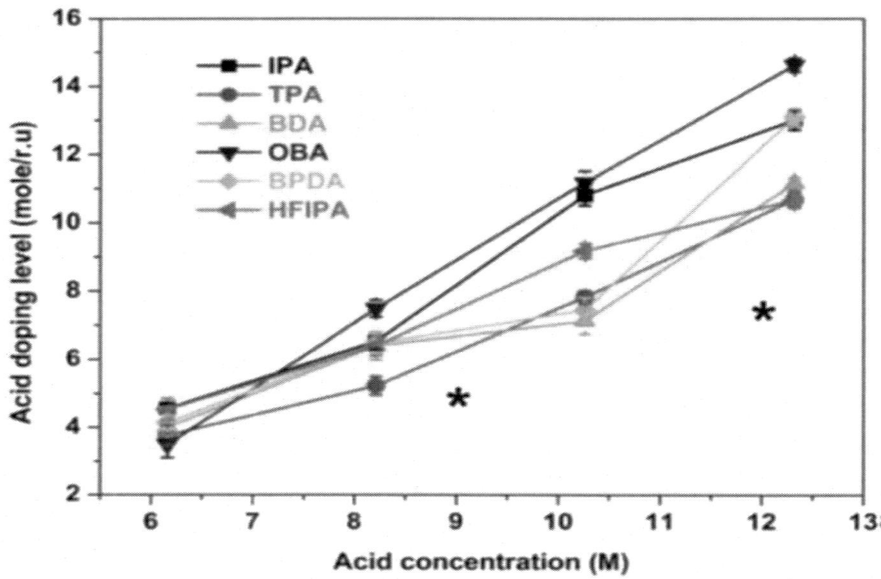

Figure 4. Acid doping level of Py-PBI membranes obtained by dipping in different phosphoric acid concentrations at room temperature.

The symbol (*) in the figure shows the PA loading data of IPA-based conventional PBI reported in the literature earlier.48. (IPA :Isophthalic acid; TPA: Terephthalic acid; BDA: Biphenyl 4,4′-dicarboxylic acid; OBA: 4,4′-oxybis(benzoic acid); BPDA: 4,4′-dicarboxy benzophenone; HFIPA: 4,4′(hexafluoroisopropylidene)bis(benzoic acid))

Poly(Vinylphosphonic Acid)-Doped PBI Membrane

The leaching of PA from high temperature PEMFC can be arrested by replacing PA with poly(vinylphosphonic acid) (PVPA) as it is a polymeric acid and is stably bound to the PBIs via multipoint acid-base reactions. PVPA can form hydrogen-bonding network with other phosphonic acid groups. The PVPA/PBI membrane is prepared by dissolving PBI and PVPA in the ratio of 1:1 (mol/mol) in respect of the monomer repeating unit of each and casting on glass plate [32]. The proton conductivity is calculated to be 0.017 S. cm^{-1} at 120°C. Marginal decrease of mechanical property occurs with PVPA. The PVPA-doped membrane electrode assembly (MEA) maintains higher cell voltage at all applied current densities, which continues even after 400,000 cycles. The maximum power density of the PVPA-doped MEA is 252 $mWcm^{-2}$.

Ionic Liquid Doped PBI Membrane

Diethylamine bisulphate (DESH)/sulphate ionic liquid has been used for high temperature fuel cell. Composite hybrid membranes of PBI and ionic liquid diethlyamine bisulphate/sulphate, PBI/xDESH are fabricated as polymer electrolyte membrane [33]. The ratio of diethylamine/sulphuric acid is varied to obtain DESH (1/1, mol/mol) or DESH2 (1/2, mol/mol). An appropriate amount of ionic liquid DESH is added to the PBI/DMSO solution (5 wt %) and stirred to get a viscous solution. After sonication for several hours and heating at 75°C for long duration, the solution is cast and heated for 12 h at 80°C. The ionic liquid incorporation in to PBI does not change its structure, which can be confirmed by using FTIR. The imine group from the imidazole ring is not protonated. There is only weak interaction of IL with PBI while remaining free inside the structure of PBI film. The specific conductivity of PBI/6DESH2 sample is slightly higher than PBI/6DESH in the low temperature zone (60°C--120°C) and becomes more prominent between 120 and 200°C. Although the PBI/6DESH2 membranes have more "free H_2SO_4 molecules" these molecules cannot move effectively for a long distance. The conductivity is thus improved via formation of an ionic channel by the ionic liquid within the polymer matrix. The PBI/xDESH membranes can operate under a higher temperature/low humidity conditions with reasonable conductivity level (>0.01 S cm^{-1}).

PBI/ABPBI Derivative Membranes

There are efforts to modify PBI by introducing additional groups so that the properties are conducive for high temperature applications, such as PA uptake, enhanced solubility, high temperature stability and so on. Few cases are discussed in the following section.

Pendant Benzimidazole Groups Based Polyimides Membrane

Polyimides with pendant benzimidazole groups are synthesized by reacting 5-(1H-benzoimidazol-2-yl)-benzene-1-3-diamine (DABI) with 4,4′-(hexafluoroisopropylidene) diphthalic anhydride (ODA) under nitrogen flow [34]. Polyimide with 40:60 mole ratio of DABI:ODA exhibits highest conductivity of 0.097 Scm^{-1} at 150 °C while at lower mole ratio of 35:65 the proton conductivity drops to 0.0073 Scm^{-1} at the same temperature.

Fluorine-Containing Polybenzimidazole Membrane

An amorphous, organosoluble, flexible, fluorine-containing PBI can be synthesized by the reaction of 3,3'-diaminobenzidine with 2,2-bis(4-carboxyphenyl) hexafluoropropane in PPA at 200°C for 24 h, which exhibits satisfactory thermal and mechanical properties [35]. The pure PBI film has a tensile modulus of 1.19 GPa, tensile a strength of 55 MPa and elongation at break of 12%, which decrease rapidly with doping. Even though the permeability increases with increasing PA doping, the permeability of PBI–3.0 H_3PO_4 is still much lower than the Nafion 117 membrane. The conductivity of PBI–3.0H_3PO_4 reaches 1.7 x10^{-4} Scm^{-1} at 160°C.

Dihydroxy PBI Membrane

Rigid-rod polymers like PBI, having nitrogen heterocycles and hydroxy groups can form strong hydrogen bonds between the polymer chains to generate network structures. The cross-links are also possible in such structures via phosphate bridges. Homopolymer is synthesized by reacting (3,3',4,4'-tetraaminobiphenyl) with 2,5-dihydroxyterephthalic acid in presence of PPA under nitrogen atmosphere with increasing temperature up to 195°C for long duration [36]. Copolymers are also synthesized with pPBI using the same procedure. The low solubility behavior and inherent viscosities of base treated 2OH-PBI as well as the copolymer indicate the occurrence of cross-linking through phosphoric acid ester bridges. The proton conductivity of 2OH-PBI is higher than that of pPBI membranes over the entire temperature range from room temperature to 180°C. For example, the proton conductivity of the 2OH-PBI membrane at 160°C is 0.35 Scm^{-1} while that of pPBI membrane is about 0.27 Scm^{-1} at the same temperature. The tensile strengths of the homo- and co-polymers are in the range of 1.49-1.84 MPa and strain at break in the range of 437-552%. The investigation of the fuel cell performance with Pt and Pt alloy cathode catalysts exhibits slow cathode kinetics that acts as a limiting factor to overall fuel cell performance. The voltage in a fuel cell application at 0.2 A cm^{-2} and 160°C is 0.642 V.

Hydroxyl Pyridine Containing PBI Membranes

PBIs with a properly tailored alkalinity by introduction of molecules, such as pyridine, imidazole or other alkaline groups, are expected to increase the basic sites of polymers to enhance the interactions between the polymer and the PA molecules. This, in turn, boosts the acid doping and proton conductivity. Poly (2,2′-m-(4-hydroxyl pyridine)-5,5′-bibenzimidazole), termed as OHPyPBI is synthesized with this concept [37]. The phenyl group of mPBI is replaced by the hydroxyl pyridine group to get OHPyPBI. The increased alkaline group (–N=) of OHPyPBI has the ability to enhance the PA doping levels. The presence of the hydroxyl groups in the backbone chains provides the hydrogen bond networks for the proton conduction. Also, it has the potential for further chemical modifications. At an acid doping level of 8.6 mol acids per molar repeat unit of the polymer, the OHPyPBI membrane exhibits a proton conductivity of 0.102 S cm^{-1} at 180°C without humidifying. The possibility of crosslinked structure via the formation of phosphate esters between the hydroxyl groups of the polymer backbone and phosphoric acid molecule at elevated temperature enhances the tensile modulus for acid doped OHPyPBI membranes. In fuel cell set up, the peak power density of OHPyPBI/8.6PA based MEA increases from 523 mW cm^{-2} to 570 mW cm^{-2} when the temperature increases from 160°C to 180°C due to increased membrane conductivity as well as faster electrode kinetics.

Ether-Containing Polybenzimidazole Membrane

Addition of a flexible group to the backbone of PBI is likely to improve both the strength and solubility. Poly[2,2'-(p-oxydiphenylene)-5,5'-benzimidazole] (PBI) is synthesized from 3,3'-diaminobenzidine and 4,4'-oxybisbenzoic acid, and the solvent casting is used to prepare membrane [38]. The ether containing PBI membrane can withstand maximum stress of 153 MPa, which reduces with H_3PO_4 doping. At 150°C, the proton conductivity is 0.083 Scm^{-1}, which is comparable to that of a sufficiently wetted Nafion membrane. In a single cell, the maximum power density is 1.17 Wcm^{-2}, and the maximum current density is 6.0 Acm^{-2}.

N-Alkylated PBI Membrane

Nanostructured PBI membranes can be obtained by carrying out N-alkylation with long (C18) alkyl chains [39]. N-alkylation of PBI is done by dissolving it in anhydrous NMP at 80°C and cooling to RT to add sodium hydride to activate it. It is then reacted with 1-bromooctadecane at 80°C for 24 h. PBI-C18-3 is soluble in THF and $CHCl_3$, which are typically not good solvents for PBI and thus the solution casting is quite easy. The polymers exhibit excellent thermal stability. It degrades above 320°C under air atmosphere. The conductivity reduces with acid doping. It happens if the PA is confined into a microphase-separated morphology. Further, the reduction of the total number of charge carriers upon grafting allows lower phosphoric acid uptake compared to plain PBI.

Fluorine-Containing Polybenzimidazole/Imidazole Hybrid Membranes

An amorphous, organosoluble fluorine-containing PBI can be synthesized from 3,3'-diaminobenzidine and 2,2-bis(4-carboxyphenyl)hexafluoropropane by condensation polymerization in PPA at 200°C [40]. The hybrid (PBI-xIm; Im: Imidazole) exhibits initial degradation at about 160°C. The moduli and strength of the PBI/Im hybrid membranes decreases with the increase of Im. The methanol permeability of the PBI/Im hybrid membranes increases with increasing amounts of Im. The proton conductivities of all the PBI/Im hybrid membranes are above $2x10^{-3}$ Scm^{-1}. Considering mechanical strength and conductivity, the appropriate Im content in the fluorine-containing PBI should be in the range of 20-30 mole per repeat unit of PBI.

PBI Membrane with N-Grafted Sulfone Group

Sulfonated PBI is synthesized via reaction of PBI anion with the ring-opening reaction of 1,3-propane/butane sultone. Pulverized LiH is slowly added to PBI solution in DMAc and the mixture stirred at 85°C for 3 h. Purified 1,3-propane/1,4-butane sultone is added with different feed ratios, the reaction mixture stirred for 24 h and then precipitated by pouring it into acetone with few drops of hydrochloric acid to obtain PBI butane/propane sulfone. (PBI-BS/PBI-PS) [41]. With increasing sulfonation level there is increase in proton conductivity, which remains reasonably high up to 140°C. PBI-BS shows higher proton conductivity up to 160°C even at low humidity conditions. In fuel cell assembly, a maximum power output of 200 $mWcm^{-2}$ is obtained from PBI-BS at 80°C and 0.7 Acm^{-2} current density, 0.3 V cell voltage and 100% RH.

By following the same procedure, the sultones can be replaced with sodium (4-bromomethyl) benzenesulfonate to graft the benzylsulfonate groups on the imidazole nitrogen atom of PBI [42]. The proton conductivity of cast films lies in the range of 10–3-10–2 Scm-1. High conductivity is due to the increased density of mobile protons together with increased water uptake that increases proton mobility. Hydrochloric or phosphoric acid doping improves the conductivity to nearly that of nafion. A limitation exists in its shrinkage during removal of sorbed water.

The directly sulfonated polybenzimidazole can be synthesized by sulfuric acid treatment on the pre-formed polybenzimidazole membranes. Membrane samples are sulfonated by immersing these in 5, 10 and 20 wt. % H_2SO_4 solution for 1 h at 50–55°C [43]. After removing the adhered acid by wiping with filter paper, the samples are heated at 475°C for 2 min under nitrogen atmosphere. The modified material is stable up to about 400°C. The proton conductivity of the sulfonated membrane is 7.5×10^{-5} Scm^{-1} at 160°C and 100% RH.

The details of PBI/ABPBI polymers are incorporated in Table 1.

Table 1. List of PBI/ABPBI based polymers with chemical structures

Membrane	Structure	Reference
Polybenzimidazole & poly(2,5-benzimidazole)	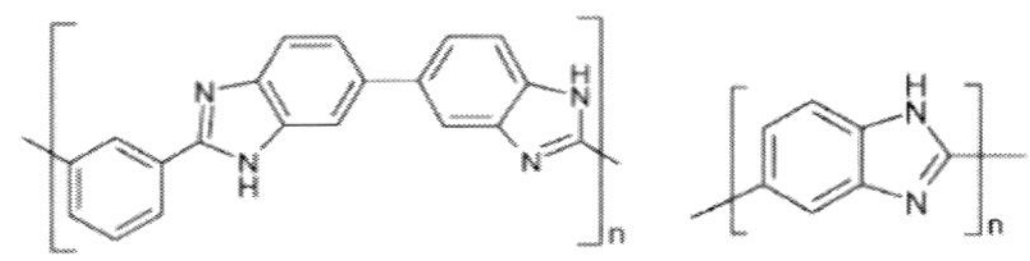	22
N-butane sulfonic acid grafted PBI & N-Benzyl sulfonic acid grafted PBI	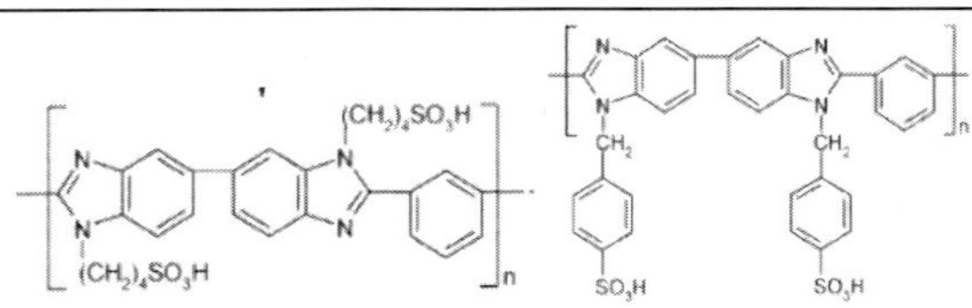	41 42
Sulfonated PBI	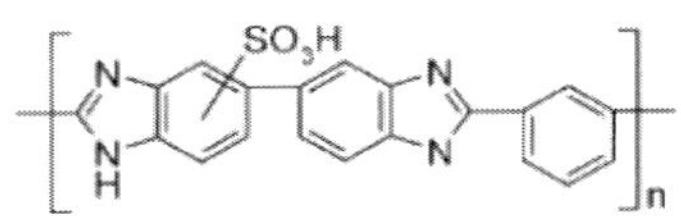	43
Fluorinated PBI	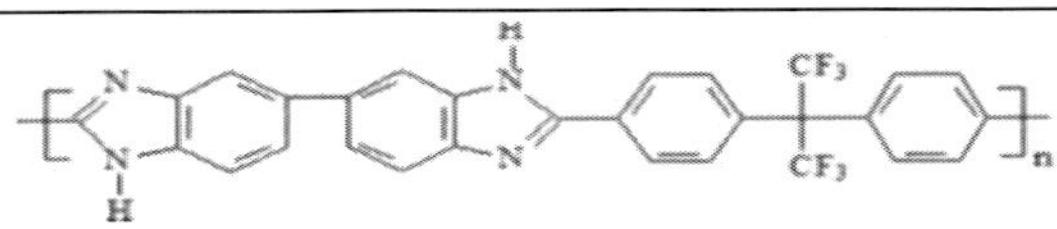	35
Pyridine bridge polybenzimidazoles	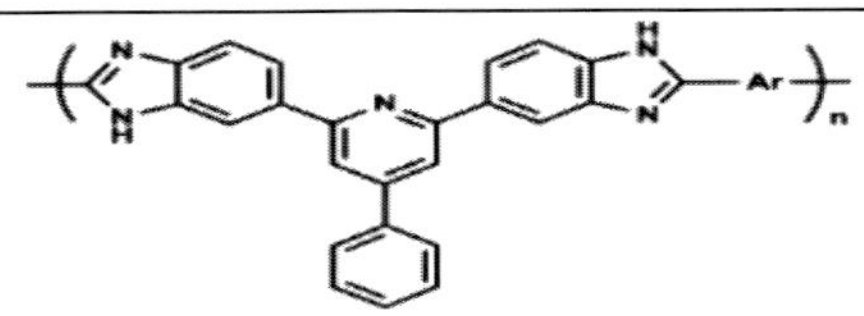	31
Polyimide benzimidazole	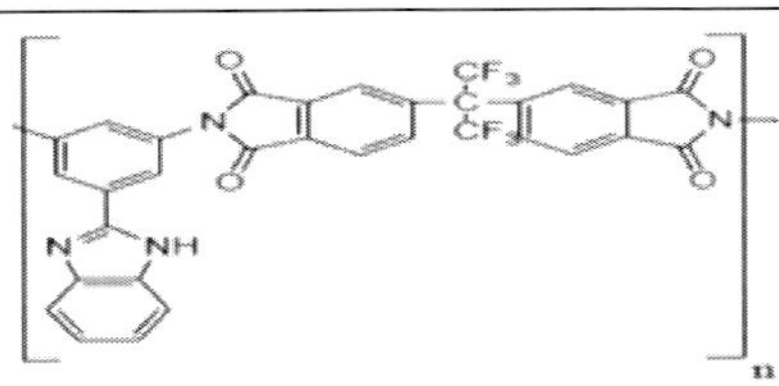	34
Hydroxypyridine PBI	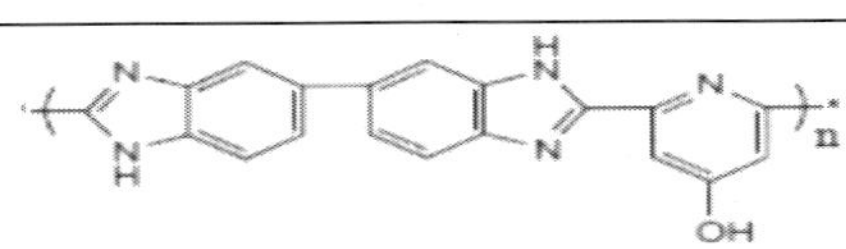	37
Dihydroxy benzimidazole	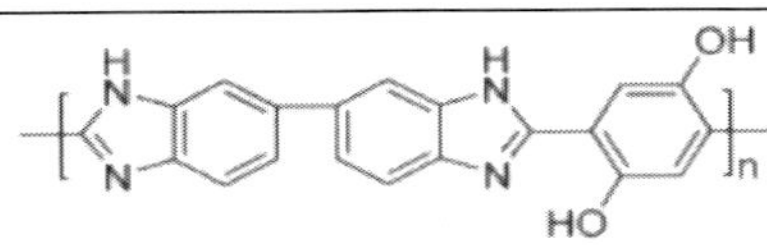	36
Ether containing PBI	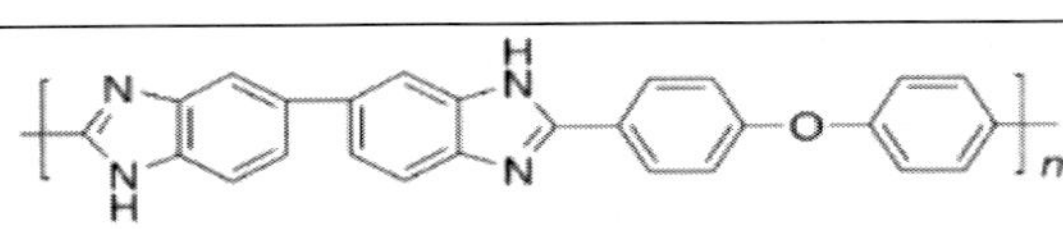	38

Cross-Linked Polybenzimidazole Membranes

Generally, the conductivity and mechanical strength of PEMFC are compromised. Different methods are adopted to improve the proton conductivity without sacrificing too much the mechanical strength or vice versa. One of the strategies, viz. the covalent cross-linking of polybenzimidazoles can be achieved by thermal treatment or an amide-type linkage through imidazole groups of the polymer. The p-xylene dichloride can be used as a crosslinking agent [44].

For preparing the crosslinked membrane, the low-molecular-weight PBI (21500) is treated with p-xylene dibromide [45]. Higher the crosslink density, higher is the resistance towards dissolution in DMAc. The cross-linked PBI membranes having acid doping levels of 8.5, 14.1, and 15.1 mol phosphoric acid, respectively, exhibit volume swellings of 62, 143, and 160% for cross-linking degrees of 13.0, 3.6, and 1.1%, respectively. By exposing the linear PBI membrane in 3% hydrogen peroxide solution containing 4 ppm Fe^{2+} at 68°C for specific duration, the weight loss of about 15% is observed, whereas 7 and 2% weight loss are observed for PBI membranes with cross-linking degrees of 3.6 and 13%, respectively.

Silane Cross-Linked Sulfonted Poly(Ether Ketone/Ether Benzimidazole) Membrane

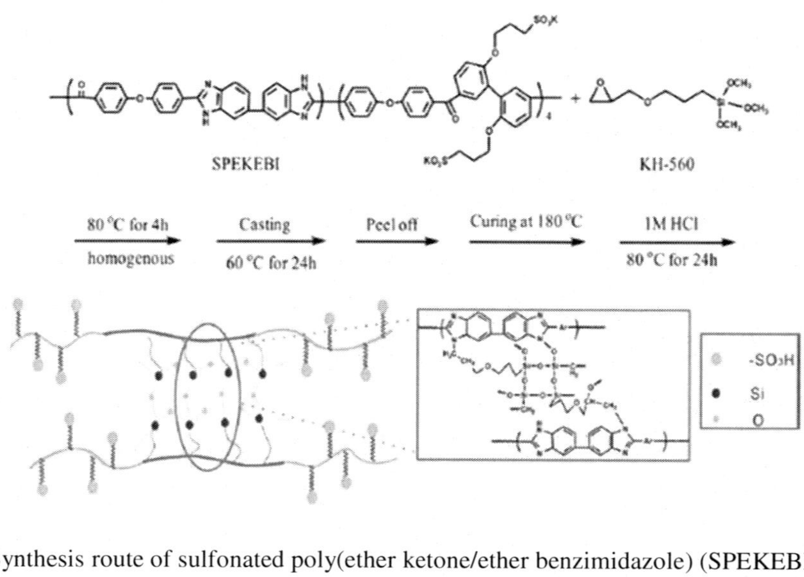

Scheme 2. Synthesis route of sulfonated poly(ether ketone/ether benzimidazole) (SPEKEBI)-x-SiO_2 membranes.

[3-(2,3-epoxypropoxy) propyl] trimethoxysilane (KH-560) can be incorporated as a crosslinker in various proportions into side-chain-type sulfonated poly(ether ketone/ether benzimidazole) (SPEKEBI), to make membranes with high ion exchange capacities and excellent performance for direct methanol fuel cells (DMFCs). 4,4'-oxydibenzoic acid (ODBA) and 3,3',4,4'-tetraaminobiphenyl (TABP) are dissolved in P_2O_5/CH_3SO_3H and heated to 120°C in nitrogen atmosphere and then disodium 2,2'-di(sulfopropyloxy)biphenyl (SBP) is reacted to form the polymer. The film is cast from a mixture of γ-(2,3-epoxypropoxy) propyltrimethoxysilane (KH-560) and SPEKEBI by following standard procedure. The reaction scheme can be seen in Scheme 2 [46]. The tensile strength and elongation at break of SPEKEBI-x-SiO_2 membranes are in the range of 10.4–18.9 MPa and 8.8–25.3%, respectively. SPEKEBI-7.5-SiO_2 membranes exhibit maximum proton conductivity of 0.244 Scm^{-1} at 80°C. The maximum power density of SPEKEBI-2.5-SiO_2 membranes is 341 mW cm^{-2} at RH of 100% and temperature of 75°C.

Copolymers Membranes

Copolymerization is done for harnessing additional properties for the PEMFC membrane so that overall performance is enhanced. A large number of copolymers are theoretically possible and are incorporated in the following discussion.

Poly Aryl Ether Bibenzimidazole Copolymer

Copolymers of poly(2,5-benzimidazole) (ABPBI) and poly[2,2'-(p-phenylene)-5,5'-bibenzimidazole] (pPBI) can be used as fuel cell membranes. The pPBI is used alone as a high-temperature PEM, having high chemical and mechanical stability originating from its stiff and linear backbone structure. However, it has very poor solubility in solvents that limits its processability. ABPBI and pPBI can be copolymerized so that the copolymers will have both improved stability of ABPBI and improved processability of stiff pPBI. The copolymer, poly{(2,5-Benzimidazole)-co-[2,2(-(p-Phenylene)-5,5(-Bibenzimidazole]}s (AB-pPBI-Xs) (X is the mol-% of DABA), can be synthesized by condensation polymerization of 3,3'-diaminobenzidine tetrahydrochloride dehydrate (DABIT) with DABA and terephthalic acid in PPA at 240°C [47]. The broader shapes of both the 11 and 26° peaks in XRD of AB-pPBI-X films confirms lower crystallinity of the copolymer films as compared to that of the homopolymers. The acid doping of AB-pPBI-X and ABPBI membranes are larger than those of pPBI, which increase with the content of ABPBI. When the acid content is larger than 65 wt. % (at 180°C and anhydrous condition), all the acid-doped membranes exhibit proton conductivities over 0.045 Scm^{-1}. The AB-pPBI-X membrane maintains higher tensile strength than that of ABPBI. For example, the tensile strength of the ABPBI membrane with 75 wt. % of acid content is 25 MPa while that of AB-pPBI-X membranes is 35 MPa.

Poly (Aryl Ether Benzimidazole) Copolymer Membrane

The presence of rigid and heterocyclic polymer backbones like PBI can be modified with aryl ether linkages to realize improved solubility and decreased Tg, while maintaining the desired thermal and mechanical properties. Poly[(2,2'-(p-phenylene)-5,5'-bibenzimidazole)-co-(2,2'-(4,4'-oxybis(1,4-phenylene))-5,5'-bibenzimidazole)]s (poPBI-X; (where X refers to the mol% of 4,4'-dicarboxydiphenyl ether (DCPE)) having varied proportion of aryl ether linkages, are synthesized in presence of PPA by varying the feed ratio of DCPE to terephthalic acid (TA) from 10 to 90 mol% [48]. When 2 wt.% polymerization solution is used, the solution for the pPBI synthesis becomes opaque with progress of the polymerization, while the corresponding solution for poPBI-Xs remains transparent throughout the synthesis indicating its solubility. The mol% of aryl ether linkage in poPBI-Xs can be determined by comparing the peak areas in ^1H NMR. The Tg of the polymer decreases from 423 to 321°C as the amount of aryl ether linkages in the polymer increases from poPBI-10 to poPBI-90. Reasonable mechanical strength (> 40MPa) of poPBI-10 and poPBI-30 is maintained even at large acid doping levels of 6–8 mol PA. The proton conductivity of poPBI-X membranes is mostly dependent on the proportion of aryl ether linkages. The poPBI-50 and poPBI-Xs (X > 50) membranes exhibit conductivity lower than that of pPBI. The flexible aryl ether linkages at low level are able to enhance the proton transfer by decreasing the Tg of the acid doped membranes. Thus, the poPBI-10 and poPBI-30 membrane exhibits proton conductivities higher than that of pPBI and comparable to mPBI membrane.

Poly[2,2'-(M-Phenylene)-5,5'-Bibenzimidazole] and Poly(2,5-Benzimidazole) Copolymer Membrane

The copolymer of PBIs (AB-mPBI) is synthesized by copolymerization of 3, 3'-diaminobenzidine tetrahydrochloride, 3,4-diaminobenzoic acid (DABA) and isophthalic acid in PPA at 200°C [49]. Microwave irradiation can be used for faster reaction. Unlike the homopolymers, this copolymer dissolves easily in DMAc making the membrane casting easier. The AB-mPBI exhibits decomposition temperature of 520°C, which is similar to that of mPBI and ABPBI. Higher acid doping is possible with the copolymers having higher content of ABPBI. The conductivity of AB-mPBI membrane with an acid doping level of ca 4.0 mol is 7.5×10^{-2} S cm^{-1} at 180°C. At 120°C, the tensile strengths at break for the AB-mPBI and mPBI membranes with an acid doping level of ca 2.6 mol are comparable (7.5 MPa and 8.0 MPa).

Poly(Benzimidazole-Co-Benzoxazole) Membranes

Poly(benzimidazole-co-benzoxazole)s (PBI-co-PBO) are synthesized by polycondensation reaction of 3,3'-diaminobenzidine, terephthalic acid with 3,3'-dihydroxybenzidine or 4,6-diaminoresorcinol in PPA [50]. PBI-co-PBO1 (0.9:0.1) membrane exhibits marginally lower proton conductivity due to the lowering of phosphoric acid doping level. The presence of BO unit in the copolymer is responsible for decrease in the doping level. The maximum tensile strength of membrane is 3.4 MPa with the same sample having 10 mol % of BO units. The single cell operated with non-humidified hydrogen and air at 150°C exhibits performance similar to that of PBI. All the PBI-co-PBO membranes maintain OCV around 0.97 V. At 0.6 V, the current density of PBI is 253 mA cm^{-2} while that of PBI-co-PBO lies in the range of 210-220 mA cm^{-2}.

Sulfonated Poly(Arylene Ether Benzimidazole) Based Copolymer

For making compromise among the sulfonated polymers and PBI, a sulfonated benzimidazole polymer/copolymer can be made. Disodium-3,3'-disulfonate-4,4'-dichlorodiphenylsulfone (SDCDPS) can be reacted with 5,5'-bis[2-(4-hydroxyphenyl) benzimidazole] (HPBI) via nucleophilic aromatic polycondensation reaction to get a sulfonated poly(arylene ether benzimidazole) copolymer [51]. The copolymer solution is cast onto petri dish and dried to get copolymer film. The transparent membranes are immersed in boiling 0.5M H_2SO_4 for 2 hours and dried to remove absorbed water. The membrane modification is also performed with oleum (65% SO_3). Proton conductivities of the membranes at 80°C are 0.0059 Scm^{-1} (RH =0%), and 0.0012 Scm^{-1} (RH = 0%) for H_2SO_4 and oleum treated membranes, respectively. The higher extent of sulfonation with oleum results in lower conductivity of the copolymer. The excess amount of sulfonic acid groups undergoes interaction with imidazole sites to lower the conductivity. The highest power density of the membrane observed in fuel cell is 23.7 $mWcm^{-2}$ at 80°C without humidification.

Cross-linked poly(2,5-benzimidazole) membrane

Cross-linked ABPBI (C-ABPBI) membranes with full aromatic groups are prepared by a reaction between three-arm PBIs with three carboxylic acid end groups (T-ABPBI) and 3,3'-diaminobenzidine (DABI) in PPA with simultaneous casting of the films. T-ABPBI is prepared by the polymerization of 3,4-diaminobenzoic acid and 1,3,5-benzenetricarboxylic acid in PPA [52].The cross-linking process improves the chemical stability of ABPBI and enhances the phosphoric acid (PA) content as compared to linear ABPBI and m-PBI. As a result, the C-ABPBI membranes show higher proton conductivity values than the linear polymer membranes. For example, the C-ABPBI-50 membrane having a PA content > 84 wt% exhibits a proton conductivity of 0.12 Scm^{-1} at 150°C under anhydrous conditions. The large concentration of PA indicates excess amount of free PA, which is responsible for higher proton conductivity [53].

Cross-Linked Benzoxazine−Benzimidazole Copolymer

Polybenzoxazine can be prepared via ring-opening of the cyclic benzoxazine monomers by heat treatment without the need of any catalysts. Cross-linked polymeric materials with enhanced thermal and dimensional stabilities are then obtained without the formation of any by-products. Cross-linked copolymer membranes based on polybenzimidazole and polybenzoxazine can be synthesized by first solution casting the solution comprising different wt % of 3-Phenyl-3,4-dihydro-6-tert-butyl-2H-1,3-benzoxazine (pBUa) and PBI, followed by stepwise heating from 60 to 220°C for 4 h and then maintaining the temperature for an hour [54]. PA doping is done by immersing dried P(pBUa-co-BI) films in 80−85 wt % PA solution at 80°C for 4 h. In the PA solution there is complete dissolution of the homopolymers (P(pBUa) and PBI) and their mixture while very low degree of swelling of P(pBUa-co-BI) membranes is observed indicating cross-linked and copolymerized structure of P(pBUa-co-BI)s. All the P(pBUa-co-BI) samples are thermally stable up to 450°C. The highest proton conductivities of (pBUa-co-BI)-65 (65 indicates weight ratios of pBUa to PBI) membrane, with the highest PA content (ca. 88 wt %), are about 0.10 and 0.12 S cm^{-1} at 120 and 150°C, respectively. When incorporated in a fuel cell, higher performance (0.71 V at 0.2 A cm^{-2}), durability (1999 cycles of accelerated life time test), and slower lifetime loss (−0.03 mV h^{-1}) are observed.

Block Copolymer Membrane of Meta-Polybenzimidazole and Para-Polybenzimidazole

The strategy of making copolymer from meta and para benzimidazole has solved solubility problem. However, the conductivity does not improve. The blocks of multiblock structure is known to phase separate easily into two domains that augments the internal connectivity within the block and hence leads to higher conductivity [55]. Segmented block copolymers of PBI (meta-polybenzimidazole-block-parapolybenzi-midazole; m-PBI-b-p-PBI) are synthesized by coupling reactions of amine (m-PBI-Am) and acid (p-PBI-Ac) terminated oligomers in PPA medium [56]. The length of the acid terminated oligomeric block is varied, while the length of the amine terminated oligomeric block kept fixed, and vice versa to get a series of block copolymers. There is a gradual decrease of copolymer molecular weight with the increase of one oligomer molecular weight due to sluggishness in the reactivity of the longer chain length oligomers. The PA doping level for all the samples varies between 8 and 11 mol per PBI repeat unit, while no trend exists with increase of block length/structure or molecular weight of the polymer. The block length influences the proton conductivity, as at 160°C the smallest block length copolymer ($Am_{0.12}$-b-$Ac_{0.11}$) conductivity is 0.05 Scm^{-1} whereas that of largest block length copolymer ($Am_{0.59}$-b-$Ac_{0.54}$) is 0.11 Scm^{-1}. The proton conductivity varies significantly from one type of block structure to another because of the variation in nanophase separation of these block copolymers.

Poly(Imide Benzimidazole) Copolymer Membrane

The poly (imide benzimidazole)s present a copolymer structure in which the basic benzimidazole groups facilitate the acid uptake, while the polyimide (PI) units improve the mechanical strength due to their relatively weak interactions with the doped acid molecules in comparison with benzimidazole units. The copolymer is synthesized by heating a mixture of biphenyl-4,4'-diyldi(oxo)-4,4'-bis (1,8-naphthalene dicarboxylic anhydride) (BPNDA), 2-(4-aminophenyl)-5-aminobenzimidazole (APABI), 4,4'diaminodiphenyl ether (ODA) in m-cresol in the presence of benzoic acid and isoquinoline at180°C for 20 h [25]. By careful control of the monomer stoichiometry (APABI:ODA molar ratio) different random copolymers can be synthesized. This random copolymer (PIBI) based membranes exhibit excellent radical oxidative stability which is comparable to that of crosslinked mPBI membrane and Nafions. With 85% orthophosphoric acid as the doping medium during doping in the autoclave at 180°C, the PIBI-1/1 membranes can still absorb considerable amount of PA. The acid uptake under these conditions reaches 300 wt% for the PIBI-1/1 membrane and 340 wt% for the PIBI-2/1 membrane. The PIBI membranes exhibit significantly higher mechanical strength as compared to PBI membrane. For example, with similar acid uptake, the PIBI-1/1 membrane has much higher tensile strength (28 MPa) than a membrane based on high molecular weight PBI (6.2 MPa). The PIBI-1/0 membrane exhibits an anhydrous proton conductivity of 0.26 S cm^{-1} at 170°C which is comparable to the highest values of the acid doped PBI membranes. The proton conductivity increases in the order PIBI-1/0 >PIBI-4/1>PIBI-2/1>PIBI-1/1 which is consistent with the order of the corresponding acid uptakes.

Table 2. List of PBI/ABPBI copolymers and their chemical structures

Membrane	Structure	Reference
Poly[(2,2'-(p-phenylene)-5,5'-bibenzimidazole)-co-(2,2'-(4,4'-oxybis(1,4 phenylene))-5,5'-bibenzimidazole)]	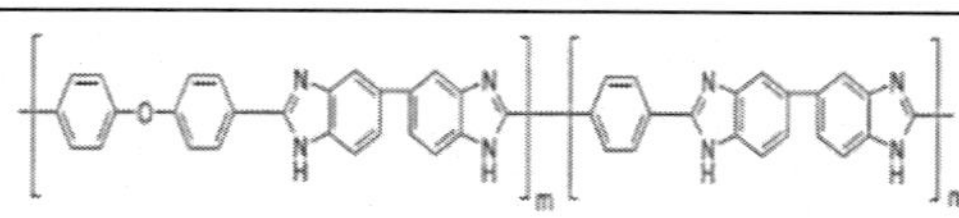	48
ABPBI & p-PBI copolymer	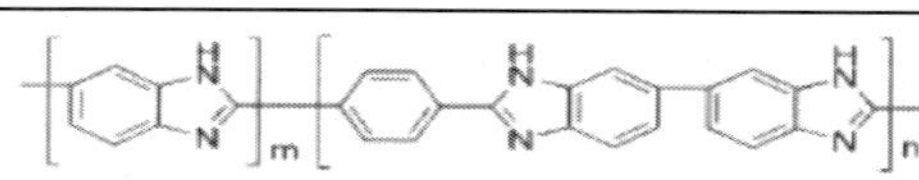	47
Copolymer (ABPBI) is a combination of ABPBI and m-PBI	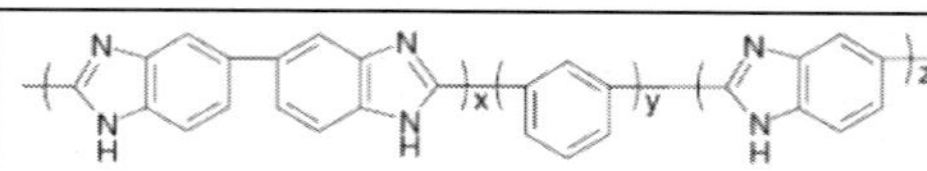	49
Poly(benzimidazole-co-benzoxazole)	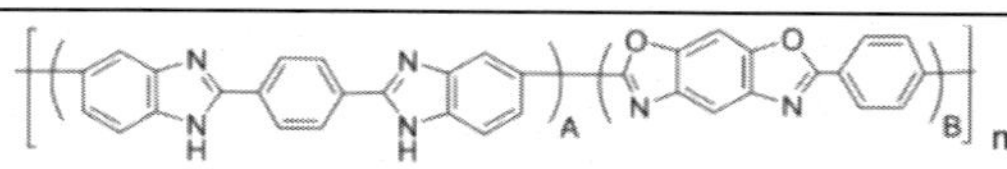	50
m-PBI & p-PBI block copolymer	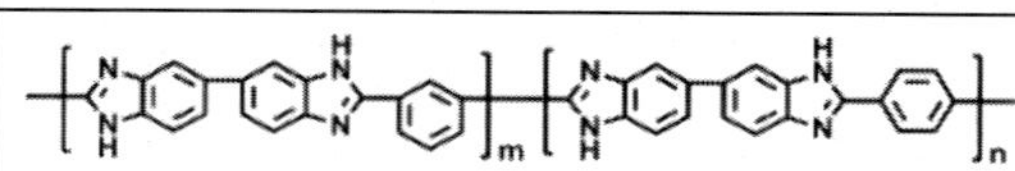	56
poly(imidebenzimidazole) copolymer	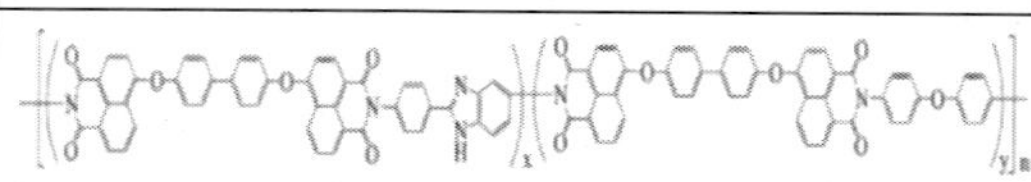	25
Cross-Linked Polybenzoxazine based copolymer	---	53
Sulfonated Poly(Arylene Ether Benzimidazole) based Copolymer	---	51

Blend/Composite Membrane

PBI's structure (rigid-rod) is such that during thermal processing internal physical stresses develop and the additives are, therefore, used to stabilize the polymer matrix. Hydrogen bonding with imidazole groups and/or possible aromatic π-π stacking controls the molecular structure of PBI. The disturbance of interactions results in defects that are pronounced after drying. Most common stabilizers are sulfuric or phosphoric acids. Concentrated acids soften and/or swell the polymer, break up crystal packing or ordering, and protonates imidazole nitrogen on the polymer backbone. The pliable polymer obtained helps in membrane formation. Among various methods, the blending methods can use pre-polymers, like polyamic acids, to give PBI blends after thermal processing. Regarding the performance, the drawbacks of PBI membrane for use in PEMFC are the too much dependence on the proton conductivities (which depends on the phosphoric acid doping percentages), rapid deterioration of mechanical properties and chemical stability of the membrane (due to serious acid leaching under continuous usage). To overcome the drawbacks various approaches have been adopted, like blending and composite making.

Sulfonated Poly(Ether Sulfone)/Sulfonated Polybenzimidazole Blend Membrane

Sulfonated polybenzimidazole homopolymer (MS-p-PBI 100) and sulfonated poly(aryl ether benzimidazole) copolymers (MS-p-PBI 50, 60, 70, 80, 90) can be blended with sulfonated poly(ether sulfone) copolymer (PES) to get membrane with enhanced boiling water stability as well as dimensional stability as compared to that of pristine PES 70 membranes [57].The water uptake of the PES 70 and amphiphilic MS-p-PBI 100 blend membranes is lower than that of the acid form PES 70 membrane. There is substantial improvement in swelling behaviour and dimensional stability of the blend membranes. The blend membrane (acid form) used for PEMFC and DMFC operation exhibits a current density of 800 mA cm^{-2} at 0.4 V with a stoichiometric flow rate (H_2/air = 1.5/2) of gases. A current density value of 230 $mAcm^{-2}$ is possible at 0.4 V for 1 M methanol feed at 70°C.

Sulfonated PVA/PBI Based Crosslinked Composite Membrane

Poly(vinyl alcohol) (PVA) has an excellent resistance to methanol crossover and several organic groups like hydroxyl, amine, carboxylate, sulfonate, and quaternary ammonium can be incorporated in to it to impart hydrophilicity and/or ionic groups for use as a polymer electrolyte membrane in fuel cell. A crosslinked sulfonated PVA/PBI membrane is made by first dissolving PVA and PBI in DMSO and then adding the chlorosulfonic acid to it. Guteraldehyde is added to crosslink PVA and then a film is cast and dried over several days at 45-50°C [58]. The methanol permeability of the membranes (10^{-6}-10^{-8}) increases along with higher water uptake at a lower degree of effective crosslinking whereas it decreases with a dense and compact morphological structure. At high level of sulfonation of PVA, the tensile strength and percent elongation of the composite PEM increase with increase in PBI content. The composite membrane, with the highest degree of sulfonation and PBI content, exhibits highest proton conductivity of 2.72 x 10^{-3} Scm^{-1}.

Nafion/Sulfonated PBI Composite Membrane

Sulfonated PBI(s-PBI) is deposited as an additional layer on a Nafion membrane to create composite polymer electrolyte that maintains a reasonable proton conductivity and reduces the crossover of methanol [59]. At low current densities, methanol crosses through the PBI membrane that creates large polarization and poisoning at the cathode. The resulting mixed potential at the cathode lowers the overall cell voltage. The presence of the s-PBI over a Nafion 112 MEA slows the methanol crossover sufficiently to increase the open circuit voltage by 150 mV.

SILOXANE BASED COMPOSITE

Trisilanolphenyl Polyhedral Oligomeric Silsesquioxane/ABPBI Composite Membrane

TriSilanolPhenyl polyhedral oligomeric silsesquioxane (SO-POSS), having large number of –Si-O-Si groups and a small number of —OH groups in its structure, is used as the nanofillers for modification of ABPBI [60]. Due to the presence of uniformly dispersed 3% SO-POSS particles in ABPBI matrix the thermal stability of the composite membranes increases. The uptake of both water and H_3PO_4 increases rapidly with the addition of SO-POSS because of the formation of hydrogen bonds between the POSS and H_2O/H_3PO_4. This has direct consequence to the conductivity, which is enhanced at temperature over 100°C. Proton conductivities of H_3PO_4 doped with 3wt% SO-POSS/ABPBI membranes reach to a maximum of 2.55×10^{-3} S·cm^{-1} at 160°C.

ABPBI and Octaammonium POSS Based Composite

ABPBI and octaammonium POSS (OA-POSS) based composite is another variant, which can be prepared by following the standard procedure using methanesulfonic acid as solvent [61]. By increasing the OA-POSS content in the membrane from 1% to 3% there is transformation of the semicrystalline nature of ABPBI to amorphous nature. 3% OA-POSS based composite membrane exhibits water uptake of 106% and PA uptake of 241%. At 160°C without any humidification, this membrane shows highest proton conductivity of 0.25 Scm^{-1}.

Octaammonium POSS nanofillers is used to modify ABPBI membrane through an in situ synthesis method via hydrophilic silicon–oxygen–silicon groups and base amino groups to stabilize H_2O and H_3PO_4 molecules [62]. This is very important for PEMFC to cover intermediate temperature with the hydrous and anhydrous conditions through different proton transfer mechanisms (so called "vehicle" mechanism under hydrous condition and "hopping" mechanism under anhydrous condition, respectively).

ABPBI/AM-POSS composite membrane is prepared by dissolving the mixture in methane sulfonic acid and casting and stepwise heating up to 200°C. After proper drying, the membrane is doped with PA by dipping in it. At PA doping levels of 60%, the proton conductivities of ABPBI/AM-POSS composites are slightly higher than those of PA doped pristine ABPBI membranes, which increases to a very high value >0.1 S cm^{-1} even at low temperature range (below 100°C), with a doping level of 250%.

PBI and Phosphonic Acid Groups-Functionalized POSS Membrane

Significant challenge exists for functionalization of molecules or macromolecular species with multiple phosphonic acid groups because of various side reactions including condensation and aggregation of phosphonic acid groups. To counter the challenges, one of the efforts is the functionalization of phenyl rings of POSS with ~7 phosphonic acid groups via bromination, followed by phosphonation. The functionalized POSS is incorporated in to PBI and used in high temperature PEMs [63]. The bromine of Br-POSS is substituted with phosphonate ester groups via a Pd(0) catalyzed P–C coupling reaction that is carried out with diethyl phosphite ($(EtO)_2POH$) to get $PO(OEt)_2$-POSS (Figure 5). The proton conductivity increases with increasing temperature and the composite maintains higher proton conductivity (Figure 6). The conductivity follows mostly proton hopping mechanism.

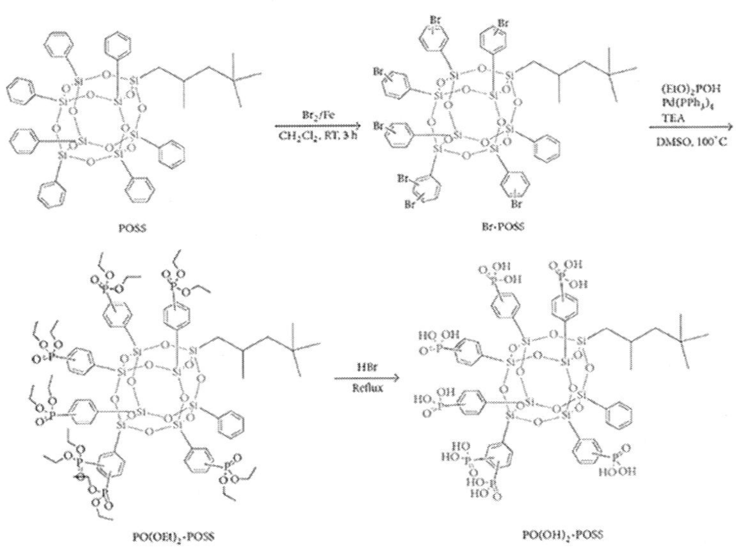

Figure 5. Synthesis of phosphonic acids-functionalized POSS.

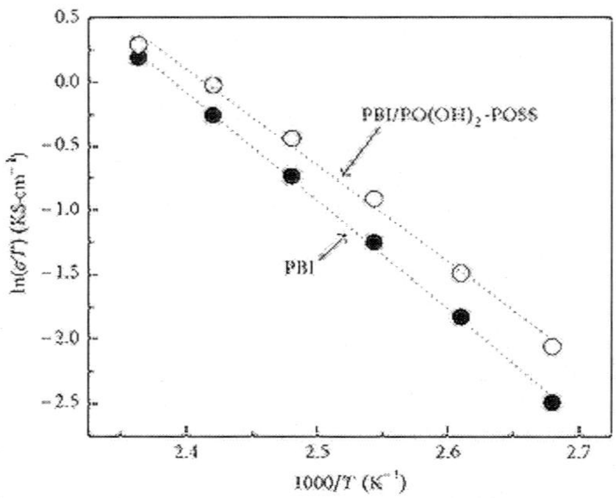

Figure 6. Proton conductivities of PA-doped PBI and PBI/PO(OH)$_2$-POSS membranes with temperature ranging from 100 to 150°C under anhydrous conditions. PA content of the membranes is ca. 50wt%. Copyright © 2016 Sung-Kon Kim. This is an open access article distributed under the Creative Commons Attribution License, which permits unrestricted use, distribution, and reproduction in any medium, provided the original work is properly cited (Figures 5 & 6).

Sulfonated Hydrocarbon Resin/PVA-Benizimidazole Based Membrane

Benzimidazole is added as the protogenic agent to a sulfonated hydrocarbon resin (SR) and poly(vinyl alcohol) (PVA) to make polyelectrolyte membranes [64]. The SR-BZ/PVA membranes are prepared as semi-interpenetrating polymer networks by crosslinking PVA with glutaraldehyde using different concentrations of benzimidazole. The proton transport in the membranes with benzimidazole occurs by the hopping mechanism because an acid–base complex is formed between the acidic sulfonic groups of the polymer electrolyte and the basic groups of the benzimidazole compound. While increasing the BZ content in the membrane the mechanism changes because of the presence of more basic groups than the acidic groups; the membrane becomes anionic. The conductivity of the SR-BZ/PVA membranes under anhydrous conditions depends on the content of BZ, and the proton conductivity is maximum for the low concentration of BZ. For the optimum BZ concentration, the interactions between the sulfonic and imidazole groups favor an easier way for the protons to exchange between these groups. In fact, the presence of 5% BZ exhibits maximum proton conductivity. By increasing the amount of BZ, the concentration of proton donor (–SO_3H) reduces that results in decreases of conductivity.

PBI/Poly(Tetrafluoro Ethylene) (PTFE) Composite Membranes

PEM with low thickness and high conductivity exhibits high proton transport efficiency and thus a high PEMFC performance. However, a thin PEM is likely to have poor mechanical strength that may cause the fuel gas crossover through the PEM, leading to a lower open circuit voltage (OCV) and a lower fuel cell performance. For increasing the mechanical strength, PTFE can be combined by designing PBI/PTFE composite membranes with Nafion as a coupling agent [65]. The perfluorocarbon main chain of Nafion is compatible with the PTFE supporting thin film, and the ether sulfonic acid side chains having $–SO_3H$ groups interact with the imidazole groups of PBI via Lewis acid–base interaction. Membrane is prepared by first treating the porous PTFE thin film with a Nafion thin layer and then impregnating in a PBI/DMAc solution. The proton conductivity of the membrane is 0.014 $S.cm^{-1}$ at 160°C. The unit cell exhibits 450 h continuous unit cell life test at 160°C with a current density of 200 mA cm^{-2} without much variation of cell voltage. The 20 cycles on/off unit cell test exhibits constant cell voltage during the first six on/off cycle operation. After the six on/off cycles operation, there is dip in cell voltage with a decay rate of $\sim 1.08 \times 10^{-4}$ V h^{-1} (from 144 h to 456 h) or 2.5×10^{-3} V/cycle (from 6th cycle to 20th cycle).

Metal Compounds/PBI Composite/Nanocomposites Membrane

Incorporation of metal compounds enhances the mechanical strength, dimensional stability, thermal stability and acid uptake etc. Various metal compounds are used for making composites and nanocomposites. Few important composites are incorporated in the following discussion.

Titania-Doped Polybenzimidazole Based Composite Membrane

By incorporating TiO_2 into the PBI membrane, the PA concentration in the membrane is improved because the adsorption capability of the inorganic phase helps in retaining the acid in the membrane at high temperatures and also manages well the by-product water produced [66]. Maximum power density of 195 $mWcm^{-2}$ is observed for pure PBI and much higher value of 434$mWcm^{-2}$ is observed for TiO_2/PBI, respectively. The open circuit voltages higher than 0.9V indicate that the membranes have low gas permeability. The composite membrane of 2wt% TiO_2/PBI maintains a constant potential throughout the measurement.

PBI/SiO_2 Composite Membrane

PBI/SiO_2 Composite membrane can be produced by first mixing PBI solution with nano SiO_2 powder in an ultrasonic bath and then casting onto a flat glass. The residual DMAc is removed by maintaining the film over long hours at high temperature [67]. The membranes are thermally stable up to 600°C. The maximum proton conductivity value of 3.9×10^{-3} Scm^{-1} is exhibited at a temperature of 180°C by the composite membrane at 15% loading of nano-SiO_2. The hydroxyl groups, present in SiO_2 nanoparticles, form hydrogen bonds with the polymer chains, which facilitate the proton transport.

ABPBI–silica nanocomposite membranes can be prepared by ex situ techniques using methane sulfonic acid (MSA) as solvent [68]. Fumed silica is first modified with (3-mercaptopropyl) methyl dimethoxysilane (MPDMS) in toluene at 145°C for 7 h. After modification the $-SO_3H$ groups are generated on the silica nanoparticles. Uniform and loosely aggregated morphology is seen in the TEM micrograph of ABPBI-2MS and ABPBI-10MS composites (MS: modified silica having $-SO_3H$ group). The hydrogen bonding between $-SO_3H$ group on MS and N-atom on the ABPBI chain is responsible for uniform morphology. The ABPBI-10MS membrane exhibits highest water uptake of 28% and 24% at 30 and 80°C, respectively, corresponding to 42% and 65% improvements as compared to the virgin ABPBI. All the doped ABPBI-MS composite membranes possess higher proton conductivities than the doped ABPBI and unmodified and doped ABPBI membranes; the highest being 0.04 Scm^{-1}.

Phosphomolybdic Acid Doped Poly(2,5-Benzimidazole) Membrane

The large molecular size of phosphomolibdic acid ($H_3PMo_{12}O_{40}$ denoted as PMo12) prevents the diffusion of the cluster into membranes of PBI polymers. Therefore, for preparing ABPBI–PMo12 hybrid membranes, ABPBI solution in MSA and variable amounts of the PMo12 acid are stirred together until a homogeneous viscous solution is obtained [69]. The solution is cast to get membrane which is further doped with PA to ensure high proton conductivity as well as stability. To maintain flexible nature of the ABPBI–PMo12 (60%) membrane, it can be doped in a bath concentration up to approximately 68% H_3PO_4. ABPBI–45% PMo12 membrane impregnated in a 68% phosphoric acid solution reaches a proton conductivity of 3×10^{-2} S cm^{-1} at 185°C under dry conditions.

Zirconium Pyrophosphate/ABPBI Composite Membranes

Zirconium pyrophosphate/ABPBI (ZPP/ABPBI) composites are made for improving the physical properties, including mechanical strength and dimensional stability of ABPBI [70]. ABPBI is synthesized in presence of ZPP and polyphosphoric acid (PPA) and directly cast on glass plate. After removing and drying, the membrane is doped with PA. The glass transition temperature of ABPBI is around 450°C, while most of ZPP/ABPBI composite membranes do not exhibit glass transitions due to the limited mobility of benzimidazole chains caused by inorganic materials. Tensile strength increases with ZPP loading. However, at higher loading it decreases. The ZPP/ABPBI membranes exhibit high proton conductivities, above 0.12 S .cm^{-1} at 180°C.

Zr/Polyimidazole (PBI-O-Pht) Derivative Based Composite Membranes

Zr precursors (zirconium tetraacetate ($Zr(OAc)_4$) or zirconium acetylacetonate ($Zr(acac)_4$) based composite PBI membrane can be prepared with very low dose of zirconium to get favourable redistribution of PA. Zirconium atoms, having valence of four, exhibit coordination numbers of up to nine and may form crosslinks between PBI chains. The composite is prepared by mixing Zr(IV) acetylacetonate ($Zr(acac)_4$) or Zr(IV) tetraacetate ($Zr(OAc)_4$) solution in NMP with the polymer, (poly[oxy-3,3-bis(4′-benzimidazol-2″-ylphenyl)phtalide-5″(6″)-diyl] (PBI-O-PhT) solution and cast on glass plate. After solvent evaporation, the films are heated in vacuum at 160°C for 2 h for removal of residual solvent, and then heated in an oven with air circulation for 1 h at 350 °C for the three-dimensional crosslinking of polymer chains. The doping of the cross-linked films is done with 77% PA at 60 °C for three days [71]. A possible mechanism of the crosslinking process of PBI by $Zr(acac)_4$ and further doping with PA is shown in Figure 7.

The labile nature of the N–Zr bonds in the excess of PA imparts dynamical behaviour to the Zr-crosslinks. The strong tendency of Zr-atoms to coordinate atoms of both PBI chains and PA is a unique advantage as it not only enhances the mechanical properties of the composite, but also has improved ability to retain the liquid electrolyte. The proton conductivity of the membrane varies in the range of 0.07-0.1 S·cm^{-1} at 160°C for 0.75-2% zirconium salts with acid uptake from 380 to 430%.

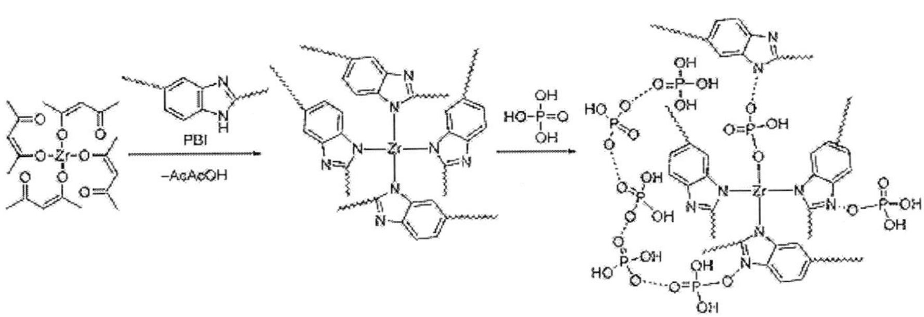

Figure 7. Possible mechanism of the crosslinking process of PBI by $Zr(acac)_4$ and further doping with PA.

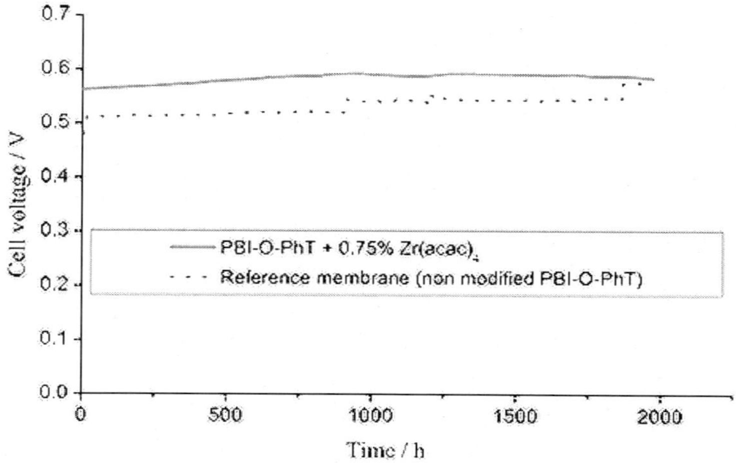

Figure 8. 2000-hour stability test of a composite PBI-O-PhT + 0.75 wt % $Zr(acac)_4$ membrane and the reference membrane in an operating fuel cell. H_2/air, T = 160 °C, current density 0.4 A·cm^{-2}.

The membranes are evaluated in fuel cells for 2000 h (Figure 8). An increase in performance of about 15–20 mV is observed, which is due to the redistribution of PA inside the MEA and the formation of the effective boundary between the three phases (electron and proton conducting phases and gas phase) during the first 1000 h. The characteristic 1000 h is the time needed to remove excessive acid from the active layer of the electrodes.

Polybenzimidazole/Strontium Cerate Nanocomposites Membrane

The perovskite-type (AB1-xMxO3-δ) ceramics exhibit p-type electronic (hole) conduction under oxidizing atmospheres and become a proton conductor in the presence of water vapor and hydrogen. The nanocomposite membranes based on polybenzimidazole/strontium cerate (PSC) nanoparticles can be prepared by solution casting method using DMAc as solvent [72]. The presence of nanoparticles in the membranes enhances the ability to pick up PA. A high acid uptake of 190% (for PSC8; 8 wt% cerate nanoparticle) is possible for PA-PSC nanocomposite membranes. The proton conductivity of the corresponding nanocomposite membrane reaches a maximum of 0.075 Scm^{-1} at RT. The large A-site cations are responsible for low activation enthalpy of proton mobility in perovskites structure. The PBI membranes exhibit a broad XRD peak at around 2θ =25° indicating its amorphous nature. In fuel cell set up, the PSC_8 nanocomposite membrane exhibits a power density of 0.44 Wcm^{-2} and a current density of 0.88 Acm^{-2} at

0.5 V and 180ºC. The OCV value for the same membrane is 0.86 at 180°C (membrane thickness: 55 μm).

Phosphosilicate Gel-Polybenzimidazole Nanocomposite Membrane

The phosphosilicate gel (PPSG) enhances the mechanical strength of the PBI membranes and reduces the leaching of the PA from the membrane when blended with PBI. The phosphosilicate gel-PBI nanocomposite membrane is fabricated by conventional solution casting method with DMAc used as a solvent. With 5 wt% PPSG loading the ion exchange and water uptake properties of the membrane enhance by 200% and 46%, respectively [73]. In fuel cell set up, 60% enhancement of peak power generation at 150°C is achieved with 5 wt% PPSG loading. Also, the maximum allowable load current enhances by more than 1.8 times.

Polyaniline and Phosphoric Acid Doped Sulfonated PBI Composite Membranes

Polyaniline (PANI) and PA doped sulfonated PBI composite membrane is prepared for use in high-temperature PEMFC without the need of humidification [74]. The monosodium salt 5-sulfoisophthalic acid and 3, 3', 4, 4'-Tetraaminobiphenyl are reacted in presence of polyphosphoric acid (PPA) at 190-200°C for 20 h and then the solution is cast to form the membrane.

Sulfonated Polystyrene Based Blend Nanocomposites

Poly(benzimidazole-amide) (PBA) is blended with sulfonated polystyrene (PS-S) and the nanocomposites is prepared by adding 0.1–2 wt.% titania nanoparticles-grafted-multi-walled carbon nanotubes [75]. For preparation of PBA/PS-S blend, the stock solutions of PBA and PS-S are prepared separately in DMAc, mixed is specific ratio and stirred for 24 h. The hybrid proton exchange membranes (poly(benzimidazole amide)/sulfonated polystyrene/titania nanoparticles-grafted-multi-walled carbon nanotubes) are prepared by mixing totania-nanoparticle grafted MWCNT (0.1–2 wt.%) with appropriate amount of the stock solution and the films are cast on glass Petri dishes at 70°C and dried at 80°C for 24 h under vacuum. PBA/PS-S/MWCNT/Titania NPs 0.1 exhibits tensile stress of 65 MPa and elongation at break of 0.72% as against the corresponding values 54 MPa and 0.55% for the blend. The Tg is also increased to a large extent. Water uptake, swelling ratio and proton conductivity increases with incorporation of nanoparticles.

Polybenzimidazole-MWCNT Composite Membranes

By mixing 1 wt. % of MWCNTs to PBI solution in DMAc and casting on glass plate, the PBI-MWCNT nanocomposites membrane can be prepared [76]. In Raman spectra, the I_D/I_G ratio of CNT dispersed in the PBI-CNT membrane is higher than that of pure CNT, which is due to the π-π interaction between the polymer structure and the carbon nanotubes side walls. This increases the sp^3 carbons on MWCNT surface and produces disorder in the structure. The disorder can also be confirmed by the Raman spectra as the G' band disappears from the PBI-MWCNT membrane sample. The PBI-MWCNT membrane maintains a Young's Modulus of 4.6 GPa and a tensile strength of about 106 MPa which reduce on PA doping to values much higher than the corresponding PBI membrane. The maximum proton conductivity of doped film PBI-MWCNT is 7.4×10^{-2} S cm^{-1} while the corresponding value for PBI membrane is 6.3×10^{-2} S cm^{-1} at 180°C under anhydrous conditions. In fuel cell set up, the highest open circuit voltages are 0.8 and 0.96 V for PBI/PA and PBI-MWCNT/PA respectively at 180°C.

Nanoclay (Cloisite-30B)-Sulfonated Polybenzimidazole Membrane

Organoclay is used for increasing mechanical and thermal properties of membranes. In addition, it can be used in nanocomposite proton conducting membrane preparation. Reactive organoclay, Cloisite-30B, having surfactant containing hydroxyl groups can act as a pseudo cross-linker. Cloisite-30B organoclay, sufonated PBI and 1,4-bis (hydroxymethyl) benzene as crosslinker are dissolved in DMAC and sonicated [77]. The homogenous dispersion is cast on glass plate. The exfoliation of clay platelets in the cross-linked SPBI is confirmed by the absence of characteristic peak. All the films exhibit elongation at break of 7–11% and the tensile strength reaching a maximum of 271 MPa. Due to the presence of organoclay loading there is decrease in the water uptake for sulfonated PBI/clay (1%) nanocomposites. The SPBI/Closite-30b (1%) exhibits lower proton conductivity compared to pure SPBI, while 5% clay loading exhibits slightly higher proton conductivity.

Sulfonated Poly(Arylene Ether Sulfone)/ ABPBI-Grafted GO Nanocomposite Membrane

The base-functionalized graphene oxides improve the proton conductivity of PEM because of acid-base interactions between the sulfonic acid groups of the polymer matrix and the basic groups of functionalized graphene oxide. Additional pathways for proton conductions are created due to this interaction [78]. Basic polymer-grafted graphene oxide is synthesized by reacting graphene oxide with 3,4-diaminobenzoic acid and used as filler to improve the proton conductivity, dimensional stability, and mechanical strength of PEM [79]. GO is dispersed in DMAc by sonication, and DABA, PPA are added into the mixture and finally refluxed at 160°C to get ABPBI-GO. Agglomerated domains are observed in sulfonated poly(arylene ether sulfone) (SPAES)/ABPBI-GO membranes. The thermal stability, mechanical strength and toughness increase with lower ABPBI-GO loading. The SPAES/ABPBI-GO 1.0 (1% ABPBI-GO) membrane exhibits higher proton conductivity of 0.152 Scm^{-1} at 80°C and 90% RH as compared to both lower and higher ABPBI-GO loaded composites.

Polybenzimidazole/Mxene Composite Membranes

MAX phase is a group of layered ceramics (ternary carbide and nitride) with metallic properties having chemical formula as $M_{n+1}AX_n$ (n=1, 2 or 3, M is a transition metal, A is a group IIIA or IVA element, and X is carbon or nitrogen atom). Excellent combination of metallic conductivity and hydrophilic surfaces are observed in the material. 2D nanomaterial like Mxene can also be used for enhancing the membrane performance in intermediate temperature (>100°C) polymer electrolyte membrane fuel cells (ITPEMFCs). Ti_3C_2Tx-MXene is incorporated into polybenzimidazole (PBI)-based membranes by using a solution blending technique [80]. More hydrophilic interfacial domains are produced by the presence of $Ti_3C_2T_x$ that provide more space for water and acid storage. The elongation at break of PA doped PBI with 3 wt% $Ti_3C_2T_x$ remains unaffected at 150°C, and the tensile strength and Young's modulus increase by ~150% and ~160%, respectively. Unlike pure PBI membrane, the $Ti_3C_2T_x$ –PBI membranes exhibits continuous increase in proton conductivity when temperature rises beyond 150°C reaching a value of 0.014 Scm^{-1}. For single fuel cell test, the maximum power densities of PBI, 3 wt% $Ti_3C_2T_x$ -PBI and 5 wt% $Ti_3C_2T_x$ -PBI -based fuel cell are 156$mWcm^{-2}$, 169$mWcm^{-2}$ and 209$mWcm^{-2}$ respectively at 150°C.

Sulfonated ABPBI/MMT/Ionic Liquids Composite Membranes

Ionic liquids doped nanoclay/sulfonated ABPBI composite membranes are prepared for use as fuel cell membrane [81]. Good dispersion and exfoliation of 3wt% cloisite Na+ is observed by incorporating it in ABPBI membrane. Simultaneously, the ABPBI crystalline peak disappears due to the hindrance of nanoclay towards crystallization. Thus, 3% clay loading exhibits maximum mechanical properties due to highest exfoliation. Additionally, the hydroxyl groups on cloisite Na^+ interact with ABPBI polymer chains via hydrogen bonding with the secondary amine groups of ABPBI that improves the compatibility between the fillers and matrix, leading to enhanced mechanical properties of the composites. The doping level reaches a maximum value of 1.4 at the nanoclay loading of 3wt% and also exhibits highest conductivity of 4.0×10^{-2} Scm^{-1}.

ABPBI/Sulfonated Sepiolite Nanocomposite Membranes

High proton conductivities are possible by additional proton transfer through the absorbed or bound water molecules in clays via vehicle and/or Grotthuss mechanisms. ABPBI/sulfonated sepiolite (S-Sep) composite membranes can be made with enhanced mechanical, thermal, and oxidative stabilities by constructing ABPBI chains arranged along the sepiolite (Sep) fibers and acid-base crosslinks formed between S-Sep fibrous particles and ABPBI chains [82]. A proton conduction multipath is generated. Sep is a type of porous fibrous clay mineral having unique physical structure including a high specific surface area, micro-porous channels (0.35-1.06 nm) that run parallel to its length and excellent water absorbability and maintenance. Initially, the enlarged internal and external surface area are derived by precisely controlled acid activation treatment, which leads to defibration and magnesium removal. Next, the nano-sized Sep particle obtained is modified by a silane coupling agent that is followed by sulfonation to get S-Sep. The S-Sep sample retains the crystalline structure after the acid activation and surface modification treatments. Finally, ABPBI/S-Sep composite membrane is obtained via solution casting method followed by PA doping. Typical brittle fracture morphologies in SEM micrograph indicate good compatibility between the S-Sep particles and the ABPBI polymer. ABPBI/2% S-Sep sample exhibits the highest value of HDT and the best mechanical properties between 25-300°C. The E' at 300°C of the ABPBI/2% S-Sep sample is 5.4 GPa, which is very close to that of ABPBI at 25°C. The ABPBI/2% S-Sep-1.92PA exhibits maximum proton conductivity of 0.075 Scm^{-1}. Fuel cell study with a single cell exhibits that the OCV values of all ABPBI/2S-Sep composite membrane at 80, 120, and 180°C are more than 0.90 V and higher than that of pure ABPBI membrane, which indicates pore free membrane with minimum crossover. The ABPBI/2% S-Sep composite membrane exhibits improved performance with peak power density of 0.13, 0.18, and 0.23 Wcm^{-2} at 80, 120, and 180°C, respectively.

Nafion/PBI Nanofiber Composite Membranes

The crosslinked PBI nano-fiber thin film can be used as a support material to impregnate Nafion solution for preparing Nafion/ PBI nano-fibre (i.e., Nafion/PBI-f) composite membranes. The PBI crosslinked nano-fiber film is prepared by first making nanofiber via electro-spinning from a PBI solution in DMAc and then treating it with glutaraldehyde to crosslink [83]. The crosslinked PBI nano-fiber film is then used as a support material for impregnating in Nafion solutions to prepare Nafion/PBI-f composite membranes. The hybrid membrane exhibits increase in tensile strength and decrease in elongation. Water uptake and proton conductivity at 80°C and 95% RH are higher for Nafion/PBI-f membrane than pure nafion/PTFE composite membrane. Nafion/PBI-f2 membrane (67% Nafion & 22 μm thickness) exhibits similar unit cell PEMFC performance to Nafion/PTFE membrane.

Zirconium/Cerium Phosphate Benzimidazole Copolymer Nanocomposite Membrane

The nanocomposite membranes, viz., zirconium phosphate-fibrous cerium phosphate/polybenzimidazole, -/polybenzimidazole-co-polyaniline, -/polybenzimidazole-co-polypyrrole, and /polybenzimidazole-co-polyindole nanocomposite membranes are prepared and characterized for fuel cell application [84]. Self-supported sheet of zirconium phosphate- fibrous cerium phosphate nanocomposite membrane, $[Zr(HPO_4)_2]_{0.25}$ $[Ce(HPO_4)_2]_{0.75}$ $3.87H_2O$ is used as polymerizing agent. $[Zr(HPO_4)_2]_{0.25}$ $[Ce(HPO_4)_2]_{0.75}.3.87H_2O$ self-supported sheet immersed in ethanol solution of benzimidazole changes to beige colour indicating formation of PBI and the incorporation is 5.7 wt%. Similarly, when the nanocomposite membrane is dipped in ethanol solution of a mixture of benzimidazole, aniline, pyrrole and indole, respectively, the colour of the sheets gradually changes to green, green, and brown, respectively leading to weight gain of membrane from 7 to17 wt%. The in-situ chemical oxidation of the benzimidazole, and its co-monomers is triggered by the reduction of Ce(IV) ions present in the inorganic matrix of nanocomposite membrane.

Zirconium Tricarboxybutylphosphonate/PBI Nanocomposite Membrane

To enhance the proton conductivity of PBI a highly proton conductive inorganic nanomaterial can be combined with it. The zirconium tricarboxybutylphosphonate (Zr(PBTC)) powder can be dispersed in a PBI solution in DMAc and cast on glass plate to get up to 50wt% loaded membrane (Zr(PBTC)P50) [85]. The Zr(PBTC)P50 membrane is further phosphorated or sulfonated by immersing into H_3PO_4 or H_2SO_4 solutions to get Zr(PBTC)P50/H_3PO_4 or Zr(PBTC) P50-S60, respectively. The diffraction patterns of Zr(PBTC)P50/ H_3PO_4 and Zr(PBTC)P50-S60 membrane show an increase in the reflection characteristics of the crystalline region at 2θ =24°. This confirms the formation of inter-chain and intra-chain hydrogen bonding originated from the introduction of $H_2PO_4^-$/HSO_3^- in the polymetric chains of PBI. The Zr(PBTC)P50 membrane shows a proton conductivity of 3.82 x 10^{-3} S cm^{-1} at 200°C, while that of Zr(PBTC)P50/H_3PO_4 and Zr(PBTC)P50-S60 are higher at 5.24 x 10^{-3} and 8.13 x 10^{-3} S cm^{-1}, respectively.

Porous Polybenzimidazole Membrane

The polymer electrolyte membranes are required to have good chemical and electrochemical stability, mechanical strength, high electrolyte transport, high proton ionic conductivity and so on. The introduction of porosity into the PBI membrane is expected to increase its ionic conductivity.

The preparation of porous membrane is done by first casting PBI solution containing porogen onto glass plates at room temperature by using a doctor-blade to get plasticized PBI films [86]. After drying the films at 110°C for 2 h, the porogen-plasticizer is extracted by immersing the membranes in methanol. Among different phthalates such as dimethyl, diethyl, dibutyl, and diphenyl, as well as triphenyl phosphate used as porogen, the dibutyl phthalate performs best. The film is dried again and doped with phosphoric acid over long duration. The porogens containing aliphatic tails, such as dibutyl and dimethyl phthalate, contain more interconnected and bigger pore size than the aromatic tails, such as diphenyl phthalate and triphenyl phosphate. The acid uptake for a porous membrane (70% dibutyl phthalate incorporated as porogen) increases up to 439% from the value of 132% for pure PBI. There is increase of conductivity from 1.5×10^{-3} Scm^{-1} for the bulk polymer to a maximum value of 5×10^{-2} Scm^{-1} for a doped porous membrane with an initial porogen content of 70 wt %. Also, there is a very small decrease in conductivity (<10%) of the membranes even after one year of storage.

Glucose and saccharose are also used as porogen to prepare porous PBI membranes having high proton conductivity that after acid doping at room temperature contain 15.7 mol H_3PO_4 per PBI repeat unit [87]. Further, the performance and stability study of the porous PBI membrane in high-temperature proton-exchange-membrane fuel cell indicates stability for 200 h under a constant current discharge of 0.5 A cm^{-2} except for the last fifty hours.

Track-Etching Process Derived Porous Membrane

In the track-etching process an irradiation of a thin polymeric film is done by energetic heavy ions to create linear damage tracks, which are then chemically etched to convert in to pores. For example, a nanoporous PBI membranes is prepared by irradiating a 40 μm thick dense PBI film with 420MeV Xe ions under vacuum [88]. For PBI films, track sensitization can be performed before etching with blue visible range irradiation for controlled pore realization. The etching of latent tracks in PBI can be successfully done by using NaClO aqueous solution (0.3-2.0 M) at pH 9.5 and temperature 65°C for more than 40 min depending on the required final pore size. TE PBI membranes with end pores diameters 15-50 nm and maximum overall membrane porosity of 10% are obtained by applying the process. The TE-PBI samples are then doped with phosphoric acid by immersing for long period.

The radiochemical PBI cross-linking in the surroundings of the tracks generates additional H-bonding interactions among neighboring chains due to smaller inter-chain distances. The conduction mechanism for TE membranes involves a new dynamic hydrogen bonding network of shorter benzimidazole fragments, cross-linked nanodomains and free amphoteric phosphoric acid molecules settled on the pore walls. The PBI membrane with 50 nm end pore diameter and $5x10^9$ cm^{-2} pore density is considered adequate for high temperature PEM applications as it exhibits conductivity values above $0.015Scm^{-1}$ at150°C after 250 h operation time.

Crosslinked Porous Polybenzimidazole Membrane

P-Xylene Dichloride Based Crosslinker

The porosity, coupled with cross-linking approaches, can be followed to eliminate the drawbacks associated with the non-crosslinked porous membranes. To the PBI solution in DMAc, DBP and p-xylene dichloride crosslinker are added and is followed by casting on glass plate. p-xylene dichloride reacts with the N–H functional groups in the imidazole ring of PBI [89]. The film is heated at 120°C for 2 h and DBP removed by methanol leaching. The tensile strength of the film is higher than that of porous PBI film and also pristine PBI film. The cross-linked PBI membranes restrict the plasticizing effect of phosphoric acid to maintain the mechanical properties of the acid-doped cross-linked porous PBI membranes. The crosslinked porous PBI membranes have good radical oxidative stability. As compared to the pristine PBI membrane, the proton conductivity of the porous crosslinked PBI membranes increases 20 times when the acid uptake increases by 2.9 times. With highest degree of porosity, the proton conductivity reaches $1.5x\ 10^{-2}$ S cm^{-1} at 80°C and the corresponding values is $3x\ 10^{-2}$ S cm^{-1} at 160°C under the condition of without humidification.

Epoxy Resin Crosslinker Based PBI Membrane

The imidazole ring and the epoxy group reacts at a comparatively low temperature, which results in a cross-linked network among the PBI backbones. Varied amount of cross-linker and DBP are added to PBI solution and continuously stirred at room temperature for several hours [90]. The solution is poured on glass plate and cured at progressively increasing temperatures. Unlike pristine porous membranes, the cross-linked structure not only maintains the pore size, but also provides sufficient strength for achieving higher acid doping levels and proton conductivity. The tensile strength of the crosslinked membrane decreases with increasing temperature. For example, for membrane with 5% crosslinker the tensile strength decreases at 130°C to about one third (3.90 MPa) of that at room temperature under dry air with ambient humidity. On the other hand, the elongation at break increases at 130°C. The oxidative stability increases with crosslinking. The proton conductivity for 50% porogen sample is 0.048 S cm^{-1} at 200°C under anhydrous conditions, which is 4 times that of pristine PBI. By introducing 10% crosslinker to this sample, a high proton conductivity of 0.046 S cm^{-1} is exhibited by membrane under identical condition.

A porous PBI membrane can also be fabricated by casting a solution of PBI and DBP in dimethyl sulfoxide and then the membrane is dried and porogen removed [91]. The dried film is then doped with 6M KOH solution. The in-plane ionic conductivity of KOH-doped 70 wt.% porous PBI at room temperature increases from 1.5 $x10^{-2}$ S cm^{-1} of bare PBI to 4.9 x 10^{-2} S cm^{-1}, while at 65°C the value increases to 1 $x10^{-1}$ S cm^{-1}. The KOH-doped porous PBIs exhibit excellent thermal stability and satisfactory mechanical properties and electrochemical stability.

Ionic Liquid Doped Porous Membrane

The ionic liquid, 1-H-3-methylimidazolium bis(trifluoro-methanesulfonyl) imide ([h-mim] Ntf2) is first melted and porous PBI support is submerged into it for 1 h [92]. By applying vacuum the air bubbles are removed for IL to enter the pores. The proton conductivity of IL filled PBI membranes is about 250-450 times higher than the values obtained for the unfilled, porous PBI support ($3x10^{-6}$ Scm^{-1} at 134°C and $7x10^{-6}$ S cm^{-1} at 190°C). Further, the PBI/IL membrane exhibits substantial increase in proton conductivity with increasing temperature and reaches a value of $1.86x10^{-3}$ S cm^{-1} at 190°C. The PBI/IL membrane outperforms Nafion ® 117 in terms of fuel cell performance at temperatures above 95°C. At 150°C, a power density of 0.039 Wcm^{-2} is observed.

Miscellaneous Membranes

Double Layer Membrane

Although the porous membranes used as PEMs in fuel cells can uptake more phosphoric acid, it suffers from the fuel crossover problem. Also, for long term fuel cell operation, the acid leakage problem exists. Possible connective pore structure in high porosity membrane may aggravate the problem. Asymmetric membranes are considered to have the potential to tackle both the problems. The asymmetric membrane comprises a dense layer and a porous layer; the former acting as a barrier for fuel penetration. The leakage is minimized if the dense layer faces the cathode, where the byproduct water is produced. To a 3.5 wt% PBI solution certain amount of ionic liquid 1-ethyl-3-methylimidazolium bis(trifluoromethyl-sulfonyl) imide ([EMIM][TFSI]) and xylene dichloride crosslinker are added, stirred to obtain homogeneous solutions and the membranes are cast onto glass plates in an oven at 80°C for 12 h [93]. The resultant film is immersed in methanol for leaching out [EMIM] [TFSI]. The proton conductivity at 160°C under anhydrous conditions reaches a value of 6.57 x 10^{-2} S cm^{-1} for Asym PBI50 (Ionic liquid: 50 wt%) and 6.26 x 10^{-2} S cm^{-1} for Asym PBI67 (Ionic liquid: 50 wt%), which are about 3 times higher than that for acid-doped dense PBI.

Hybrid and Porous Membrane

Generally, the inorganic–organic hybrid membranes are promising systems, since the synergic combination of both organic and hydrophilic inorganic components produce special physicochemical properties. Higher water uptake capabilities are normally expected in a composite membrane as compared to acid–base membranes, which is highly beneficial to tackle the limitation of phosphoric acid auto-dehydration for operation above 140°C. The undesirable effect of higher rigidity of membrane due to presence of inorganic component may lead to formation of voids. Therefore, sufficient surface characteristics and a low particle size are two of the main requirements for the inorganic loading for designing electrolyte membranes with suitable transport selectivity and mechanical properties. In one of the efforts in this direction, the hybrid membranes based on polybenzimidazole (PBI) and ETS-10 titanosilicate type materials functionalized with sulfonic groups is prepared for high temperature PEMFC applications. The synthesis of ETS-10 zeotype crystals is carried out by sequential addition of sodium silicate, KF and TiO_2 (anatase) to an aqueous solution of KCl and NaCl in double-distilled water [94]. The homogenized gel is poured into Teflon lined autoclaves and crystallization takes place under hydrothermal conditions at 230°C during 24 h. The ETS-10 obtained is sulphonated by first introducing a mercapto group by using 3-(mercaptopropyl) trimethoxysilane. The mercapto group is then converted to a sulfonic acid group by treating with H_2O_2 solution. By using sulfonated ETS-10 and PBI solution mixture, the composite membranes are made by following the conventional techniques. The relatively high proton conductivities e.g., 2.2 folds for SO_3H-ETS-10 (3 wt.%) at 170°C is due to the presence of the sulfonic acid groups grafted onto ETS-10 crystals, which contribute to a most effective acid-doping of PBI polymer providing a new proton conduction pathways. The non-selective defects are formed at the organic-inorganic interface and also pronounced membrane swelling takes place in this porous systems.

Sandwiched-Porous Anion Exchange Membrane

Alkaline doped PBI membrane with its high hydroxide ion conductivity and excellent chemical stability in alkaline media, has found application as an anion exchange membrane (AEM) in alkaline direct alcohol fuel cells [95]. Long-term operation of alkaline doped PBI membranes deteriorates due to progressive release of the doped alkali. By making composite with the alkaline-stable inorganic additives the alkali retention ability can be improved. However, this is not practised due to incompatibility issues. The issues can be addressed by fabricating a sandwiched-porous PBI (sp-PBI) membrane and designing a new catalyst structure using PBI-decorated reduced graphene oxide (rGO) as the supporting material [96]. A dense PBI thin film is fabricated on either side of porous PBI to ensure high mobility of electrolyte in the interconnected pores as well as to reduce hydrogen permeability and improved mechanical properties. When the porosity reaches 60%, the alkali uptake of PBI (35%) becomes 1.5-times higher than that of dense PBI. Also, the alkali uptake of sp-PBI is much higher than that of dense PBI. Following the trend, the ionic conductivity of p-PBI with 60% porosity reaches 0.142 S cm^{-1} at 90°C as compared to 0.096 Scm^{-1} for dense PBI. By using the as-prepared Pd nanowires (NWs)/PBI/rGO catalysts in the catalyst layer, a triple phase boundary is established. The PBI-based membrane electrode assembly (MEA) yield a peak power density of 544 mW cm^{-2} at 90°C in a H_2/O_2 AEM fuel cell.

Hyperbranched Poly(Benzimidazole-Co-Benzene) Membrane

A hyperbranched honeycomb structure can be introduced in PBI membrane to improve its mechanical strength. Hyperbranched PBI-type polymer, poly(benzimidazole-co-benzene (PBIB) with a honeycomb structure, is prepared by the condensation polymerization of 3,3'-diaminobenzidine (DAB) and trimesic acid (TMA) in mole ratio of 3:2 at around 275°C for long duration [97]. Membrane is prepared by placing a mixture of PBIB and H_2SO_4 between two teflon sheets and heating the entire assembly to 230°C under pressure in a carver hot press. The whole FE-TEM image appears like a net having highly ordered and uniform six-membered units, which confirms the presence of hyperbranched honeycomb structure. The thermal decomposition of the PBIB powder starts at 250°C. The mechanical properties of both undoped and doped PBIB films are quite good, having tensile strength of 32-35 MPa. The proton conductivity of the doped PBIB membrane is much higher than that of PBI and sulfonated PBI membranes due to the highly ordered and compact honeycomb structure of the PBIB. This facilitates hoping of protons as the amine and imine groups are expected to be closer as compared to that of a linear PBI chain. In a fuel cell set up, when the temperature is increased from 70 to 100°C, the voltage and power output increase and with further increase of temperature to 160°C, the performance deteriorates. The power output from the doped PBIB membrane at different temperatures is higher under the humid conditions than that under the dry conditions.

CHARACTERIZATION OF SOLID POLYMER ELECTROLYTE

Membrane Preparation and Acidification

The PBI/ABPBI/modified PBI etc. membranes are prepared from their 5-10% solutions in solvents like DMAc. The solutions are heated to high temperature under inert atmosphere for fast dissolution of polymer. The clear solutions obtained are filtered and spread on glass plates using a doctor's blade, and the solvent is evaporated at high temperature for extended period, normally ranging from 12 to 48h. After cooling to room temperature, the glass plate with polymer film is immersed in deionized water bath and the membranes are slowly removed from the glass plates. For making polymer/polymer blends and composites, the same procedure is followed. For making porous films porogens are added to polymer solution and cast. Later the porogens are leached out by dipping in a leachant solvent like methanol.

Phosphoric Acid Doping

The dried polymer film is immersed in excess of phosphoric acid of specific strength at specific temperature, ca. 80°C for extended period of 4-24 h. The PA-doped membrane is then taken out of the PA solution and then excess acid is removed by soaking with filter paper. The membrane is then dried at high temperature approx. 60-80°C under vacuum for several days and weighed to find the weight gain for calculating extent of PA doping.

The weight difference, $(W_2 - W_1)$ is the weight of the absorbed PA. The PA content of the membrane is then calculated as the weight percent (wt %) of PA absorbed in the membrane using the equation,

$$PA\ content\ (wt\%) = \frac{W2-W1}{W2} \times 100 \qquad (1)$$

Proton Conductivity

The proton conductivity is obtained by using a four-platinum wire cell. The general set condition is the galvanostat mode with a frequency range of 1 Hz - 1 MHz. The membrane is generally cut into 1×4 cm^2 pieces and put into a temperature controlled chamber (30-170°C) under N_2 atmosphere to achieve an anhydrous condition. Ion conductivity is calculated by the following equation:

$$\sigma = L/ AR$$

L is the distance between the electrodes which is fixed as 1 cm. R is the resistance and A is the cross-sectional area of the membrane.

Through plane ionic conductivities of the membranes are also determined by means of complex impedance analysis in the temperature range of interest, ca. 25-200°C. Samples are sandwiched between two stainless steel disks, separated by a Teflon holder. The pressure is kept constant and defined by the maximum limits of screws used to close the probe arrangement. Films are then painted on both sides with electrically conductivity silver paint. Thus, a perfect contact between the film and the steel electrodes is established, which helps in avoiding impedance noise in the Nyquist plot due to micropore defects at the polymeric film surface.

Oxidative Stability

One of the most important requirements of PEMFC is its stability under fuel cell condition. During fuel cell operation, trace amount of hydrogen peroxide is also generated, which can destroy even a highly performing membrane. Therefore, the oxidative stability determination is a must for any membrane developed for fuel cell application. The oxidative stabilities are investigated by immersing the membranes in Fenton's reagent (3% H_2O_2 solution containing 4 ppm Fe^{2+}) at 70°C. After every 24 h, the samples are collected by filtering and repeated cleaning with deionized water. The sample is then dried at 120°C for more than 20 h in a vacuum oven. The degradation of the membranes is calculated by the weight loss and visual observation.

Thermal Stability

Thermal stability is determined by using thermogravimeytric analyzer (TGA). Normally, two environmental conditions are used during heating; the first is the nitrogen atmosphere and the second one is the air. The former gives information on thermal stability while the latter gives thermo-oxidative stability. The standard heating rate is used.

Mechanical Properties

Mechanical properties are commonly determined by using universal testing machine. Important parameters, such as tensile modulus, tensile strength, compressive modulus, compressive strength, flexural strength and elongation at break are measured for characterizing the PEMFC membrane.

Current-Voltage Characteristics

The polarization curve, which represents the current-voltage characterizes of the membrane electrode assembly, gives information on the performance of fuel cell. A typical polarization curve is presented in Figure 9. The shape of the polarization curve originates from four over potentials or irreversibility.

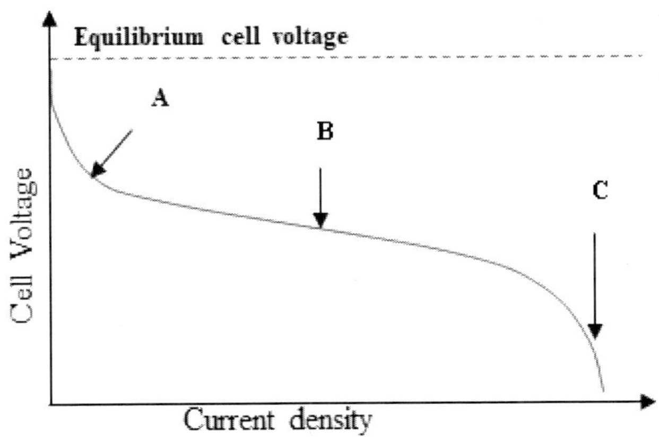

Figure 9. Representative polarization curve for fuel cell: (A) Activation losses (oxygen reduction reaction); (B) Ohmic losses; (C) Mass transport losses (Diffusion and concentration of O_2).

References

[1] Authayanun, S., Im-Orb, K., Arpornwichanop, A. 2015. "A Review of the Development of High Temperature Proton Exchange Membrane Fuel Cells." *Chinese J. Catal.* 36: 473-83.

[2] Costamagna, P., Srinivasan, S. 2001. "Quantum jumps in the PEMFC science and technology from the 1960s to the year 2000: Part II. Engineering, technology development and application aspects." *J. Pow. Sources* 102: 253–69.

[3] Appleby, A. J., Foulkes, F. R. 1989. *Fuel Cell Handbook*, Van Nostrand Reinhold, New York.

[4] Hogarth, M., Glipa, X. 2001. "*High Temperature Membranes for Solid Polymer Fuel Cells,* John Matthey Technology Centre Report" (ETSU F/02/00189/REP).

[5] Wainright, J. S., Litt, M. H., Savinell, R. F. 2003. "High-Temperature Membranes." *In Handbook of Fuel Cell, Fundamentals, Technology, and Applications*. edited by W. Vielstich, A. Lamm and H. A. Gasteiger. vol. 3, p. 436. New Jersey, John Wiley & Sons.

[6] Li, Q., He, E., Jensen, J. O., Bjerrum, N. J. 2003. "Approaches and Recent Development of Polymer Electrolyte Membranes for Fuel Cells Operating above 100°C." *Chem. Mater*. 15: 4896–4915.

[7] Liu, Y. H., Nguyen, T. H., Kristian, N., Yu, Y. L., Wang, X. 2009. "Reinforced and self-humidifying composite membrane for fuel cell applications." *J. Membr. Sci.,* 330: 357–62.

[8] Pan, C., He, R., Li, Q., Jensen, J. O., Bjerrum, N. J., Hjulmand H. A., Jensen, A. B. 2005. "Integration of high temperature PEM fuel cells with a methanol reformer." *J. Pow. Sources* 145: 392–98.

[9] Wahome, J. A., 2008. In *Abstracts*, 64th Southwest Regional Meeting of the American Chemical Society, Little Rock, AR, United States, October 1–4.

[10] Endoh, E. 2006. "Proton Exchange Membrane Fuel Cells 6." *ECS Trans*. 3 (1): 9–18.

[11] Kerres, J. A. 2001. "Development of ionomer membranes for fuel cells." *J. Membr. Sci.* 185: 3–27.

[12] Rikukawa, M., Sanui, K. 2000. "Proton-conducting polymer electrolyte membranes based on hydrocarbon polymers." *Prog. Polym. Sci.* 25: 1463–1502.

[13] Savadogo, O. 1998. "Emerging membranes for electrochemical systems: (I) solid polymer electrolyte membranes for fuel cell systems." *J. New Mater. Electrochem. Systems* 1: 47–66.

[14] Jones, D. J., Roziere, J. 2001. "Recent advances in the functionalisation of polybenzimidazole and polyetherketone for fuel cell applications." *J. Membr. Sci., 185: 41–58.*

[15] Roziere, J., Jones, D. J. 2003. "Non-Fluorinated Polymer Materials for Proton Exchange Membrane Fuel Cells." *Annu. Rev. Mater. Res*. 2003, 33, 503–55.

[16] Li, Q., Jensen, J. O., Savinell, R. F., Bjerrum, N. J. 2009. "High temperature proton exchange membranes based on polybenzimidazoles for fuel cells." *Prog. Polym. Sci.* 34: 449–47.

[17] Wainright JS, Wang JT, Weng D, Savinell RF, Litt M. 1995. "Acid-doped polybenzimidazoles —a new polymer electrolyte." *J. Electrochem. Soc* 142:L121–L123.

[18] Li, Q., Hjuler, H. A., Bjerrum, N. J. 2001. "Phosphoric acid doped polybenzimidazole membranes: Physiochemical characterization and fuel cell applications." *J. Appl. Electrochem.*, 31: 773–79.

[19] Membranes and electrodes for fuel cells as PEMEAs merges with E-TEK. *Fuel cell bulletin* 2006, no. 2. p. 12–15.

[20] Vogel, H., Marvel, C. S. 1961 "Polybenzimidazoles, new thermally stable polymers." *J. Polym. Sci,* 50: 511–539.

[21] Cassidy, P. E. 1980. *Thermally Stable Polymers*, ch. 6, pp. 141–221, NY, Marcel Dekker, Inc.

[22] Asensio, J. A., Sánchez, E. M., Gómez-Romero, P. 2010. "Proton-conducting membranes based on benzimidazole polymers for high-temperature PEM fuel cells. A chemical quest." *Chem. Soc. Rev.* 39: 3210–39.

[23] Zhang, H., Shen, P. K. 2012. "Recent development of polymer electrolyte membranes for fuel cells." *Chem. Rev.* 112: 2780–2832.

[24] Lu, W., Zhang, G., Li, J., Hao, J., Wei, F., Li, W., Zhang, J., Shao, Z. G., Yi, B. 2015. "Polybenzimidazole-crosslinked poly(vinylbenzylchloride) with quaternary1,4-diazabicyclo 2.2.2) octanegroupsashigh-performanceanion exchange membrane for fuel cells." *J. Pow. Sources* 296: 204–14.

[25] Yuan, S., Guo, X., Aili, D., Pan, C., Li, Q., Fang, J. 2014. "Poly(imide benzimidazole)s for high temperature polymer electrolyte membrane fuel cells." *J. Membr. Sci.* 454, 351–58.

[26] Herranz, D., Ocón, P. 2019. "Benzimidazole as Solid Electrolyte Material for Fuel Cells." *Intech Open*. doi: http://dx.doi.org/10.5772/intechopen.85430.

[27] Modestov, A. D., Tarasevich, M. R., Leykin, A. Y., Filimonov, V. Y. 2009. "MEA for alkaline direct ethanol fuel cell with alkali doped PBI membrane and non-platinum electrodes." *J. Pow. Sources* 188: 502–06.

[28] Xiao, L., Zhang, H., Jang, T., Scanlon, E., Chen, R., Choe, E. W., Ramanathan, L. S., Yu, S., Benicewicz, B. C. 2005. "Synthesis and characterization of pyridine-based polybenzimidazoles for high temperature electrolyte membrane fuel cell applications." *Fuel Cells* 5: 287–95.

[29] Molleo, M., Schmidt, T., Benicewicz, B. (2012). "Polybenzimidazole fuel cell technology: theory, performance, and applications." In *Encyclopedia of sustainability, science and technology,* edited by Meyers, R. A., 391–431, New York, Springer.

[30] Kim, T. H., Lim, T. W., Lee, J. C. 2007. "High-temperature fuel cell membranes based on mechanically stable para-ordered polybenzimidazole prepared by direct casting." *J. Pow. Sources* 172: 172–79.

[31] Maity, S., Jana, T. 2013. "Soluble Polybenzimidazoles for PEM: Synthesized from Efficient, Inexpensive, Readily Accessible Alternative Tetraamine Monomer." *Macromolecules* 46: 6814-823.

[32] Berber, M. R., Fujigaya, T., Sasaki, K., Nakashima, N. 2013. "Remarkably Durable High Temperature Polymer Electrolyte Fuel Cell Based on Poly(vinylphosphonic acid)-doped Polybenzimidazole." *Sci. Rep.* 3:1764.

[33] Mamlouk, M., Ocon, P., Scott, K. 2014. "Preparation and characterization of polybenzimidzaole/diethylamine hydrogen sulphate for medium temperature proton exchange membrane fuel cells." *J. Pow. Sources* 245: 915-26.

[34] Lee, M., Khan, S. B., Akhtar, K., Han, H., Seo, J. 2013. "Fuel cell: Synthesis and Properties of Polyimide for PEMFC at High Temperature." *Int. J. Electrochem. Sci.* 8: 4225 – 33.

[35] Chuang, S. W., Hsu, S. L. C. 2006. "Synthesis and Properties of a New Fluorine-Containing Polybenzimidazole for High-Temperature Fuel-Cell Applications." *J. Polym. Sci.: Part A: Polym. Chem.* 44: 4508–13.

[36] Yu, S., Benicewicz, B. C. 2009. "Synthesis and Properties of Functionalized Polybenzimidazoles for High-Temperature PEMFCs." *Macromolecules* 42: 8640–48.

[37] Yang, J., Xu, Y., Zhou, L., Che, Q., He, R., Li, Q. 2013. "Hydroxyl pyridine containing polybenzimidazole membranes for proton exchange membrane fuel cells." *J. Membr. Sci.* 446: 318–25.

[38] Li, J., Li, X., Zhao, Y., Lu, W., Shao, Z., Yi, B. 2012. "High-Temperature Proton-Exchange-Membrane Fuel Cells Using an Ether-Containing Polybenzimidazole Membrane as Electrolyte." *Chem Sus Chem* 5: 896 – 900.

[39] Dominguez, P. H., Grygiel, K., Weber, J. 2014. "Nanostructured poly(benzimidazole) membranes by N-alkylation." *eXPRESS Polym. Lett.* 8: 30–8.

[40] Chuang, S. W., Hsu, S. L. C., Yang, M. L. 2008. "Preparation and characterization of fluorine-containing polybenzimidazole/imidazole hybrid membranes for proton exchange membrane fuel cells." *Eur. Polym. J.* 44: 2202–06.

[41] Bae, J. 2002. "Properties of selected sulfonated polymers as proton-conducting electrolytes for polymer electrolyte fuel cells." *Solid State Ion.* 147: 189–94.

[42] Glipa, X., Haddad, M. E., Jones, D. J., Rozière, J. 1997. "Synthesis and characterisation of sulfonated polybenzimidazole: a highly conducting proton exchange polymer." *Solid State Ion.* 97: 323-31.

[43] Staiti, P., Lufrano, F., Aricò, A. S., Passalacqua, E., Antonucci, V. 2001. "Sulfonated polybenzimidazole membranes — preparation and physico-chemical characterization." *J. Membr. Sci.* 188: 71–8.

[44] Wang, K. Y., Xiao, Y. C., Chung, T. S. 2006. "Chemically modified polybenzimidazole nanofiltration membrane for the separation of electrolytes and cephalexin." *Chem. Eng. Sci.* 61: 5807-17.

[45] Li, Q., Pan, C., Jensen, J. O., Noye, P., Bjerrum, N. J. 2007. "Cross-Linked Polybenzimidazole Membranes for Fuel Cells." *Chem. Mater.* 19: 350-52.

[46] Yao, Z., Cui, M., Zhang, Z., Wu, L., Xu, T. 2017. "Silane Cross-Linked Sulfonted Poly(Ether Ketone/Ether Benzimidazole)s for Fuel Cell Applications." *Polymers* 9:631.

[47] Kim, S. K., Kim, T. H., Jung, J. W., Lee, J. C. 2008. "Copolymers of Poly(2,5-benzimidazole) and Poly[2,2'-(p-phenylene)-5,5'-bibenzimidazole] for High-Temperature Fuel Cell Applications." *Macromol. Mater. Eng.* 293: 914–21.

[48] Kim, T. H., Kim, S. K., Lim, T. W., Lee, J. C. 2008. "Synthesis and properties of poly(aryl ether benzimidazole) copolymers for high-temperature fuel cell membranes." *J. Membr. Sci.* 323: 362–70.

[49] Yang, J., He, R., Che, Q., Gao, X., Shi, L. 2010. "A copolymer of poly[2,2'-(m-phenylene)-5,5'-bibenzimidazole] and poly(2,5-benzimidazole) for high-temperature proton-conducting membranes." *Polym Int* 59: 1695–170.

[50] Lee, H. J., Lee, D. H., Henkensmeier, D., Jang, J. H., Cho, E. A., Kim, H. J., Kim, H. 2012. "Synthesis and Characterization of H3PO4 Doped Poly(benzimidazole-co-benzoxazole) Membranes for High Temperature Polymer Electrolyte Fuel Cells." *Bull. Korean Chem. Soc.* 33: 10 3279.

[51] Gerçel, H. F., Tosun, C. G., Akyalcm, L. 2016. "Polymer Electrolyte Membrane Fuel Cell Performance of a Sulfonated Poly(Arylene Ether Benzimidazole) Copolymer Membrane." *Adv. Mater. Sci. Eng.* 2016: 6123213, 8 pages.

[52] Kim, S. K., Kim, T. H., Ko, T., Lee, J. C. 2011. "Cross-linked poly(2,5-benzimidazole) consisting of wholly aromatic groups for high-temperature PEM fuel cell applications." *J. Membr. Sci.* 373: 80–8.

[53] Ma,Y.-L., Wainright, J.S., Litt, M.H., Savinell, R.F. 2004. "Conductivity of PBI membranes for high-temperature polymer electrolyte fuel cells." *J. Electrochem. Soc.* 151: A8–A16.

[54] Kim, S. K., Choi, S. W., Jeon, W. S., Park, J. O., Ko, T., Chang, H., Lee, J. C. 2012. "Cross-Linked Benzoxazine−Benzimidazole Copolymer Electrolyte Membranes for Fuel Cells at Elevated Temperature." *Macromolecules* 45: 1438−46.

[55] Elabd, Y. A., Hickner, M. A. 2011. "Block Copolymers for Fuel Cells." *Macromolecules* 44: 1−11.

[56] Maity, S., Jana, T. 2014. "Polybenzimidazole Block Copolymers for Fuel Cell: Synthesis and Studies of Block Length Effects on Nanophase Separation, Mechanical Properties, and Proton Conductivity of PEM." *ACS Appl. Mater. Interfaces* 6: 6851−64.

[57] Krishnan, N. N., Lee, H. J., Kim, H. J., Kim, J. Y., Hwang, I., Jang, J. H., Cho, E. A., Kim, S. K., Henkensmeier, D., Hong, S. A., Lim, T. H. 2010. "Sulfonated poly(ether sulfone)/sulfonated polybenzimidazole blend membrane for fuel cell applications." *Eur. Polym. J.* 46: 1633–41.

[58] Anis, A., Al-Zahrani, S. M. 2012. "Sulfonated PVA/ PBI Based Crosslinked Composites Towards Anhydrous Proton Conductive Polymer Electrolyte Membranes for Fuel Cells." *Int. J. Electrochem. Sci.* 7: 9174 – 85.

[59] Deluga, D. A., Kelley, S. C., Pivovar, B., Shores, D. A., Smyrl, W. H. 2000. "Composite membranes to reduce crossover in PEM fuel cells," Paper presented at the *Fifteenth Annual Battery Conference on Applications and Advances* (Cat. No.00TH8490), Long Beach, CA, USA, January 11-14. (pp. 51-53).

[60] Qingting, L., Na, N., Quan, S., Xiaoxue, W., Xujin, B., Zhang, F., Rong, Z., Shengfei, H., Feng, Z., Xiao, L. 2018. "Poly(2,5-benzimidazole)/TriSilanolPhenyl POSS Composite Membranes for Intermediate Temperature PEM Fuel Cells." *J. Wuhan Univ. Technol.-Mat. Sci. Edit.* 33: 212.

[61] Bao, X. J., Liu, Q. T. 2010. "Poly(2,5-benzimidazole) Based Polymer Electrolyte Membranes for High Temperature Fuel Cell Applications." In Proceeding; Chapter 9: Power Generation, Storage & Novel Energy Sources. pp. 434 – 437, Paper presented at Clean Technology Conference and Trade Show. *Clean Technology*. Anaheim, June 21-24.

[62] Liu, Q., Sun, Q., Ni, N., Luo, F., Zhang, R., Hu, S., Bao, X., Zhang, F., Zhao, F., Li, X. 2016. "Novel octopus shaped organic-inorganic composite membranes for PEMFCs." *Int. J. Hydrog. Energy*, 41:16160-166.

[63] Kim, S. K. 2016. "Polybenzimidazole and Phosphonic Acid Groups-Functionalized Polyhedral Oligomeric Silsesquioxane Composite Electrolyte for High Temperature Proton Exchange Membrane." *J. Nanomater.* 2016: 2954147, 7 pages.

[64] Silva, B. B. R., Soares, J. B., Malfatti, C. F., M. C., M. 2011. "Forte, Benzimidazole effect on the performance of polyelectrolyte membranes based on sulfonated hydrocarbon resin." *J. Membr. Sci.* 374: 12–19.

[65] Lin, H. L., Huang, J. R., Chen, Y. T., Su, P. H., Yu, T. L., Chan, S. H. 2012. "Polybenzimidazole/poly(tetrafluoro ethylene) composite membranes for high temperature proton exchange membrane fuel cells." *J Polym Res* 19:9875.

[66] Ya, K., Kumazawa, K., Kawamura, G., Muto, H., Matsuda, A. 2018. "Cell performance enhancement with titania-doped polybenzimidazole based composite membrane in intermediate temperature fuel cell under anhydrous condition.' *J. Ceramic Soc. Jpn.* 126: 789-93.

[67] Ioana-Maria, N., Aurora, J., Victoria, C., Cristian, B., Bucuresti, I. 2017. "Advanced Polymeric Materials Based on PBI/SiO_2 Composite with High-Performances designated for PEM-Fuel Cells." Presented at the *Electrical Vehicles Int. Conference & Show*, Bucuresti, Romania, October 5-6.

[68] Linlin, M., Mishra, A. K., Kim, N. H., Lee, J. H. 2012. "Poly(2,5-benzimidazole)–silica nanocomposite membranes for high temperature proton exchange membrane fuel cell." *J. Membr. Sci.* 411– 412: 91– 98.

[69] Gomez-Romeroa, P., Asensio, J. A., Borros, S. 2005. "Hybrid proton-conducting membranes for polymer electrolyte fuel cells Phosphomolybdic acid doped poly(2,5-benzimidazole)—(ABPBI-$H_3PMo_{12}O_{40}$)." *Electrochim. Acta* 50: 4715–20.

[70] Kim, T. H., Lim, T. W., Park, Y. S., Shin, K., Lee, J. C. 2007. "Proton-Conducting Zirconium Pyrophosphate/Poly(2,5-benzimidazole) Composite Membranes Prepared by a PPA Direct Casting Method." *Macromol. Chem. Phys.* 208: 2293–2302.

[71] Kondratenko, M. S., Ponomarev, I., Gallyamov, M. O., Razorenov, D. Y., Volkova, Y. A., Kharitonova, E. P., Khokhlov, A. R. 2013. "Novel composite Zr/PBI-O-PhT membranes for HT-PEFC applications." *Beilstein J. Nanotechnol.* 4: 481–92.

[72] Shabanikiaa, A., Javanbakht, M., Amoli, H. S., Hooshyari, K., Enhessari, M. 2015. "Polybenzimidazole/strontium cerate nanocomposites with enhanced proton conductivity for proton exchange membrane fuel cells operating at high temperature." *Electrochim. Acta.* 154: 370-78.

[73] Ghosh, P., Halder, D., Ganguly, S., Banerjee, D., Kargupta, K. 2014. "Phosphosilicate gel-polybenzimidazole nanocomposites novel membrane for fuel cell application." *Int J Plast Technol.* 18: 403-08.

[74] Zhang, Y., Zou, J., Wang, L., Chen, H., Si, J. 2014. "Preparation of Novel Polyaniline and phosphoric acid doped sulfonated polybenzimidazole composite membranes for Fuel Cells." *Appl. Mech. Mater.* 584-586: 1669-72.

[75] Kausar, A. 2016. "Proton exchange fuel cell membranes of poly (benzimidazoleamide)/sulfonated polystyrene/titania nanoparticles grafted-multi-walled carbon nanotubes." *J. Plastic Film & Sheeting* 31 (1): 27-44.

[76] Moreno, N. G., Gervasio, D., García, A. G., Francisco, J., Robles, P. 2015. "Polybenzimidazole-multiwall carbon nanotubes compositemembranes for polymer electrolyte membrane fuel cells." *J. Pow. Sources* 300: 229-37.

[77] Ahmadizadegan, H. 2017. "Effect of adding Nanoclay (Cloisite-30B) on the Proton Conductivity of Sulfonated Polybenzimidazole." *Nanochem Res* 2: 96-108.

[78] Cao, L., Sun, Q., Gao, Y., Liu, L., Shi, H. 2015. "Novel acid-base hybrid membrane based on amine-functionalized reduced graphene oxide and sulfonated polyimide for vanadium redox flow battery." *Electrochim. Acta.* 158: 24-34.

[79] Ko, T., Kim, K., Lim, M., Nam, S. Y., Kim, T., Kim, S., Lee, J. 2015. "Sulfonated poly(arylene ether sulfone) composite membranes having poly(2,5-benzimidazole)-grafted graphene oxide for fuel cell applications." *J. Mater. Chem. A,* 3: 20595-20606.

[80] Fei, M., Lin, R., Deng, Y., Xian, H., Bian, R., Zhang, X., Cheng, J., Xu, C., Cai, D. 2018. "Polybenzimidazole/Mxene composite membranes for intermediate temperature polymer electrolyte membrane fuel cells." *Nanotechnology* 29: 035403 (6pp).

[81] Bao, X., Zhang, F., Liu, Q. 2015. "Sulfonated poly(2,5-benzimidazole) (ABPBI)/ MMT/ ionic liquids composite membranes for high temperature PEM applications." *Int. J. Hydrog. Energy,* 40: 16767-774.

[82] Zhang, X., Liu, Q., Xia, L., Huang, D., Fu, X., Zhang, R., Hu, S., Zhao, F., Li, X., Bao, X. 2019. "Poly(2,5-benzimidazole)/Sulfonated Sepiolite Composite Membranes with Low Phosphoric Acid Doping Levels for PEMFC Applications in a Wide Temperature Range," *J. Membr. Sci.* 574:282-98.

[83] Yu, T. L. L., Liu, S. H., Lin, H. L., Su, P. H. 2010. "Nafion/PBI Nanofiber Composite Membranes for Fuel Cells Applications." Proc. ASME 2010. Presented at the 8th Int. *Fuel Cell Sci., Eng. Technol. Conference Fuel Cell* 2010. New York, USA. June 14-16.

[84] Shakshooki, S. K., Elakari, F. A., Shaabani, A. M. 2018. "Zirconium Phosphate/Fibrous Cerium Phosphate Nanocomposite Membrane Self Supported Benzimidazole, Its Co-Aniline, Co-Pyrrole and Co-Indole Polymerization Agent." *Acad. J. Chem.* 3: 88-100.

[85] Yamazaki, Y., Jang, M. Y., Taniyama, T. 2004. "Proton conductivity of zirconium tricarboxybutylphosphonate/PBI nanocomposite membrane." *Sci. Technol. Adv. Mater.* 5: 455–59.

[86] Mecerreyes, D., Grande, H., Miguel, O., Ochoteco, E., Marcilla, R., Cantero, I. 2004. "Porous Polybenzimidazole Membranes Doped with Phosphoric Acid: Highly Proton-Conducting Solid Electrolytes." *Chem. Mater.* 16: 604–07.

[87] Li, J., Li, X., Yu, S., Hao, J., Lu, W., Shao, Z., Yi, B. 2014. "Porous polybenzimidazole membranes doped with phosphoric acid: Preparation and application in high-temperature proton-exchange-membrane fuel cells." *Energ. Convers. Manage* 85: 323–27.

[88] Eguizábal, A., Sgroi, M., Pullini, D., Ferain, E., Pina, M. P. 2014. "Nanoporous PBI membranes by track etching for high temperature PEMs." *J. Membr. Sci.* 454: 243–52.

[89] Shen, C. H., Jheng, L. C., Hsu, S. L. C., Wang, J. T. W. 2011. "Phosphoric acid-doped cross-linked porous polybenzimidazole membranes for proton exchange membrane fuel cells." *J. Mater. Chem.* 21: 15660-65.

[90] Wang, S., Zhao, C., Ma, W., Zhang, G., Liu, Z., Ni, J., Li, M., Zhang, N., Na, H. 2012. "Preparation and properties of epoxy-cross-linked porous polybenzimidazole for high temperature proton exchange membrane fuel cells." *J. Membr. Sci.* 411–412: 54– 63.

[91] Zarrin, H., Jiang, G., Lam, G. Y. Y., Fowler, M., Chen, Z. 2014. "High performance porous polybenzimidazole membrane for alkaline fuel cells." *Int. J. Hydrog. Energy* 39: 118405-415.

[92] van de Ven, E., Chairuna, A., Merle, G., Benito, S. P., Borneman, Z., Nijmeijer, K. 2013. "Ionic liquid doped polybenzimidazole membranes for high temperature Proton Exchange Membrane fuel cell applications." *J. Pow. Sources* 222: 202-09.

[93] Jheng, L. C., Hsu, S. L. C., Tsai, T. Y., Chang, W. J. Y. 2014. "A novel asymmetric polybenzimidazole membrane for high temperature proton exchange membrane fuel cells." *J. Mater. Chem.* A. 2: 4225-233.

[94] Eguizábal, A., Lemus, J., Urbiztondo, M., Garrido, O., Soler, J., Blazquez, J. A., Pina, M. P. 2011. "Novel hybrid membranes based on polybenzimidazole and ETS-10 titanosilicate type material for high temperature proton exchange membrane fuel cells: A comprehensive study on dense and porous systems." *J. Pow. Sources* 196: 8994– 9007.

[95] Modestov, A. D., Tarasevich, M. R., Leykin, A. Y., Filimonov, V. Y. 2009. "MEA for alkaline direct ethanol fuel cell with alkali doped PBI membrane and non-platinum electrodes." *J. Pow. Sources* 188: 502–06.

[96] Zeng, L., Zhao, T. S., An, L., Zhao, G., Yan, X. H. 2015. "A high-performance sandwiched-porous polybenzimidazole membrane with enhanced alkaline retention for anion exchange membrane fuel cells." *Energy Environ. Sci.* 8: 2768-774.

[97] Bhadra, S., Kim, N. H., Choi, J. S., Rhee, K. Y., Lee, J. H. 2010. "Hyperbranched poly(benzimidazole-co-benzene) with honeycomb structure as a membrane for high-temperature proton-exchange membrane fuel cells." *J. Pow. Sources* 195:2470–477.

BIOGRAPHICAL SKETCH

Dr Asit Baran samui

Affiliation: Institute of Chemical Technology, N. Mehta Marg, Mumbai 400019, India

Education: Ph.D. (Polymer Science & Technology)

Business Address: Institute of Chemical Technology, N. Mehta Marg, Mumbai 400019, India

Research and Professional Experience:

- Carried out research and development work as `Scientist' in `Defence Research & Development Organization', India over 35 years
- Teaching in universities for last 6 years

Professional Appointments:

Scientist 'G' in DRDO
Visiting Professor, DIAT
Visiting Faculty, ICT
Technical Consultant to 2 Nos. Chemical Industries

Honors: Scientist of the Year Awards for DRDO Laboratory (Twice)

Publications from the Last 3 Years:

1) Versatility of polyethylene glycol (PEG) in designing solid-solid phase change materials (PCMs) for thermal management and their application to innovative technologies, Swati Sundararajan, Asit B. Samui*, Prashant S. Kulkarni, *J. Mater. Chem. A. 5,* 18379-18396 (2017).-Review
2) Poly(ethylene glycol) (PEG)-modified epoxy phase-change polymer with dual properties of thermal storage and vibration damping, Swati Sundararajan, Amit Kumar, Bikash C. Chakraborty, Asit B. Samui, Prashant S. Kulkarni, *Sustainable Energy Fuels*, 2018, 2, 688,-697.
3) Ionic liquid (IL) capped MnO_2 nanoparticles as an electrode material and IL as electrolyte for supercapacitor application, B. G. Bharate, Pankaj E. Hande, A. B. Samui, P. S. Kulkarni, *Renewable Energy,* 126, 437-444 (2018).
4) Selective nanomolar detection of mercury using coumarin based fluorescent Hg(II)-ion imprinted polymer, Pankaj E. Hande, Asit B. Samui, Sanjay Kamble, Prashant S. Kulkarni, *Sensors and Actuators B: Chemical,* 246, 2017, 597-605.
5) Shape-stabilized poly(ethylene glycol) (PEG)-cellulose acetate blend preparation with superior PEG loading via microwave-assisted blending, Swati Sundararajan, Asit B. Samui*, Prashant S. Kulkarni, *Sol. Energy,* 144, 2017, 32-39.
6) Thermal energy storage using poly(ethylene glycol) (PEG) incorporated hyperbranched polyurethane as solid-solid phase change material (PCM), Swati Sundararajan, Asit B. Samui, Prashant S. Kulkarni, *Ind. Eng. Chem. Res., 2017, 56 (49), 14401–14409.*
7) Synthesis and characterization of poly(ethylene glycol) (PEG) based hyperbranched polyurethanes as thermal energy storage materials, Swati Sundararajan, Asit B. Samui, Prashant S. Kulkarni, *Thermochimica Acta,* 650, 2017, 114-122.
8) An efficient method for determination of the Diphenyl Amine (stabilizer) in Propellants by Molecularly Imprinted Polymer based Carbon Paste Electrochemical Sensor, Pankaj E. Hande, Asit B. Samui, Prashant S. Kulkarni, *Propellants, Explosives, Pyrotechnics,* 42 (4), 2017, 376–380.

9) Tailoring of Energetic groups in Acroyloyl Polymers, Deepak Kumar, K. Durga Bhaskar Yamajala, Asit B. Samui and Shaibal Banerjee, (2017) *Tailoring of energetic groups in acroyloyl polymers, Designed Monomers and Polymers,* 20:1, 332-343, doi: 10.1080/15685551.2016.1258977.
10) Phase changing dicationic polymeric ionic liquid with CO2 capture abilities, S Sundararajan, AB Samui, PS Kulkarni, *International Journal of Environmental and Ecological Engineering* 11 (3), 2017.
11) Amit Kumar, P.S Kulkarni, A.B Samui, Phase Change Cotton Tape Development for Thermal Management, *International Journal of Advances in Science Engineering and Technology* (IJASEAT), 5 (1), 2017, 51-54.
12) Asit Baran Samui, Tapan Kanai, Polyhydroxyalkanoates based copolymers, *International Journal of Biological Macromolecules* 140 (2019) 522–537.

Book Chapters

1) “Applications of Carboxymethylcellulose in soaps and detergents” by A. B. Samui in *Carboxymethylcellulose: Properties, Applications and Effectiveness*. Eds.: Md. Ibrahim H. Mondal, Nova Science Publishers-Submitted, 2017.
2) “Main chain photoresponsive liquid crystalline polymers” by A. B. Samui & Balakrishna, K. in *Liquid Crystalline Polymers*, Eds. V. Mittal, Central West Publishing, Australia-2019.
3) “Polymers for Marine Applications” Taylor & Francis LLC *Encyclopedia Program*—2019.

In: Benzimidazole
Editor: Aksel A. Vestergaard

ISBN: 978-1-53616-986-7

Chapter 9

SUBSTITUTED BENZIMIDAZOLES AS LIGANDS: RECENT DEVELOPMENTS IN DESIGNING AND APPLICATIONS

Aayushi Arora[1], Preeti Oswal[1], Gyandshwar Kumar Rao[2], Arun Kumar[1] and Ajai Kumar Singh[3,*]

[1]Department of Chemistry, School of Physical Sciences, Doon University, Dehradun, Uttarakhand, India
[2]Department of Chemistry, Amity School of Applied Sciences, Amity University Haryana (AUH), Manesar, Gurugram, Haryana, India
[3]Department of Chemistry, Indian Institute of Technology Delhi, New Delhi, India

ABSTRACT

Benzimidazole framework has got considerable attention due to their importance in biology and ligand designing. The substitution at one or both "N" with functional groups such as –SR, $–PPh_2$, –SeR, –TeR, –OR, $–NR_2$ etc., results in the formation of multi-dentate ligands suitable for catalysis and transition metal complexation. Several benzimidazole based ligand systems have been designed by using this strategy and their transition metal complexes (including those of group VIII) have been investigated. The single-crystal structures of the metal complexes have many interesting features due to a variety of secondary interactions resulted from ligand itself. Diverse applications of benzimidazole based ligands and their metal complexes have been reported in the literature. The present

* Corresponding Author's Email: aksingh@chemistry.iitd.ac.in.

chapter covers a critical analysis of the recent developments in this area and future scope of benzimidazole framework in ligand designing.

Keywords: ligand, metal, biological activity, catalysis, luminescence

INTRODUCTION

Benzimidazole is an important heterocyclic aromatic organic molecule. It has drawn interest of many researchers because it behaves as a versatile pharmacophore in medicinal chemistry [1]. It is a bicyclic compound that contains a benzene ring and an imidazole ring fused together (Figure 1). In benzimidazole, benzene acts as a deactivating group which results in the transfer of electron density towards benzene ring from the imidazole ring. This electron transfer plays a crucial role in deciding the reactivity of benzimidazole [2].

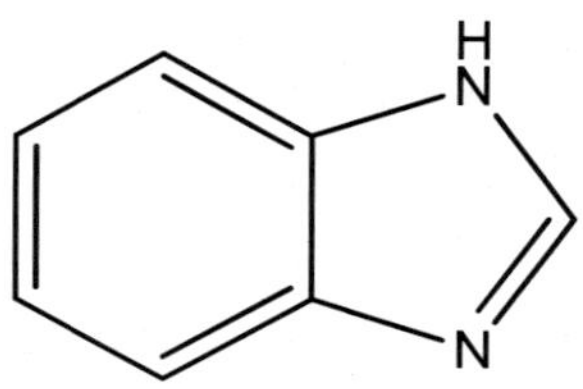

Figure 1. Benzimidazole.

Benzimidazole exists in two tautomeric forms (A and B) which are interconvertible only through a hydrogen shift between first and third nitrogen atom in the molecule (Figure 2). Due to this phenomenon, the compounds that were considered to be isomers initially, are tautomers in reality. The nitrogen atoms positioned at 1 and 3 show amphoteric nature i.e., they possess both acidic and basic properties [3]. Benzimidazoles are less basic imidazoles and are generally soluble in dilute acids. The acidic properties exhibited by benzimidazole result due to the stabilization of the anion by resonance [4].

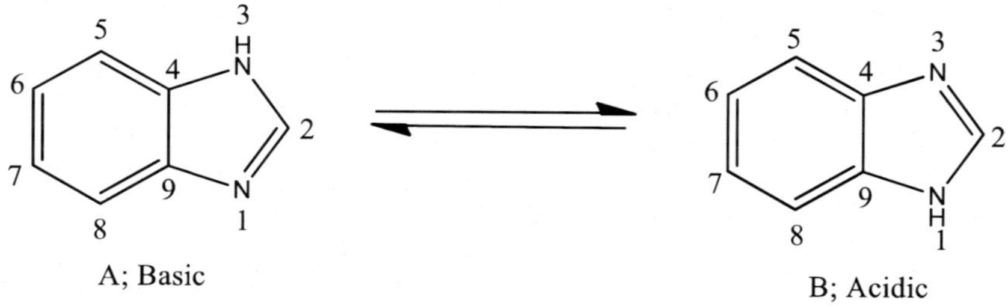

Figure 2. Tautomers of benzimidazole.

Benzimidazole ring plays a crucial role in a variety of biological activities namely antitumor, anticancer, antifungal, antiviral, anti-inflammatory, antimicrobial, anticonvulsant and many others [5]. N-ribosyl-dimethylbenzimidazole is a component of Vitamin B12, acts as an axial ligand for cobalt center [6].It forms an integral part of the drug discovery and serves as a chief and important scaffold for the designing and preparation of new molecules pertaining to medicinal value [7].

BENZIMIDAZOLE DERIVATIVES WITH AMINE FUNCTIONALITY

Numerous benzimidazole derivatives with amine functionality and their metal complexes have been designed and reported in the recent past. Such ligands and their metal complexes have found significant biological such as anti-microbial, antitumour, and catalytic applications such as in coupling reactions, hydrogenation of carbonyl compounds etc. Compound **1**, synthesized via reaction of Tm-1,4,7,10-tetraazacyclododecane-(*N*,*N*',*N*''-triacetic acid)-(*N*'''-orthoaminoanilide (Tm-DO$_3$A-oAA) with NO and O_2(Scheme 1), is used as a responsive Magnetic Resonance Imaging (MRI) contrast agent. MRI contrast agents are the compounds that allow a chemical exchange of labile protons between itself and bulk water that leads to a change in the chemical shift. This causes a faster relaxation rate (R_2) ($R_2 = 1/T_2$, where T_2 is relaxation time) leaving the other R_1 (relaxation rate $R_1 = 1/T_1$ where T_1 is second relaxation time)unchanged. The presence of amine-substituted benzimidazole in the contrast agent increases the exchange rate, therefore increases the sensitivity of a biomarker [8].

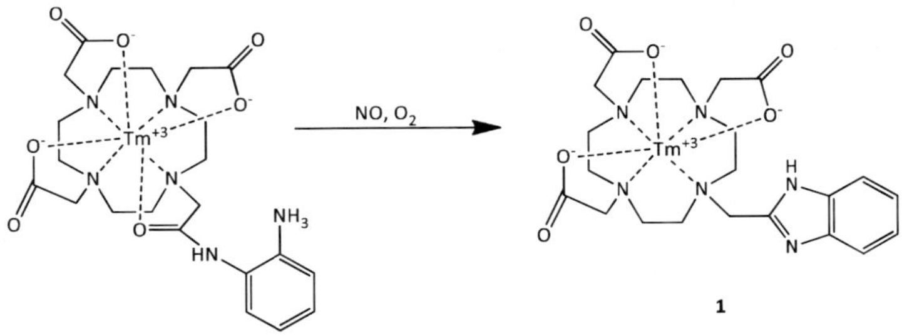

Scheme 1. Synthesis of biomarker 1.

Benzimidazole unit has also been investigated as an integral part of the hepatitis C virus inhibitor. Two hepatitis C virus (HCV) translation inhibitors **2** and **3** (Scheme 2) containing benzimidazole derivatives have been utilized in gaining insight into the internal ribosomal entry site (IRES) of hepatitis C virus RNA genome. It has been evaluated that both the benzimidazole ligands have efficient binding with the highly conserved IRES pocket. They widen the interhelical angle of IRES subdomain that further leads to its undocking and thereby inhibiting the IRES driven translation in HCV–infected cells. These inhibitors have been recognized as a new class of anti–HCV drugs that would bear considerably higher tolerance to resistance [9].

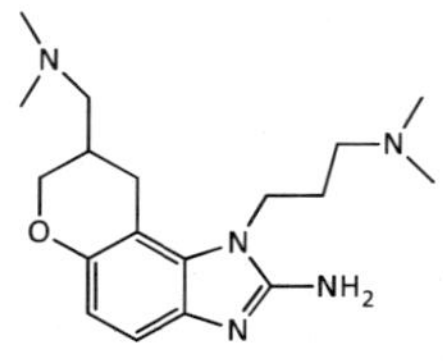

2

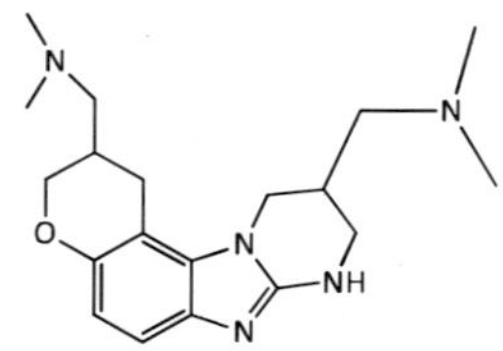

3

Scheme 2. "N" substituted compounds **2** and **3**.

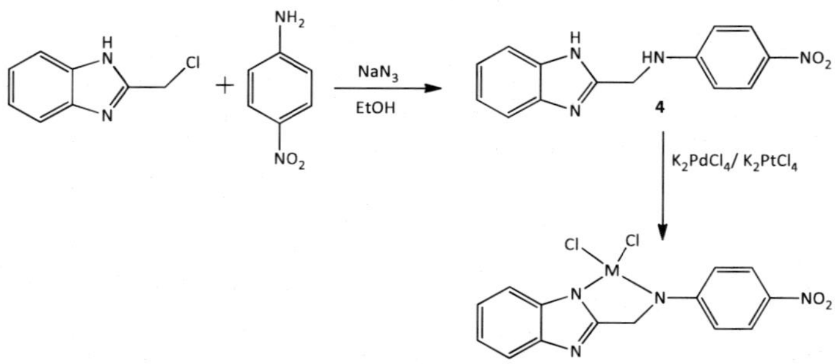

5: M: Pd
6: M: Pt

Scheme 3. Synthesis of ligand **4** and complexes **5**, **6.**

The ligand **4** (Scheme 3) and its Pd(II) and Pt(II) complexes **5**, **6**have been synthesized [10]. The molecular structure of the ligand was confirmed by mass spectrometry. The complex **6** displayed cytotoxic activity against both Gram-positive and Gram-negative bacteria while complex **5** was found show cytotoxicity only against *E. coli* and *N. Gonorrhoeae*, Gram-negative bacteria. The antitumor activity of both the complexes was investigated for three cell lines of different origin namely *breast cancer*(MCF−7), *Colon carcinoma* (HCT) and *Human hepatocellular carcinoma* (Hep−G2). It was observed that both the complexes showed higher activity at 100 μM against all three cell lines as compared to cis-platin. In addition, complex **5** showed higher antitumor properties in comparison to its platinum analog since Pt(II) has strong thermodynamic preference for S-donor such as glutathione, methionine present in cytosol serve as competing ligands [10].

Some interesting aminobenzimidazole derivatives (**7−20**) have been reported. The condensation of dimethyl N-cyanodithioimino carbonate with 2-iodoaniline in the presence of Cs_2CO_3 in DMF at 100ºC gave an adduct that was further reacted with the secondary amines infusion to afford the corresponding aryl guanidines. These guanidines, in the presence of CuI and 2,2'-biimidazole in DMF resulted in aminobenzimidazole derivatives **7−20** (Scheme 4). These 1-cyano-2-aminobenzimidazole derivatives were explored for their cytotoxic activities against three cell lines namely A549, K562 and PC−3. Most of the derivatives showed cytotoxic activities against tumour cells of A549 and K562 in the micromolar range. Derivative **20** showed significant inhibition of K562 cell cycle [11].

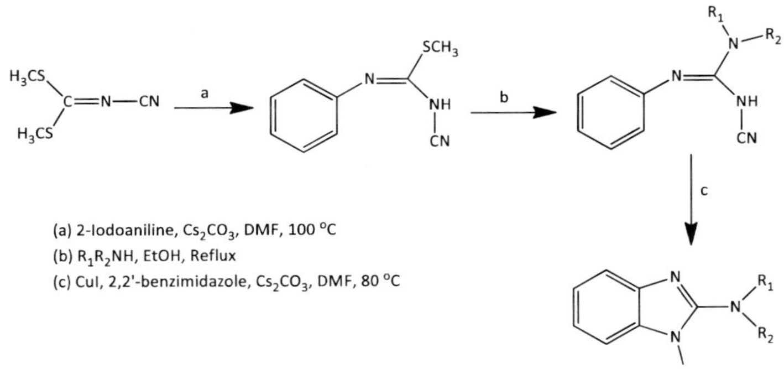

Scheme 4. Synthesis of aminobenzimidazole derivatives **7–20**.

Table 1: Various substitutions on benzimidazole framework

No.	R_1R_2NH	No.	R_1R_2NH
7	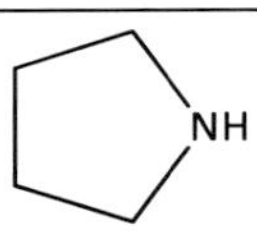	**14**	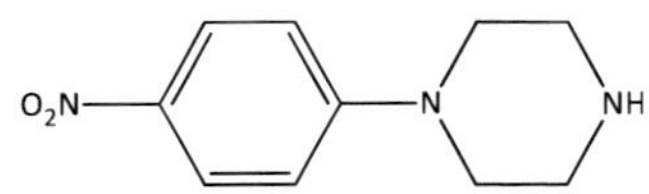
8	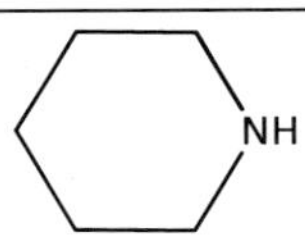	**15**	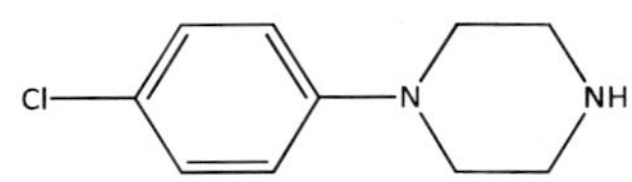
9	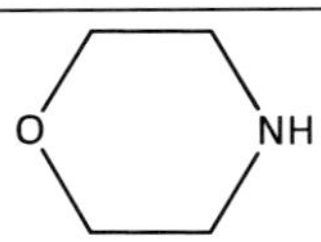	**16**	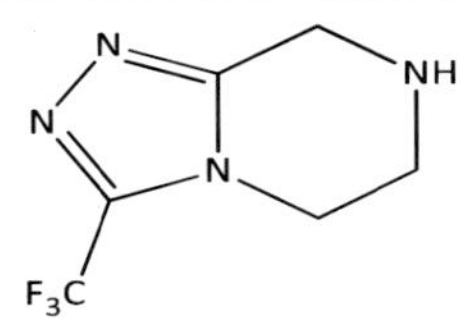
10	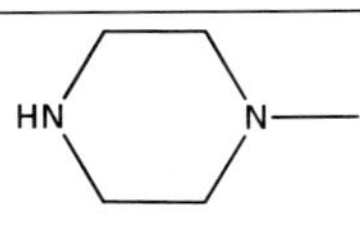	**17**	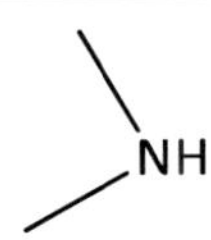
11	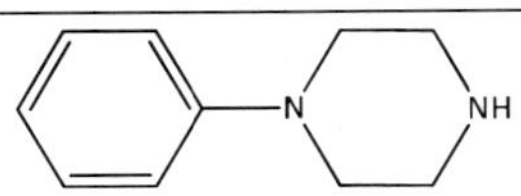	**18**	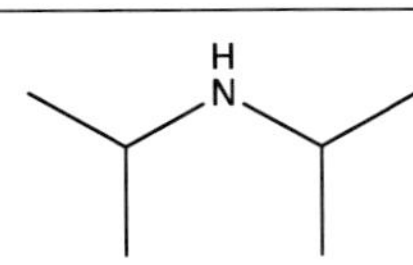
12	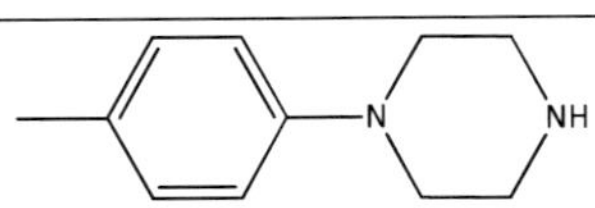	**19**	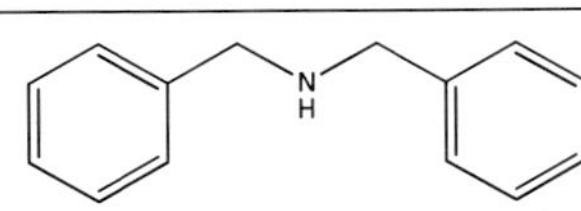
13	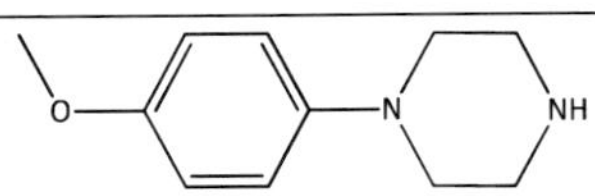	**20**	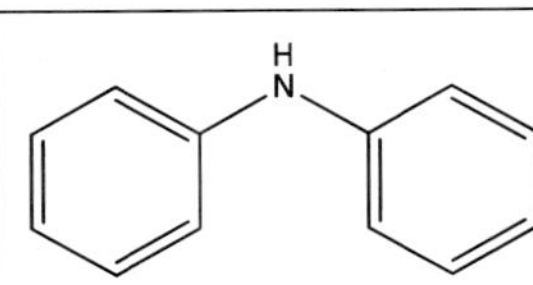

Three cobalt(II) complexes **21–23** (Scheme 5) have been reported for their antimicrobial activity against*Pseudomonas Fluorescence* (Gram-negative) and *Bacillus subtilis* (Gram-positive) bacteria. All the three cobalt complexes were more active towards the Gram-positive bacteria. It wasconcluded that activity of all the three complexes has been related to the cell wall structure of the bacteria [12].

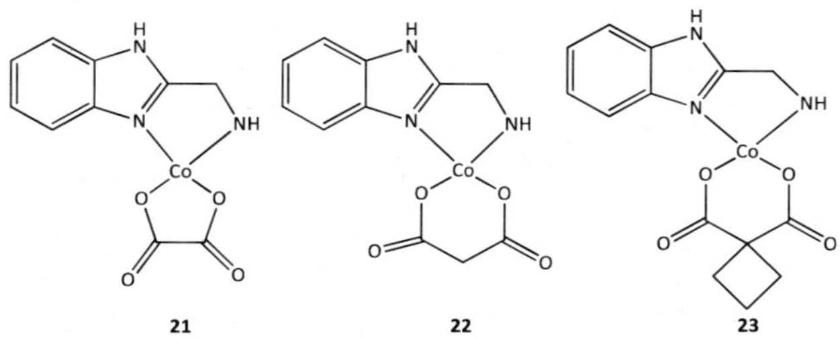

Scheme 5. Co(II) complexes of 2-aminomethylbenzimidazole.

In addition to biological applications, aminobenzimidazole derivatives have also been explored for various catalytic applications. Three aminobenzimidazole derivatives **25–27** (Scheme 6) containing *N*-(1-phenylethyl)-1*H*-benzo[d]imidazole-2-amines have been prepared and explored for their efficiency as organocatalysts in α–chlorination of ethyl-2-oxocyclopentanecarboxylate with *N*-chlorosuccinimide (NCS) in toluene. It was observed that 10 mol% of **26** showed 99% conversion with the highest enantiomeric excess (ee) of 40% at −50ºC followed by **27** that gave ee of 12%. These organocatalysts played a bifunctional role as they activate both the dicarbonyl compound as well as the chlorinating agent (NCS) [13].

Various derivatives of chiral 2-aminobenzimidazoles **28–29**(Scheme 7) have been synthesized. **29** has been explored as organocatalyst for the direct conjugate addition of dicarbonyl compound to maleimide and N-substituted maleimides in order to produce Micheal adducts that possess high enantiomeric control. Density functional theory (DFT) studies confirmed that hydrogen bonding of an organocatalyst with maleimide results in the formation of a ternary structure that facilitates the formation of final product [14].

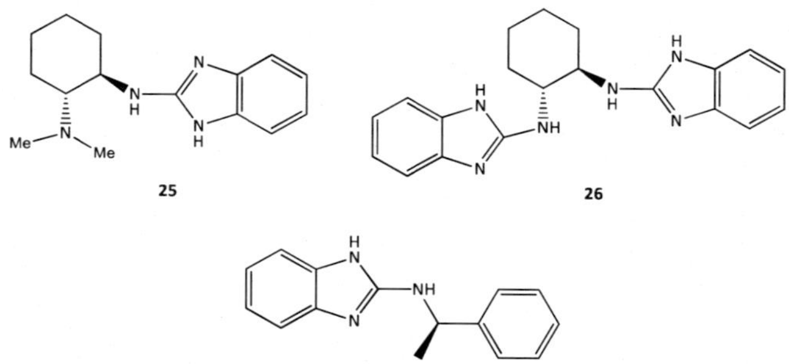

Scheme 6. Derivatives of 2-aminobenzimidazole.

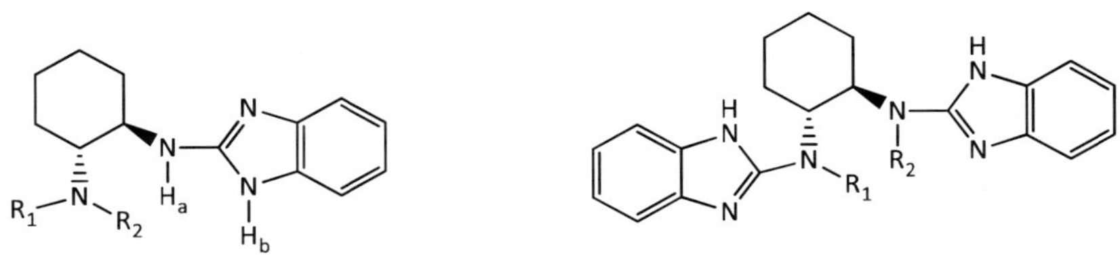

28a; R_1= R_2 = H
28b; R_1= R_2= Et
28c; R_1= R_2=2,5-Dimethylpropyl-1yl
28d; R_1= R_2=piperdin-1-yl

29: R_1=R_2=Me

Scheme 7. 2-Aminobenzimidazole based organocatalysts **28** and **29.**

Substitution of aminobenzimidazoleswith an electron withdrawing group (−CF_3) leads (Scheme 8) to the formation of **31** that behave as a bifunctional organocatalyst for Micheal addition of dimethyl malonate to nitro-olefins. Ligand behaves as base to abstract a hydrogen from malonate and also as Brønsted Lewis acid to activate nitro-olefin. The investigation also revealed that the catalytic mechanism is sensitive to the presence of co-catalyst. When TFA is used as a co-catalyst, the dimethylamine moiety is protonated and it formed hydrogen bonds with nitro-olefin. Thus, the guanidine moiety functions as a weak base to abstract a hydrogen of malonate that results in the formation of an ion pair between the guanidinium cation and malonate anion. In other words, dimethylamine and guanidine moieties reverse their role in two situations accordingly [15].

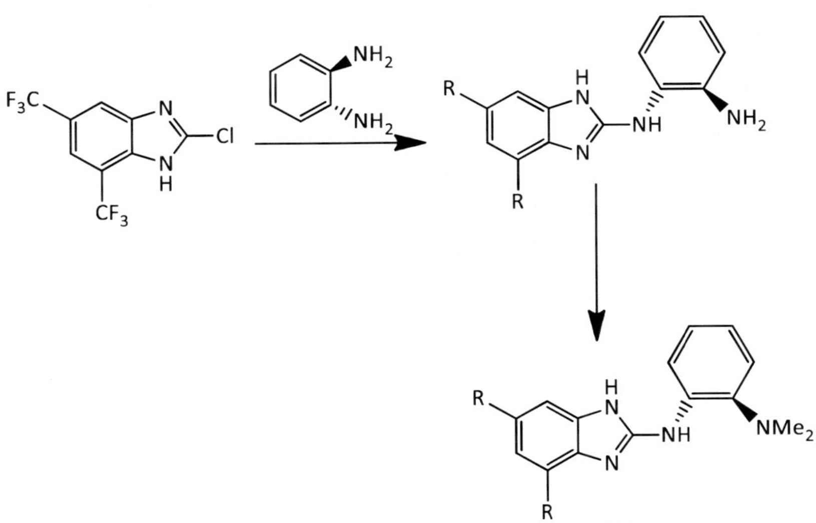

30; R: H
31; R: CF_3

Scheme 8. Synthesis of compounds **30** and **31.**

A Ru complex **33** (Scheme 9)of hybrid benzimidazole ligand **32** was analyzed for asymmetric hydrogenation of ketones. It was observed that the outcome of Ru-based catalysts is depend on the use of aminobenzimidazole derivative during asymmetric hydrogenation of aryl ketones. The benzimidazole functionality reverses the typically observed chiral induction and allows the feasibility of the hydrogenation to be carried out in non-protic solvents up to 99% enantiomeric excess [16].

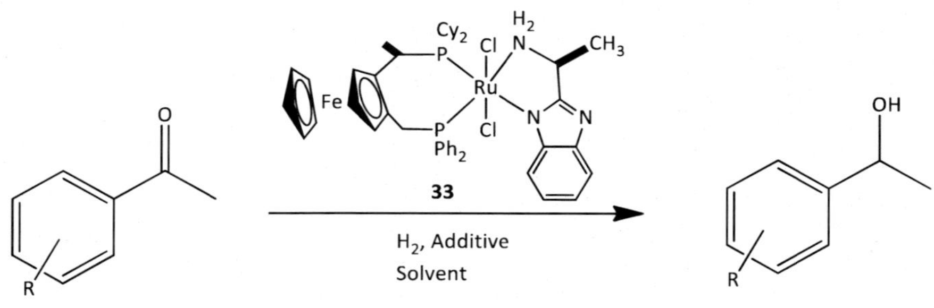

Scheme 9. Transfer hydrogenation catalyzed by **33.**

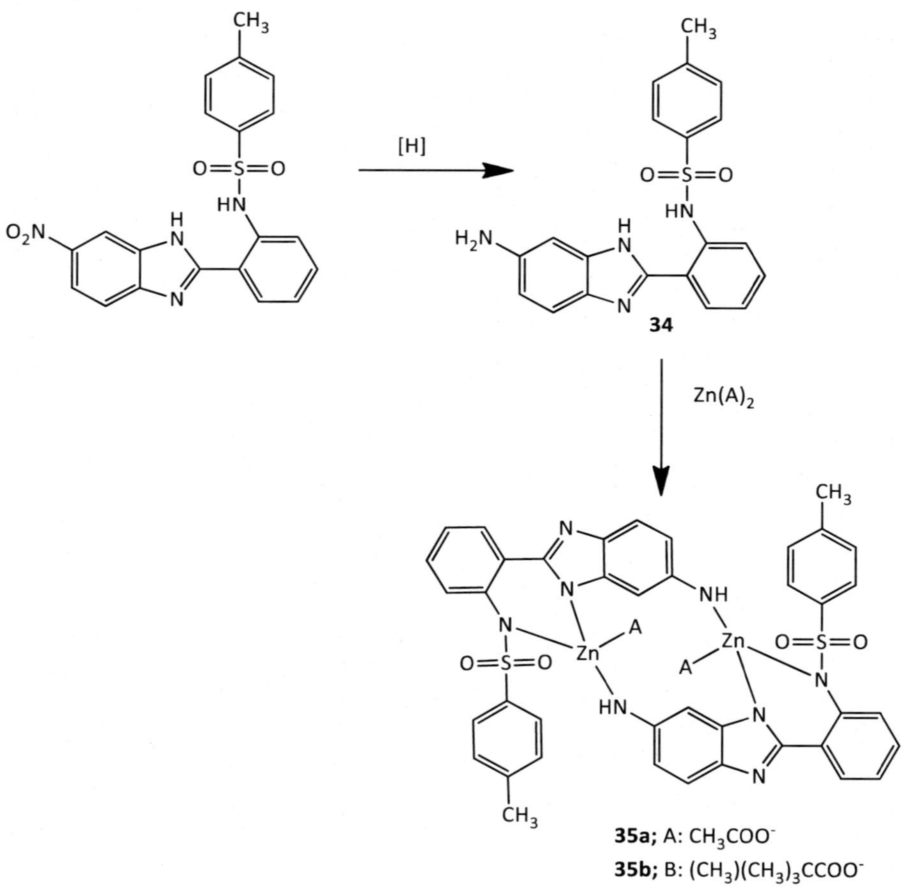

Scheme 10. Synthesis of Zn(II) complexes of ligand **34.**

The 1-propyl-2-(2-tosylaminophenyl)-5-aminobenzimidazole **34** (Scheme 10) has been synthesized by the reduction of its corresponding nitro derivative. Zn (II) complexes **35a** and **35b** have been synthesized and its structure elucidation by X-ray crystallography confirmed the presence of bridging of Zn atom by two ligand units. Every Zn atom has distorted tetrahedral geometry resulted due to the coordination of monodentate pivalate groups. The anomalously high Stokes shift 8405 cm^{-1} nm in solution was observed in the ligand that is attributed to intramolecular proton transfer by N–H······N in the excited state to form quinoid tautomer that demonstrates radiative deactivation of excited electron energy. In solid state, the fluorescence is hypsochromically shifted to 440nm. The Stokes shift in case of Zn(II) complex **35b** and ligand **34** was found to be 3940 and 4300 cm^{-1} respectively. The decrease in the Stokes shift was caused primarily due to the replacement of sulphonamide ion in **34** by the zinc ion **35**. This replacement resulted in the interruption of intramolecular proton transfer [17].

The reaction of iminodiacetic acid / (2-hydroxyethyl)-iminodiacetic acid with 1,2-phenylenediamine resulted in the formation of bis-benzimidazole compounds **36** and **37** (Scheme 11). These ligands have been tested for their anti-corrosion properties. It was observed that tertiary amine substituted benzimidazole (**36**) served as better corrosion inhibitors in comparison to secondary amines as the adsorption of former over the Fe metal surface was higher [18].

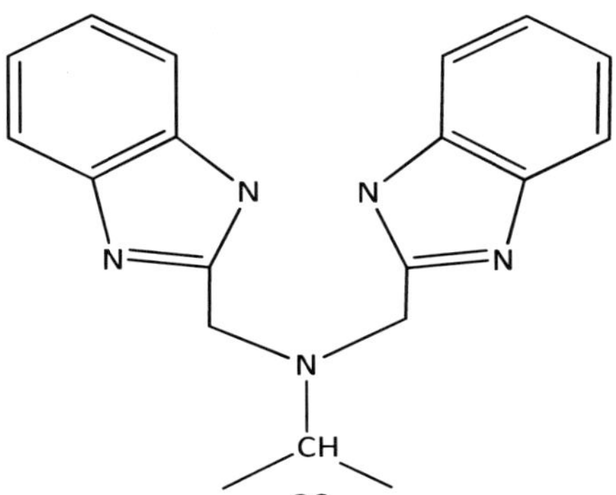

36

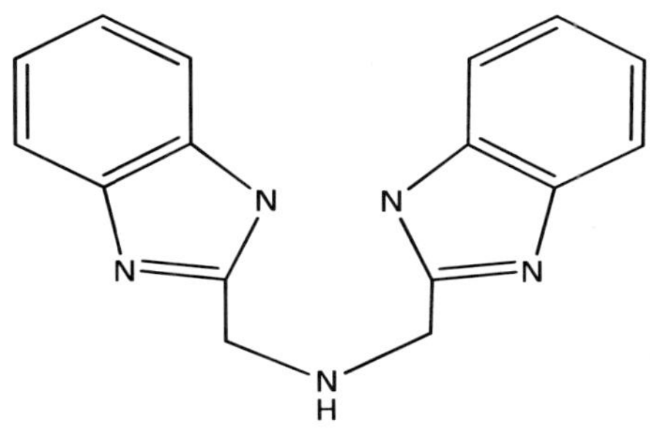

37

Scheme 11. Tertiary and secondary amine-containing benzimidazole ligands **36** and **37**.

BENZIMIDAZOLE DERIVATIVES WITH PHOSPHINE FUNCTIONALITY

Diverse benzimidazole scaffolds have been designed containing phosphines donor group as a part due to the tuneable donor/acceptor properties of phosphorous and thus their versatile nature as a catalyst. Some of the features of phosphine ligated benzimidazole ligands are, (i) a tunable framework to provide diversification, (ii) hemilability due to the presence of weak coordinating ligands that result in dynamic interaction of metal centres which increases the catalyst longevity and (iii) the presence of phosphine group which donor/acceptor ability could be fine-tuned by changing substituent's at "P" [19]. The phosphorous donor containing ligands **38** and **39**[1-aryl-2-(dialkylphosphino)-benzimidazole] (Scheme 12) have been synthesized in order to tune the properties of the framework. The ligands in combination with $Pd(OAc)_2$has been explored as a catalyst for Suzuki coupling and Buchwald Hartwig amination. Both the ligands were found to be superior in comparison to other phosphine ligands lacking benzimidazole framework. $[Pd(CH_3CN)_2Cl_2]$ along with **38** or **39** excellently catalyze the Buchwald Hartwig amination of several aryl halides at 0.5 mol% palladium loading. Similarly, the Suzuki coupling reaction was also catalyzed by the two ligands in the presence of Pd at 0.01 mol% of $[Pd(CH_3CN)_2Cl_2]$ only [20].

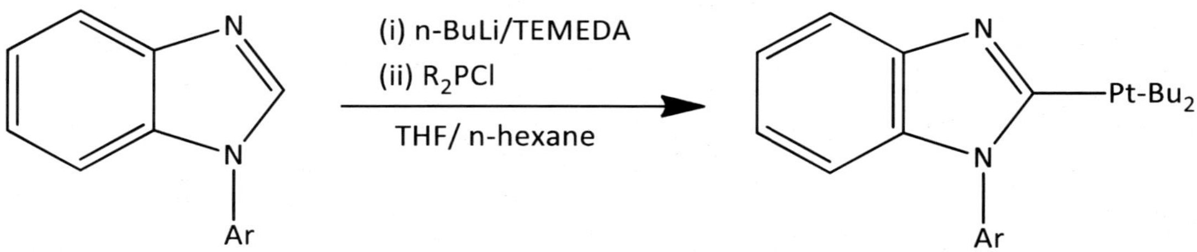

38: R: Phenyl
39: R: Naphthyl

Scheme 12. Synthesis of compounds **38** and **39.**

A simple condensation between *o*-phenylenediammine and 2-bromobenzoic acid in the presence of polyphosphoric acid yielded methylated precursors. These precursors after treatment with n-butyllithium and $ClPR_2$ yielded corresponding benzimidazolyl phosphines **40–42** (Scheme 13). An additional advantage of these ligands lies in its easy preparation strategy as well as inexpensive benzimidazole raw material. Ligand **41**in combination with several Pd precursors was tested for amination reaction in order to optimize the best Pd source. Several aryl chlorides including heteroaryl chlorides were successfully coupled with various amines at 0.2–0.5 mol% of $Pd_2(dba)_3$using **41** as a ligand [19]. The ligand **41** has also been explored as pre-catalyst with $Pd(dba)_2$ as a palladium source in Suzuki-Miyaura coupling reaction of aryl chloride with potassium phenyltrifluoroborate at 135 °C. Various aryl chlorides and heteroaryl chlorides have been successfully coupled to yields substituted biphenyls in excellent yields when the metal to ligand ratio was 1:2 [21].

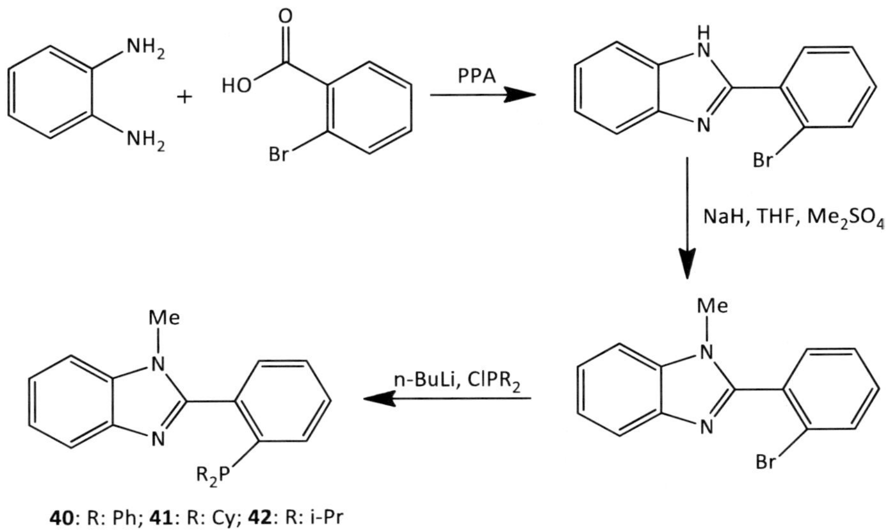

Scheme 13.Preparation of **40–42**.

Ruthenium complexes **45–46** of phosphine substituted benzimidazole derivatives **43–44** have also been prepared (Scheme 14) and characterized using several analytical techniques including X-ray crystallography. The Ru atom in both the complexes has piano-stool geometry and Cl lies trans to C^- donor group. Another interesting feature is the presence of a five-membered chelating ring in complex**46**whereas, in complex **45**,a six-membered ring has been formed by the coordination of P and C^- to Ru center. This makes the complex **45** more efficient than complex **46** [22].

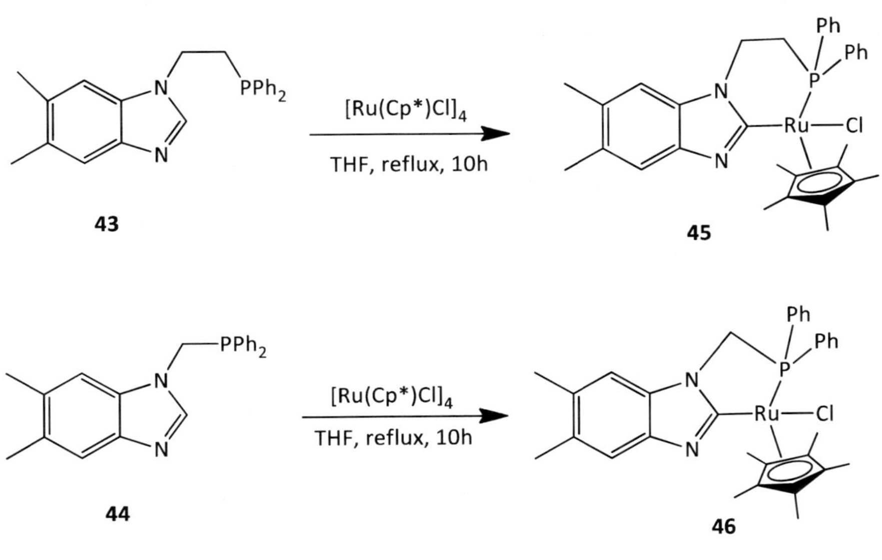

Scheme 14. Synthesis of Ru(II) complexes of phosphine ligated benzimidazoles.

Many phosphines ligated benzimidazole derivatives **40–41** and **47–57** (Scheme 17)in combination with Pd precursors were studied for their efficiency in decarboxylative coupling of aryl mesylates. Various substitutions on the P and benzimidazole ring were found to play a key role in the determination of the catalytic activity of these ligands. The activity of ligand**56** was best among them where "N" was having a methyl group and "P" center was derived with cyclohexyl rings. The crystal structure of the Pd complex of ligand **56** showed a distorted square planar geometry around Pd center and a bidentate coordination mode through nitrogen and phosphorus donor atoms of ligand. The decarboxylative cross-coupling of aryl and heteroaryl carboxylates with mesylates was carried out using 5 mol% $Pd(dba)_2$ and 12 mol% **57**. Moderate conversions ranging from 27–73% were obtained. The cross-coupling reaction was also performed using $PdCl_2$-**56** complex prepared and it was observed that the in situ generated palladium species is superior to the complex in catalyzing the reaction. Various sterically hindered vinyl mesylateshave has also been successfully coupled to give moderate yields using phosphine substituted benzimidazole derivative **56** [23].

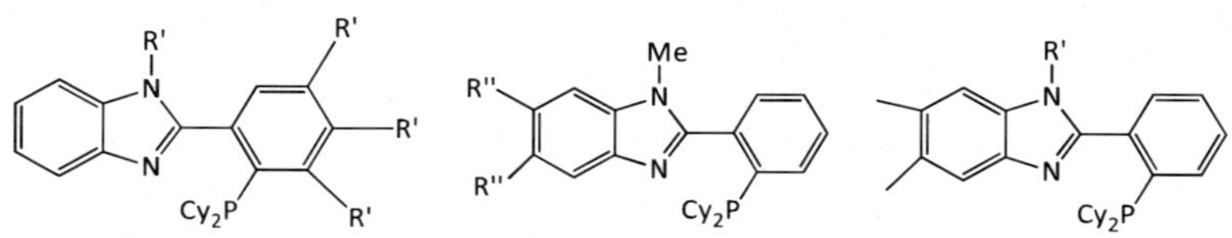

47; R: H; R': iPr
48; R: H; R': Ph
49; R: H; R': CH_2OMe
50; R: OMe; R': Me

51; R'': Cl
52; R'': OMe
53; R'': Me

54; R': i-Pr
55; R':n-Bu
56; R': n-octyl
57; R': n-$C_{18}H_{37}$

Scheme 15. Various phosphorus ligated benzimidazole derivatives.

SCHIFF BASES WITH BENZIMIDAZOLE CORE

Schiff bases are an important class of compounds owing to their important applications in a variety of fields such as analytical, inorganic and biological chemistry. Schiff bases consisting of a heterocyclic framework such as benzimidazole have found potent applications in biology and medicine because of their activity against various viruses and microbes [24]. A number of methods have been employed in the synthesis of benzimidazole derived Schiff bases. These methods suffered serious limitations of (i) long reaction time (ii) low yield (iii) need of specific catalyst and, (iv) need of recrystallization for obtaining pure compound. Compounds **58–69** have been synthesized using $Ni(NO_3)_2.6H_2O$ at room temperature in short reaction time (Scheme 16). An extensive investigation using a variety of catalysts namely $Cu(NO_3)_2.6H_2O$, $Ni(NO_3)_2.6H_2O$, $Mn(NO_3)_2.6H_2O$, $Fe(NO_3)_2.6H_2O$ and $Co(NO_3)_2.6H_2O$ at varied mol% was explored for the reaction of 2-aminobenzimidazole and *o*-nitrobenzaldehyde. It was observed

$Ni(NO_3)_2.6H_2O$ was the most efficient catalyst for the synthesis of Schiff base ligands at room temperature among the metal precursor [24].

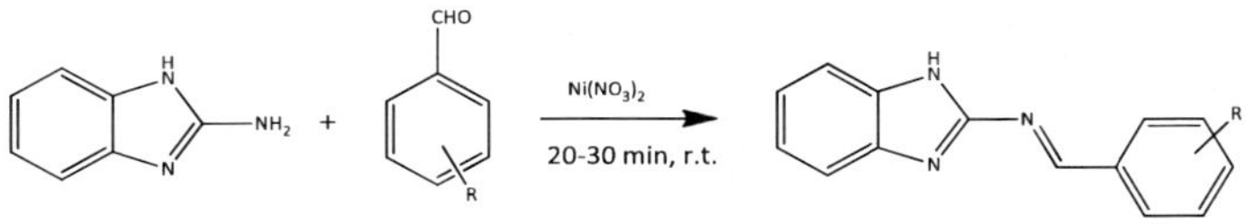

58: R: H; **59**: R: 2-Cl; **60**: R: 3-Cl; **61**: R: 4-Cl
62: R: 4-Br; **63**: R: 3-NO_2; **64**: R: 4-NO_2
65: R: 4-Me; **66**: R: 4-OMe; **67**: R: 2-OH
68: R: 3-CHO; **69**: R: 4-CHO

Scheme 16. Various Schiff base ligated benzimidazole derivatives **58–69**.

1,3-Bis(2'-benzimidazolylimino)iso-indoline **70** (Scheme 17) was synthesized by the coupling of 2-aminobenzimidazole and 1,3-diiminoisoindoline. The ligand **70** has three protons out of which one is present on isoindoline nitrogen and the other two are on each of the benzimidazole rings. Five metal complexes **71–75** have been synthesized and characterized including single-crystal X-ray structuretechnique. The ligand binds with metals in monoanionic *N,N,N*-tri-coordination mode forming a square planer geometry.It also forms octahedral metal complexes when M:**L** ratio was 1:2.The Fe(III) complex **71**has trigonal bipyramidal geometry where one chloride ion occupies the additional coordinative. Synthesis of cobalt complex was achieved via the reaction of $Co(C_2H_3O_2)_2.4H_2O$ with the ligand **70**. Complex **73** was synthesized by the reaction of **70** with $Ni(C_2O_3H_2)_2.4H_2O$ and forms a distorted tetrahedral geometry with the ligand, a DMF molecule, and an acetate ion. Copper(II) complex **74** forms a distorted square planar geometry and a weak interaction (2.5 Å) at an axial coordination site with the oxygen atom of acetate ion has been observed. The zinc(II) **75** has a trigonal bipyramidal geometry in which a solvent molecule occupies the fifth coordination site [25].

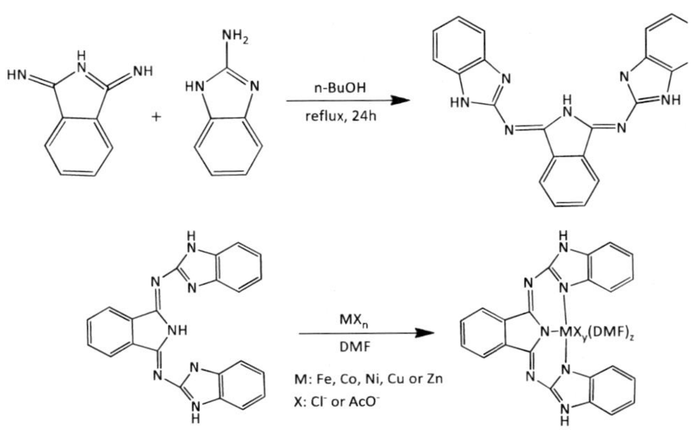

70

71: M: Fe; X: Cl^-; y=2; z=0
72: M: Co; X: AcO^-; y=1; z=
73: M: Ni; X:AcO^-; y=1; z=1

Scheme 17. Preparation of ligand **70** and its metal complexes.

A series of four Cu(II) complexes **76–79** differ in the R group on phenyl ring have been synthesized from their corresponding Schiff base ligands containing benzimidazole unit (Scheme 18). All four complexes exhibited distorted square planar geometry around Cu(II) center. The antibacterial activity of the free ligands and metal complexes has been investigated against both Gram-positive and Gram-negative bacteria.Complexes showed higher antimicrobial activity than that of free ligand. This was explained in the term of coordination and chelating property of metal complexes. It was observed that the metal complexes were more active against Gram-positive bacteria than Gram-negative bacteria. This was due to the presence of impermeable peptidoglycan layer in case of Gram-negative bacteria. The Cu(II) complexes **76–79** showed inhibition of the reduction of nitroblue tetrazolium (NBT) to formazan which were used for detection of the SOD mimetic catalytic activity. The higher biological activities of copper complexes were explained on the basis of presence of flexible ligands that can accommodate geometrical change involving Cu(II) to Cu(I) and the labile water molecule which can be easily substituted by oxygen in the catalytic pathway [26].

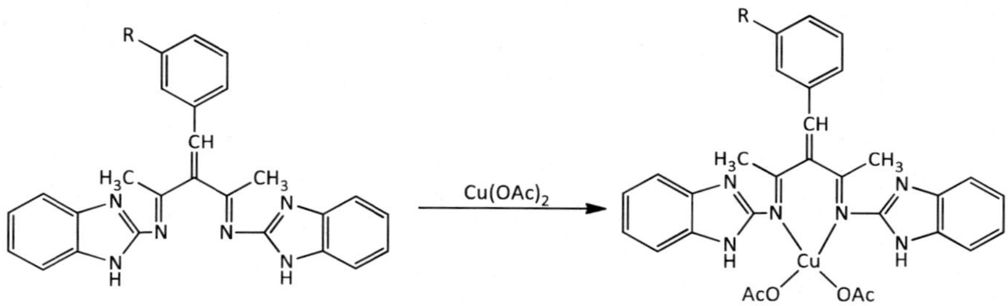

76: R: Br; **77**: R: F
78: R: Cl; **79**: R: CN

Scheme 18. Preparation of Cu(II) complexes **76–79.**

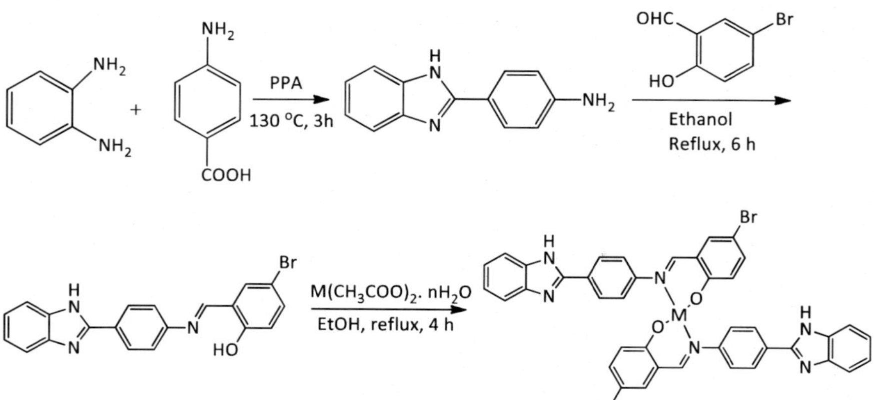

80

81: M: Zn; **82**: M: Ni
83: M: Cu; **84**: M: Pd

Scheme 19. Preparation of ligand **80** and its Zn(II), Ni(II), Cu(II), Pd(II) complexes.

A Schiff base ligand **80** (Scheme 19) containing a benzimidazole framework along with 'O' and 'N' donor atoms and their metal complexes **81–84** have been synthesized and characterized by different spectroscopic techniques. DNA binding activity of these compounds has been explored by thermal denaturation experiments as well as UV–Vis absorption spectroscopy titrations. It was observed that the ligand and metal complexes have the ability to bind to (st) DNA by intercalation (**80** and **81**). The free ligand **80** showed stronger DNA binding ability whereas **84** showed the weakest. Antimicrobial activity activities have also been performed on both Gram-positive and Gram-negative bacteria. Complexes **81–83** showed good to moderate antibacterial activity with no toxicity in the brine shrimp assay with the exception of the Cu(II) which was slightly toxic [27].

BENZIMIDAZOLE DERIVATIVES WITH PYRIDINE MOTIF

A large number of pyridine substituted benzimidazole ligands and their metal have been reported. The choice of a pyridine based benzimidazole ligand is based upon the fact that the large pyridyl-benzimidazole rings provide recognition sites for stacking interactions as well as exhibit hydrogen bonding that result in the supramolecular compounds. This leads to one, two and three-dimensional structures that have several applications in biology [28], dye-sensitized solar cells (DSSC) [29], fluorescence sensing [30] and catalysis [31]. The tetranuclear Cu(II) complex $[Cu_4Cl_6O(C_{13}H_{11}N_3)_4]$ **86** (Scheme 20) has been prepared by the reaction of $CuCl_2$ with the benzimidazole-pyridine ligand **85**. The Cu(II) ion is tetrahedrally coordinated with the central oxygen atom. The N atom of the monodentate 1-(pyridine-2-ylmethyl)-1*H*-benzimidazole **85** occupies the fifth coordination position on the outside of the cluster located on the central Cu–O axis. This results in a distorted trigonal bipyramid geometry around Cu(II) with the O and N atoms occupying axial positions [32].

Ligands 2,6-bis(benzimidazole-2´-yl)pyridine (**87**), 2,6-bis(1-ethylbenzimidazol-2´-yl)pyridine (**88**) and 2-(2-pyridyl)benzimidazole (**89**) has been used for the synthesis of three Mn(I) complexes namely fac-[MnBr(**87**-κ^2-N^1,N^2)$(CO)_3$] (**90**), fac-[MnBr(**87**-κ^2-N^1,N^2)$(CO)_3$] (**91**) and fac-[MnBr(**89**)$(CO)_3$] (**92**) (Scheme 21). These complexes have been studied for blue light photo-induced carbon monoxide release properties in DMSO and myoglobin solution. The single-crystal structure of **90** and **91** showed the facial carbonyl ligands and Br atom occupying the axial position. It was found that the presence of the uncoordinated group in the vicinity of metal centre accelerates the release of CO molecule when the Br·····H–N interactions are absent [33].

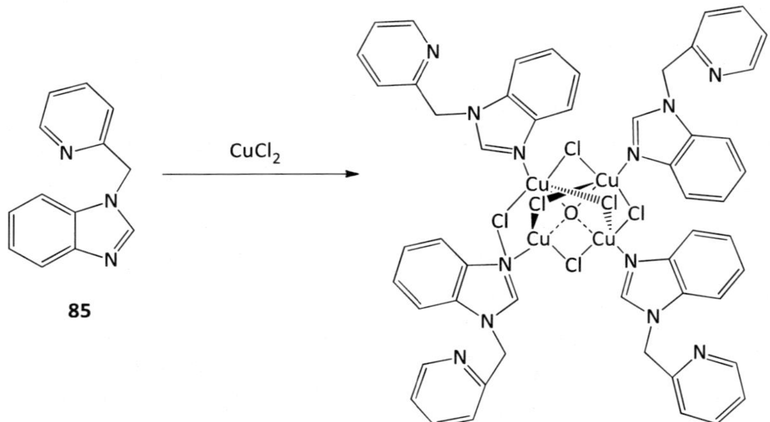

Scheme 20. Synthesis of Cu(II) complex **86.**

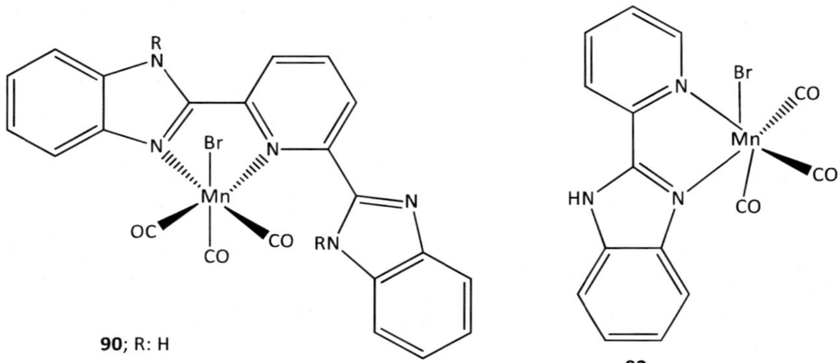

Scheme 21. Mn(II) complexes **90–92**.

Two Cu(II) complexes [Cu(**93**)$_2$](ClO$_4$)$_2$] (**94**) and [Cu(**93**)(bpy)](ClO$_4$)$_2$ (**95**) have been reported (**93**; N-butylbenzimidazole, bpy; 2,2′-bipyridine) (Scheme 22). A distorted square pyramidal geometry around Cu(II) was observed with a M:**L** ratio 1:2 in the case of **95**. Both the metal complexes have been investigated for their antifungal and antibacterial properties against Gram-positive bacteria (*Bacillus subtilis* and *Bacillus cereus*) and Gram-negative bacteria (*E. coli*, *Enterobacteraerogenes*, *Klebsiellapnuemoniae*). Cyctotoxic activities were also tested against various yeasts such as *Saccharomyces cerevisiae*, *Candida utilis,* and *Candida albicans*. The complexes exhibited considerable antibacterial activity when compared with ampicillin as standard (10μg). However, the cytotoxic activity against *K. pnuemoniae* was least among all bacteria studied. In case of antifungal activity, the complexes were found to be more active against *S. cerevisiae* than *C. utilis* and least against *C. albicans*. Both the complexes were found better against *E. coli* and *B. subtilis*as compare to precursor metal salt CuCl$_2$ and Cu(ClO$_4$)$_2$ [34].

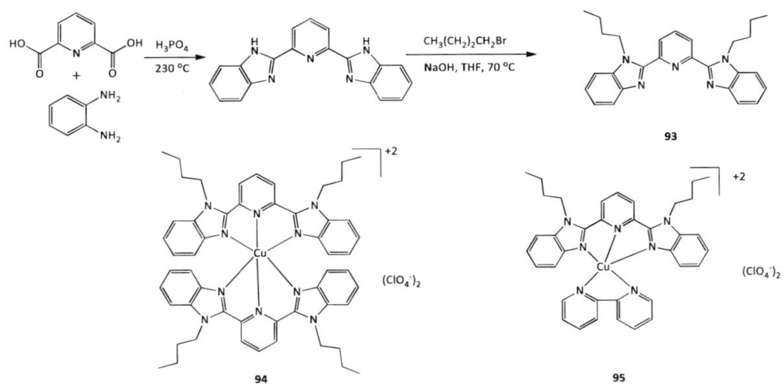

Scheme 22. Synthesis of ligand **93** and Cu(II) complexes **94** and **95**.

The Mn(II) complexes [Mn(L)(NCS)$_2$], [Mn(L)Cl$_2$], [Mn(bzimpy)Cl$_2$] and [Mn(L)(NO$_3$)$_2$] **96–99** have been reported (Scheme 23).Complexes **96** and **97** contain Mn(II) in five coordinated state resulting in a highly distorted square pyramidal geometry where Cl and NCS occupy an axial position. Both complexes exhibit π–π stacking interactions. In **97**, one benzimidazole ring is stacked over the other benzimidazole ring of the adjacent molecule under symmetry operation −x, 1−y, 1−z. However, two sets of stacking interactions occur in complex **96**. One is the stacking of pyridine benzimidazole unit of the same section with that of neighboring under the symmetry 2−x, 1−y, −z. The second set of stacking interaction is the edge of the benzene ring with the benzene ring of same section in neighboring molecule resulting from a symmetry operation 2−x, 1−y, 1−z [28].

The seven coordinated Mn(II) complex **99** exhibits a pentagonal bipyramidal geometry. The π–π stacking interactions in **99** are the same as that of **96**. The cation of complex **98** is linked via π–π edge-edge stacking with that of the same section of a neighboring molecule. The **97** was found to be efficient in catalase mimic activity with a turn over frequency of 68 s^{-1} [28].

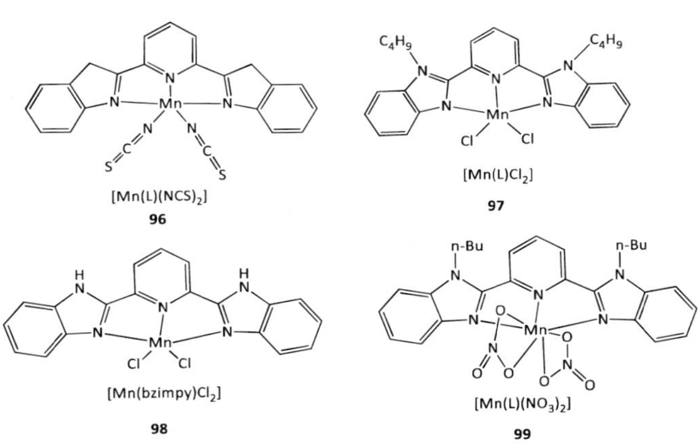

Scheme 23. Mn(II) complexes **96–99**.

Europium complex Eu(tta)$_3$.BBIMP (**101**) has been synthesized by reacting 2,6-bis(benzimidazole-2´-yl)pyridine ligand (**100**) and Eu(tta)$_3$.2H_2O (Scheme 24). The complex **101** exhibited remarkable luminescence properties i.e., sharp emission and large stimulated emission cross-section which are of the same order of Nd–glass laser. The Eu(III) complex showed high intrinsic luminescent quantum yield and hence promising alternatives for optical applications such as high power laser media [35].

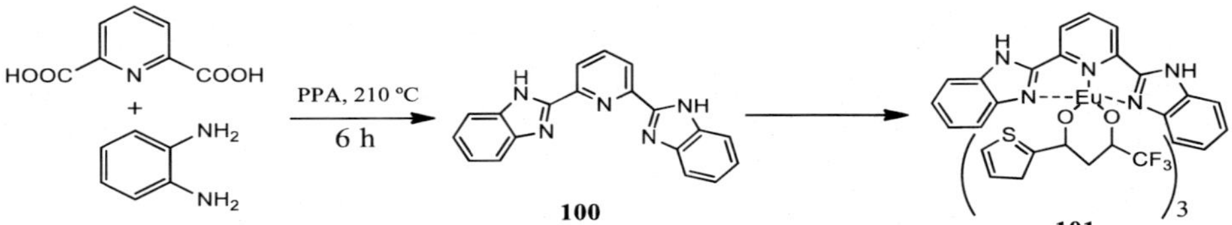

Scheme 24. Synthesis of europium complex **101**.

The ligand 1-(4-pyridylmethyl)-1*H*-benzimidazole (**102**) (Scheme 25), and their complexes Ag(**102**)(ClO_4)]$_n$ (**103**), {[Cu(**102**)$_2$(ClO_4)$_2$](H_2O)$_2$}$_n$ (**104**) have also been reported to exhibit photoluminescence activity. The compound **103** was observed to show double helix chain structure that results in the formation of a two-dimensional network connected by Ag·····Ag and Ag·····O secondary interactions. A two-fold impenetrating three dimensional network exists in case of **104**. These complexes exhibited excellent luminescent properties with d^{10} metals. A solid-state luminescence spectrum of **103** was recorded and showed an emission band at 314 nm upon excitation at 280nm. This is attributed to intra ligand transition which was in good agreement with values reported earlier [36].

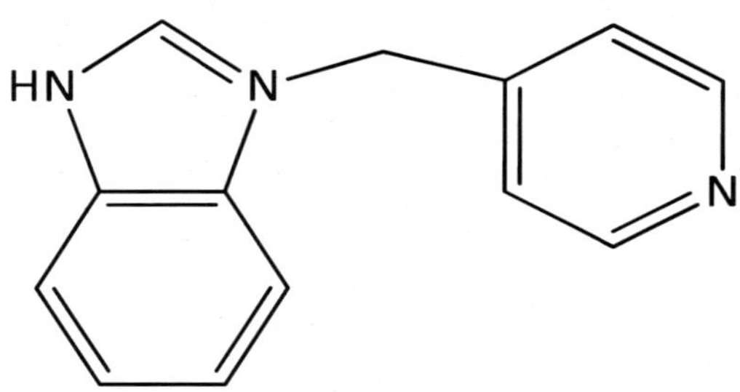

Scheme 25. Structure of ligand **102**.

Four Zn(II) tetrahedral complexes (Scheme 26), *viz.* {[Zn(pmbm)$_2$(tpa)].H_2O}n (**104**), {[Zn(pmbm)(phda)].2(H_2O)}$_n$ (**105**), [Zn(pmbm)(aze)]$_n$ (**106**), {[Zn(pmbm)(1,4-ndc)].2(CH_3OH)}$_n$ (**107**) have been synthesized by solution-phase ultrasonic reactions of Zn(AC)$_2$.2H_2O with 1-(4-pyridylmethyl)-benzimidazole (**102**) and in presence of various dicarboxylates ligands under the ammoniacal condition (H_2tpa = terephthalic acid, H_2phda = phenylenediacetic acid, H_2aze = azelaic acid, 1,4-ndcH_2 = 1,4-naphthalenedicarboxylic acid). Intense emission band at 436nm for **104**, 380 nm and 477 nm for **105**, 386 nm for **106** and 408 nm for **107**respectively, have been observed by excitation at 300 nm. The free ligand exhibited photoluminescence at 408 nm. The red-shift in emission and luminescence of the **104** is mainly due to π to π* transition. It is due to two reasons (i) the linkage of pmbm ligand to the metal center thereby increasing the rigidity of the ligand and reducing the energy loss by radioactive decay and, (ii) the emission of ligand to metal charge transfer. Complexes **105** and **106** exhibit a blue shift because of the intra ligand emission state which is ~~are~~ common in the case of N-donor ligands [37].

Eight Zn(II) and Cd(II) containing coordination polymers of ligand **108**viz. [Zn(m-bdc)(**108**)] (**109**), [Zn(NH_2-bdc)(**108**)] (**110**), [Zn_3(p-bdc)$_3$(**108**)].2H_2O (**111**), [Zn_3(btc)$_2$(**108**)(H_2O)].H_2O (**112**), [Cd(o-bdc)(**108**)(H_2O)] (**113**), [Cd(m-bdc)(**108**)] (**114**), [Cd(p-bdc)(**108**)(H_2O)] (**115**) and [Cd_2(btda)(**108**)].H_2O (**116**) have been hydrothermally synthesized (*m*-H_2bdc = 1,3-benzenedicarboxylic acid, NH_2–H_2bdc = 5-NH_2-1,3-benzenedicarboxylic acid, *p*–H_2bdc = 1,4-benzenedicarboxylic acid, H_3btc = 1,3,5-benzenetricarboxylic acid, *o*–H_2bdc = 1,2-benzenedicarboxylic acid, btda = 3,3',4,4'-benzophenonetetracarboxylic dianhydride. The *N*–donor ligand 1,1'-(1,4-butanediyl)-bis[2-(3-pyridyl)benzimidazole]) **108** (Scheme 27) and *O*-donor anions may show contributions to the fluorescent emissions simultaneously on the basis of emission peaks and the luminescent lifetimes for complexes **109–111**, **113** and **114**. The complexes **112** and **115** showed emission peaks λ = 370 nm which is attributed to π–π* transitions of the ligand **108**. A similar peak at λ = 370 nm was also observed in case of free ligand [38].

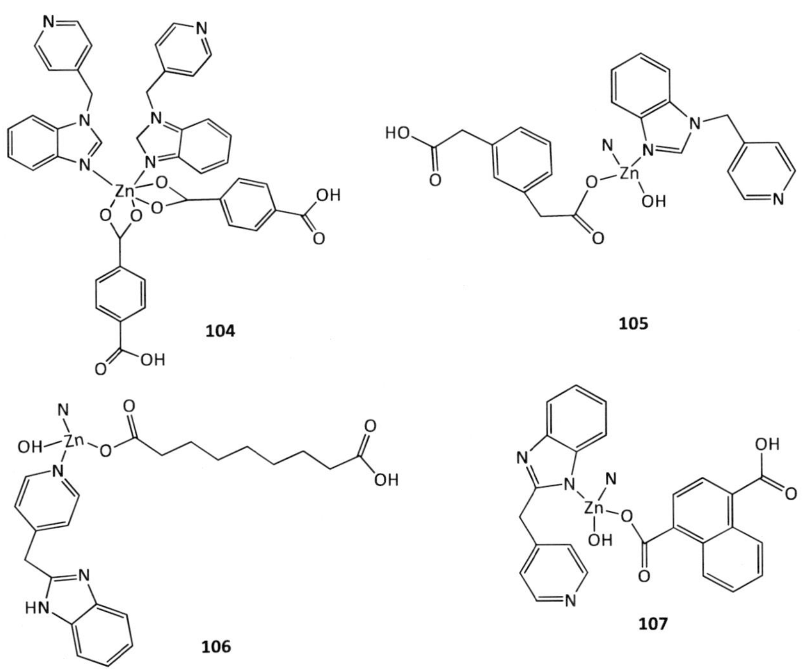

Scheme 26. Structure of Zn(II) complexes **104–107**.

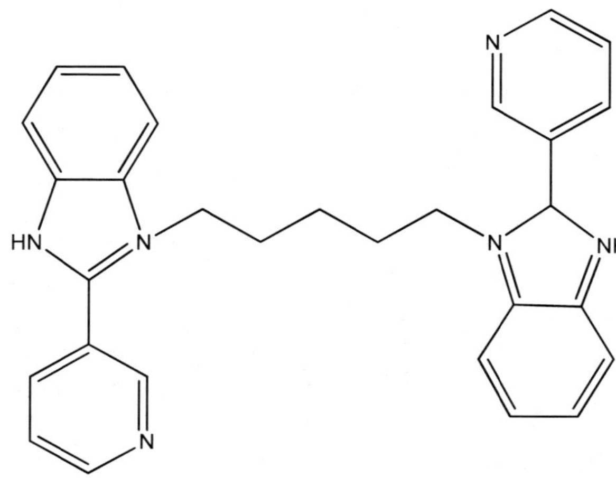

Scheme 27.Structure of ligand **108**.

4-(1-*H*-benzimidazol-2-yl)-5-(hydroxymethyl)-2-methylpyridin-3-olhydrochloride (BIMIPY) (**117**) has been synthesized by the reaction of pyridoxal hydrochloride with *o*-phenylene diamine in ethanol. The molybdenum complex $[(MoO_2)_2(H_2O)_2(\mu\text{-}O)(BIMIPY\text{–}H^+)_2].4DMSO.2H_2O$ (**119**) has been prepared by the reaction of **118** with bis(acetylacetonato)-dioxidomolybdenum (Scheme 28). Complex **119**exhibited distorted octahedral geometry around metal center. Each molybdenum atom is coordinated to a water molecule and the two nitrogen atoms present on imadazolic moiety (C=N). The two dioxymolybdenum ions (MoO_2^{+2}) are chelated to the oxygen atoms of the deprotonated hydroxyl groups. It was observed that ligand **117** exhibited intense blue photoluminescence at 471 nm along with a shoulder at 410 nm whereas **118** did not show any fluorescence in methanol [39].

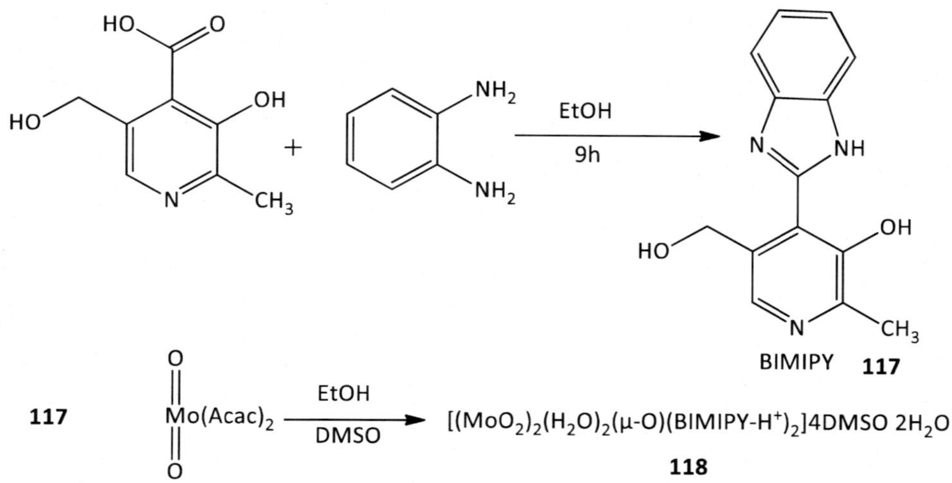

Scheme 28. Synthesis of **117** and **118**.

Ligand **119** [1-(pyridine-2-ylmethyl)-2-(3-(1-(pyridine-2-ylmethyl)benzimidazol-2-yl) propyl) benzimidazole]has been synthesized by the reaction of 2-(3-(benzimidazol-2-yl)propyl)benzimidazole and 2-(chloromethyl)pyridine (Scheme 29). The Cu(II) complex (**120**) of ligand **119** showed a distorted trigonal bipyramidal geometry around Cu center resulted due to the coordination of two nitrogen atoms of benzimidazole groups, two oxygen atoms of the nitrate anions and one oxygen atom of a water molecule. The free ligand **119** and its Cu(II) **120** complex have been explored as a fluorescent sensor for 4-(2-aminoethyl)benzene-1,2-diol. It was observed that **119** functions as an efficient sensor for 4-(2-aminoethyl)benzene-1,2-diol via host-guest interaction. The Cu(II) complex **120** showed weaker interaction in comparison to the parent ligand **119** [30].

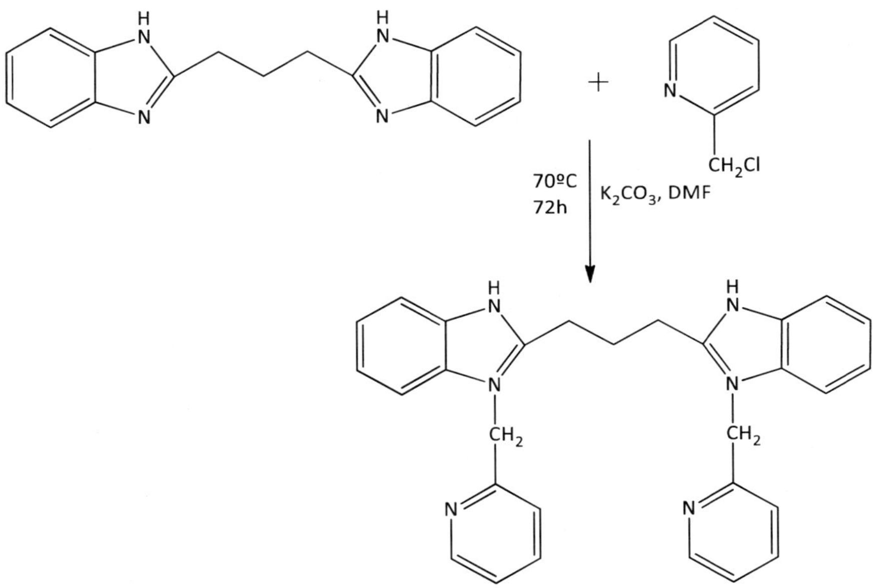

119

Scheme 29. Synthesis of ligand **119**.

Gel-like materials (GLs) have been prepared via the reaction of 2,6-bis(2-benzimidazyl)pyridine ligand (**87**) with or without squaric acid in the presence of Zn(II) or Ni(II) precursors (Scheme 30). It has been observed that a mixture of **87**, squaric acid and Zn(II) or Ni(II) gelate N,N-dimethylformamide to generate hybrid gels. Gels G2 and G3 have been found better than G1 in terms of dye absorption due to the reason that the addition of squaric acid has lead to a 3-D network in G2 and G3 which can trap dye molecules in its porous structure [40].

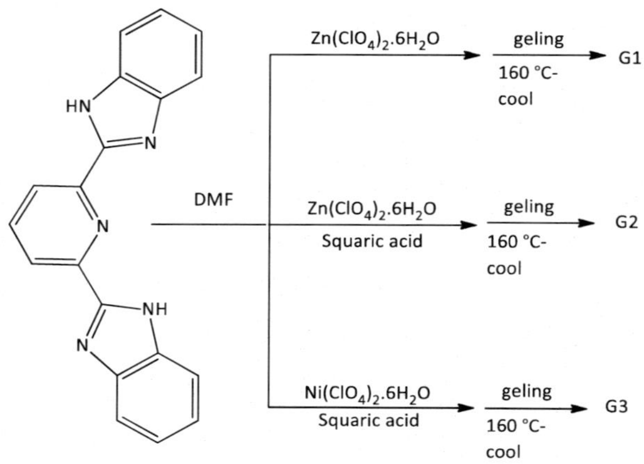

Scheme 30. Preparation of Zn and Ni gels G1−G3.

Ruthenium complexes have emerged as light-harvesting devices for dye-sensitized solar cells (DSSC). Inclusion of benzimidazole and pyridine in the framework is opted because of their electron mobilities. The benzimidazole derivatives can be developed as layer materials for electron transport and hole blocking in organic light-emitting diode devices [29]. Heteroleptic ruthenium complexes **122−125** (Scheme 31) containing pyridine substituted benzimidazole moiety namely [Ru(dcbpy)(1-methyl-2-(pyridine-2-yl)benzoimidazole)-(NCS)$_2$] (**122**), [Ru(dcbpy)(1-benzyl-2-(pyridine-2-yl) benzoimidazole)(NCS)$_2$] (**123**), [Ru(dcbpy)(1-decyl-2-(pyridine-2-yl)-benzoimidazole) (NCS)$_2$] (**124**) and [Ru(dcbpy)(1,10-dimethyl-2,2'-bibenzoimidazole)(NCS)$_2$] (**125**), have been prepared by the replacement of one 4,4'-dicarboxylic-2,20-bipyridine (dcbpy) ligand **121** by suitable benzimidazole ligand. The order of performances of the four complexes was found to be **123>124>122>125**. Photocurrent and photovoltage decay studies denote a systematic trend of electron lifetimeswhich follows the order **123>124>122>125** [29].

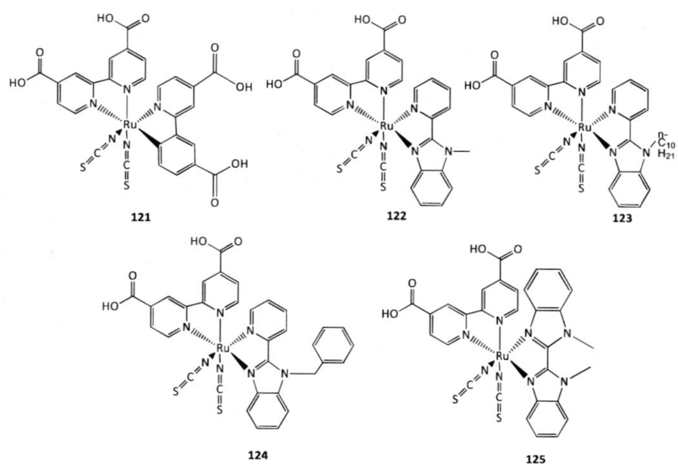

Scheme 31. Ru(II) complexes **121−125**.

Heteroleptic Ru(II) complexes **126–129** (Scheme 32) [Ru(4,4′-dicarboxylic-2, 2′-bipyridine)(1-(2,4-difluorobenzyl)-2-(4-(thiophen-2-yl)pyridine-2-yl)-benzimidazole) $(NCS)_2$] (**127**), [Ru(4,4′-dicarboxylic-2,2′-bipyridine)(1-(2,4-difluorobenzyl)-2-(4-(5-hexylthiophen-2-yl)pyridine-2-yl)benzimidazole)$(NCS)_2$] (**128**), and [Ru(4,4′-dicarboxylic-2,2′-bipyridine)(1-(2,4-difluorobenzyl)-2-(4-(5-(5-hexylthiophen-2-yl) thiophen-2-yl)pyridine-2-yl)benzimidazole)(NCS)] (**129**) have been synthesized and tested for their efficiency for dye-sensitized solar cells. The substitution of fluorine on benzimidazole has increased the open-circuit voltage drastically. The light-harvesting capabilities of the complexes have also been enhanced upon the substitution of thiophene ring on pyridine. However, the long hexyl chain on thiophene (**129** and **130**) retards charge recombination to some extent [41].

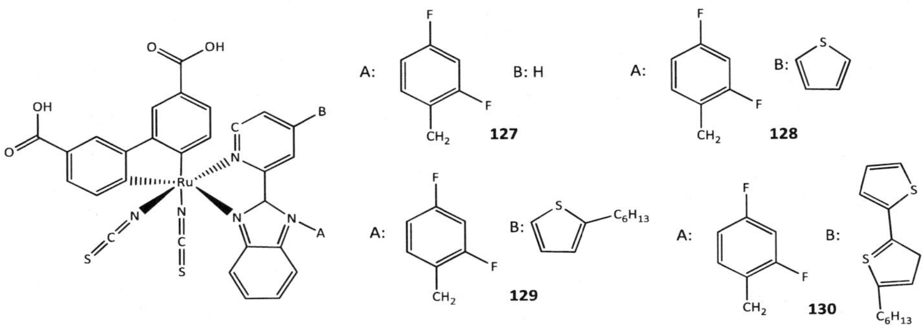

Scheme 32. Ru(II) complexes **126–129**.

Three d[10] coordination polymers $\{[Zn_3(PyBIm)_3(PyHBIm)_2(tma)(H_2O)].H_2O\}_n$ (**130**), $[Zn(PyHBIm)(oba)]_n$ (**131**), $[Cd(PyHBIm)(oba)]_n$ (**132**) (oba=4,4′-oxybis-(benzoate)), having one dimensional network have been synthesized via hydrothermal reaction. Compound **130** exhibit 1–D dimensional network-like structure which further assembles into three- dimensional network through π–π stacking interaction and hydrogen bonding. The complex**130** contains three zinc atoms in a single frame. The structure of **130** exhibited a distorted octahedral geometry around Zn atoms while the third Zn atom has distorted trigonal bipyramidal geometry. The two neighboring zinc atoms in **130** bridged by interval PyBIm (2-(2-pyridyl)benzimidazole) and tma (trimesate) ligands which form a 1-D helical chain. Compounds **131** and **132** exhibit 1-D-meso helical structure that assembles itself through aromatic π–π stacking interaction and hydrogen bonding resulted in 3-D networks. The neighboring zinc atoms in compounds **132** and **133** are bridged by long flexible spacers in a chelate coordination mode resulting in a meso-helical chain. All three complexes showed considerable thermal stability and photoluminescence at room temperature [42].

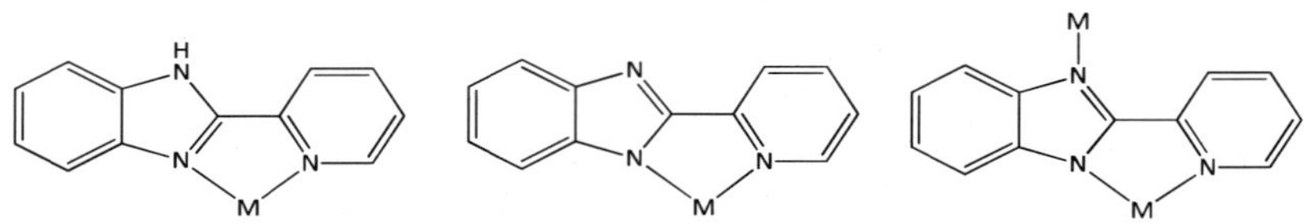

Scheme 33. Different modes of binding of ligand utilized in the synthesis of coordination polymers **130–132**.

A flexible N-containing polydentate ligand i.e., 1,10-(1,4-butanediyl)bis[2-(4-pyridyl)benzimidazole] (**L**, **134**) (Scheme 34), that consist of both pyridyl groups and benzimidazole groups into a single framework and its Mo and Cu complexes H_4**134**(γ–Mo_8O_{26}).2H_2O (**135**), [Cu(H_2**134**)(β–Mo_8O_{26})(H_2O)$_2$].3H_2O (**136**) and [Cu(H_2**134**)(γ–Mo_8O_{26})].3H_2O (**137**), Cu(H**134**)$_2$(β–Mo_8O_{26}) (**138**) has been synthesized. It was found that ligand opts a cis conformation in **136** while it a trans configuration in **137** has been observed. Compound **136** exhibits a two-dimensional structure while **137** results in three-dimensional network. The versatility in structural changes may be explained by the different conformations of the **134**. In **136**, the ligand **134** coordinates to Cu(II) ions in cis-conformations and produce meso-helical chains. These chains are linked by the (β–Mo_8O_{26})$_4$ anions to generate a 2-D layer. In compound **138**, the ligand arranges itself in a distorted cis-conformation and acts as a monodentate ligand resulting in a 1-D chain structure. Owing to the distorted cis-conformation of the ligand **134**, the chains in **138** are connected by hydrogen bonds between N atoms of benzimidazole groups and protonated pyridine groups to form a 3-D supramolecular framework [43].

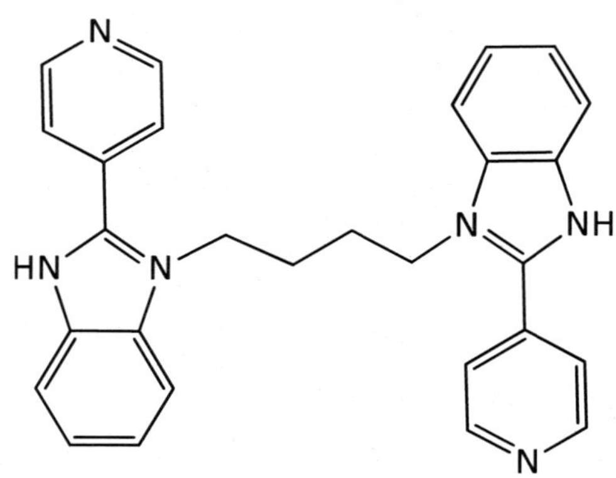

134

Scheme 34. Structure of ligand **134**.

2,6-Bis(benzimidazolyl)pyridine containing framework was also utilized to prepare Co(II) complexes **139–141**of the general formula: [Py(Bm-R)$_2$]CoCl_2 (where Py = pyridyl, Bm= benzimidazolyl) (Scheme 35). The geometry around Co(II) atom in all the three complexes was distorted square pyramidal. These complexes were explored in the stereospecific polymerization of 1,3-butadiene (BD). Upon activation with alkylaluminium, these catalysts were found to show high efficiency in the generation of high molecular weight polymers containing cis conformation up to 98%. The variation in the structure of the catalyst through N-alkylation results in the change in the activity of the catalyst. It is because of the electronic effect that is imposed on the active centers while the stereospecificity remains unchanged [44].

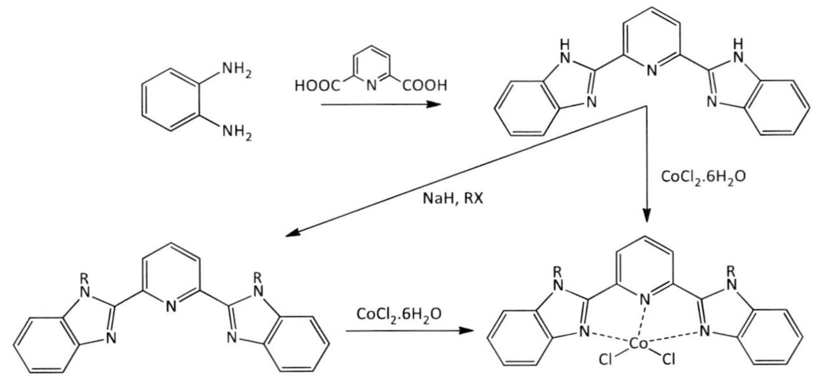

139: R: H; **140**: R: Me; **141**: R: Bz

Scheme 35. Synthesis of Co(II) complexes **139**–**141**.

A series of Iron(III) complexes **143**–**148** ligated with 2-(benzimidazole)-6-(1-aryliminoethyl)pyridines ligand (**142**) have been prepared (Scheme 36) and tested for their efficiency in ethylene oligomerization and polymerization using methylaluminoxalane (MAO) as a co-catalyst. The highest catalytic activity reported was of the order of 5.38 x 10^6 g $mol^{-1}(Fe)h^{-1}$ in the generation of producing α-olefins (C4–C28) with high selectivity. Some significant differences were observed when these reported Fe(III) catalysts were compared with their Fe(II)-analogues which was resulted due to a change in the mode of coordination. The Fe(II) analog of 6-difluoro-N-(1-(6-(1-methyl-1H-benzo[d]imidazol-2-yl)pyridine-2-yl)ethylidene)benzenamine was reported to yield the molecular $L_2Fe.FeCl_2$. The complexes bearing less bulky substituents on ligands showed better activity due to enhanced ethylene insertion [45].

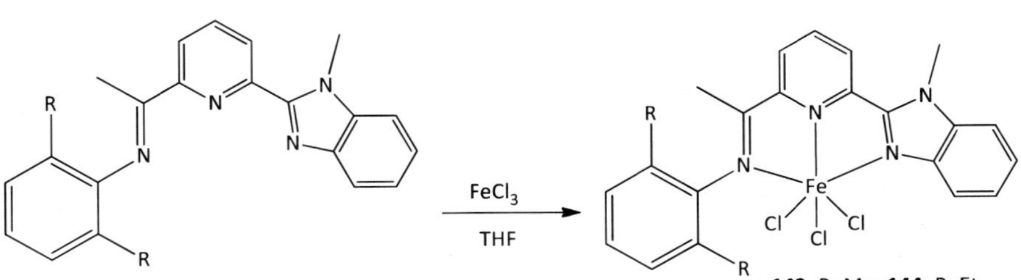

142

143: R: Me; **144**: R: Et
145: R: i-Pr, **146**: R: F
147: R: Cl, **148**: R: Br

Scheme 36. Synthesis of Fe(II) complexes **143**–**148**.

Six heteroleptic ruthenium (II) complexes **149–154** have been synthesized by the reaction of bidentate pyridyl-benzimidazole ligands with [Ru(*p*-cymene)Cl_2]$_2$ in methanol (Scheme 37).The Ru(II) complex **154** forms a half-sandwich ''three-legged piano-stool'' geometry with the η^6 bound arene ring forming the seat, the two nitrogen atoms of the ligand and one terminal chloride ligand a the legs of the piano-stool. The Ru ion showed a pseudo-octahedral coordination geometry, with the arene formally occupying three facial coordination sites. All the complexes were tested in the transfer hydrogenation of acetophenone in 2-propanol using KOH. Among all the complexes,**149** showed the best efficiency in transfer hydrogenation of acetophenone as the maximum conversion was achieved in 1h. The efficiency of the catalyst follows the order **152** >**150** >**153** >**154** >**151** [31].

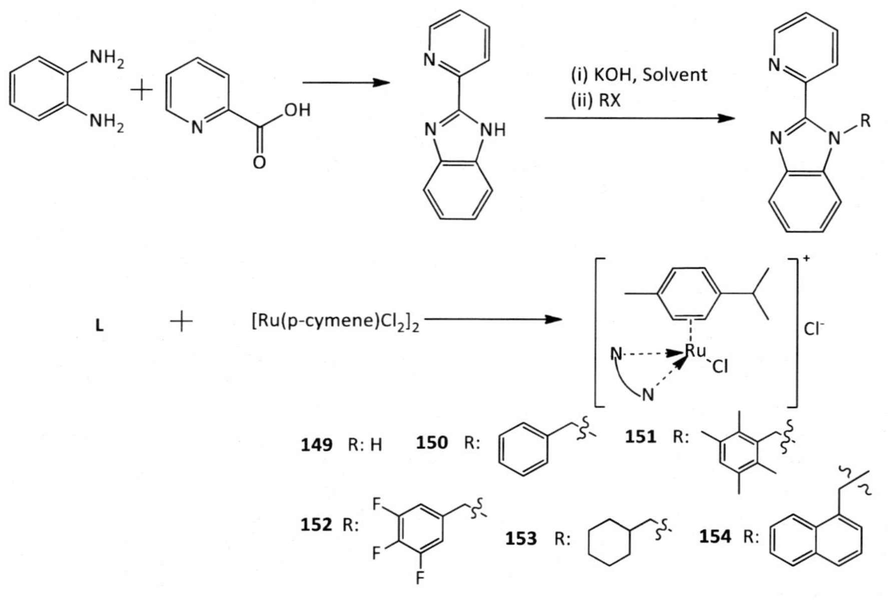

Scheme 37. Synthesis of Ru(II) Complexes **149–154**.

Another series of Ru(II) complexes **159–165** have been reported containing bis(benzimidazol-2-yl)-pyridine and (N-propylbenzimidazol-2-yl)pyridine derivatives **87**, **155–158** (Scheme 38) and have been explored for transfer hydrogenation of acetophenone. The catalyst amount of 0.5 mol% of 5 was found enough to catalyze hydrogenation reaction. Significant enhancement of the catalytic activity was witnessed in the case of **161–163** for hydrogenation reaction which is attributed to the establishment of an unsymmetrical environment around the Ru(II) atom via two different coordinating donor arms at 2,6-positions of the pyridyl backbone. Only 0.1 mol% of catalyst **164**or **165** was efficient in catalyzing the asymmetric transfer hydrogenation of acetophenone and **165** give 79% *ee* while **164** gave only 56%. The steric effect of the substituent on the chiral oxazolyl moiety has an influence on the enantioselectivity of the products [46].

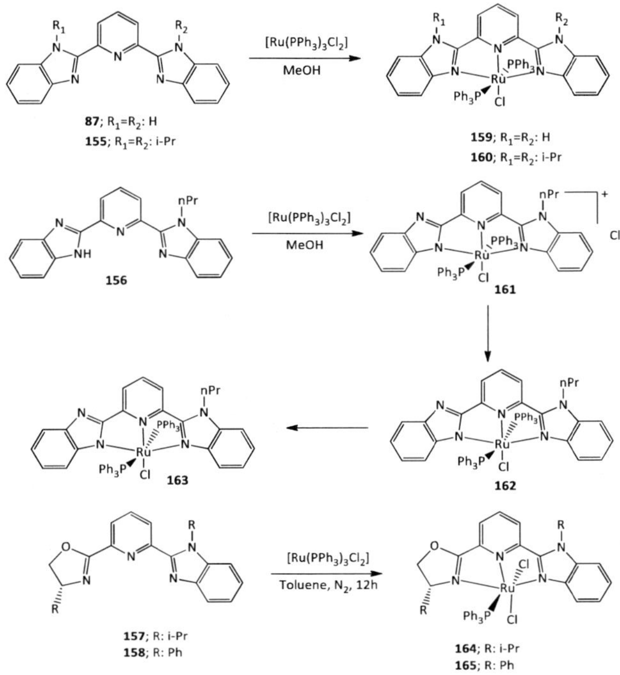

Scheme 38. Synthesis of Ru(II) complexes **159–165**.

The synthesis of the ligand 6,6'-bis[((1-methylbenzimidazol-2-yl)thio)methyl]-2,2'-bipyridine (bmb–bpy) (**166**) was carried out by the coupling of 6,6'-bis(bromomethyl)-2,2'-bipyridine with 2-benzimidazolethiol in the presence of KOH in toluene. The subsequent methylation of thioether with methyl iodide and anhydrous potassium carbonate in acetonitrile afforded bmb-bpy. The Cu(II) complex [Cu(bmb-bpy)(H_2O)]$(OTf)_2$ (**167**) was synthesized by the reaction of $Cu(OTf)_2$ with the ligand in equimolar ratio (Scheme 39) [47].

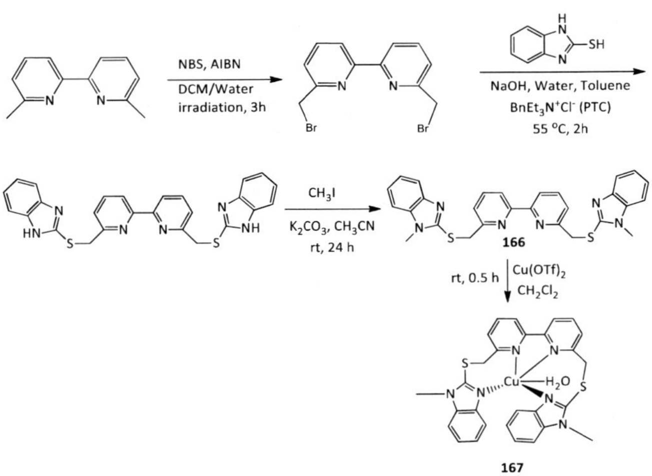

Scheme 39. Synthesis of **166** and Cu(II) complex **167**.

BENZIMIDAZOLE DERIVATIVES WITH OXYGEN DONOR SUBSTITUENTS

The benzimidazole derivatives containing oxygen donor atom have been reported and characterized including their crystal structures [48–50]. Oxygen donor in such ligands behaves as Z-donor and metal complexes are formed as a result of (N, O) bidentate chelation [51, 52]. The reaction of 2-(2-Hydroxy-5-methylphenyl)-1*H*-benzimidazole (**168**) with $Zn(NO_3)_2$ and $Cu(NO_3)_2$ afforded [Cu(**168**)$_2$].3H_2O (**169**) and [Zn(**168**)(H_2O)$_2$]NO_3 (**170**) respectively (Scheme 40). The crystal structure of the ligand **168** showed the intramolecular and intermolecular hydrogen bonding interactions N–H·····Cl and C–H·····Cl respectively [53].

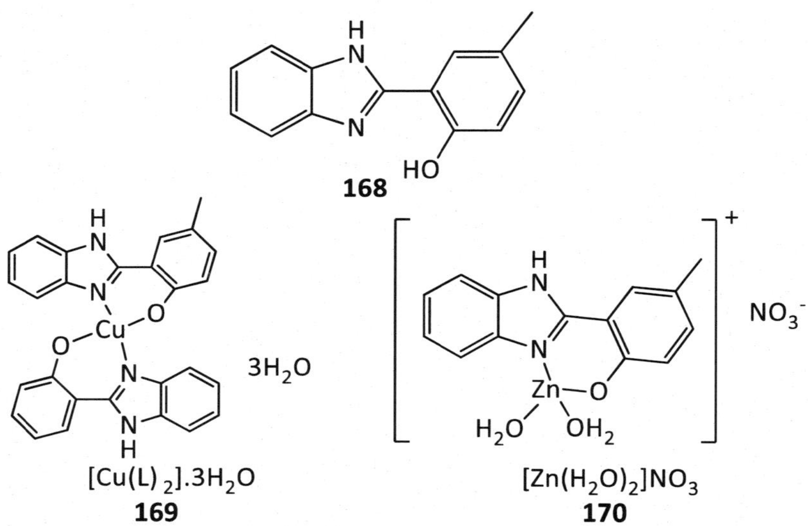

Scheme 40. Structure of **169–170**.

A new bis-benzimidazole ligand 1,3-bis(1*H*-benzimidazol-2-yl)-propan-1-one (**171**) has been reported by Tavman and coworkers. It was further utilized for the synthesis of six different Zn(II) complexes [Zn(**171**)$_2$](NO_3)$_2$.2H_2O (**172**), [Zn(**171**)$_2Cl_2$].H_2O (**173**), [Zn(**171**)(OAc)$_2$] (**174**), [Zn(**171**)$_2Br_2$].2H_2O (**175**), [Zn(**171**)$_2I_2$].4H_2O (**176**) and [Zn(**171**)(H_2O)$_2$](ClO_4)$_2$ (**177**). The imidazole ring in benzimidazole derivative is involved in chelation with the metal in the complexes.The ligand (**171**) exhibits keto-enol tautomerism in polar solvents (Scheme 41) and the keto form dominates both at room temperature and at higher temperatures. All the compounds **171–177** show fluorescence in the visible region and have been explored for their potential in light-emitting devices. It was found that a redshift occurred in **175**, **176** and **177** in comparison to the ligand. Complex **174** differs slightly in its fluorescence property due to the presence of acetate ion in complex [54].

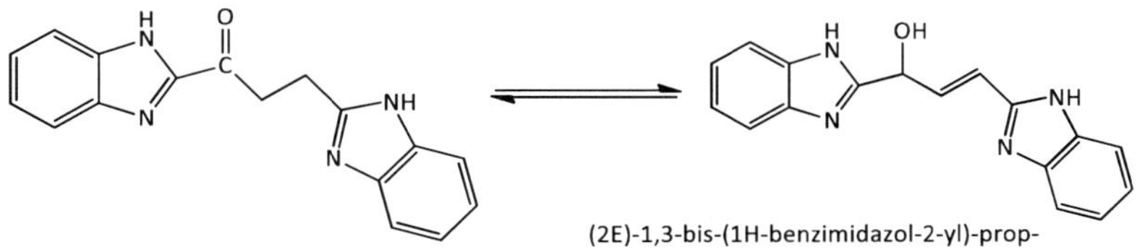

1,3-bis(1H-benzimidazol-2-yl)-propan-1-one **171** ⇌ (2E)-1,3-bis-(1H-benzimidazol-2-yl)-prop-2-en-1-ol **171**

Scheme 41. Keto-enol tautomerism in **171**.

The reaction of triphenylborane/dibromobenzoborane with oxygen substituted benzimidazole derivative [2-(2-hydroxyphenyl)benzimidazole;**178**] afforded two boron ligated compounds **179** and **180** (Scheme 42). Both the ligands contain boron bridged π-framework. The **179** has a more rigid and flat structure than **180**. Chelation of boron atom with nitrogen and oxygen atom present on the ligand framework results in π-conjugated structure in which boron atom adopts a four-coordinated geometry. In addition to good thermal stability, both the boron-containing derivatives show deep blue emissions in solution as well as in solid-phase with narrow band gaps. Ligand**180**showed better thermal stability than **179** because of the presence of fused biphenyl π−skeleton that also leads to the enhancement in its mobility. The **180** serves as a better luminescent material in organic light-emitting diodes (OLEDs) [55].

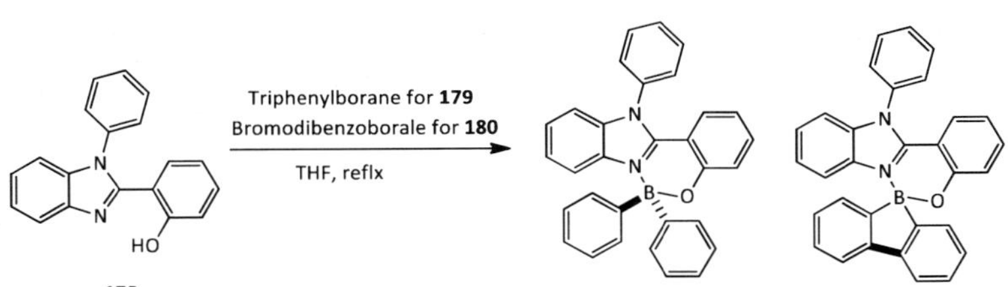

178 **179** **180**

Scheme 42. Synthesis of **179** and **180**.

Cu(II) complex (**182**) of 3-(1H-benzo[d]imidazol-2-yl)-[1,1′-binaphthalene]-2,2′-diol (**181**) has been reported (Scheme 43). It was found that the ligand is sensitive to the basic medium results in a significant change in the color from orange to green. The Cu(II) complex (**182**) behaved as a turn-on sensor for the detection of H_2S at various pH conditions [56].

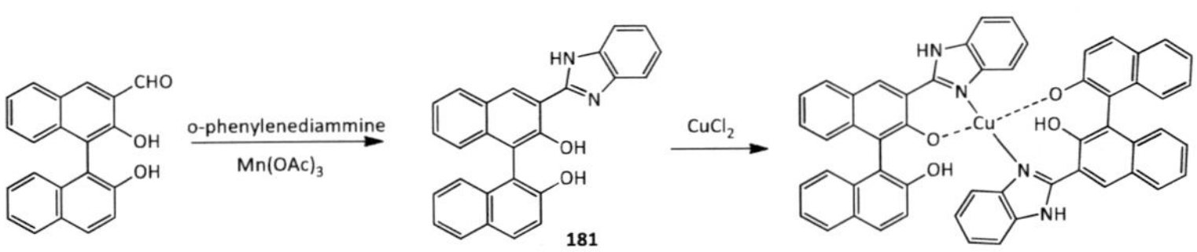

Cu(**181**)$_2$ **182**

Scheme 43. Sunthesis of **181** and Cu(II) complex of **181**.

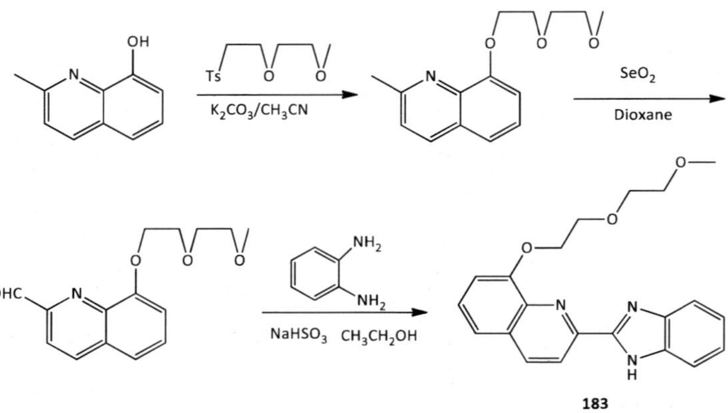

Scheme 44. Synthesis of **183**.

An oxygen-containing derivative (**183**) based on 2-(2-(8-hydroxyquinolin)-yl)benzimidazole has been prepared (Scheme 44) and reported to sense Hg^{+2}. The skeleton containing benzimidazole and quinolone has been reported to be coplanar. The oxygen and nitrogen atoms present on 8-hydroxyquinoline moiety in combination with nitrogen atom present on benzimidazole unit sense Hg^{+2}in 1:1 stoichiometry. Theoretical studies demonstrate that ligand to metal charge transfer phenomenon (LMCT) is responsible for the quenching when Hg^{+2} is bound. The usable pH range of 5−11 reported in the case of this sensor **183** is extremely beneficial for its utility in determining Hg^{+2} in biological samples [57].

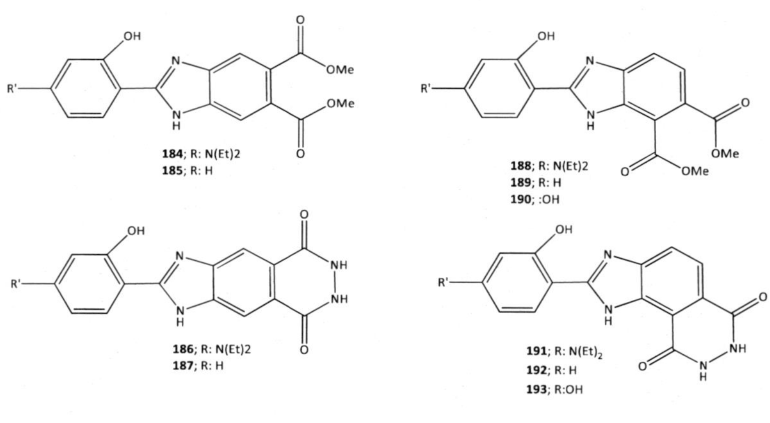

Scheme 45. Structure of ligands **184-193**.

The solvatofluorism studies of ligands **184−193** (Scheme 45) revealed that the compound **193** exhibit a dual emission in the solvent DMF while all other compounds **184−192** showed only a broad single emission in all the solvents. This broad emission is assigned to the unrestricted transfer of protons in the excited states. The theoretical keto-enol values obtained from DFT calculations agreed well with the experimentally observed emission value. The distance between –OH proton and the nitrogen atom of the benzimidazole ring was found to be below 2Å. It is explained on the basis that hydrogen present on the hydroxyl group comes in the vicinity of the nitrogen atom present on benzimidazole ring due to hydrogen bonding to facilitate transfer of proton. The chemiluminescence of **186–187** and **191–193** was also studied against luminol and isoluminol. It was observed that the chemiluminescence generation was nearly 3.5–7.3 times higher when compared against isoluminol. The initiation of this chemiluminescence occurred via the addition of the alkaline solution of potassium hexacyanoferrate(III) and hydrogen peroxide to the stock solution. The chemiluminescence intensity reached a maximum in 1.5seconds and thereafter decreased rapidly.The species formed upon excitation i.e., excited 3-aminophthalate and 4-aminophthalate ion have been reported to behave as light emitter in the chemiluminescence of isoluminol(**186–187**) and luminal (**191–193**) respectively [58].

Metal-organic frameworks (MOFs) have gained considerable interest in gas adsorption [59], catalysis [60], sensing [60] and biomedical applications [61]. Among biomedical applications, MRI image coupled with fluorescence imaging is a field that requires specially designed MOFs. In this context, two metal-organic frameworks (**195** and **196**) have been synthesized via incorporating benzimidazole derivative 1,4-bis(5-carboxy-1H-benzimidazole-2yl)benzene (**194**) (Scheme 46) along with gadolinium (Gd). MOF **195** contains an impenetrating network of 1D open channel in which series of hydrogen-bonded water molecules rests. Both the Gd-MOFs showed an intense fluorescence upon excitation at 405nm. The cytotoxicity these two MOFs was explored against MCF7 cells. A bright blue colored merged image of fluorescence was reported which showed the presence of latter MOF in the cytoplasm. Owing to the non-agglomerating property and enhanced fluorescence, **196** was preferred as MRI contrast agent. Both **195** and **196** exhibited good MRI imaging along with enhanced longitudinal relaxivity [61].

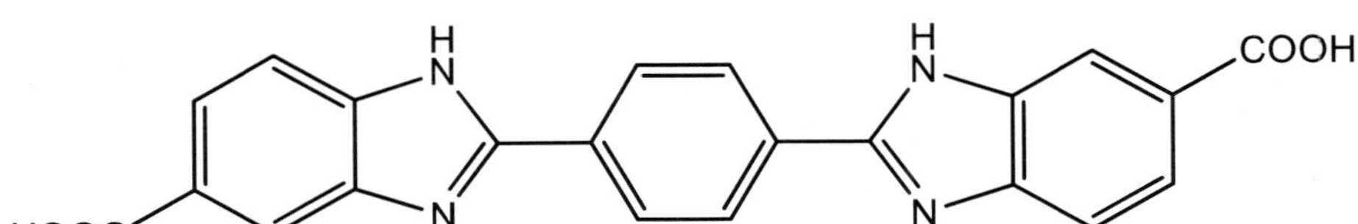

194

Scheme 46. Structure of **194**.

A series of homoleptic and heteroleptic Co(II), Ni(II), Cu(II), Zn(II) and Cd(II) complexes of ligand 1*H*-anthra[1,2-d]imidazol-6,11-dione-2-[2-hydroxyphenyl] (**197**) namely [Co(**197**)] (**198**), [Co(**197**)(bpy)] (**199**), [Ni(**197**)] (**200**), [Ni(**197**)(bpy)] (**201**), [Ni(**197**)(phen)] (**202**), [Cu(**197**)] (**203**), [Cu(**197**)(bpy)] (**204**), [Cu(**197**)(phen)] (**205**), [Zn(**197**)] (**206**), [Zn(**197**)(bpy)] (**207**), [Zn(**197**)(phen)] (**208**), [Cd(**197**)] (**209**), [Cd(**197**)(bpy)] (**210**) and[Cd(**197**)(phen)] (**211**), have been synthesized. The single-crystal structure revealed that the metal is in penta-coordinated in case of **199**, **200**, **201**, **202**, **203**, and **204**. However, a hexa-coordination was observed in case of **200**, **201** and **204** resulting in octahedral geometry. All the complexes have been reported to have N–H·····O, O–H·····N and O–H·····N interactions as well as certain non-classic C–H·····O and C–H·····N interactions and π–π stacking interactions. The ligands along with their metal complexes were tested for their cytotoxic activities against Gram-positive *Bacillus subtilis* and Gram-negative bacteria *Escherichia coli* and *Heamophilus influenza*. The Cd(II) complexes **209–211** and [Ni(**197**)(bpy)]$_2$.2H_2O (**201**) were reported to show good cytotoxic activity against Gram-positive bacteria. It was found that the in vivo antimicrobial activity of homoleptic and heteroleptic complexes were nearly equal indicating no effect of the incorporation of co-ligands [62].

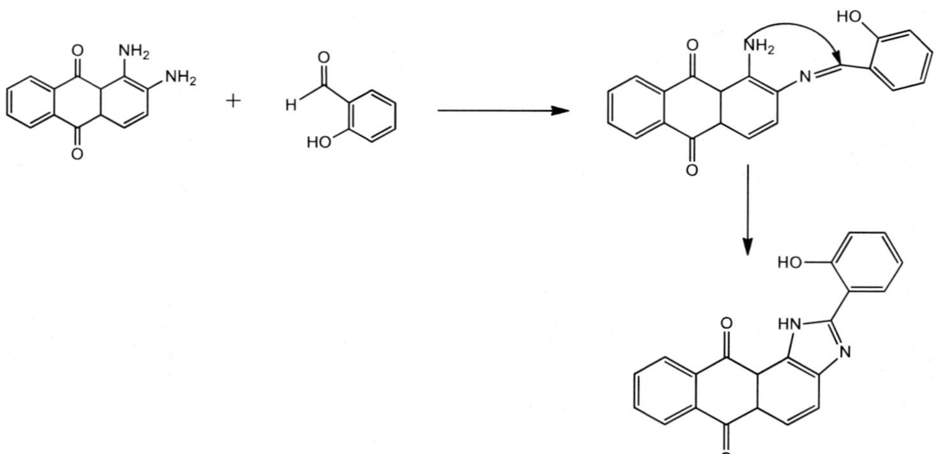

197

Scheme 47. Synthesis of ligand **197**.

The reaction of chiral benzimidazole ligand, 1,2-Bis(1*H*-benzimidazol-2-yl)ethane-1,2-diol (**212**) with dimethyltin(IV)chloride, diphenyltin(IV)chloride and triphenyltin(IV)chloride resulted in the formation of [$C_{18}H_{19}N_4O_2SnCl$] (**213**), [$C_{28}H_{23}N_4O_2SnCl$] (**214**) and [$C_{52}H_{42}N_4O_2Sn_2$] (**215**) respectively (Scheme 48). All three complexes were investigated for their binding with CT-DNA. Studies suggested that **213** and **214** binds with CT-DNA through multifaceted mode i.e., electrostatic mode as well as partial intercalative interactions. However, complex **215** binds only via one mode i.e., electrostatic interactions only. The complex **214** binds with CT-DNA more strongly than **213** and **215**since more conformational changes occurred in the helical structure of CT-DNA [63].

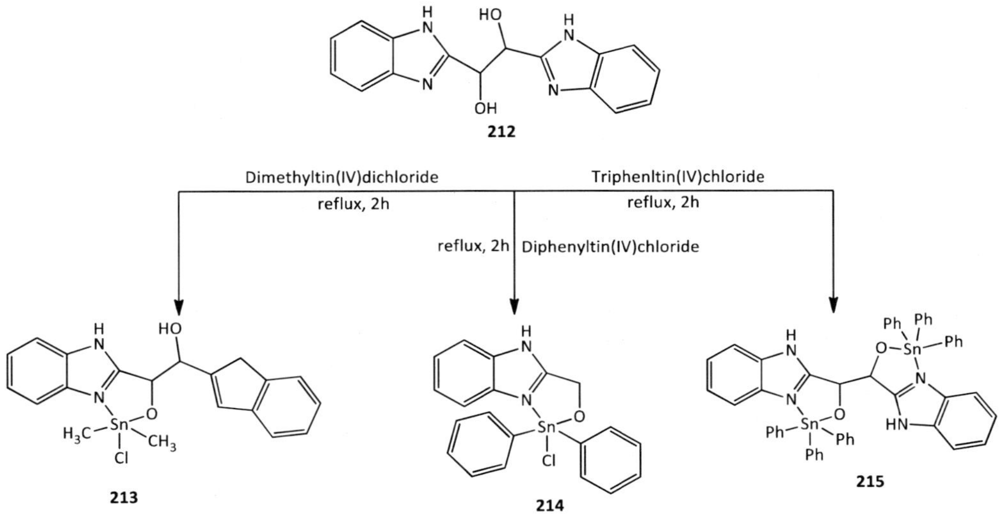

Scheme 48. Synthesis of tin complexes **213–215**.

A tetranuclear [Ni_4(**216**)$_6$Cl]·Cl (**217**) complex has been prepared using 2-(1*H*-benzo[d]imidazol-2-yl)-6-methoxyohenol (**216**) as a ligand (Scheme 49) and explored as a catalyst in the polymerization of methyl methacrylate. Even though, **216** exhibits a cationic tetranuclear host structure, there are two types of coordination environment around four nickel atoms and interaction of two Ni center resulted in the creation of active sites during the polymerization [64].

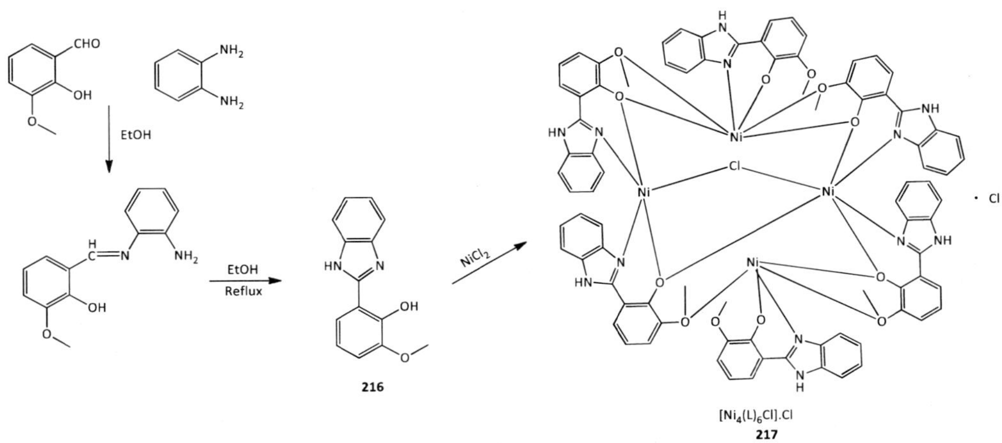

Scheme 49. Synthesis of Ni(II) complex **217**.

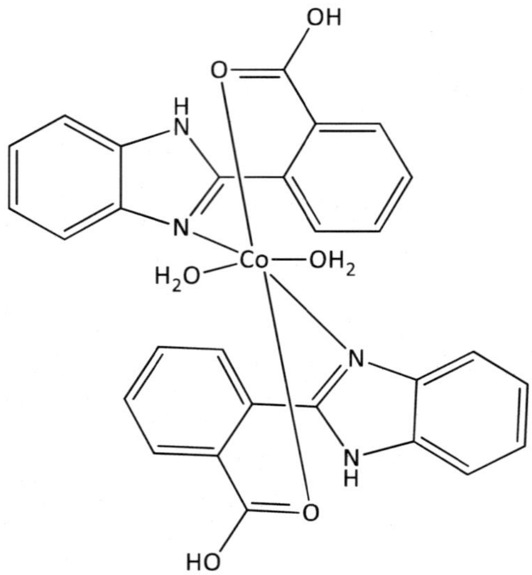

218

Scheme 50. Co(II) complex **218**.

Co(II) complex (**218**) (Scheme 50) of oxygen-containing benzimidazole derivative has been used as a catalyst for carbonylation reaction i.e., synthesis of phenylacetic acid derivatives. The catalyst showed high regioselectivity for various substituted benzyl halides to the subsequent product. Large scale experiments have also been set up to establish a wide scope, cheap availability and high efficiency of the catalyst. It was also reported that the catalyst (**218**) is recyclable up to four reaction cycles with subsequent conversions 84%, 80%, 73%, and 60% respectively [65].

CHALCOGENO–SUBSTITUTED BENZIMIDAZOLES

The chalcogen ligated benzimidazole derivatives possess unique properties due to the presence of soft donor (S/Se/Te) and hard donor (N) in a single framework. Complexes of such ligands have been studied for their catalytic [66] and magnetic behavior [67]. Chalcogen ligated benzimidazole derivatives have been explored in biology and medicine. A complex $[CuCl_2(C_{10}H_7N_3S)(C_2H_5OH)]$ (**219**) (Scheme 51) containing N and S donor atom was prepared using thiabendazole and cupric chloride. The Cu(II) atom in the complex exhibits distorted square pyramidal geometry resulted by ligation of two nitrogen atoms of ligand, one oxygen atom from ethanol and two Cl^- donors. Hydrogen bonding O–H·····Cl and N–H·····Cl and π–π stacking interactions between thiazole rings in the crystal structure resulted in three structures. The ligand **218**showed anthelmintic properties. However, its insolubility in water limits its potential use. The coordination of metal would make it more water-soluble while retaining the biological activity of the ligand [68].

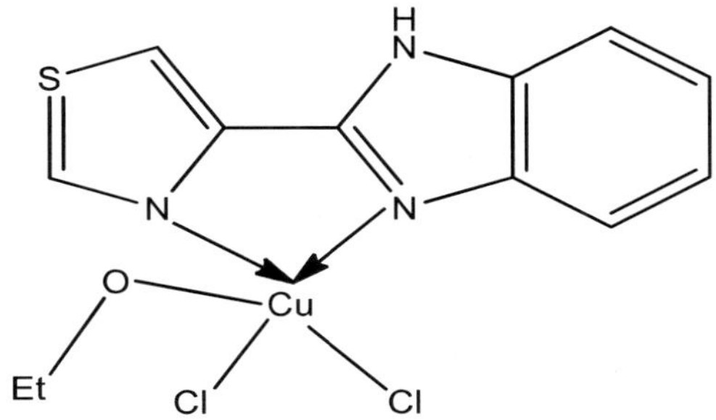

219

Scheme 51. Cu(II) complex **219**.

The ligand **220** was prepared by alkylating 2-(4,5-(4,5-bis(propylthio)-tetrathiafulvalenyl)-1*H*-benzimidazol-2-yl)-pyridine with 2,6-di(pyrazol-1-yl)-4-(bromomethyl)pyridine (Scheme 52). Two dysprosium complexes [Dy_2(hfac)$_6$(**220**)].(CH_2Cl_2)$_2$.C_6H_{14} (**221**) (hfac$^-$: 1,1,1,5,5,5-hexafluoroacetylacetonate) and [Dy_2(hfac)$_3$(tta)$_3$(**220**)] (**222**) of the ligand **220** (tta$^-$ = 2-thenoyltrifluoroacetonate) have been synthesized. The geometry around Dy1 is square antiprism(D4d) and Dy2 is spherical tricapped trigonal prism (D3h).The latter exhibited exchange between ancillary ligands in solution. Both complexes behave as single-molecule magnets (SMM) and showed multi-relaxation mode upon application of external magnetic field. Every single mode is attributed to a different Dy(III) atom. The Dy(III) atom present in N_2O_6 (D_4d environment) relaxes slower than the Dy(III) in an N_3O_6 (D_3h environment). Authors suggested that it is possible to tune the SMM property by suitable ligands designing around Dy(III) center [67].

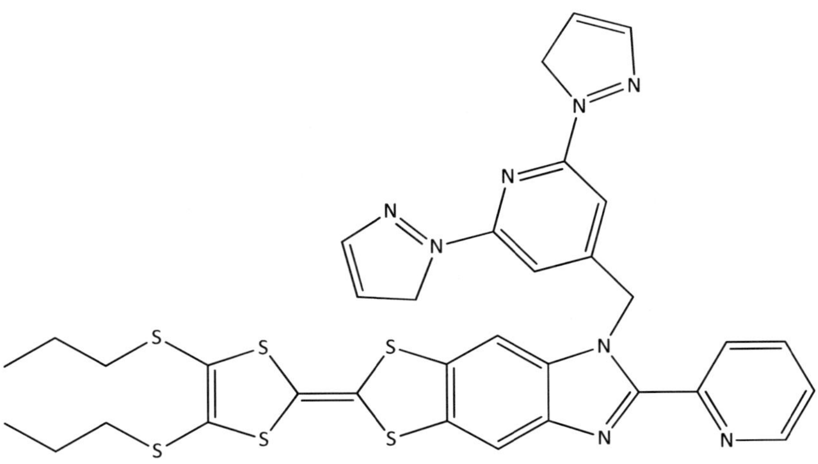

220

Scheme 52. Structure of ligand **220**.

2-(2'-(Methylseleno/thio)phenyl)benzimidazole (**223–224**) were synthesized by refluxing 2-(methylseleno/thio)benzaldehyde with an equimolar amount of 1,2-diamminobenzene. The Pd(II) / Pt(II) complexes (**225–226**) of these ligands were synthesized by the reaction of $Pd(C_6H_5CN)_2$-Cl_2/$Pt(cod)_2Cl_2$ (Scheme 53). The intramolecular C–H·····π interactions and N–H·····N, N–H·····Se interactions results in the 2D planar structure. Both the Pd(II) (**225–226**) and Pt(II) (**227–228**) complexes adopt square planar geometry. In addition to that, both these complexes incorporate the coordination of benzimidazole to Pt(II)/Pd(II) that resulted in *cis*-isomer [69].

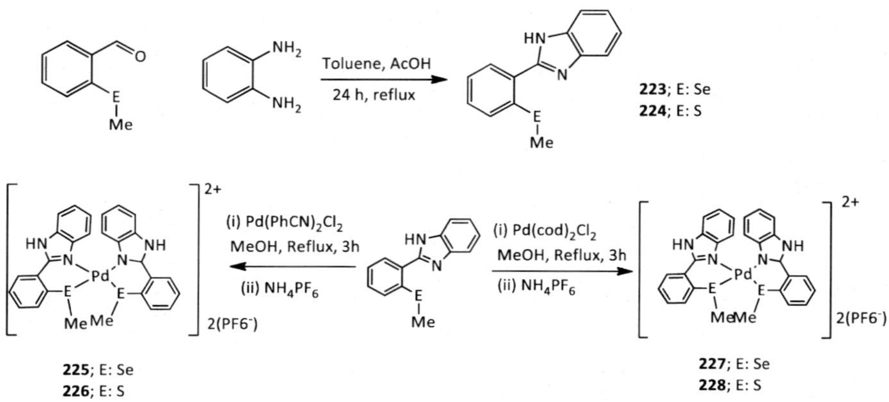

Scheme 53. Synthesis of Pd(II) and Pt(II) complexes.

Metal complexes (**230–234**) of the general formula $M[Se(BIA)]_2$ containing benzimidazole and a Se^- donor atom (Scheme 54) have been explored as anti-tumour agents for *Ehlirchasicites carcinoma* (EAC) cells induced albino mice. The monitoring of the drug effect was done by calculating total packed cell volume (TPCV) and the survival time. It was observed that $Zn[Se(BIA)]_2$ (**230**) was the most effective and potent antitumor drug followed by $Ni[Se(BIA)]_2$ (**231**) and $Cu[Se(BIA)]_2$ (**233**). The complexes $Cd[Se(BIA)]_2$ (**232**) and $Hg[Se(BIA)]_2$ (**234**) were found inactive [70].

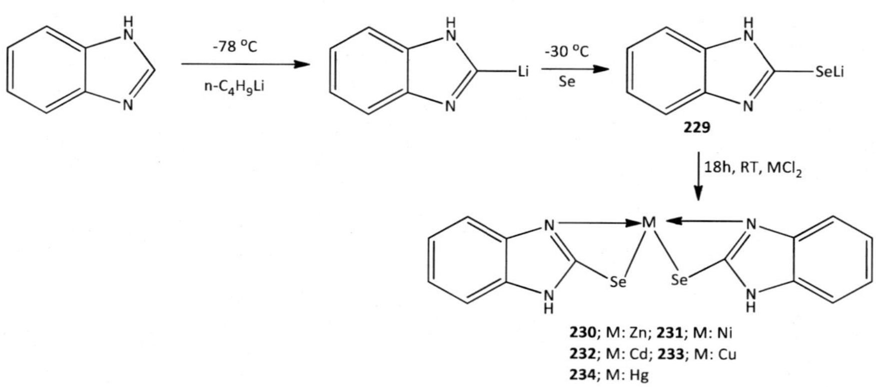

Scheme 54. Synthesis of M(II) complexes **230–234**.

Metal complexes (**236–241**) synthesized by incorporating diethyl S-(1-*H*-benzimidazol-2-yl)methyl dithiocarbamate (**235**) as ligand have been reported (Scheme 55) to exhibit anti fungicidal activity against *R. solani* and *S. rolfesii* fungi. The fungi toxic effect of all the complexes is higher than that of ligands which is due to the chelation effect. The chelation decreases the polarity of the metal complex which in turn resulted in increased lipophilicity and thus facilitating the complexes to cross the cell membrane easily. The Co(II) complex (**237**) exhibited highest anti fungicidal activity at lowest EC_{50} value among all the complexes [71].

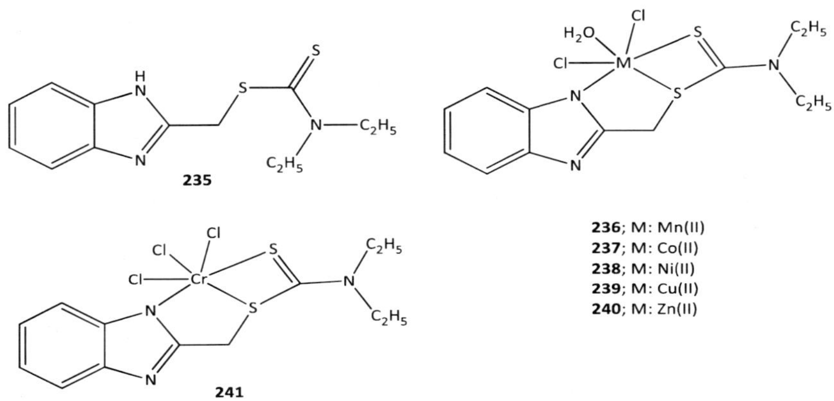

Scheme 55. M(II) complexes **236–241**.

The reaction of ZnX_2/CdX_2(X; halide) and 1-methyl-1,3-dihydro-2*H*–benzimidazole-2-selone (**241**) resulted in the formation of[H(sebenzimMe)]$_2$MX_2(**242–247**) (Scheme 56). The crystal structures of the complexes showed the involved in hydrogen bonding with N–H of imidazole and the nature of the interactions depend upon the nature of the halide present.Complexes [H(sebenzimMe)]$_2$ZnX_2 (**242** and **243**) and [H(sebenzimMe)]$_2$CdX_2(**245–246**)(X = Cl, Br), exhibit two intramolecular N–H·····X interactions, whereas the iodide derivatives, [H(sebenzimMe)]$_2$ZnI_2 (**244**) and [H(sebenzimMe)]$_2$CdI_2 (**247**), exhibit only one intramolecular N–H·····I interaction [72].

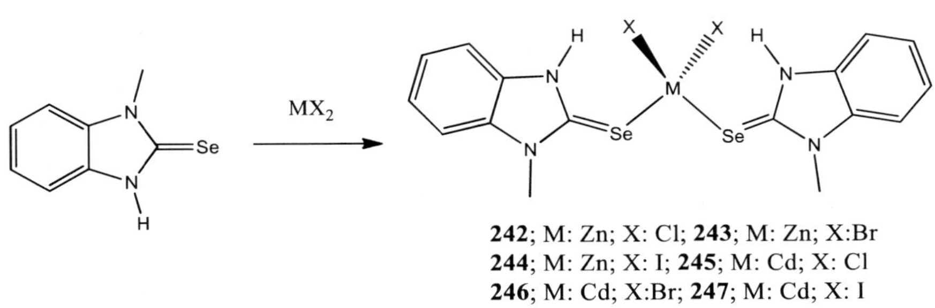

Scheme 56. Synthesis of M(II) complexes **242–247**.

Bis(pincer)chalcogens ligands (**248–253**) have been synthesized by the reaction of their corresponding chalcogen precursor (Se powder for **248** and Na_2E_2 for **249–253**). The ligands **250** and **251** were utilized for the synthesis of bis(dihaloselones) **254–256** (Scheme 57). The single-crystal structure of **248** and **249** revealed that the chalcogen donors are cis to each other with respect to the central benzene ring. It has been also observed that the planes of two benzimidazole rings lie perpendicular to the benzene ring in **253**. Compounds **248** and **249** exhibited π–π stacking interactions that result in the formation of one-dimensional network while **253** has intermolecular secondary interactions between selenium atom and bromine atom [73].

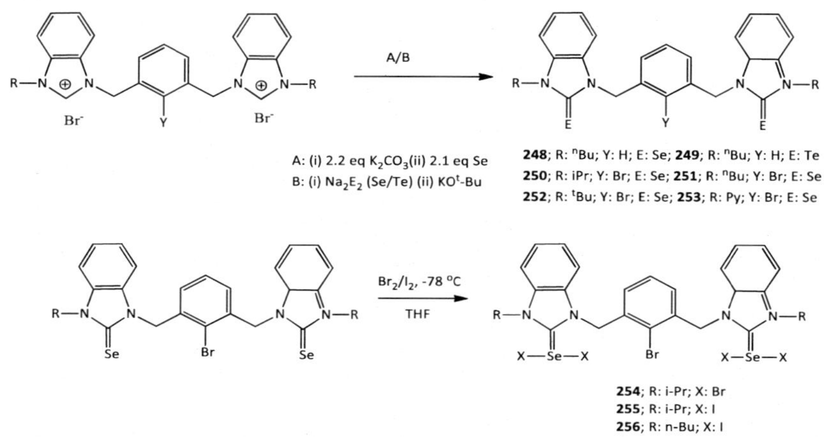

Scheme 57. Synthesis of pincer ligands **248–256**.

The synthetic strategy for the syntheses of the benzimidazolium salt (**257**) and Pd(II) complex [Pd(**257**)–$HClCl_2$] (**258**), is shown in Scheme 58.The geometry around the Pd center in complex is square planar and formation of a six-membered chelate ring resulted from bidentate coordination mode was observed. The catalytic potential of **258** was explored for Suzuki Miyaura C–C coupling reaction of aryl/heteroaryl halide and phenylboronic acid using potassium carbonate as a base. Good conversions (>80%) were achieved in two hours of reaction time using only 0.02 g of the catalyst. In addition, the catalyst was easily separable and recyclable up to seven reaction cycles [74].

The reaction of 1-benzyl-3-phenylthio/selenomethyl-1,3-dihydrobenzoimidazole-2-thione/selenone [**259–262**] with $[(\eta^5\text{-Cp*})IrCl(\mu\text{–}Cl)]_2$ (Scheme 59) followed by addition of NH_4PF_6 resulted in the formation of$[(\eta^5\text{-Cp*})Ir(\mathbf{L})Cl][PF_6]$ (**263–266**). Half sandwich pseudo octahedral piano stool type geometry around Ir atom was observed in these complexes. Compounds (**263–266**) have been used as potential catalysts for the synthesis of 1,2-disubstituted benzimidazoles. Moderate to good yields were observed at 0.1 mol% of catalyst loading. Complex **263** was found more efficient in comparison to other complexes. This complex was also utilized for catalytic oxidation of alcohols to ketones and aerobic oxidation of benzylic alcohol [66].

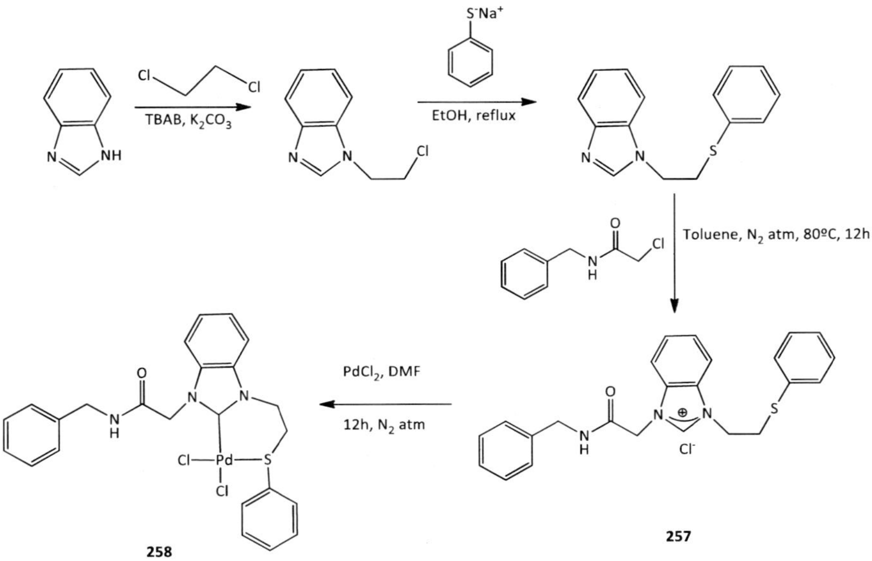

Scheme 58. Preparation of Pd(II) complex **258**.

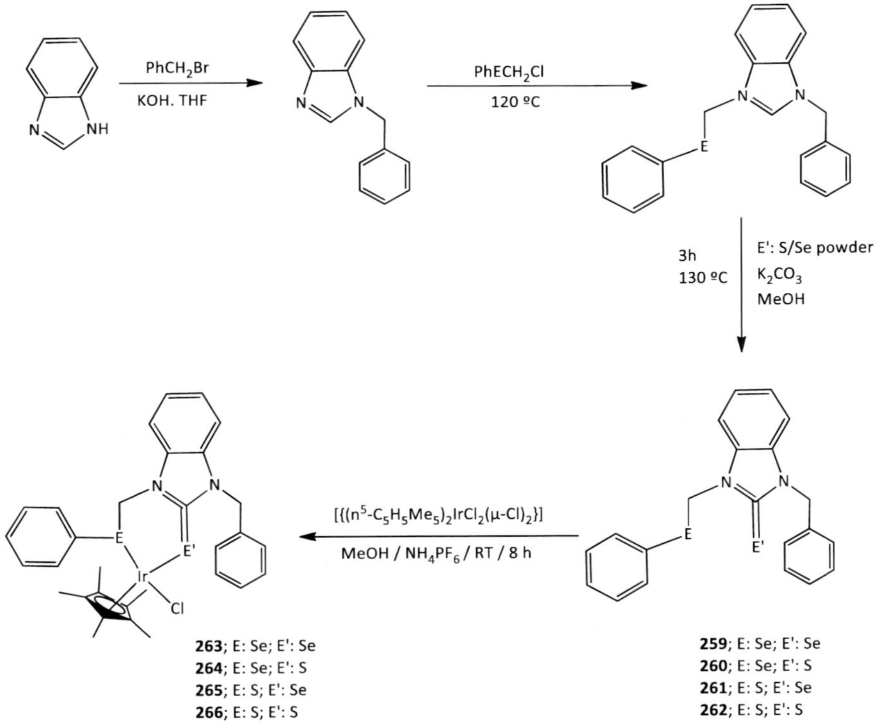

Scheme 59. Synthesis of Ir(II) complexes **263–266**.

BENZIMIDAZOLE DERIVATIVES WITH *N*–HETEROCYCLIC CABENE FUNCTIONALITY

Carbenes derivatives of benzimidazole have gained attraction owing to their flexibility and conformational freedom that allow them to exhibit greater structural diversity. A flexible multidentate N-heterocyclic ligand 2-(1-*H*-imidazol-1-yl-methyl)-1-*H*-benzimidazole (imb, **267**) and its two Cd(II) complexes {[Cd(imb)(μ_2-Cl)Cl].CH_3OH}$_n$ (**268**) and {[Cd(imb)(μ_2-Cl)Cl].DMF}$_n$ (**269**) have been reported. **268** exhibits a ladder chain-type structure while the incorporation of DMF into the second (**269**) leads to an infinite loop-like structure. The inter chain Cd·····Cd interactions and hydrogen bonding between O–H·····Cl uncoordinated methanol and chlorine atom and N–H·····O interaction between benzimidazole and uncoordinated methanol exist in **268**. In **269**, N–H·····O hydrogen bonding existed between the benzimidazole and DMF molecule along with π–π stacking interactions apart from Cd·····Cd interactions [75].

The complexes **271** [(η^6-*p*-Cymene)$_2$Ru(HL)(Cl)](Cl), **272** [(η^6-p-Cymene)$_2$Ru(HL)(Br)](Br) and **273** [(η^6-*p*-Cymene)$_2$Ru(HL)(I)](I) were synthesized by the reaction of a NHC derivative of benzimidazole (bis(3,5-dimethyl-1*H*-pyrazol-1-yl)methyl)-1*H*benzo[d]imidazole (HL, **270**) with [(η^6-*p*-Cymene)$_2$Ru(Cl)$_2$] for **271**, [(η^6-*p*-cymene)$_2$Ru(Br)$_2$]for **272** and [(η^6-*p*-cymene)$_2$Ru(I)$_2$] for **273**, respectively (Scheme 60). These Ru(II) complexes were tested for their cytotoxic activity and exhibited low affinity towards DNA bases even though they were cytotoxic in nature with IC_{50} values below 15μM. The ligand **270** dissociate from the complexes **271–273** in the presence of glutathione (GSH). However, the Ru(II) complexes retain their activity in the presence of GSH because they impair mitochondrial function and reduce cellular GSH pool. Complex **271** showed the best cytotoxic activity against prostate cancer cell line. These complexes were also reported to show anti-proliferative activity by the mitochondria-mediated intrinsic apoptotic pathway [76].

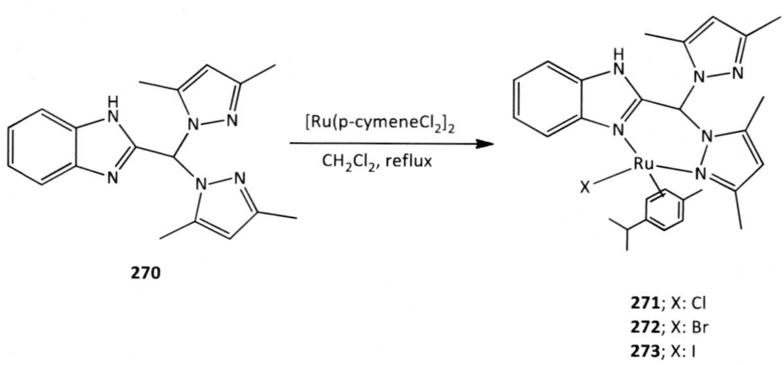

Scheme 60. Synthesis of Ru(II) complexes **271–273**.

Six new NHC containing benzimidazole derivatives (**274–279**) have been synthesized (Scheme 61) and their silver and palladium complexes have been reported. The Pd(II) complexes (**280–285**) have been used as catalysts for the Suzuki Miyaura cross-coupling reaction. $Pd(OAc)_2$ with free ligand have also been explored for Suzuki coupling reaction and it was found that the synthesized complexes showed lesser activity than $Pd(OAc)_2$ with free ligand. This might attribute due to in situ generations of Pd–NHC complex which showed better activity that the pre-formed Pd–NHC complex since the induction period involved in $Pd(OAc)_2$/LHX. During this period, Pd(II) NHC complex is formed and then it gradually decomposes to catalyze the cross–coupling reaction [77].

Palladium(II) complexes (**286–287**) of benzimidazole based NHC ligands (Scheme 62) have been synthesized and explored as a catalyst for Suzuki coupling as well as coupling of acid chlorides with aryl boronic acids. The heteroaryl halides that usually require harsh reaction conditions underwent reaction smoothly at 40 °C yielding 80–90% cross-coupled product. The multistep Suzuki coupling reaction involving dihalobenzene with various derivatives of diphenylbenzene have also been carried out and moderate yields were reported in 1.5 h even for bulkier reagents. Acid chlorides have also been coupled with aryl boronic acid using **287** as catalyst and 2 h was sufficient enough to catalyze the reaction at 50 °C and at a catalyst loading of 2 mol% [78].

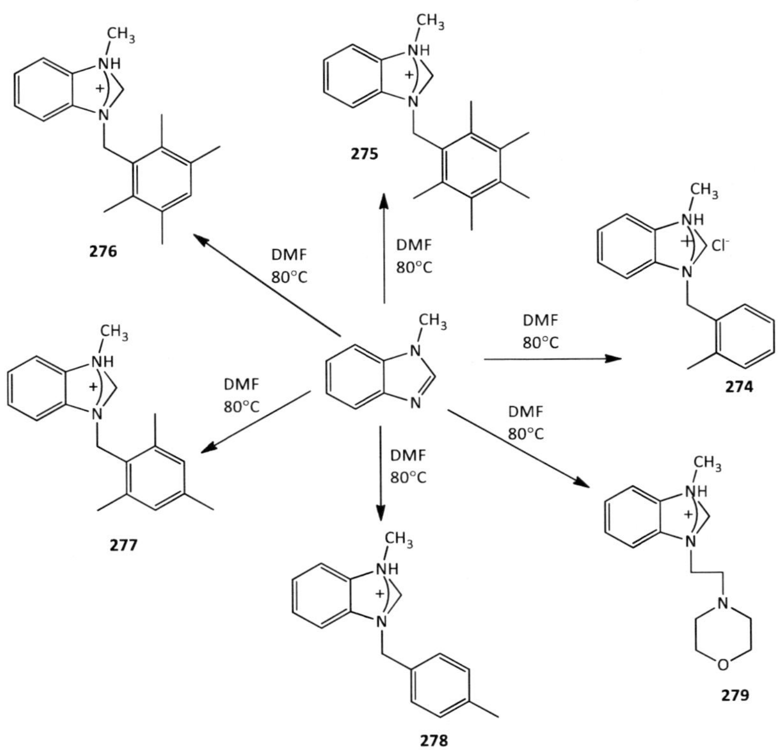

Scheme 61. Synthesis of ligands **274–279**.

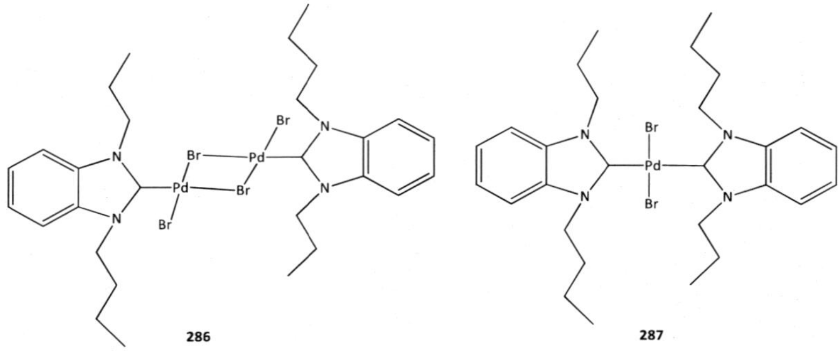

Scheme 62. Pd(II) complexes **286** and **287**.

Pd(II) and Ni(II) complexes namely **291** and **292** of ligand **288**, **293** and **294** of ligand **289**, **295** and **296** of ligand **290** have been reported by Li and coworkers (Scheme 63). Complexes **286** and **288** have been tested for their efficiency towards Heck coupling reaction and excellent conversions >92% were obtained in 24 h at 130 °C when **294** was used a catalyst. It was also observed that the NHC substituted benzimidazole were superior to the pyridine functionalized benzimidazole derivatives. Catalyst **293** and **295** were explored for their catalytic potential in Kumada coupling and former was found to be slightly better at 0.5 mol% of catalyst loading [79].

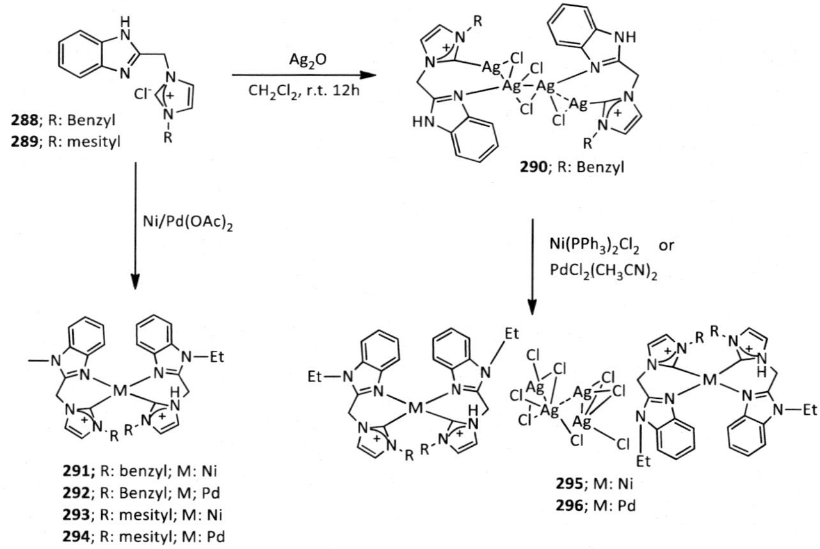

Scheme 63. Synthesis of Pd(II) complexes **292, 294** and **296** and Ni(II) complexes **291, 293** and **295**.

The N-alkylation of the 2,6-diisopropyl-phenyl-benzimidazole with substituted benzyl bromides yielded unsymmetrical benzimidazolium salts which were used for the synthesis of three robust Pd-NHC complexes (**297–300**) (Scheme 64). Catalytic amination of chloroarenes with primary and secondary amines have been carried out at 100 °C at varied mol% of **297**. The benzimidazole based NHC catalysts (**297–300**) are better in comparison to the previously reported NHC catalysts because of the presence of σ-donating property of benzimidazole and the flexibility of the methylene group and the less congestic steric hindrance around the Pd center [80].

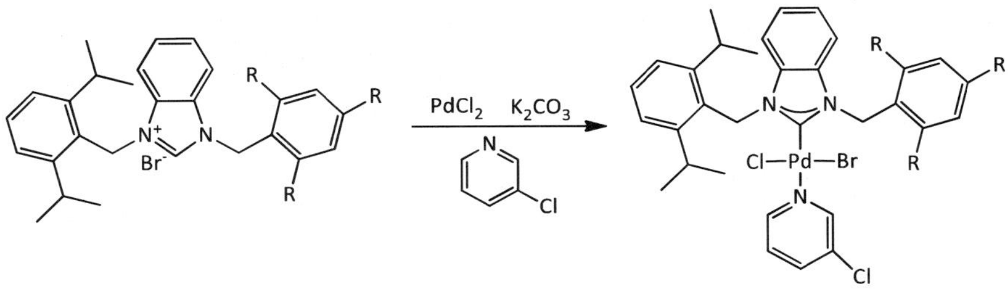

297; R: i-Pr
298; R: Me
300; R: H

Scheme 64. Synthesis of Pd–NHC complexes **297–300**.

The reaction of 2-(2-bromoethyl)-1,3-dioxane with 1-alkylbenzimidazole derivatives lead to the synthesis of the new benzimidazolium salts **301–303**. The reaction of 1,3-dialkylbenzimidazolium salts with Pd(OAc)$_2$ palladium-*N*-heterocycliccarbene (NHC) complexes (**304–306**) were formed (Scheme 65). The amination of bromobenzene with various primary amine derivatives has been performed using all three Pd–NHC catalysts. Catalyst**306** was found to show significant higher activity than other two for amination reaction [81].

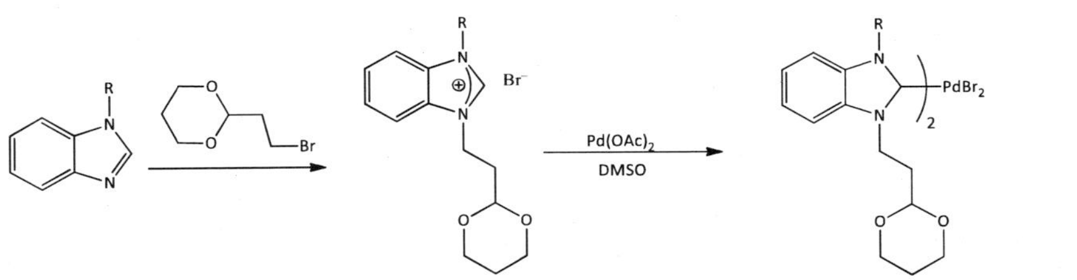

301; R: CH_2CH_2OMe
302; R: $CH_2C_6H_2(CH_3)_3$-2,4,6
303; R: $CH_2C_6H_2(OCH_3)_3$-3,4,5

304; R: CH_2CH_2OMe
305; R: $CH_2C_6H_2(CH_3)_3$-2,4,6
306; R: $CH_2C_6H_2(OCH_3)_3$-3,4,5

Scheme 65. Synthesis of Pd–NHC based catalysts **304–306**.

[Ni(**307**)$_2$](PF$_6$)$_2$ (**309**), [Hg(**307**)$_2$(CH$_3$CN)$_2$](PF$_6$)$_2$ (**310**) and [Ni(**308**)$_2$ (CH$_3$CN)](PF$_6$)$_2$ (**311**) and of 3-(1-ethyl-1*H*-benzimidazol-2-yl)-methyl)-1-((6-methylpyridin-2-yl)methyl)imidazolyl-idene) (**307**), and 3-(1-ethyl-1*H*-benzimidazol-2-yl)methyl)-1-((6-methylpyridin-2-yl)-methyl)benzimidazolylidene) (**308**) has been synthesized and reported (Scheme 66). These complexes have been tested as potential catalysts in Friedel Crafts alkylation.It was observed that **311** was highly efficient for alkylation of indoles with b-nitrostyrenes at room temperature yielding moderate-to-excellent yields [82].

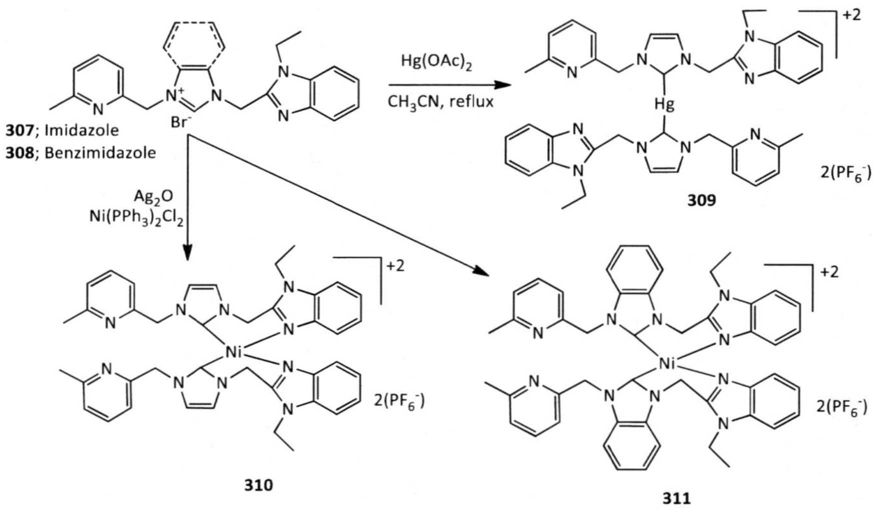

Scheme 66. Preparation of complexes **309–311**.

REFERENCES

[1] Singh, V. K. & Parle, A. (2019). "The Intriguing Benzimidazole: A Review." *International Journal of Pharmaceutical Sciences and Research*, *10*, 1540–1552.

[2] Barker, H. A., Smyth, R. D., Weissbach, H., Toohey, J. I., Ladd, J. N. Z & Volcani, B. E. (1960). "Isolation and Properties of Crystalline Cobamide Coenzymes Containing Benzimidazole or 5,6–Dimethylbenzimidazole." *Journal of Biological Chemistry*, *235*, 480–488.

[3] Bhatnagar, A., Sharma, P. K. & Kumar, N. (2011). "A Review on "Imidazoles": Their Chemistry and Pharmacological Potentials." *International Journal of PharmTech Research*, *3*, 268–282.

[4] Shukla. (2012). "Synthesis and Biological Screening of benzimidazole derivatives, *International Journal of Pharmaceutical Sciences and Research*, *3*, 922–927.

[5] Rao, G. E., Babu, P. S., Koushik, O. S., Sharmila, R., Vijayabharathi, M., Maruthikumar, S., Prathyusha, R. & Pavankumar, P. (2016). "A Review on Chemistry of Benzimidazole Nucleus and Its Biological Significance." *International Journal of Pharmaceutical, Chemical and Biological Sciences*, *6*, 227–232.

[6] O'Neil & Maryadele, J. (2001). "The Merck index an encyclopedia of chemicals, drugs, and biologicals." *The Merck Edition, 13 Edition*, Merck & Corporation Inc., p–1785, 10074.

[7] Goswami, B. (2012). "Pharmacological Activities of Benzimidazole Derivatives: Overview." *International Journal of Science, Innovations and Discoveries*, *2*, 121−136.

[8] Daryaei, I, Randtke, E. A. & Page, M. D. (2017). "A Biomarker−Responsive T_{2ex} MRI Contrast Agent." *Magnetic Resonance in Medicine*, *77*, 1665−1670.

[9] Dibrova, S. M., Dinga, K., Brunna, N. D., Parkerb, M. A., Bergdahl, B. M., Wylesc, D. L. & Hermanna, T. (2012). "Structure of a Hepatitis C Virus RNA Domain in Complex with a Translation Inhibitor Reveals a Binding Mode Reminiscent of Riboswitches." *Proceedings of National Academy of Sciences of United States of America*, *109*, 5223−5228.

[10] Ghani, N. T. A. & Mansour, A. M. (2011). "Novel Pd(II) and Pt(II) Complexes of *N,N*−Donor Benzimidazole Ligand: Synthesis, Spectral, Electrochemical, DFT Studies and Evaluation of Biological Activity." *Inorganica Chimica Acta*, *373*, 249–258.

[11] Hu, Z., Ou, L., Li, S. & Yang, L. (2014). "Synthesis and Biological Evaluation of 1-Cyano-2-aminobenzimidazole Derivatives as a Novel Class of Antitumor Agents." *Medicinal Chemistry Research*, *23*, 3029−3038.

[12] El−Sherif, A. A. (2010). "Synthesis, Solution Equilibria and Antibacterial Activity of Co(II) with 2−(Aminomethyl)−Benzimidazole and Dicarboxylic Acids." *Journal of Solution Chemistry*, *39*, 1562−1581.

[13] Sánchez, D. S., Baeza, A. & Alonso, D. A. (2016). "Organocatalytic Asymmetric Chlorination of 1,3-Dicarbonyl Compounds Catalyzed by 2−Aminobenzimidazole Derivatives." *Symmetry*, *8*, 3−12.

[14] Gómez−Torres, E., Alonso, D. A., Gómez−Bengoa, E., & Nájera, C. (2013). "Enantioselective Synthesis of Succinimides by Michael Addition of 1,3-Dicarbonyl Compounds to Maleimides Catalyzed by a Chiral Bis(2-aminobenzimidazole) Organocatalyst." *European Journal of Chemistry*, 1434−1440.

[15] Lee, M., Zhang, L., Park, Y. & Park, H. G. (2012). "Chiral 2−Aminobenzimidazole Bifunctional Organocatalysts: Effect of di−CF_3 and TFA on Catalytic Mechanisms." *Tetrahedron*, *68*, 1452−1459.

[16] Li, Y., Ding, K. & Sandoval, C. A. (2009) "Hybrid NH_2−Benzimidazole Ligands for Efficient Ru-Catalyzed Asymmetric Hydrogenation of Aryl Ketones." *Organic Letters*, *11*, 907−910.

[17] Nikolaevskiia, S. A., Koshchienko, Y. V., Chernyshev, A. V., Burlov, A. S., Cheprasov, A. S., Aleksandrov, G. G., Kiskinb, M. A. & Metelitsa, A. V. (2014). "Zinc Complexes of 1−Propyl−2−(2−Tosylaminophenyl)−5−Aminobenzimidazole: Synthesis, Structure, and Luminescence Properties." *Russian Journal of Coordination Chemistry*, *40*, 468−472.

[18] Garcia−Ochoa, E., Guzmán−Jiménez, S. J., Hernández, J. G., Pandiyan, T., Vásquez−Pérez, J. M. & Cruz−Borbolla, J. (2016). "Benzimidazole Ligands in the

Corrosion Inhibition for Carbon Steel in Acid Medium: DFT Study of its Interaction on Fe_{30} Surface." *Journal of Molecular Structure*, *1119*, 314–324.

[19] Chung, K. H., So, C. M., Wong, S. M., Luk, C. H., Zhou, Z., Lau, C. P. & Kwong, F. Y. (2012). "Buchwald-Hartwig Amination of Aryl Chlorides Catalyzed by Easily Accessible Benzimidazolyl Phosphine-Pd Complexes." *Synlett*, *8*, 1181–1186.

[20] Wong, K., So, C. M. & Kwong, F. Y. (2012). "The Recent Development of Phosphine Ligands Derived from 2–Phosphino–Substituted Heterocycles and Their Applications in Palladium Catalyzed CrossCoupling Reactions." *Synlett*, *23*, 1132–1153.

[21] Wong, S. M., So, C. M., Chung, K. H., Luk, C. H., Lau, C. P. & Kwong, F. Y. (2012). "P,N–Type Benzimidazolyl Phosphine Ligands for the PalladiumCatalyzed Suzuki Coupling of Potassium Aryltrifluoroborates and Aryl Chlorides." *Tetrahedron Letters*, *53*, 3754–3757.

[22] Hahn, F. E., Naziruddin, A. R., Hepp, A. & Pape, T. (2010). "Synthesis, Characterization, and H–Bonding Abilities of Ruthenium(II) Complexes Bearing Bidentate NR,NH–Carbene/Phosphine Ligands." *Organometallics*, *29*, 5283–5288.

[23] Song, B., Knauber, T. & Gooßen, L. J. (2013). "Decarboxylative Cross–Coupling of Mesylates Catalyzed by Copper/Palladium Systems with Customized Imidazolyl Phosphine Ligands." *Angewandte Chemistry International Edition*, *52*, 2954–2958.

[24] Mobinikhaledi, A., Foroughifar, N. & Kalhor, M. (2010). "An Efficient Synthesis of Schiff Bases Containing Benzimidazole Moiety Catalyzed by Transition Metal Nitrates." *Turkey Journal of Chemistry*, *34*, 367–373.

[25] Engle, J. T., Martić, G. & Ziegler, C. J. (2013). "Investigations into the Coordination Chemistry of 1,3–Bis(2'–benzimidazolylimino)isoindoline." *Macroheterocycles*, *6*, 353–359.

[26] Joseph, J., Suman, A., Nagashri, K., Joseyphus, R. S. & Balakrishnan, N. (2017). "Synthesis, Characterization and Biological Studies of Copper(II) Complexes with 2-Aminobenzimidazole Derivatives." *Journal of Molecular Structure*, *1137*, 17–26.

[27] Mahmood, K., Hashmi, W., Ismail, H., Mirza, B., Twamley, B., Akhter, Z., Rozas, I. & Baker, R. J. (2019). "Synthesis, DNA Binding and Antibacterial Activity of Metal(II) Complexes of a Benzimidazole Schiff Base." *Polyhedron*, *157*, 326–334.

[28] Kose, M. & McKee, V. (2014). "Mn(II) Complexes of a Tridentate Benzimidazole Ligand: Synthesis, Structural Characterisation and Catalase Mimetic Studies." *Polyhedron*, *75*, 30–39.

[29] Huang, W. K., Cheng, C. W., Chang, S. M., Lee, Y. P. & Diau, E. W. G. (2010). "Synthesis and Electron-Transfer Properties of Benzimidazole-Functionalized Ruthenium Complexes for Highly Efficient Dye–Sensitized Solar Cells." *Chemical Communications*, *46*, 8992–8994.

[30] Khattar, R. & Mathur, P. (2013). "1-(Pyridin-2-ylmethyl)-2-(3-(1-(pyridine-2-ylmethyl)benzimidazole-2-yl)propyl)benzimidazole and its Copper(II) Complex as a

New Fluorescent Sensor for Dopamine (4-(2-aminoethyl)benzene-1,2-diol)." *Inorganic Chemistry Communications*, *31*, 37–43.

[31] Dayan, O., Demirmen, S. & Özdemir, N. (2015). "Heteroleptic Ruthenium(II) Complexes of 2-(2-pyridyl)benzimidazoles: A Study of Catalytic Efficiency Towards Transfer Hydrogenation of Acetophenone." *Polyhedron*, *85*, 926–932.

[32] Li, H., Jiang, H. & Sun, H. (2011). "Hexa-l-chlorido-l4-oxido-*tetrakis*({1-[(pyridine-2-yl)methyl]-1*H*-benzimidazole-κN^3}copper(II))." *Acta Crystallohraphica Section E*, *E167*, m1372.

[33] Mansour, A. M., & Friedrich, A. (2017). "The CO Release Properties of κ2 *N1,N2*-Mn(I) Tricarbonyl PhotoCORMs with Tridentate Benzimidazole Coligands." *Inorganic ChemistryFrontiers*, *4*, 1517–1524.

[34] Kose, M. (2014). "Synthesis, Characterization, and Antimicrobial Properties of Two Cu(II) Complexes Derived from a Benzimidazole Ligand." *Journal of Coordination Chemistry*, *67*, 2377–2392.

[35] Tao, C., Yuan, X., Yin, Q., Yan, H., Ni, W. Yan, L. & Zhang, L. (2016). "Synthesis, Characterization and Photoluminescent Properties of Europium(III) Complexes with Ligands Bearing Benzimidazole Groups." *Journal of Materials Science: Materials in Electronics*, *27*, 5715–5722.

[36] Shen, T. L., Zhu, X. L., Lu, X., Du, J. L. & Li, L. J. (2014). "Ag(I) and Cu(II) Coordination Polymers of Pyridylmethyl-Benzimidazole Ligand: Syntheses, Crystal Structure, and Influence of Metal Ion." *Molecular Crystals and Liquid Crystals*, *605*, 197–205.

[37] Hao, H. J., Du, M. Y., Wang, D. F., Sun, C. J., Wang, Z. H., Huang, R. B. & Zheng, L. S. (2013). "Syntheses, Structures, Photoluminescence of Four Dicarboxylate-Controlled Zn(II) Coordination Complexes Incorporating Flexible 1–(4–pyridylmethyl)–benzimidazole Ligand." *Journal of Molecular Structure*, *1048*, 124–129.

[38] Liu, H. Y., Ma, J. F., Liu, M. M. & Yang, J. (2013), "A Series of Zn(II) and Cd(II) Coordination Polymers Based on Flexible Bis–[(pyridyl)–benzimidazole] Ligand and Different Carboxylates: Syntheses, Structures, and Photoluminescent Properties." *Crystal Engineering and Communications*, *15*, 2699–2708.

[39] Pereira, M. B., Kopp, C. R., Fontana, L. A., Oliveira, G. M. D., Back, D. F., Piquini, P. C. & Villetti, M. A. (2014). "Synthesis, X–ray Structural Features, DFT Calculations and Fluorescence Studies of a New Pyridoxal–benzimidazole Ligand and its Respective Molybdenum Complex." *New Journal of Chemistry*, *38*, 3092–3101.

[40] Hu, A. B., Zhou, H., Pan, Z. Q. & Cheng, Q. R. (2016). "Hybrid Gels of Bis(benzimidazole)-base Ligand with Perchlorate Salts Driven by Squaric Acid: Synthesis, Characterization and Dye Adsorption." *Journal of Porous Materials*, *23*, 663–669.

[41] Huang, W. K., Wu, H. P., Lin, P. L. & Diau, E. W. G. (2013). "Design and Characterization of Heteroleptic Ruthenium Complexes Containing Benzimidazole Ligands for Dye-Sensitized Solar Cells: The Effect of Thiophene and Alkyl Substituents on Photovoltaic Performance." *The Journal of Physical Chemistry*, *117*, 2059–2065.

[42] Wang, C. J., Yue, K. F., Zhang, W. H., Jin, J. C., Huang, X. Y., Wang, Y. Y. (2010). "Three New d^{10} Coordination Polymers Based on 2-(2-pyridyl)benzimidazole Ligand: Synthesis, Structures and Properties." *Inorganic Chemistry Communications*, *13*, 1332–1336.

[43] Liu, H. Y., Wu, H., Yang, J., Liu, Y. Y., Liu, B., Liu, Y. Y., & Ma, J. F. (2011). "pH-Dependent Assembly of 1D to 3D Octamolybdate Hybrid Materials Based on a New Flexible Bis-[(pyridyl)-benzimidazole] Ligand." *Crystal Design and Growth*, *11*, 2920–2927.

[44] Appukuttan, V., Zhang, L., Ha, J. Y., Chandran, D., Bahuleyan, B. K., Ha, C. S., Kim, I. L. (2010). "Stereospecific Polymerizations of 1,3–Butadiene Catalyzed by Co(II) Complexes Ligated by 2,6–Bis(benzimidazolyl)pyridines." *Journal of Molecular Catalysis A: Chemical*, *325*, 84–90.

[45] Hao, P., Chen, Y., Xiao, T. & Sun, W. H. (2010). "Iron(III) Complexes Bearing 2-(benzimidazole)-6-(1-aryliminoethyl)pyridines: Synthesis, Characterization and Their Catalytic Behaviours Towards Ethylene Oligomerization and Polymerization." *Journal of Organometallic Chemistry*, *695*, 90–95.

[46] Ye, W., Zhao, M., Du, W., Jiang, Q., Wu, K., Wu, P. & Yu, Z. (2011). "Highly Active Ruthenium(II) Complex Catalysts Bearing an Unsymmetrical *NNN* Ligand in the (Asymmetric) Transfer Hydrogenation of Ketones." *Chemistry: A Europeon Journal*, *17*, 4737–4741.

[47] Zhao, S., Ito, S., Ohba, Y. & Katagiri, H. (2014) "Determination of the Absolute Configuration and Identity of Chiral Carboxylic Acids Using a Cu(II) Complex of Pyridine-benzimidazole Based Ligand." *Tetrahedron Letters*, *55*, 2097–2100.

[48] Pınar, S., Akkurt, M., Küçükbay, H., Şireci, N. & Büyükgüngör, O. (2006). "1,3-Bis(4-methyl-benz-yl)benzimidazolium Chloride Monohydrate." *Acta Crystallographica Section E*, *E62*, o2223–o2225.

[49] Hu, B., Deng, X. T., Wang, C. G. & Wang, X. Y. (2006). "2,2'-(Propane-1,3-diyl)bis(benzimidazolium)dichlorideDehydrate." *Acta Crystallographica Section E*, *E62*, m3477–m3478.

[50] Zheng, S. R., Cai, Y. P., Zhang, X. L. & Su, C. Y. (2005) "2,2′-(Iminodimethylene) bis(1*H*-benzimidazolium)(1+) chloride." *Acta Crystallographica Section C*, C61, o642–o644.

[51] Wahlgren, C. G., Addison, A. W., Burman, S., Thompson, L. K., Sinn, E. & Rowe, T. M. (1989). "Oxo-bridged Complexes of Iron(III) Derived from 2-(2′-

Hydroxyphenyl)-benzothiazole and 2-(2′-Hydroxyphenyl)benzimidazole Ligands." *Inorganic Chimica Acta*, *166*, 59–69.

[52] Tavman, A., Agh–Atabay, N. M., Guner, S., Gucin, F. & Dulger, B. (2007). "Investigation of Raman, FT–IR, EPR Spectra and Antimicrobial Activity of 2-(5-H/Me/Cl-1*H*-benzimidazol-2-yl)-phenol Ligands and Their $Fe(NO_3)_3$ Complexes." *Transition Metal Chemistry*, *32*, 172–179.

[53] Tavman, A. (2010). "Zn(II) and Cu(II) Complexes of 2-(2-Hydroxy-5-methylphenyl)-1*H*-benzimidazole and Crystal Structure of 2-(2-Hydroxy-5-methylphenyl)-1*H*-benzimidazolium Chloride." *Russian Journal of Inorganic Chemistry*, *55*, 377–383.

[54] Tavman, A. (2012). "Preparation and Spectral Characterization of 1,3-bis(1H-benzimidazol-2-yl)-propan-1-one and Its Various Zn(II) Complexes." *Main Group Metal Chemistry*, *35*, 81–89.

[55] Zhang, Z., Zhang, H., Jiao, C., Ye, K., Zhang, H., Zhang, J. & Wang, Y. (2015). "2-(2-Hydroxyphenyl)benzimidazole-Based Four-Coordinate Boron-Containing Materials with Highly Efficient Deep–Blue Photoluminescence and Electroluminescence." *Inorganic Chemistry*, *54*, 2652–2659.

[56] Sun, M., Yu, H., Li, H., Xu, H., Huang, D. & Wang, S. (2015). "Fluorescence Signaling of Hydrogen Sulfide in Broad pH Range Using a Copper Complex Based on BINOL–Benzimidazole Ligands." *Inorganic Chemistry*, *54*, 3766–3772.

[57] Zhong, K., Zhou, X., Hou, R., Zhou, P., Hou, S., Bian, Y., Zhang, G., Tang, L. & Shang, X. (2014). "A Water-Soluble Highly Sensitive and Selective Fluorescent Sensor for Hg^{2+} Based on 2-(2-(8-hydroxyquinolin)-yl)benzimidazole via Ligand to Metal Charge Transfer (LMCT)." *RSC Advances*, *4*, 16612–16617.

[58] Deshmukh, M. S., Sekar, N. (2015). "Chemiluminescence Properties of Luminol Related *o*-Hydroxybenzimidazole Analogues: Experimental and DFT Based Approach to Photophysical Properties." *Dyes and Pigments*, *113*, 189–199.

[59] Furukawa, H., Cordova, K. E., O'Keeffe, M. & Yaghi, O. M. (2013). "The Chemistry and Applications of Metal-Organic Frameworks." *Science*, *341*, 974–987.

[60] Lee, J. Y., Farha, O. K., Roberts, J., Scheidt, K. A., Nguyena, S. B. T. & Hupp, J. T. (2009). "Metal-Organic Framework Materials as Catalysts." *Chemical Society Reviews*, *38*, 1450–1459.

[61] Kundu, T., Mitra, S., Díaz, D. D. & Banerjee, R. (2016). "Gadolinium(III)-Based Porous Luminescent Metal-Organic Frameworks for Bimodal Imaging." *ChemPlusChem*, *81*, 1–6.

[62] Casanova, I., Durán, M. L., Viqueira, J., Sousa–Pedrares, A., Zani, F., Realc, J. A. & García–Vázquez, J. A. (2018). "Metal Ccomplexes of a Novel Heterocyclic Benzimidazole Ligand Formed by Rearrangement-Cyclization of the Corresponding Schiff Base. Electrosynthesis, Structural Characterization and Antimicrobial Activity." *Dalton Transactions*, *47*, 4325–4340.

[63] Arjmand, F., Sayeed, F. & Parveen, S. (2011). "In Vitro Binding Studies of Organotin(IV) Complexes of 1,2-bis(1H-benzimidazol-2-yl)ethane-1,2-diol with CT−DNA and Nucleotides (50−GMP and 50−TMP): Effect of the Ancillary Ligand on the Binding Propensity." *Journal of Organometallic Chemistry*, *696*, 3836−3845.

[64] Shi, Q., Chen, L., Liu, D, Wu, J., Wu, L., Zhu, L., Cheng, Q., Lu, R., Fan, D. & Lü, X. (2013). "A New Tetranuclear Nickel Complex Based on the Benzimidazole Ligand for the Controllable Polymerization of Methyl Methacrylate." *Inorganic Chemistry Communications*, *29*, 22−26.

[65] She, M. Y., Xiao, D. W., Yin, B., Yang, Z., Liu, P., Li, J. L. & Shi, Z. (2013). "An Efficiently Cobalt-Catalyzed Carbonylative Approach to Phenylacetic Acid Derivatives." *Tetrahedron*, *69*, 7264−7268.

[66] Sharma, A. K., Joshi, H., Bhaskar, R. & Singh, A. K. (2017). "Complexes of (η^5-Cp*)Ir(III) with 1-benzyl-3-phenylthio / selenomethyl-1,3-dihydrobenzoimidazole-2-thione/selenone: Catalyst for Oxidation and 1,2-Substituted Benzimidazole Synthesis." *Dalton Transactions*, *46*, 2228−2237.

[67] Feng, M., Pointillart, F., Lefeuvre, B., Dorcet, V., Golhen, S., Cador, O. & Ouahab, L. (2015). "Multiple Single-Molecule Magnet Behaviors in Dysprosium Dinuclear Complexes Involving a Multiple Functionalized Tetrathiafulvalene-Based Ligand." *Inorganic Chemistry*, *54*, 4021−4028.

[68] Li, L., Diao, K. S., Ding, Y. Q., Tanga, J. N. & Wang, D. Y. (2012). "Dichlorido(ethanol-*tO*)[2-(1,3-thiazol-4-yl-*N*)-1*H*-benzimidazole-*jN*3]copper(II)." *Acta Cystallographica Section E*, *E62*, m508−m509.

[69] Dutta, P. K., Panda, S. & Zade, S. S. (2014). "Synthesis, Characterization and Coordination Properties of 2-(2'-(methylseleno)phenyl)benzimidazole." *Inorganic Chimica Acta*, *411*, 83−89.

[70] Bhardwaj, M., Bansal, M., Gupta, S. K. & Das, A. K. (2008). "Synthesis of Benzoimidazole Selenato Complexes and Determination of Its Antitumor Activity." *International Journal of Chemical Sciences*, *6*, 908−916.

[71] Mohamed, G. G., Ibrahim, N. A. & Attia, H. A. E. (2009). "Synthesis and Anti−Fungicidal Activity of Some Transition Metal Complexes with Benzimidazole Dithiocarbamate Ligand." *Spectrochemica Acta Part A*, *72*, 610−615.

[72] Quinlivan, P. J., Chaijan, M. R., Palmer, J. H., Shlian, D. G. & Parkin, G. (2019). "Coordination of 1-Methyl-1,3-dihydro-2*H*-benzimidazole-2-selone to Zinc and Cadmium: Monotonic and Non−Monotonic Bond Length Variations for $[H(sebenzimMe)]_2MCl_2$ Complexes (M = Zn, Cd, Hg)." *Polyhedron*, *164*, 185−194.

[73] Ghavale, N., Manjare, S. T., Singh, H. B. & Butcher, R. J. (2015). "Bis(chalcogenones) as Pincer Ligands: Isolation and Heck Activity of the Selone−Ligated Unsymmetrical C,C,Se–Pd Pincer Complex." *Dalton Transactions*, *44*, 11893−11900.

[74] Sharma, K. N., Satrawala, N. & Joshi, R. K. (2018). "Thioether-NHC Ligated Pd(II) Complex in Crafting of Filtration-Free Magnetically Retrievable Catalyst for Suzuki-Miyaura Coupling in Water." *Europeon Journal of Chemistry*, *2018*, 1743−1751.

[75] Su, X., Li, T., Xiu, Y. & Meng, X. (2012). "Syntheses, Structures and Properties of Two Cd(II) Complexes Based on the 2-(1*H*-Imidazol-1-yl-methyl)-1*H*-benzimidazole Ligand." *Verlag der Zeitschrift für Naturforschung*, *67*, 678−684.

[76] Bhattacharyya, S., Purkait, K. & Mukherjee, A. (2017). "Ruthenium(II) *p*-Cymene Complexes of a Benzimidazole-Based Ligand Capable of VEGFR2 Inhibition: Hydrolysis, Reactivity and Cytotoxicity Studies." *Dalton Transactions*, *46*, 8539−8554.

[77] Pereira, M. B., Kopp, C. R., Fontana, L. A., Oliveira, G. M. D., Back, D. F., Piquinib, P. C. & Villettic, M. A. (2014). "Synthesis, X-ray Structural Features, DFT Calculations and Fluorescence Studies of a New Pyridoxal−Benzimidazole Ligand and Its Respective Molybdenum Complex." *New Journal of Chemistry*, *38*, 3092−3101.

[78] Gupta, S., Basu, B. & Das, S. (2013). "Benzimidazole-Based Palladium *N*-Heterocyclic Carbene: A Useful Catalyst for C−C Cross-Coupling Reaction at Ambient Condition." *Tetrahedron Letters*, *69*, 122−128.

[79] Li, F., Hu, J. J., Koha, L. L. & Hor, T. S. A. (2010). "Substituent-Dependent Structures and Catalysis of Benzimidazole-Tethered *N*-Heterocyclic Carbene Complexes of Ag(I), Ni(II) and Pd(II)." *Dalton Transactions*, *39*, 5231−5241.

[80] Fang, W., Jiang, J., Xu, Y., Zhou, J. & Tu, T. (2013). "Novel Robust Benzimidazolylidene Palladium Complexes: Synthesis, Structure, and Catalytic Applications in Amination of Chloroarenes." *Tetrahedron*, *69*, 673−679.

[81] Doğana, O., Demira, S., Özdemira, I. & Çetinkaya, B. (2011). "Palladium(II)-NHC Complexes Containing Benzimidazole Ligand As a Catalyst for C–N Bond Formation." *Applied Organometallic Chemistry*, *25*, 163−167.

[82] Huang, G., Sun, H., Qiu, X., Shen, Y., Jiang, J. & Wang, L. (2011). "Synthesis, Structural Characterization of Benzimidazole-Functionalized Ni(II) and Hg(II) *N*-Heterocyclic Carbene Complexes and Their Applications as Efficient Catalysts for Friedel-Crafts Alkylations." *Journal of Organometallic Chemistry*, *696*, 2949−2957.

In: Benzimidazole
Editor: Aksel A. Vestergaard
ISBN: 978-1-53616-986-7

Chapter 10

BENZIMIDAZOLE DERIVATIVES: CHELATING PROPERTIES TOWARDS TRANSITION METALS IONS AND THEIR APPLICATIONS

Karim Chkirate** and El Mokhtar Essassi
Laboratory of Heterocyclic Organic Chemistry,
Pole of Competence Pharmacochemistry, Mohammed V University, Faculty of Sciences, Rabat, Morocco

ABSTRACT

Benzimidazole derivatives are known as heterocyclic compounds involved in a several biological processes and have been proved to possess biological activity and pharmacological properties. In this chapter, their coordination behavior towards transition metal ions such as copper(II), zinc(II), cobalt (II), cadmium(II) and cobalt (II) are investigated. The structures of the resulted compounds have been elucidated using crystallographic studies.

The synthesis and the biological properties of the coordination compounds containing benzimidazole derivatives are also presented and discussed.

Keywords: benzimidazole, coordination, transition metal, biological activity

* Corresponding Author's Email: kchkirate@gmail.com.

1. INTRODUCTION

The interest of nitrogenous heterocycles is important because of their presence in natural and unnatural molecules with biological activity and wide application in the field of coordination chemistry. Benzimidazoles are present in many pharmaceutical products with a wide range of biological activities [1-2].

Aromatic N-heterocycles such as benzimidazoles (Figure 1) contain, as part of the aromatic ring, a nitrogen atom with an isolated pair of electrons, located on an orbital in the plane. They also present empty orbitals at low energy, the orbitals π * antistickers of the aromatic system, perpendicular to the molecular plane. Therefore, electronically, these heterocycles are both ligands σ-donor and π-acceptor [3]. Their steric characteristics and their flatness make these aromatic molecules relatively unencumbered.

However, the exact steric profile strongly depends on the presence of substituents, their size and their position on the ring.

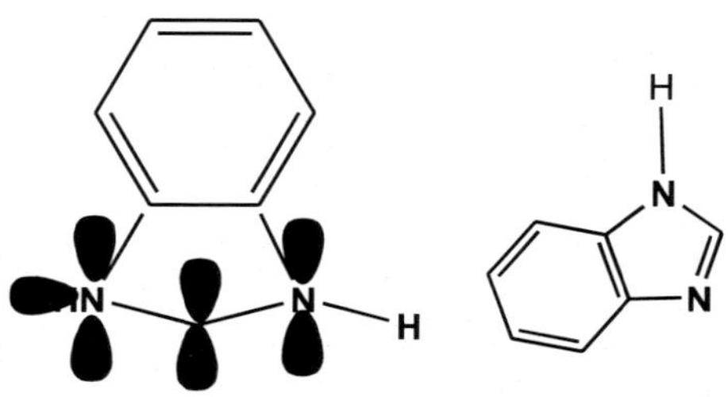

Figure 1. Position of orbitals in the benzimidazole ring.

Given the richness of the chemistry of benzimidazoles and the diversity of their applications, we present in this chapter the complexing properties of heterocyclic systems derived from benzimidazole with respect to metals.

2. COMPLEXING PROPERTIES OF BENZIMIDAZOLE DERIVATIVES TOWARDS METALS

Benzimidazole presenting two tautomeric forms Ia and Ib, is a heterocyclic aromatic organic compound. Its electronic doublet free of nitrogen atom at position 3 does not enter the aromaticity because orbital p is already used to make a π bond, while that of nitrogen atom at position 1 hybridized sp2 is included.and place its doublet in a pure orbital p to participate in the phenomenon of conjugation.

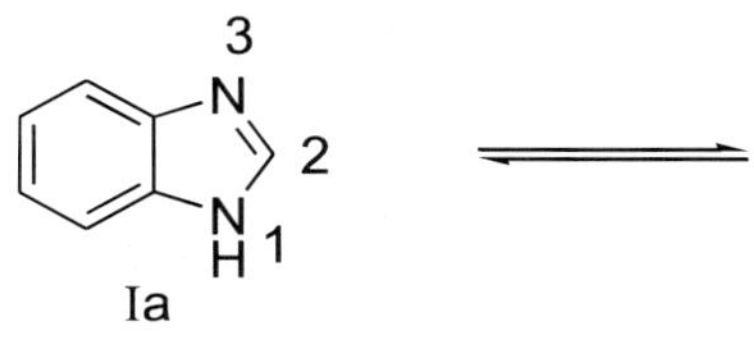

Figure 2. Tautomerism of the benzimidazole nucleus.

2.1. Monodentate or Bidentate Ligands Based on Benzimidazole

Małecki et al. [4], have developed a simple synthesis of two *p*-cymene ruthenium (II) complexes with 2-(2-aminophenyl)-1H-benzimidazole and 2-aminobenzimidazole as ligands (Scheme 1); their luminescent properties were examined. Experimental studies on complexes have been accompanied by calculations of the functional density theory (DFT).

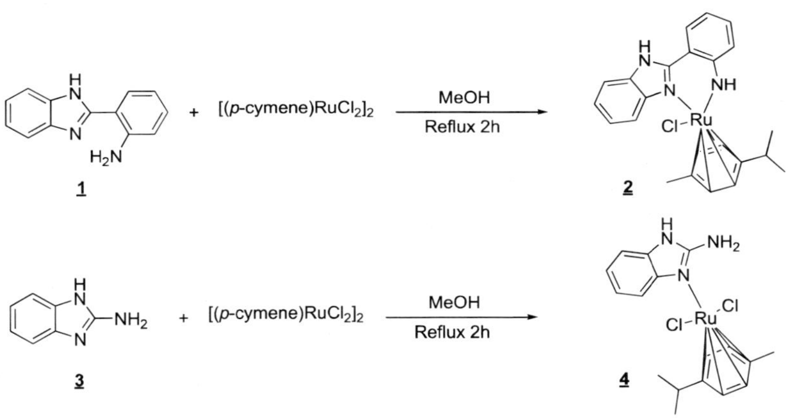

Scheme 1. Synthesis of two p-cymene ruthenium (II) complexes with 2-(2-aminophenyl)-1H-benzimidazole and 2-aminobenzimidazole as ligands.

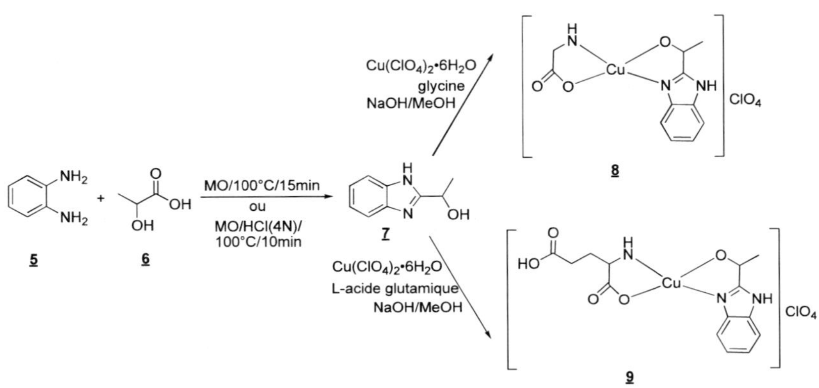

Scheme 2. Preparation of two complexes of mixed ligands [Cu(BzOH)(Gly)]ClO4 (8) and [Cu(BzOH)(Glu)]ClO4 (9).

Two complexes of mixed ligands, namely, $[Cu(BzOH)(Gly)]ClO_4$ (**8**) and $[Cu(BzOH)(Glu)]ClO_4$ (**9**) where BzOH = 1-benzimidazolylethanol, Gly = glycine and Glu = acid L-glutamic, were synthesized by Ibrahim et al. [5] (Scheme 2). Electrochemical measurements, including cyclic voltammetry and electrical molar conductivity, were also performed. The spectral characteristics indicate square planar CuN_2O_2 coordination geometries for the two complexes. Spectroscopic and electrochemical studies were performed to correlate the structural characteristics of the complexes with their catecholase and catalase activities. Complex (**8**) showed high catalase activity, and nuclease activity was also evaluated.

Recently, the design of small molecules, as potential therapeutic agents for Alzheimer's disease, is a field of study that has attracted much attention. Among the works in this direction, there are those described by Yellol et al. [6], who have synthesized a series of new complexes of iridium (III), ruthenium (II) and platinum (II) 2-pyridylbenzimidazole based on a new ligand of methyl 1-butyl-2-pyridylbenzimidazole carboxylate structure (Scheme 3). All new compounds have been shown to be active by fluorescence and transmission electron microscopy.

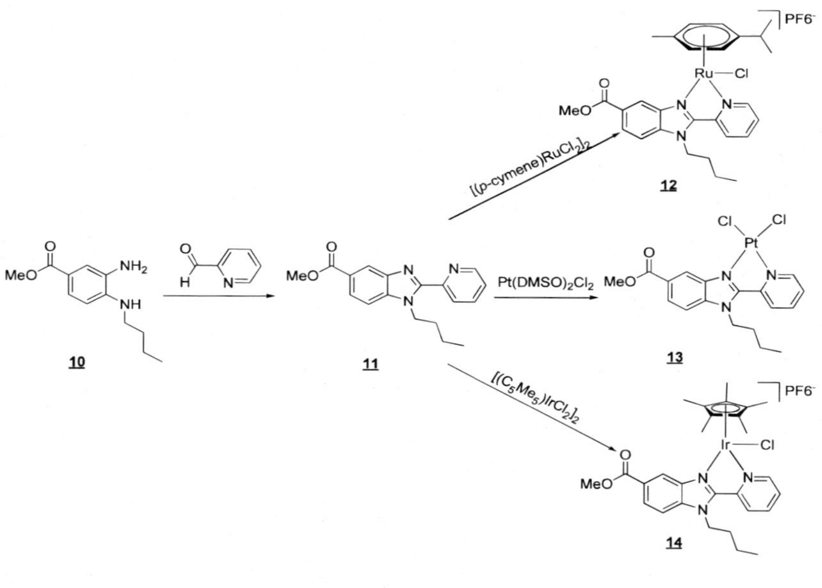

Scheme 3. Synthesis of a series of new complexes of iridium (III), ruthenium (II) and platinum (II) 2-pyridylbenzimidazole based on a new ligand of methyl 1-butyl-2-pyridylbenzimidazole carboxylate.

Using 1-(2-methoxybenzyl)-2-(2-methoxyphenyl)-1H-benzimidazole as a monodentate ligand, a new copper-II complex was synthesized by Chowdhury et al. [7] (Scheme 4). The X-ray crystalline structure of the complex reveals that it adopts a cis arrangement with respect to the ligands. The solid state structure of the complex is stabilized by an offset π-π stack which is most likely the source of a blue fluorescence. At room temperature, the complex is paramagnetic. In cyclic voltammetry, the complex shows a Cu (II) / Cu (III) oxidizing pair. DFT calculations confirm the observed experimental results.

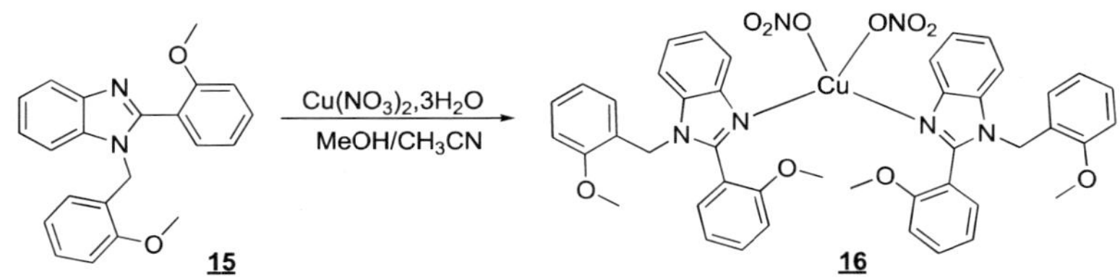

Scheme 4. Synthesis of a new copper-II complex from 1-(2-methoxybenzyl)-2-(2-methoxyphenyl)-1H-benzimidazole.

Two copper complexes derived from quinolyl-benzimidazole were prepared and characterized by Zhao et al. [8]. The interactions of the two complexes with calf thymus DNA were detected by absorption and emission spectroscopy methods; the results showed that both complexes significantly increased free radical dependence on DNA cleavage activity. The analyzes revealed that complex **20** exhibited a good inhibitory activity of cell proliferation against four different human cancer cell lines (SMMC7721, BGC823, HCT116 and HT29), in particular with HCT116 cells. These findings should promote the development of metal-based complexes for use as novel chemotherapeutic agents.

The synthesis of 4-(2-benzimidazolylazo)proline and its complexes derived from [Cu (II), Ag (I) and Au (III)] salts were carried out by Abbas et al. [9]. The binding mode for the complexes was performed on the basis of elemental analysis, spectroscopy, magnetic measurement and molar conductivity. The ligand has been shown to behave as an N, N-neutral bidentate that formed chelates with molar stoichiometry (1: 2) (M: L) for [Cu (II) and Ag (I)], whereas Au (III) forms a complex with a molar ratio (1: 1) (Scheme 6). The stability constant and the Gibbs free energy for the prepared complexes are evaluated by spectroscopic methods.

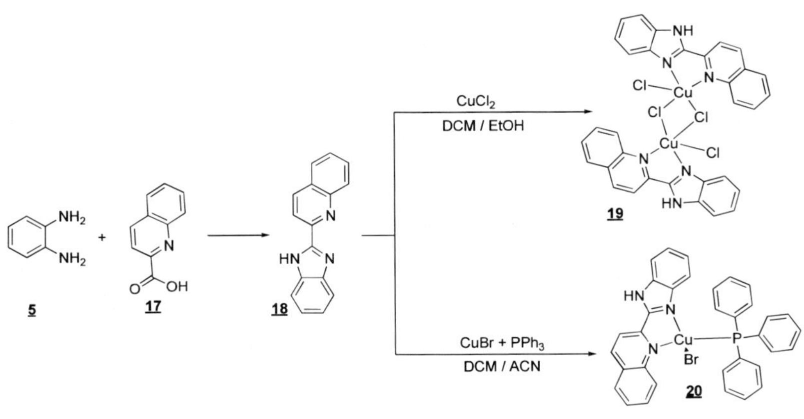

Scheme 5. Synthesis of two copper complexes derived from quinolyl-benzimidazole.

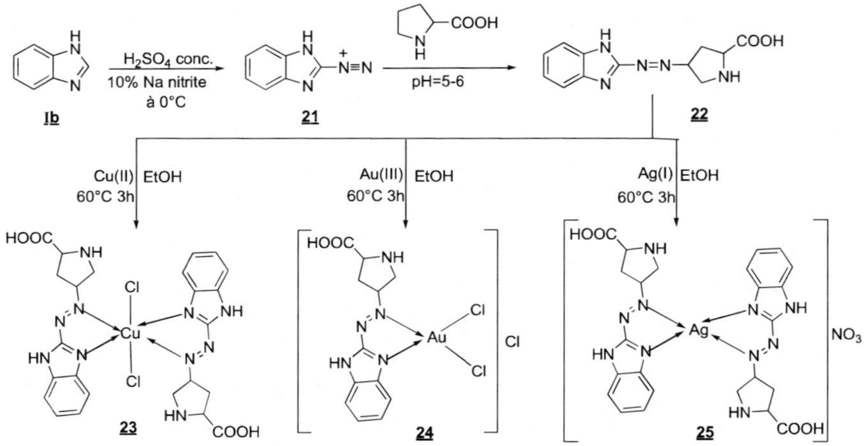

Scheme 6. Synthesis of 4-(2-benzimidazolylazo)proline and its complexes derived from [Cu (II), Ag (I) and Au (III)] salts.

Novel ruthenium complexes (II / III) of Schiff's bases, containing benzimidazole moieties, were isolated by Booysen et al. [10] (Scheme 7). With pyridine, benzimidazole forms a cis-complex of ruthenium II which is diamagnetic; while with phenol, benzimidazole forms a ruthenium III trans complex which is paramagnetic. These metal complexes were characterized and their redox properties were sounded by voltammetry. The structural characteristics of the 1st complex **27** were studied by multinuclear NMR spectroscopy (^{1}H and ^{31}P). The presence of the paramagnetic metal centers of the 2nd complex **29** was confirmed by ESR spectroscopy (electron spin resonance) in X band.

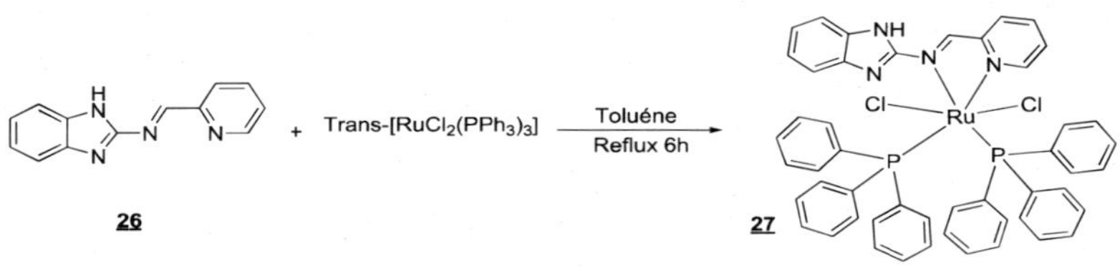

Scheme 7. Synthesis of novel ruthenium complexes (II / III) of Schiff's bases, containing benzimidazole moieties.

A coordination compound based on copper (II) halides with 4-methyl-1,2,4-triazolo[1,5-a]benzimidazole, of composition [Cu(L)$_2$Cl$_2$H$_2$O], was synthesized by Lavrenova et al. [11] (Figure 8). The compound has been characterized using spectroscopy and static magnetic susceptibility measurements. By means of structural X-ray diffraction analysis, the crystal structure was determined for the complex. The cytotoxic effects of the complex and the ligand were then studied, and it was found that the copper (II) complex led to a considerable increase in cytotoxicity.

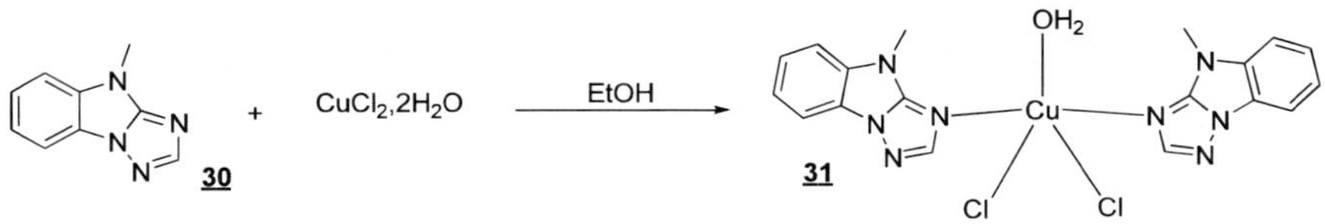

Scheme 8. Synthesis of [Cu(L)2Cl2H2O] from 4-methyl-1,2,4-triazolo[1,5-a]benzimidazole.

Ahmad et al. [12] synthesized a series of mixed ligand-metal complexes of Zn (II), Ni (II), Co (II), Cu (II), omeprazole and 8-hydroxyquinoline (scheme 9). These complexes have been characterized by various physicochemical techniques and have also been studied for their alkaline phosphatase inhibitory (ALP) activities, their antioxidant potential and their antimicrobial activities. Conductance measurement suggested that the metal complexes were non-electrolytes. The Ni (II) complex had a higher antioxidant activity than the other complexes.

All mixed ligand complexes exhibit higher antioxidant, inhibitory, and antimicrobial activity than their corresponding ligands. This indicates their therapeutic potential as future drugs for the diseases concerned.

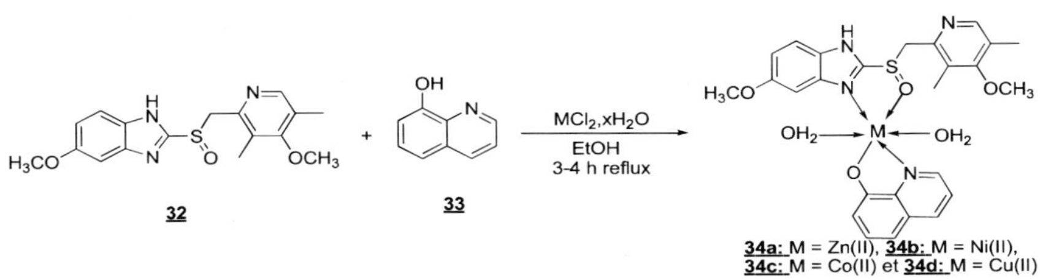

Scheme 9. Synthesis of a series mixed ligand-metal complexes of Zn (II), Ni (II), Co (II), Cu (II), omeprazole and 8-hydroxyquinoline.

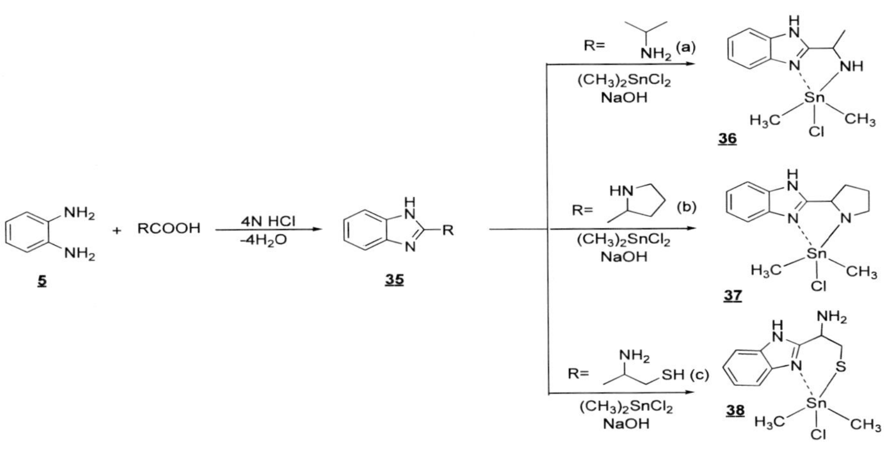

Scheme 10. Synthesis of the five-coordinate heteroaryl-tin (IV) compounds.

A similar synthesis is reported by Akremi et al. [13], concerning the synthesis of the five-coordinate heteroaryl-tin (IV) compounds, using three chiral N-heterocyclic substituents, 2-(1-aminoethyl)benzimidazole **35a**, 2-(pyrrolidin-2-yl)benzimidazole **35b** and 2-(1-amino-2-mercaptopropyl)benzimidazole **35c**. It has been found that the ligands **35a** and **35b** behave as a bidentate N, N donor; however, the **35c** ligand behaves as N, S bidentate. Characterization of these compounds by various spectroscopic methods and by elemental analyzes provided insight into the complex formation diversity (scheme 10).

Another benzimidazole derivative namely (E)-2-((4-(1H-benzimidazol-2-yl)phenylimino)methyl)-4-bromophenol and their corresponding complexes with Zn (II), Ni (II), Cu (II) and Pd (II) were prepared by Mahmood et al. [14] (Scheme 11). The ligand and synthesized complexes were fully characterized and the ligand structure confirmed by single crystal X-ray diffraction analysis. The ability of this compound and its complexes to bind to DNA was first investigated using DNA thermal denaturation experiments, generally showing weak interactions. In addition, UV-vis absorption spectroscopy was used to evaluate DNA binding and the corresponding binding constants (Kb) were calculated. All compounds were screened for their antibacterial activity and the Ni (II) complex gave promising results against all bacterial strains (Gram positive and Gram negative), while the rest of the compounds showed activity against selective strains.

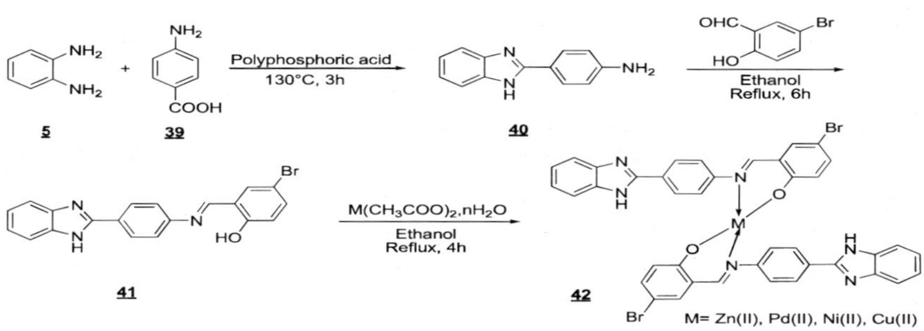

Scheme 11. Synthesis of (E)-2-((4-(1H-benzimidazol-2-yl)phenylimino)methyl)-4-bromophenol and their corresponding complexes with Zn (II), Ni (II), Cu (II) and Pd (II).

Nitrogenous heterocyclic carbene complexes of mercuramacrocycles {Hg_4Cl_4(R-Bim)$_4$} (BimH = benzimidazole, R = Ph, Py) were prepared by Yu et al. [15] either from the simple reaction of N-substituted benzimidazoles with $HgCl_2$ or from the reaction of N-substituted benzimidazoles with $Hg(OAc)_2$ and NaCl (Scheme 12). The first reaction is unique in that no additional base is required, although the yield of the reaction is low. Under similar reaction conditions, only nitrogen-bonded benzimidazole complexes were obtained when $HgBr_2$ or HgI_2 instead of $HgCl_2$ was used. The different reactivity may be due to the acid-base property of the mercury (II) and halides, which facilitates the cleavage of the Hg-Cl bond and promotes the C-metallation of $HgCl_2$ and benzimidazole. During the protonation of {Hg_4Cl_4(R-Bim)$_4$} with HBF_4,OEt, complexes of [HgCl(R-Bim-H)] [BF_4]

(R = Ph, Py) were obtained. A preliminary study on the transmetallation reaction of {Hg_4Cl_4(Ph-Bim)$_4$} with $PdCl_2(CH_3CN)_2$ and LiCl gave a C, N-linked palladium complex.

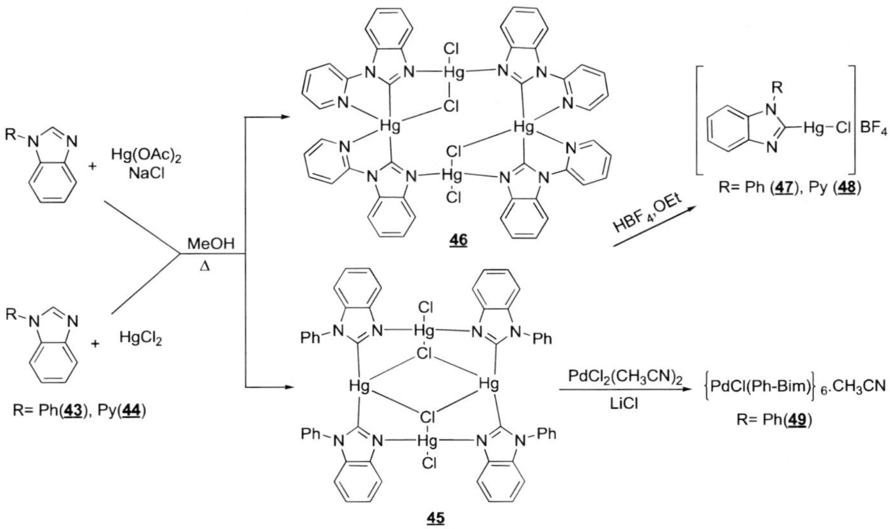

Scheme 12. Synthesis of nitrogenous heterocyclic carbene complexes of mercuramacrocycles {Hg_4Cl_4(R-Bim)$_4$} (BimH = benzimidazole, R = Ph, Py).

Su et al. [16] synthesized a zinc (II) complex carrying a benzimidazole derivative, namely [Zn(L1)Cl_2],DMF (L1 = 2-[2-(benzimidazol-2-yl)phenyl]-1-methyl-benzimidazole which has been characterized by X-ray monocrystalline structure analyzes.The complex crystallizes in the monoclinic space group P21/n.The metal center of Zn^{2+} is coordinated by two N atoms and two Cl atoms, forming deformed tetrahedron geometry The complex emits blue luminescence in a solution of DMSO. The effects of chemo-cytotoxicity and apoptosis rates of complexes on MB-MDA-231 cells of breast cancer in humans The results showed that the cytotoxic effect of the complex is dose and time dependent.

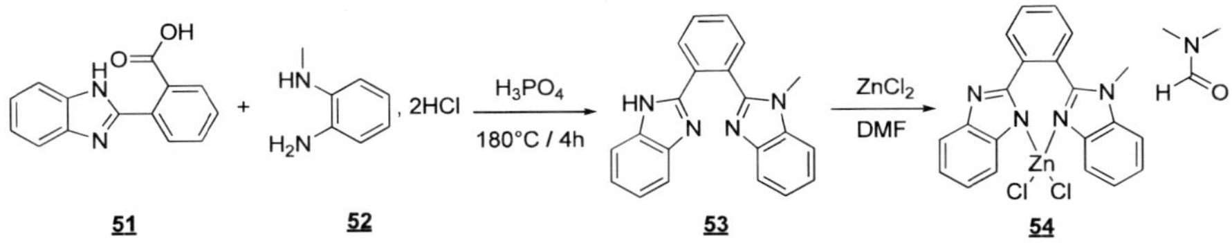

Scheme 13. Synthesis of a zinc (II) complex namely [Zn(L1)Cl_2],DMF (L1 = 2-[2-(benzimidazol-2-yl)phenyl]-1-methyl-benzimidazole.

Three new complexes formulated in the form [Cu(mbmb)$_2Cl_2$] (**56**), [Cu(mbmb)$_2(NO_3)_2$] (**57**), [Zn(mbmb)$_2Cl_2$] (**58**) (mbmb = 1-[(2-methyl-1H-benzoimidazol-1-yl)methyl]-1H-benzotriazole) were synthesized by Wang et al. [17].

Complexes **56** to **58** exhibited mononuclear structures and their central metal ions exhibited a four coordinate environment. The authors evaluated the antioxidant activity against DPPH and the anti-diabetic activity against α-amylase complexes **56**-**58** and ligand (mbmb) *in vitro*. Results from the DPPH free radical scavenging assay showed that complexes **56** and **57** showed higher antioxidant activity among synthesized compounds and their IC50 values were 43.72 μg/mL and 43.76 μg / mL, respectively. The IC50 values of the *in vitro* antidiabetic studies of complex **56** (0.74 mg/ml) and complex **57** (1.00 mg/ml) were much lower than those of standard acarbose (2.34 mg/ml) and mbmb ligand (39.87 mg / ml) that showed effective inhibition of α-amylase. In addition, the solid-state fluorescent properties of these three complexes and the mbmb ligand were studied at room temperature.

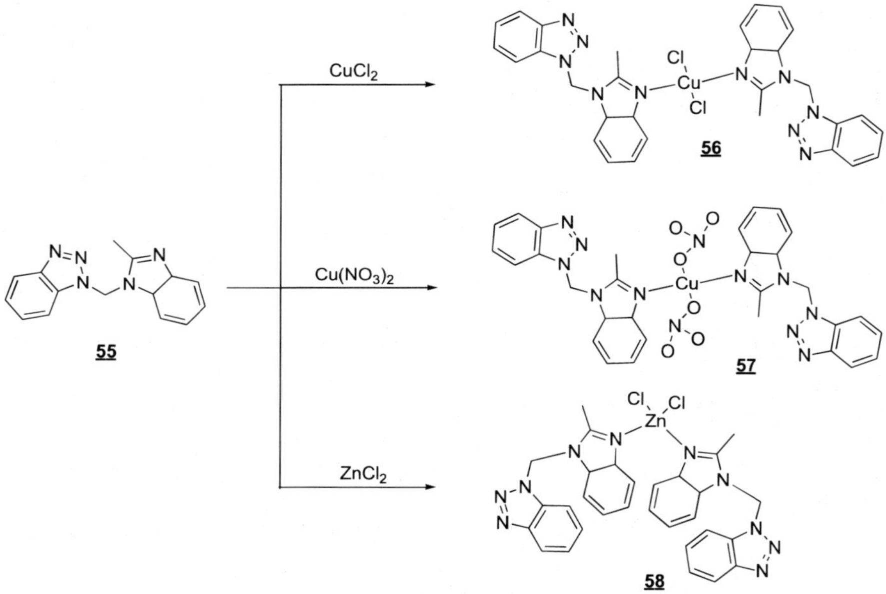

Scheme 14. Synthesis of three new complexes formulated [Cu(mbmb)$_2$Cl$_2$] (**56**), [Cu(mbmb)$_2$(NO$_3$)$_2$] (**57**), [Zn(mbmb)$_2$Cl$_2$] (**58**) (mbmb = 1-[(2-methyl-1H-benzoimidazol-1-yl)methyl]-1H-benzotriazole).

Milani et al. [18] synthesized a novel zinc (II) complex of formula [ZnCl$_2$(L1)$_2$] (**61**) [L1 = 2-(2-thienyl)-1-(2-thienylmethyl)-1H-benzimidazole] which was fully characterized by NMR, infrared, elemental analysis, electrospray ionization high resolution mass spectrometry and thermogravimetric analysis. The molecular structure was confirmed by X-ray diffraction on a single crystal. The complex **61** consists of mononucleated units of lattice-locked tetrahedral zinc (II) resulting from intramolecular interactions S...π and π-π interligands. The benzimidazole ligand and its zinc (II) complex were easily obtained by a simple synthetic route. The catalytic activity of **61** was studied in the coupling of carbon dioxide with epoxides to produce cyclic carbonates, and a series of parameters was evaluated. The complex has efficiently catalyzed the transformation of various epoxides under solvent-free conditions.

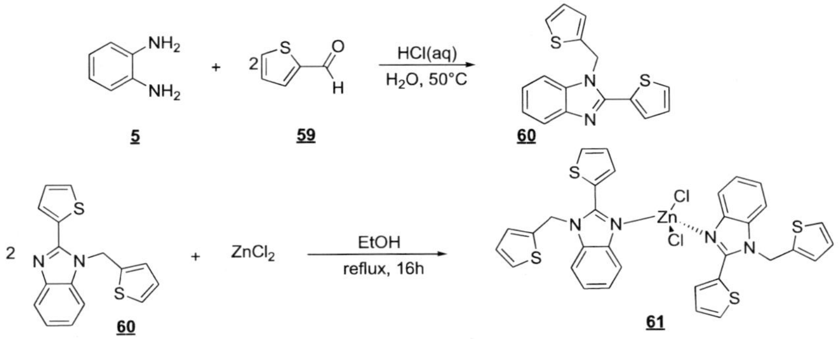

Scheme 15. Synthesis of a novel zinc (II) complex of formula [$ZnCl_2(L1)_2$] (**61**) [L1 = 2-(2-thienyl)-1-(2-thienylmethyl)-1H-benzimidazole].

Benzimidazole and sulfonamide moieties are found in a number of pharmaceutically active molecules. By incorporating the sulfonamide pharmacophore into the benzimidazole scaffold, Ashraf et al. [19] prepared 2-(o-sulfamoylphenyl) benzimidazole **63** from saccharin as a precursor. Ligand **63** has been coordinated with the divalent transition metals Mn (II) **64a**, Co (II) **64b**, Ni (II) **64c**, Cu (II) **64d** and Zn (II) **64e** to give complexes of the general formula [$ML_2(H_2O)n$] (n = 2 for 2 and n = 0 for **64b**-**64e**). All compounds were characterized by elemental analysis, conductivity measurements, magnetic susceptibility, FT-IR and NMR spectroscopy, and the molecular structures of **63**, **64b** and **64e** were determined by X-ray diffraction analysis. In all metal complexes, **63** behaved as a bidentate chelating ligand through the nitrogen sulfonamidate and the endocyclic nitrogen of benzimidazole. The molecular structures of **64b** and **64e** showed a tetrahedral geometry around the Co (II) and Zn (II) centers, respectively. The molar conductivity data revealed that the metal complexes were non-electrolytes. The benzenesulfonamide derivative and its metal complexes were evaluated for their potential antimicrobial activity against a range of bacterial and fungal strains. [Co(2-(o-sulfamoylphenyl)benzimidazolate)$_2$] **64b** was identified as the most active antibacterial compound, while none of the compounds exhibited antifungal activity.

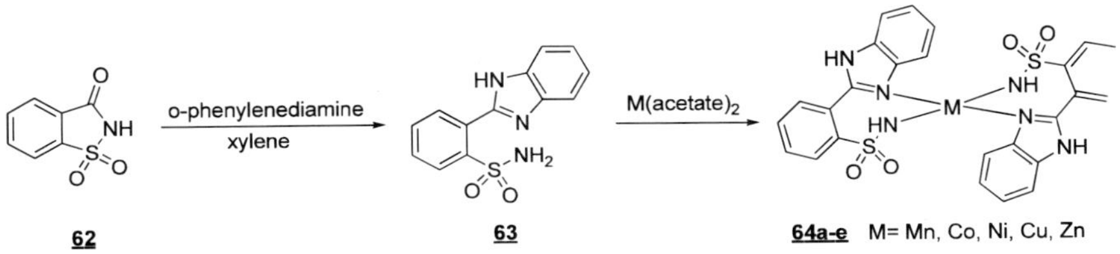

Scheme 16. Synthesis of complexes of the general formula [$ML_2(H_2O)n$] using as ligand 2-(o-sulfamoylphenyl) benzimidazole.

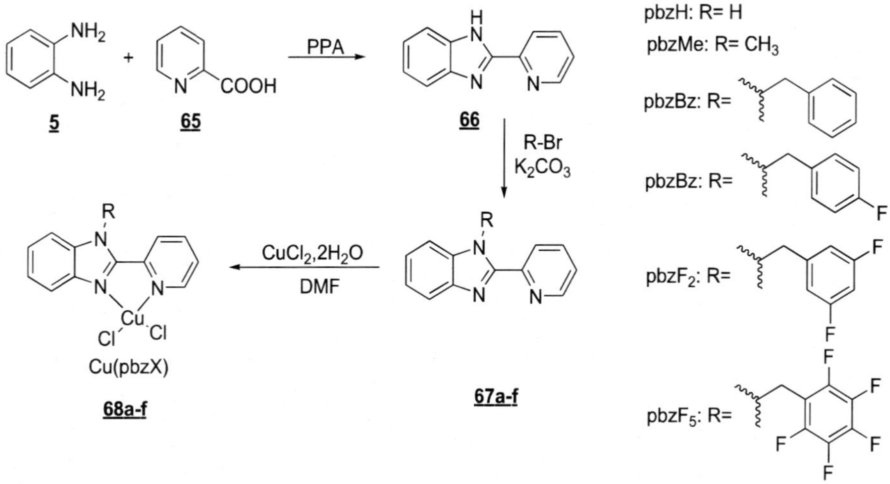

Scheme 17. Preparation of Cu (II) $CuCl_2$(pbzH), pbzH = 2-(2-pyridinyl)benzimidazole complex.

Prosser et al. [20] studied the Cu (II) $CuCl_2$(pbzH), pbzH = 2-(2-pyridinyl)benzimidazole complex and its uncoordinated nitrogen-modified derivatives of the benzimidazole moiety as anticancer agents. These compounds show promising cytotoxicity against adenocarcinoma A549 alveolar basal epithelial cells with IC50 values between 5 and 10 μM. It is important to note that this activity is greater than $CuCl_2.2H_2O$ or individual ligands, demonstrating that ligand coordination with the Cu (II) complexes is necessary for complete activity. Electron paramagnetic resonance spectroscopy (EPR) and UV-Vis spectroscopy were used to characterize the solution behavior of the complexes. DNA studies using agarose gel electrophoresis demonstrate cleavage of strands by complexes in the presence of ascorbate, which is mediated by reactive oxygen species (ROS). Through an *in vitro* fluorescence-based assay, the generation of intracellular ROS in the A549 cell line was observed; indicating that the damage caused by the ROS is responsible for the observed activity of the complexes.

2.2. Tridentate Ligands Based on Benzimidazole

Wałęsa-Chorab et al. [21] describe the synthesis of new mononuclear complexes of transition metals with (E)-2-(1-methyl-2-(1-(pyridin-2-yl) ethylidene)hydrazinyl)-1H-benzoimidazole (Scheme 18). The structures of these complexes can be divided into three groups, depending on the coordination number and the metal center environment: the first of the general formula $[ML_2]X_2$, the second $[MLX_3]$ and the third $[MLX_2]$, where X is an anion or solvent molecule. The electrochemical and electrochromic properties of the complexes have been studied. The Fe (II), Cu (II) and Co (II) complexes have been shown to be electroactive and their colors depend on the oxidation state of the metal center and the ligand molecule. The Fe (II) complex has a purple color in the neutral state and its color becomes yellow after Fe (II) → Fe (III) oxidation, then pale yellow after further oxidation of the ligand molecule and becomes green when the metal center is reduced to Fe (I). The original color can be restored after Fe (I) electrochemical oxidation to Fe (II). These complexes have been shown to exhibit high color stability in solution during multiple oxidation / reduction cycles. These complexes can therefore be considered as very interesting materials for the construction of multi-electrochromic devices.

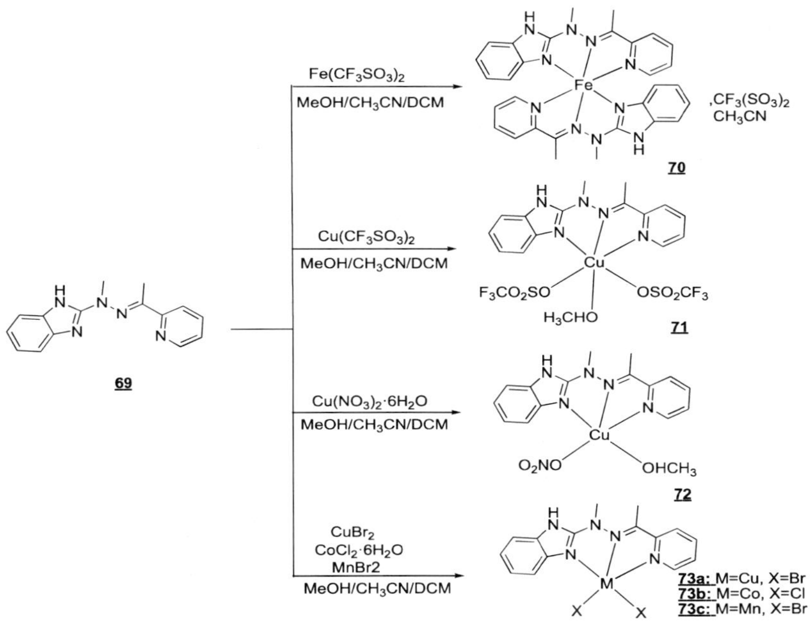

Scheme 18. Synthesis of new mononuclear complexes of transition metals with (E)-2-(1-methyl-2-(1-(pyridin-2-yl) ethylidene)hydrazinyl)-1H-benzoimidazole.

2.3. Bis-Benzimidazole Ligands

The crystalline structure of bis (benzimidazol-2-yl)methane dichloro zinc was determined by Lamkaddem et al. [22]. The metal complex is monoclinic and the metal is tetra-coordinated, forming a non-regular tetrahedral distortion with the ligand (Figure 19).

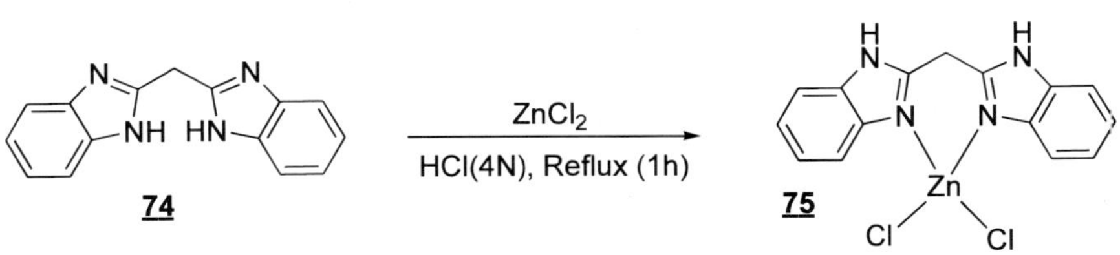

Scheme 19. Synthesis of of bis (benzimidazol-2-yl)methane dichloro zinc.

Another highly efficient synthetic method was developed by Salhaji et al. [23] in which they synthesized 2-mercapto(2-mercaptobenzimidazolyl)benzimidazole (L) and studied its complexing properties with respect to copper (II) and silver (I). A theoretical study, carried out through the AM1 program of MOPAC contained in Chem. Office 2002, allowed to determine the geometry, the dimensions, the charges of the different atoms of the ligand (L) and to predict the M-L coordination sites. The study carried out by IR and UV spectroscopies on the complexation of 2-mercapto(2-mercapto benzimidazolyl) benzimidazole by copper (II) and silver (I) has shown that the complexes formed are of ML type in the case copper and ML_2 for silver (Scheme 20). Their training constants have been determined. It was noted later that the coordination geometry around the copper and silver atoms is tetrahedral.

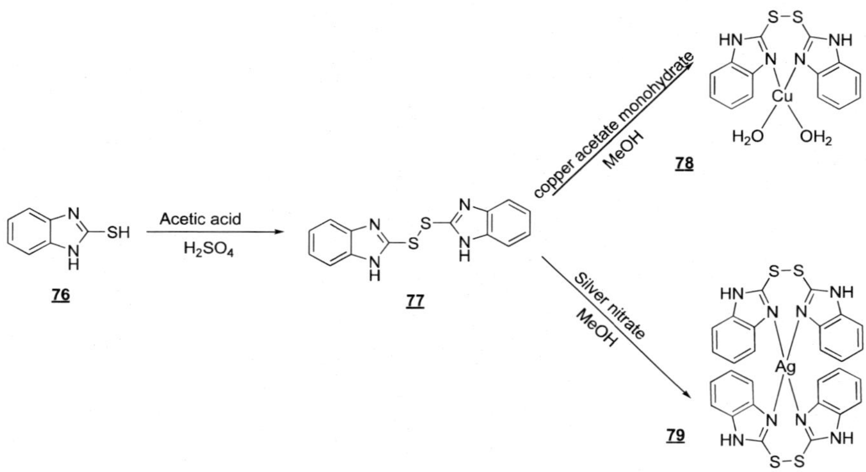

Scheme 20. Synthesis and coplexation of 2-mercapto(2-mercaptobenzimidazolyl)benzimidazole.

Cu (II) and Cu (I) monomer complexes bound to a tetradentate bis-benzimidazole diamide N, N'-bis (N-octylbenzimidazolyl-2-yl) methyl pentane diamide ligand were isolated by Mohapatra et al. [24] (Scheme 21). The RPE-Band spectra of copper (II) complexes in CH_2Cl_2 were recorded in a frozen solution as a solvent at the temperature of liquid nitrogen, thus indicating a distorted tetragonal geometry involving the binding of benzimidazole imine nitrogen to the copper ion.

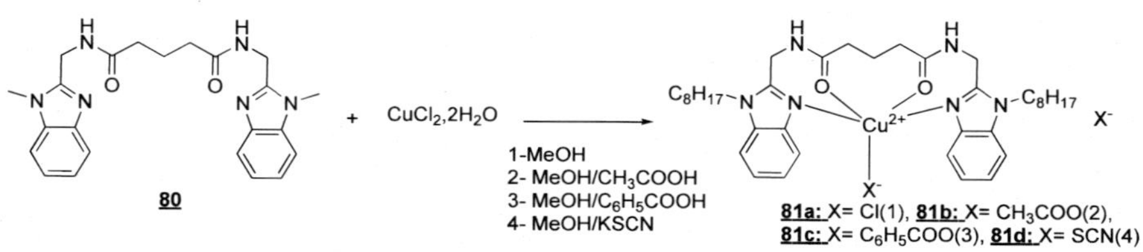

Scheme 21. Synthesis of Cu (II) and Cu (I) monomer complexes bound to a tetradentate bis-benzimidazole diamide N, N'-bis (N-octylbenzimidazolyl-2-yl) methyl pentane diamide ligand.

The exploration of new chemotherapeutic agents is a big challenge in cancer medicine. To this end, Qiao et al. [25] synthesized the 2-substituted benzimidazole-copper (II) complex, $[Cu(BMA)Cl_2].(CH_3OH)$ [BMA = N,N'-bis(benzimidazol-2-yl)methyl)amine], and its cytotoxicity has been characterized (Scheme 22). The interaction between the complex and the calf thymus DNA was detected by spectroscopic methods. Cytotoxicity studies have shown that this complex can inhibit the proliferation of the human cervical carcinoma cell line as a function of time and dose.

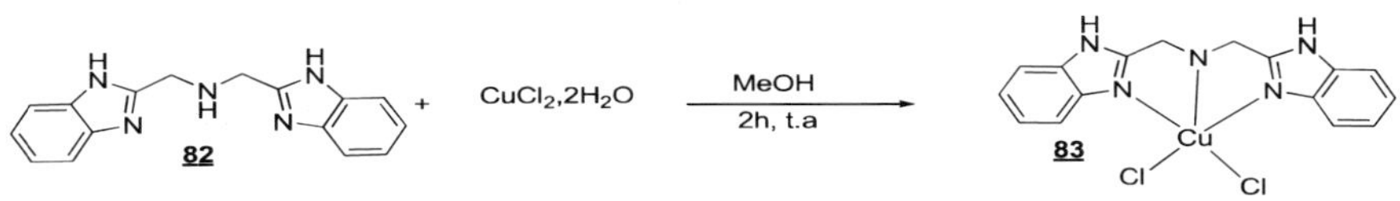

Scheme 22. Synthesis of 2-substituted benzimidazole-copper (II) complex, $[Cu(BMA)Cl_2].(CH_3OH)$ [BMA = N,N'-bis(benzimidazol-2-yl)methyl)amine].

The europium (III) complexes have attracted the interest of scientists for their excellent properties and potential applications in the fields of organic light-emitting diodes, biological labeling and laser media. Tao et al. [26] synthesized two novel europium (III) complexes having 2-thienoyltrifluoroacetylacetone (tta) as the first ligand and second ligand bearing benzimidazole groups (Scheme 23). The photoluminescent properties of solid state complexes in ethanol solution and incorporated in poly (methyl methacrylate) were studied. All europium (III) complexes had a high luminescent intensity, high quantum yields and long luminescent lifetimes. These europium (III) complexes can find potential applications in the field of high power laser carriers.

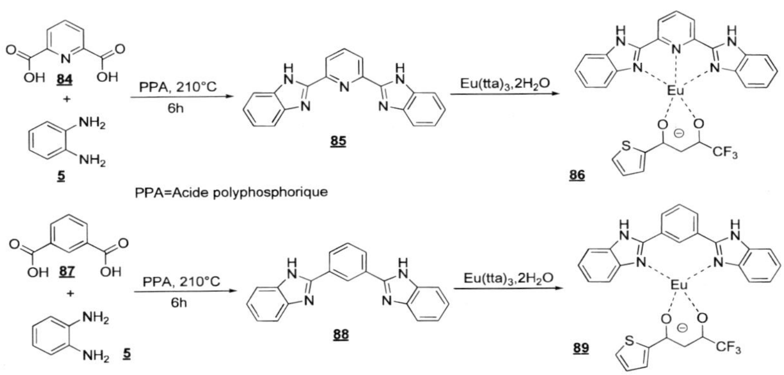

Scheme 23. Synthesis of two novel europium (III) complexes.

Rodriguez-Cordero et al. [27] prepared zinc (II) complexes ($L1ZnBr_2$, $L2ZnBr_2$) in good yields using bis-benzimidazole ligands with olefinic bridges and various substituents on nitrogen atoms in benzimidazole rings (L1 = H, L2 = Bz). The crystalline structure of L2 was determined, showing that the two benzimidazole rings and the olefinic bridge carbon atoms are coplanar with a slight deviation and that the benzyl substituent is perpendicular to this plane. The crystalline structure of L1ZnBr2 has also been determined, showing that in coordinated L1 the benzimidazole rings are out of plane with respect to the olefinic bridge, which allows their coordination through the nitrogen atoms to form a seven-membered ring with deformed tetrahedral geometry. The two crystal structures of L2 and L1ZnBr2 exhibit π-π stacking interactions that maintain their supramolecular arrangement. In addition, DNA binding studies using spectral absorption assays allowed for mild hypochromism and bathochromism when ligands and complexes interacted with calf thymus DNA in pure DMSO and buffer at 3% DMSO / Tris-HCl.

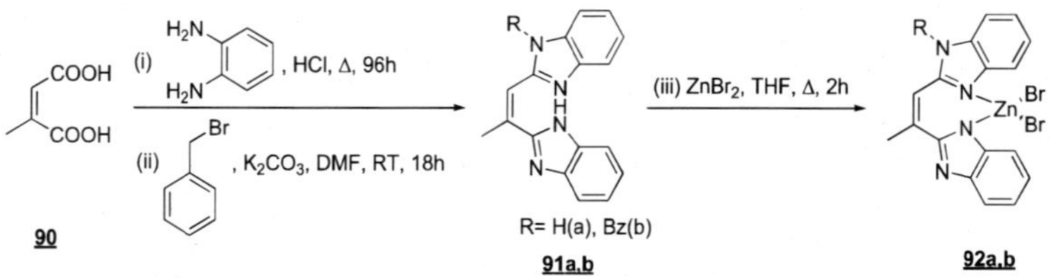

Scheme 24. Preparation of zinc (II) complexes ($L1ZnBr_2$, $L2ZnBr_2$).

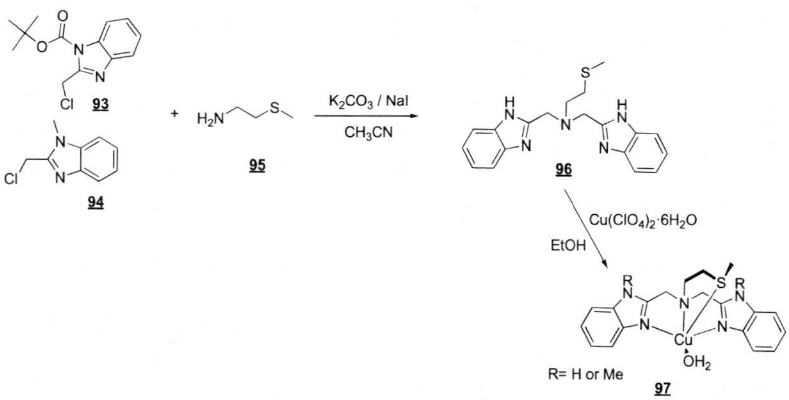

Schéma 25. Synthesis of copper (II) complexes with NH and N-methylated benzimidazole ligands (Cu-L1 and Cu-L1Me; L1 = bis(2-methylbenzimidazolyl)(2-methylthioethyl) amine, L1Me = bis (1-methyl)2-methylbenzimidazolyl)(2-methylthioethyl) amine.

Two cytotoxic copper (II) complexes with NH and N-methylated benzimidazole ligands (Cu-L1 and Cu-L1Me; L1 = bis(2-methylbenzimidazolyl)(2-methylthioethyl) amine, L1Me = bis (1-methyl)2-methylbenzimidazolyl)(2-methylthioethyl) amine were synthesized and exposed to human erythrocytes and molecular models of its membrane (Suwalsky et al. [28]). Scanning electron microscopy (SEM) of erythrocytes incubated with solutions of both Cu (II) complexes showed morphological changes in normal cells to echinocytes and haemolysis at higher concentrations. The real-time observation of dose-dependent effects of the complexes on Live erythrocytes by defocusing microscopy (DM) confirmed the results obtained by SEM X-ray diffraction studies showed that only Cu-L1 interacted with bilayers consisting of dimyristoylphos This difference is explained by the fact that the Cu-L1Me complex is larger than Cu-L1 because it has two additional methyl groups. Thus, by steric hindrance, the Cu-L1Me molecules cannot be intercalated in a DMPC bilayer, which is also present in the gel phase. These results, as well as the increase in the antiproliferative capacity of the N-methyl Cu-L1Me complex relative to that of Cu-L1, are mainly justified by its superior lipophilicity.

3. COMPLEXING PROPERTIES OF PYRAZOLYL BENZIMIDAZOLES WITH RESPECT TO METALS

3.1. Ligands Based on Pyrazolyl-Benzimidazole

Sbai et al. [29] carried out the complexation of 2-[(5-methylpyrazolyl)methyl] benzimidazole by copper and cadmium chlorides (Scheme 26).

The structures of dichloro{2-[(5-methyl-1H-pyrazol-3-yl)methyl]-1H-1,3-benzimidazole}copper(II), $[CuCl_2(C_{12}H_{12}N_4)]$ (**99**), and the dichloro-bis(chloro{2-[(5-methyl-1H-pyrazol-3-yl)methyl]-1H-1,3-benzimidazole}cadmium(II)), $[Cd_2Cl_4(C_{12}H_{12}N_4)_2]$ (**100**), show that these compounds have the structural formula $[ML(Cl)_2]$ n, where L is 2- [(5-methylpyrazolyl)methyl]benzimidazole. When M is copper, the complex is a monomer (n = 1), with a tetrahedral coordination for the Cu atom. When M is cadmium (n = 2), the complex is located around a center of inversion giving rise to a centrosymmetric dimer in which the Cd atoms are bridged by two chloride ions and are pentacoordinated.

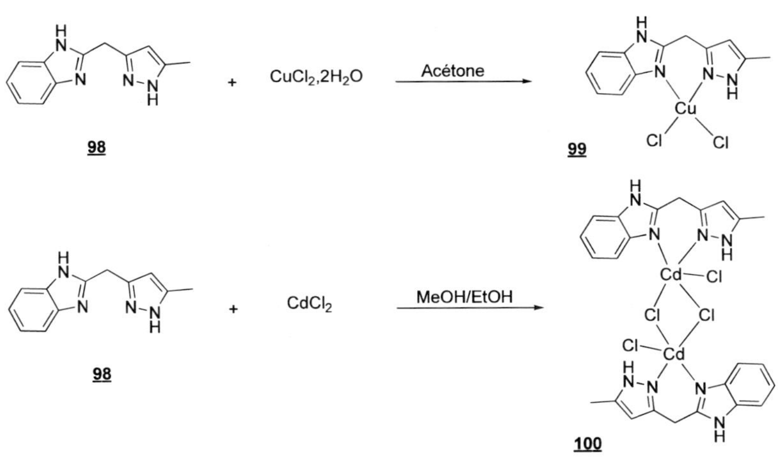

Scheme 26. Complexation of 2-[(5-methylpyrazolyl)methyl] benzimidazole by copper and cadmium chlorides.

The same authors synthesized diaquabis{2-[(5-methylpyrazol-3-yl)methyl] benzimidazole}cobalt(II) dichloride dihydrate acetone [30]. The cobalt ion, bound to two water molecules and four nitrogen atoms of the ligand, has slightly distorted octahedral coordination (scheme 27).

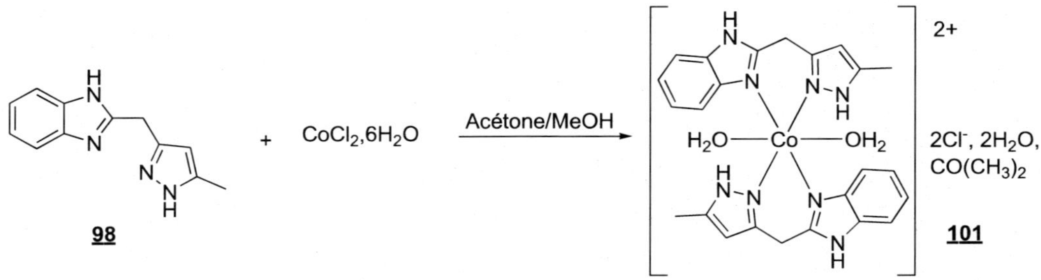

Scheme 27. Synthesized diaquabis{2-[(5-methylpyrazol-3-yl)methyl] benzimidazole}cobalt(II) dichloride dihydrate acetone.

A similar synthesis is reported by Chkirate et al. [31] leading to an asymmetric unit of the complex **102** which contains two independent molecules having different conformations. Coordination around Zn is tetrahedral distorted with the geometric constraints of the chelating ligand responsible for the majority of the distortion. Each of the independent molecules forms chains in the crystal through N-H ... Cl hydrogen bond pairs.

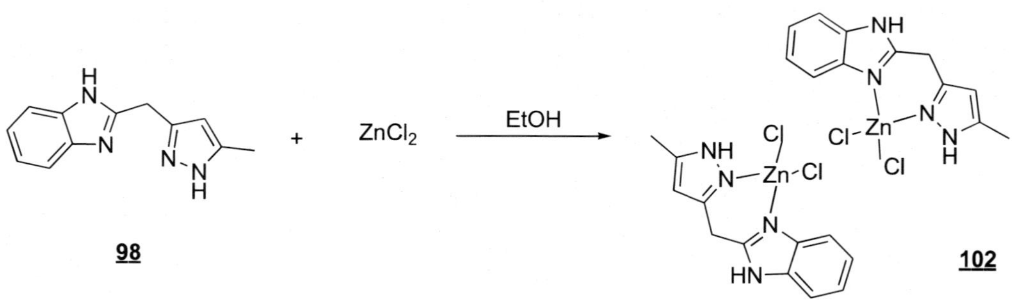

Scheme 28: Complexation of pyrazolylmethyl benzimidazole **98** by zinc dichloride.

Using the same ligand, Sbai et al. [32] synthesized the complex $[Ni(C_{12}H_{12}N_4)_2(H_2O)_2]Cl_2,2H_2O$ where the Ni atom is bonded to four nitrogen atoms and two water molecules in one deformed octahedral coordination geometry. Coordinated water molecules are involved in hydrogen bonds with two other water molecules.

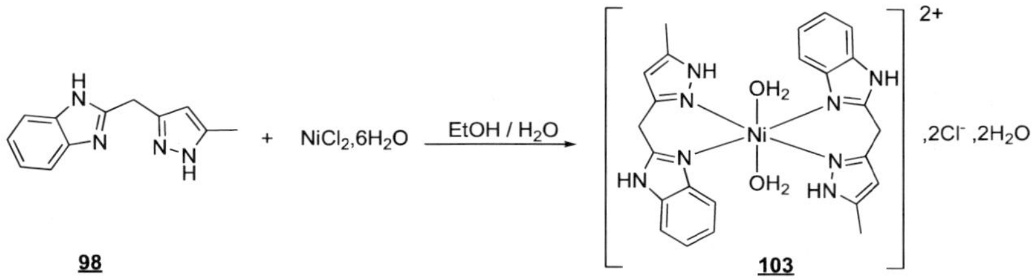

Scheme 29. Synthesis of the complex $[Ni(C_{12}H_{12}N_4)_2\text{-}(H_2O)_2]Cl_2,2H_2O$.

A novel macrocyclic polyether 1-pyrazolylbenzimidazol-2-ones **107**-**108** has been synthesized by the reaction of dichloro (poly) ethylene glycol under phase transfer catalysis conditions (Ghomsi et al.) [33]. The formation of Li^+, Na^+ and K^+ complexes with the compounds **107**-**108** was studied by UV-visible spectroscopy. Ligand **107** has been shown to form a preferred complex with a Li^+ cation (Scheme 30).

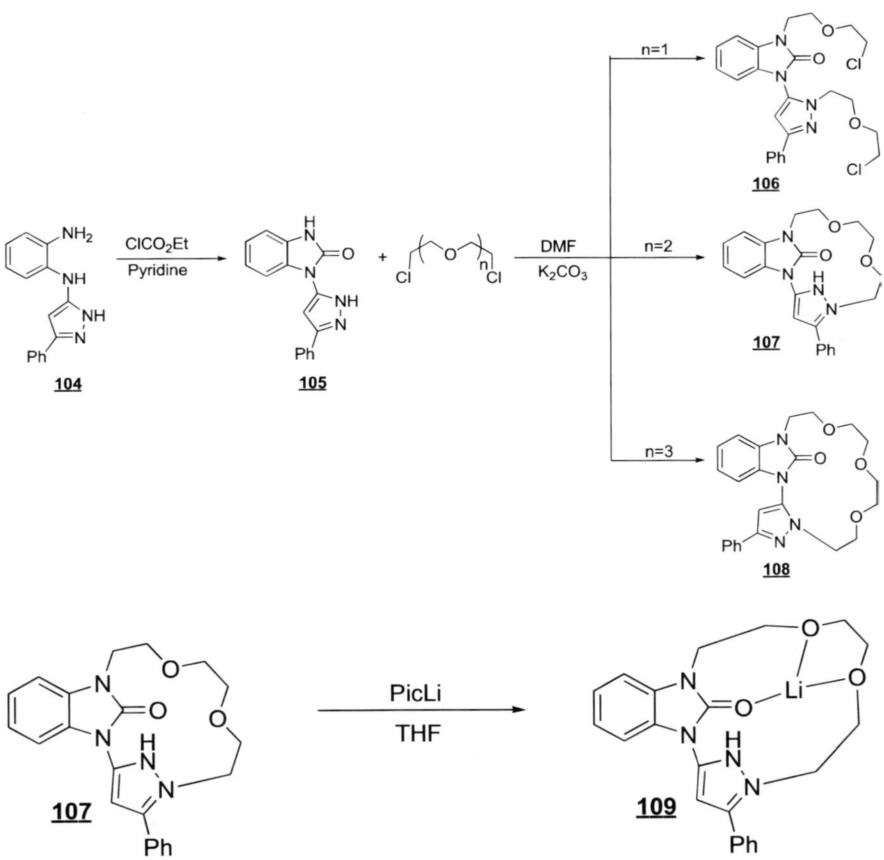

Scheme 30. Complexation of a novel macrocyclic polyether 1-pyrazolylbenzimidazol-2-ones **107**-**108** by Li^+, Na^+ and K^+.

Four new monodentate ligands of 2-(3'-(4"-sub-aryl)-1'-phenyl-1H-pyrazol-4-yl)-1H-benzimidazole structure were synthesized by Shah et al. [34] by cyclocondensation of *o*-phenylenediamine with 1,3-disubstituted pyrazole aldehydes using cerium oxide nitrate as a catalyst in the presence of hydrogen peroxide in methanol. Cu (II) and Ni (II) complexes with the 2-substituted monodentate benzimidazole ligands were prepared in 1: 2-metal: ligand stoichiometry (Scheme 31) The monodentate behavior of the ligand was confirmed on the basis of spectral studies. Ni (II) complexes have a tetrahedral geometry, while the Cu (II) complexes have a square planar geometry around the metal center.

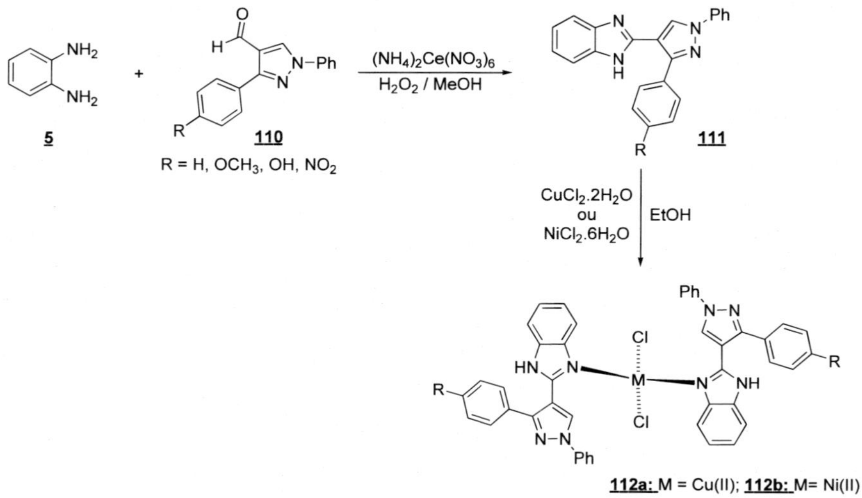

Scheme 31. Synthesis of Cu (II) and Ni (II) complexes with 2-(3'-(4"-sub-aryl)-1'-phenyl-1H-pyrazol-4-yl)-1H-benzimidazole.

In the course of their work on the synthesis of a new Mannich base of structure 4-((2-mercapto-1H-benzoimidazol-1-yl)methylamino)-1,5-dimethyl-2-phenyl-1H-pyrazol-3(2H)-one, Bhava et al. [35], reacted 2-mercaptobenzimidazole, formaldehyde and 4-aminoantipyrine (Scheme 32). In addition, the new ligand was reacted with the metal chlorides (II) to obtain the metal complexes [Cu (II), Ni (II), Co (II) and Zn (II) complexes]. Elemental analysis and spectral studies indicate octahedral geometry for all complexes. The electrochemical property of the ligand and its metal complexes has been studied by cyclic voltammetry. Biological screening of the ligand and metal complexes (II) was performed on *Streptococci*, *Staphylococcus aureus*, *Pseudomonas*, *Klebsilla Pneumoniae*, *Chromo bacterium*, *Candida albicans* by the well diffusion method which indicates that the metal complexes (II) have activity superior to the ligand.

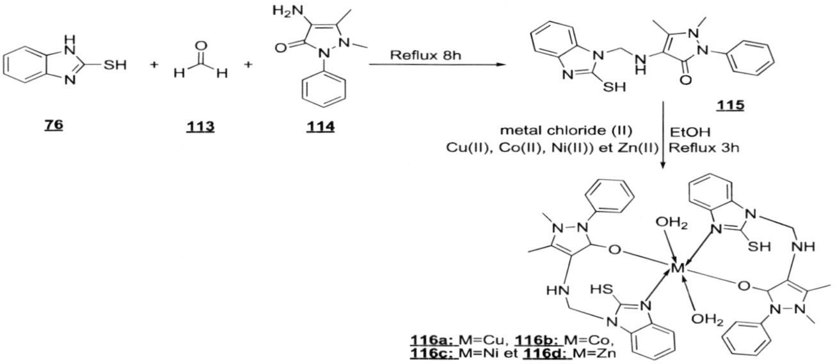

Scheme 32. Synthesis of metal complexes [Cu (II), Ni (II), Co (II) and Zn (II) complexes]using as ligand 4-((2-mercapto-1H-benzoimidazol-1-yl)methylamino)-1,5-dimethyl-2-phenyl-1H-pyrazol-3(2H)-one.

Hawes et al. [36] reported an easy and highly efficient synthesis of two ligands **118** and **119** containing the 2-(pyrazol-1-yl)-benzimidazole chelating group, and studied their structural chemistry when reacted with Cu (II) and Zn (II) ions. The mononuclear complexes $[Zn(L1)_2(NO_3)]NO_3$, **120** and $[Cu(L2)_2(NO_3)]NO_3$, **121** exhibit similar metal coordination environments, although their intermolecular interactions differ in presence or absence of an NH group of benzimidazole, which allows the intermolecular hydrogen bond to be established in **120**, whereas the complex **121** associates only through π-π interactions (scheme 33).

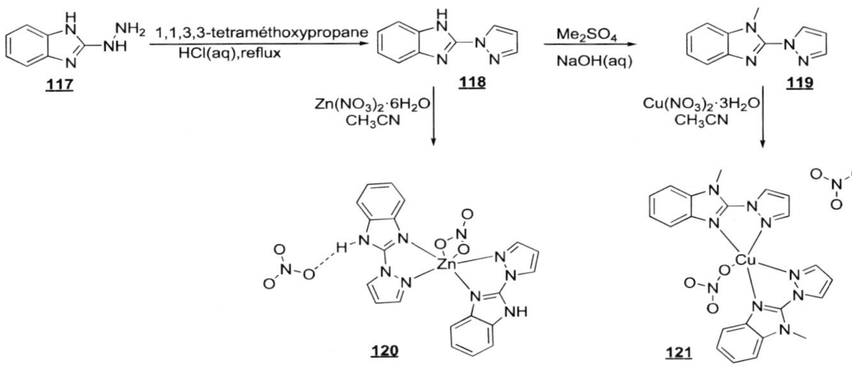

Scheme 33: Chelation of Cu (II) and Zn (II) ions with two ligands **118** and **119** containing 2-(pyrazol-1-yl)-benzimidazole.

3.2. Pyrazolyl-Pyridyl-Benzimidazole Ligands

Novel heterocycles with benzimidazole, pyrazole and pyridine units in the same molecule were prepared by El Azzaoui et al. [37]. The complexing properties with divalent copper salts were determined by mass spectrometry (FAB) and a very interesting oxidation reaction with molecular oxygen was demonstrated (Scheme 34).

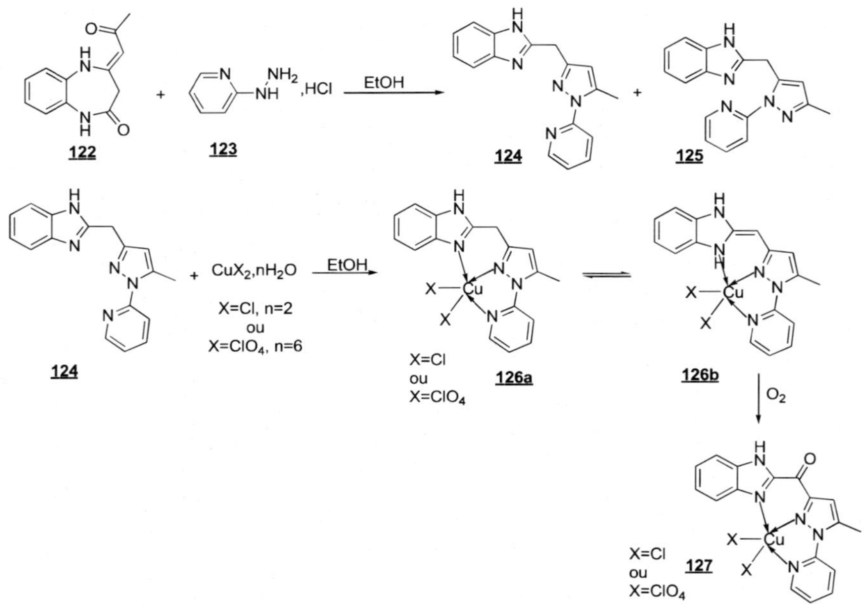

Scheme 34. Complexation of pyridylpyrazolylmethyl benzimidazole with CuX2,nH2O.

Bouhfid et al. [38] have synthesized a crystalline structure of 2-((5-methyl-1-(pyridin-2-yl)-1H-pyrazol-3-yl) methyl)-1H-benzimidazole dichloride cadmium (II) where the cadmium is in a five-link arrangement, engaging the three nitrogens of the benzimidazole-pyrazole-pyridine units describing a deformed trigonal bipyramid (scheme 35).

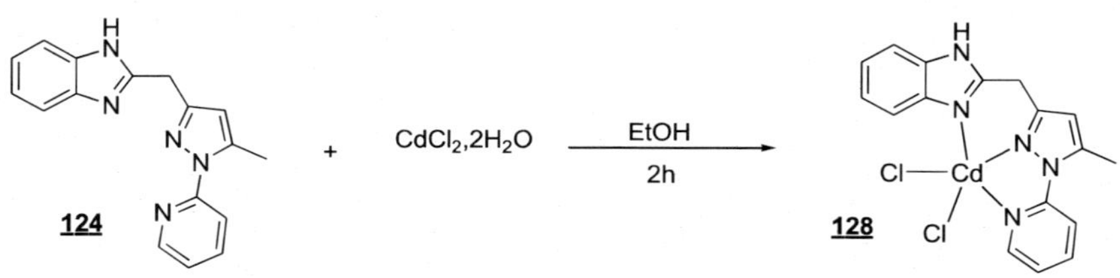

Scheme 35. Synthesis of 2-((5-methyl-1-(pyridin-2-yl)-1H-pyrazol-3-yl)methyl)-1H-benzimidazole dichloride cadmium (II).

Chromium (III) complexes, $LCrCl_3$ (L = 2-(N-R2-benzimidazol-2-yl)-6-(3,5-R1-pyrazol-1-yl) pyridine), were synthesized and characterized by Luo et al. [39] (Scheme 36). X-ray diffraction studies revealed distorted octahedral geometries around the chromium center. Various reaction parameters, including the amount of catalyst, the reaction temperature and the time were evaluated. It was evident that these complexes had good thermal stability and the substituents on the imidazole-N ligands significantly influenced the catalytic activities.

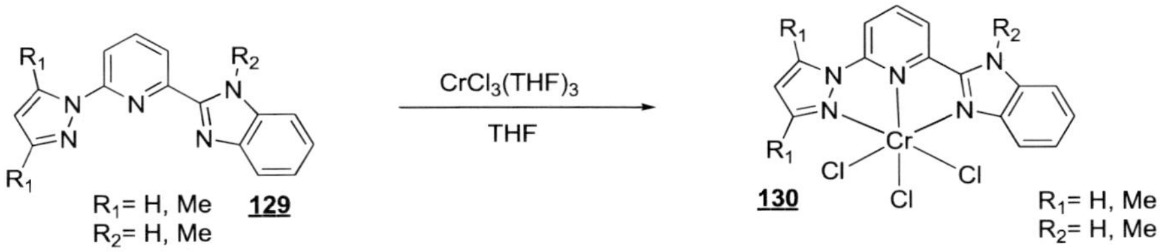

Scheme 36. Synthesis of chromium (III) complexes, $LCrCl_3$ (L = 2-(N-R2-benzimidazol-2-yl)-6-(3,5-R1-pyrazol-1-yl) pyridine).

In its work for the synthesis of a 2-(6-pyrazol-1-yl-pyridin-2-yl)-1H-benzimidazole tridentate ligand (L), Rajnak et al. [40] have prepared from this ligand, two complexes, one is a complex $[Fe(L)_2](BF_4)_2$, and the other is a complex with a neutral charge $[Fe(L)_2]$ (scheme 37). The deformed octahedral coordination of the 1st complex was confirmed by X-ray diffraction analysis. Both complexes show a thermally induced spin crossing with the critical temperature above room temperature. The photomagnetic investigation confirmed the LIESST (light-induced excited spin state) effect for both compounds.

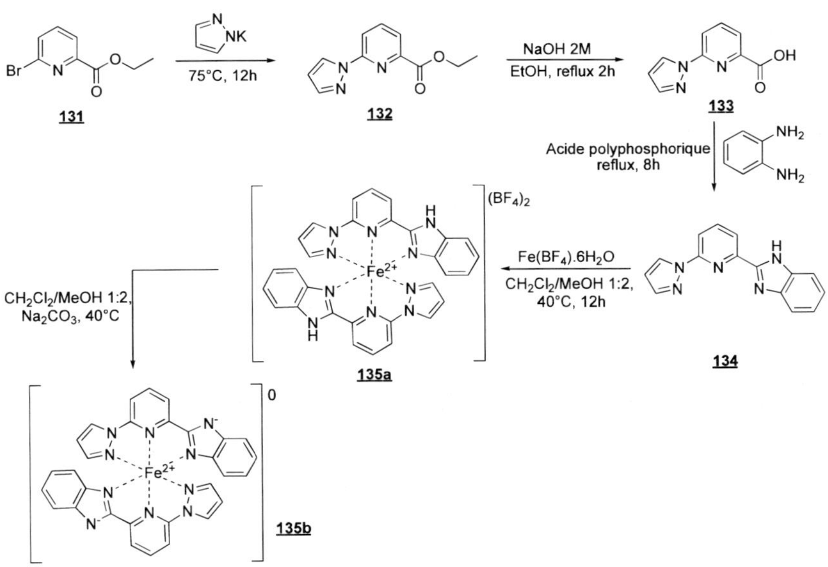

Scheme 37. Complexation of 2-(6-pyrazol-1-yl-pyridin-2-yl)-1H-benzimidazole with $Fe(BF4),6H_2O$.

Conclusion

This reported literature review on the coordination of benzimidazole derivatives with metal ions has attracted the interest of several research groups, partly because of their very interesting chemical composition, as well as their potential applications. Benzimidazole derivatives have been shown to coordinate with metal ions through imidazole nitrogen, and nitrogen bearing groups present in the molecule, resulting in a variety of metal-ligand coordination modes. In some compounds the inclusion of a metal ion gives rise to a delocalized system, while for others, different geometries are adopted, given the stereochemistry and the existence of tautomeric forms which stabilize the complexes formed.

REFERENCES

[1] Starcevic K., Kralj M., Ester K., Sabol I., Grce M., Pavelic K., Karminski-Zamola G. *Bioorg. Med. Chem.* 15 (2007) 4419-4426.

[2] Gupta S. K., Pancholi S. S., Gupta M. K., Agarwal D., Khinchi M. P. *J. Pharm. Sci. & Res.* 2 (2010) 228-231.

[3] Reedijk J. *Comprehensive Coordination Chemistry* (Eds.: G. Wilkinson, R. D. Gillard, J. A. McCleverty), Pergamon, Oxford. 2 (1987).

[4] Małecki J. G. *Struct. Chem.* 23 (2012) 461–472.

[5] Ibrahim M. M., Mersal G. A. M., Ramadan A. M., El-Shazly S. A., Amin M. A. *Int. J. Electrochem. Sci.* 9 (2014) 5298 – 5314.

[6] Yellol G. S., Yellol J. G., Kenche V. B., Liu X. M., Barnham K. J., Donaire A., Janiak C. et Ruiz *J. Inorg. Chem.,* 54 (2015) 470–475.

[7] Chowdhury S., Bhattacharya A., Saha P., Majumder S., Suresh E. & Naskar J. P. *J. Coord. Chem.* 69 (2016) 3664-3676.

[8] Zhao J., Zhi S., Yu H., Mao R., Hu J., Song W. and Zhang J. *RSC Adv.,* 7 (2017) 51162–51174.

[9] Abbas A. K. et Kadhim R. S. *Orient. J. Chem.* 33 (2017) 402-417.

[10] Booysen I. N., Adebisi A., Munro O. Q., Xulu B. *Polyhedron* 73 (2014) 1–11.

[11] Lavrenova L. G., Kuz'menko T. A., Ivanova A. D., Smolentsev A. I., Komarov V. Y., Bogomyakov A. S., Sheludyakova L. A. et Vorontsova E. V. *New J. Chem.* 41 (2017) 4341-4347.

[12] Ahmad W., Khan S. A., Munawar K. S., Khalid A. et Kawanl S. *Trop. J. Pharm. Res.* 16 (2017) 1137-1146.

[13] Akremi A., Noubigh A. et Abualreish J. A. *Orient. J. Chem.* 34 (2018) 764-770.

[14] Mahmood K., Hashmi W., Ismail H., Mirza B., Twamley B., Akhter Z., Rozas I., Baker R. *J. Polyhedron* (2018).

[15] Yu M. H., Yang H. H., Gu Y. C., Wang B. H., Liu F. C., Lin I. J. B., Lee G. H. *J. Organomet. Chem.* 887 (2019) 12-17.

[16] Su W. Y., Pan R. K., Song J. L., Li G. B., Liu S. G. *Polyhedron* 161 (2019) 268–275.

[17] Wang X., Ling N., Che Q. T., Zhang Y. W., Yang H. X., Ruan Y., Zhao T. T. *Inorg. Chem. Commun.* 105 (2019) 97-101.

[18] Milani J. L. S., Oliveira I. S., Dos Santos P. A., Valdo A. K. S. M., Martins F. T., Cangussu D., Das Chagas R. P. *Chinese. J. Catal.* 39 (2018) 245–249.

[19] Ashraf A., Siddiqui W. A., Akbar J., Mustafa G., Krautscheid H., Ullah N., Mirza B., Sher F., Hanif M., Hartinger C. G. *Inorg. Chim. Acta.* 443 (2016) 179-185.

[20] Prosser K. E., Chang S. W., Saraci F., Lê P. H., and Walsby C. J. *J. Inorg. Biochem.* 167 (2017) 89-99.

[21] Wałęsa-Chorab M., Banasz R., Marcinkowski D., Kubicki M. et Patroniak V. *R. S. C. 7* (2017) 50858–50867.

[22] Lamkaddem A., Capitelli F., Essassi E. M., Harcharras M., Larena A. A. et Ennaciri A. *J. Mar. Chim. Heterocycl.* 9 (2010) 9-14.

[23] Salhaji A., El Kacemi K., Moussaif A., Qafas Z., Brohmi L., Zniber R., El Ghoul M. et Achour R. *J. Soc. Chim. Tunisie.* 13 (2011) 51-60.

[24] Mohapatra S. C., Mathur P. *Spectroc. Acta A.* 78 (2011) 612–616.

[25] Qiao X., Ma Z. Y., Shao J., Bao W. G., Xu J. Y., Qiang Z. Y., Lou J. S. *Biometals.* 27 (2014) 155–172.

[26] Tao C., Yuan X., Yin Q., Yan H., Ni W., Yan L., Zhang L. *J. Mater. Sci. Mater. Electron.* 27 (2016) 5715–5722.

[27] Rodriguez-Cordero M., Cigüela N., Llovera L., González T., Briceño A., Landaeta V. R., Pastrán J. *Inorg. Chem. Commun.* 91 (2018) 124–128.

[28] Suwalsky M., Castillo I., Sánchez-Eguía B. N., Gallardo M. J., Dukes N., Santiago-Osorio E., Aguiñiga I., Rivera-Martínez A. R. *J. Inorg. Biochem.* 178 (2018) 87-93.

[29] Sbai F., Regragui R., Essassi E. M., Kenz A. et Pierrot M. *Acta Cryst.* C59 (2003) m334–m336.

[30] Sbai F., Regragui R., Essassi E. M., Kenz A. et Pierrot M. *Acta Cryst.* E59 (2003) m571–m573.

[31] Chkirate K., El Bakri Y., Essassi E. M. and Mague J. T. *IUCrData* 2 (2017) x170116.

[32] Sbai F., Chkirate K., Regragui R., Essassi E. M. et Pierrot M. *Acta Cryst.* E58 (2002) m337–m339.

[33] Ghomsi J. N. T., Bouhfid R., Ahabchane N. H., Essassi E. M. *Frontiers in Science and Engineering an International Journal* Edited by Hassan II Academy of Science and Technology.

[34] Maru M. et Shah M. K. *J. Chem. Pharm. Res.* 4 (2012) 1638-1643.

[35] Bhava P. S., Tharmaraj P., Raja S. J. *Int. J. Innov. Res. Dev.* 2 (2013) 378-396.

[36] Hawes C. S. et Kruger P. E. *Supramol. Chem.* 27 (2015) 757–771.

[37] El Azzaoui B., Fifani J., Tjiou E. M., Essassi E. M., Jaud J., Lopez L. et Bellan J. *Tetrahedron Lett.* 40 (1999) 4677-4680.

[38] Bouhfid R., Essassi E. M., Zerzouf A. et Reuter H. Z. *Kristallogr. NCS* 224 (2009) 339-340.

[39] Luo W., Li A., Liu S., Ye H. et Li Z. *Organometallics.* 35 (2016) 3045–3050.
[40] Rajnák C., Schäfer B., Šalitroš I., Fuhr O., Ruben M. et Boča R. *Polyhedron.* 135 (2017) 189-194.

INDEX

#

A

B

C

T

X

Z

Related Nova Publications

A Comprehensive Guide to Chemiluminescence

Editor: Luís Pinto da Silva

Series: Chemistry Research and Applications

Book Description: This book provides a detailed overview of the basic mechanisms and principles of the most relevant chemiluminescent systems, as well as describing the most recent advances and applications.

Hardcover ISBN: 978-1-53616-170-0

Retail Price: $230

An Introduction to Vanadium: Chemistry, Occurrence and Applications

Editor: Robert Bowell, PhD

Series: Chemistry Research and Applications

Book Description: Vanadium has, in the last two years, developed new uses, particularly in the field of redox batteries. The expanded use of vanadium reflects the complex redox-related chemistry of the element that allows it to form several oxidation states and to stabilise various chemical compounds.

Hardcover ISBN: 978-1-53616-119-9

Retail Price: $230